T0138684

Praise for *Understanding Cancer*

McIntosh has written a wonderful introductory textbook on cancer that covers everything from cancer diagnosis and treatment to living with cancer, while skillfully inserting the fundamentals of chemistry and cell biology where needed. Beautifully illustrated—and peppered with highly original, thought-provoking "side questions" (with answers at chapter end)—this is a great resource for any college course aimed at producing a new generation of adults prepared to make wise, scientifically based decisions for themselves and their families.

Bruce Alberts
Lead author of *Molecular Biology of the Cell*
Former Editor-in-Chief of *Science* magazine (2008–2013)
President Emeritus, U.S. National Academy of Sciences (1993–2005)

A remarkable text that explores the molecular, cellular, and medical aspects of cancer with stunning clarity. McIntosh has fashioned a compelling and enlightening narrative that will take students from the most basic aspects of cell growth and division to the cutting edge of cancer diagnosis and treatment. Not only a must read for college courses on cancer, this book is an ideal entry point for anyone interested in learning more about this disease and its causes. The illustrations are superb, the prose engaging, and the overall organization a testament to the expansive scientific vision of an internationally known biologist and educator.

Kenneth R. Miller
Professor of Biology, Brown University
Lead author of *Biology* by Miller & Levine
Recipient of an AAAS Award for Advancing the Public Understanding of Science (2009)

This is an unusual and exciting text book aimed at students with no scientific background. They will learn about science and how it is done through the study of cancer: what cancer is, how it comes about, and how it is diagnosed, treated and prevented. All of this is presented in an accessible way, conveyed with a human touch that connects the reader to stories about cancer patients. Given the book's focus on principles and clarity of writing, it is an ideal introduction to both science and cancer for the nonspecialist.

Sir Paul Nurse
Nobel Laureate in Physiology or Medicine
Director of the Francis Crick Institute
Former President of the Royal Society (2010–2015)

UNDERSTANDING CANCER

UNDERSTANDING CANCER

AN INTRODUCTION TO THE BIOLOGY, MEDICINE, AND SOCIETAL IMPLICATIONS OF THIS DISEASE

J. Richard McIntosh

Distinguished Professor Emeritus, Department of Molecular, Cellular, and Developmental Biology, and the Biophysics Group, University of Colorado, Boulder, Colorado, USA

CRC Press
Taylor & Francis Group
Boca Raton London New York

CRC Press is an imprint of the
Taylor & Francis Group, an **informa** business

CRC Press
Taylor & Francis Group
6000 Broken Sound Parkway NW, Suite 300
Boca Raton, FL 33487-2742

© 2019 by Taylor & Francis Group, LLC
CRC Press is an imprint of Taylor & Francis Group, an Informa business

No claim to original U.S. Government works

Printed in the UK by Severn, Gloucester on responsibly sourced paper

International Standard Book Number-13: 978-0-815-34535-0 (Paperback)
International Standard Book Number-13: 978-0-367-19012-5 (Hardback)

**Visit the Taylor & Francis Web site at
http://www.taylorandfrancis.com**

**and the CRC Press Web site at
http://www.crcpress.com**

Dedicated to the memory of
Robert K. McIntosh, 1966–1999,
a talented and energetic man whose promising life
was cut short by lung cancer

Contents

Preface

My reasons for writing about cancer are several. When my son Rob died of lung cancer at the age of 32, I became aware of new dimensions of this disease, complementing what I had learned as a scientist doing basic cancer research for more than 40 years. As I tried to describe Rob's illness to his friends, who had studied music, art history, and literature, they found it very hard to understand what was going on. During my last eight years of teaching at the University of Colorado, I therefore developed an undergraduate course on the biology of cancer in which I tried to convey some of the things I had learned about the best ways to explain the nature of cancer to everyone. The readings, lectures, and conversations of those years led me to believe that the essential features of cancer could be taught in a way that did not require a significant background in biology. It also convinced me that many aspects of cancer were sufficiently important that everyone should know them, not just doctors and biology majors.

This book is designed for anyone who wants to learn something about cancer, whether or not they have had previous training in biology. Readers who already know about DNA, genes, proteins, and cell biology will find they can skim Chapters Three and Four, simply refreshing their knowledge of these subjects. Others, for whom biology is new, should find these chapters informative. Understanding cancer is not a trivial undertaking, so as the chapters progress, you will be exposed to quite a lot of information about cells and how they work, both when things are normal and when they go astray during the development of cancer. However, even as the material gets more involved, I have tried to make all the explanation and description understandable, saying no more than is necessary and introducing complexity only as it is relevant for grasping some aspect of cancer or its treatment.

Cancer is a tremendously interesting subject that spans multiple fields of science, history, philosophy/religion, and relationships among people. As such, it is an excellent platform from which to discuss some important human issues that are worth serious thought. A difficulty, however, is that some of the underlying science is complicated. To learn about a scientific phenomenon, you commonly cannot understand B until you have first learned about A. For example, one cannot understand the importance of genetic instability in cancerous progression without knowing something about genes and the ways their expression is controlled. A certain amount of basic science is therefore a prerequisite for understanding cancer. To make such learning as straightforward as possible, I have kept both vocabulary and details to a minimum, while still making the explanations sufficiently thorough to reflect our current level of knowledge. Cancer is now well enough described and understood that there is a bewildering amount of information and detail available; that depth is not presented here. Instead, references are given to more complete treatments of each subject, as well as to reliable internet addresses that provide almost unlimited information about most of the topics discussed.

The primary goal of this book is to provide sufficient information to help you understand the fundamentals of cancer and the medical issues that surround its screening, diagnosis, prognosis, and treatment. Other goals include helping you to see: how the process of science works, how medical scientists have been able to make so much progress in understanding cancer, how you can minimize your own risk of getting cancer, and how to live with cancer if you or a loved one should happen to get it. You will also come to understand why no simple prevention or cure for cancer has emerged, in spite of all the money and effort that has been spent in trying to find them. Although I have tried to keep this book's language as simple as possible, some scientific terms are useful because they are precise and save many words. Such terms are defined at first use and presented in bold-faced type; a glossary at the back of the book will remind you of the meanings of these words and some others you may find useful.

Each chapter is built around medical and scientific issues, but cancer is as much a human problem as a medical one. To put a human face on the disease, I have included interviews with people who have dealt with cancer in various ways. I introduce Mary, the mother of Scott, who was diagnosed at age 3 with a Wilms Tumor, a kind of kidney cancer. Mary is a resourceful and cheerful woman who saw her son through a very hard time. Now, about 20 years later, she can look back on that time with relief and a sense of accomplishment, as well as with a clear-eyed recollection of how hard it all was. Rosemary is a senior professor at a large research university. About 18 years ago, she was diagnosed with breast cancer. She elected to have rigorous treatment and is now fine, another cancer success story. John was a 60-year-old construction worker at the time I interviewed him; he owned his own roofing business. About seven years before our conversations, he had been diagnosed with a fatal kidney cancer and given only a few months to life. Thanks to good cancer medicine and a remarkably strong spirit, John continued to live a full and useful life, succumbing to cancer only after about eight years of full and energetic life. Woods was a man who had to deal with cancer while living below America's poverty line. He acquired a bladder cancer, probably as a result of a badly designed medication given to him about 20 years earlier. His lack of money and health insurance led to serious difficulties in getting proper treatment, which points out another dimension of cancer medicine. These examples and the story of my son's cancer are woven into the drier descriptions of diagnosis, treatment, and living with cancer in the hope that they will illustrate some of the complexities of cancer medicine.

J. Richard McIntosh

Student and Instructor e-Resources

Please visit www.crcpress.com/cw/mcintosh to access:

- The answers to the "Questions for Further Thought" posed in the book.
- The video content cited in the book (see below).
- A library of images from the book in various formats for use in presentations and lectures.

Together, these materials bring additional utility to this beautifully prepared and presented textbook. We hope you will find them useful.

LIST OF VIDEO RESOURCES

Chapter 1 Cancer and the biology of human cells

MOVIE 1.1 *Exponential growth of mammalian cells grown in the lab.*

MOVIE 1.2 *Neutrophil chase.* (From Alberts B, Johnson A, Lewis J et al. [2014] Molecular Biology of the Cell, 6th ed. Garland Science. With permission from W. W. Norton & Company.)

Chapter 4 How normal cells reproduce and differentiate

MOVIE 4.1 *Breast cancer cells versus normal breast cells grown in the lab.* (From Weinberg R [2013] The Biology of Cancer, 2nd ed. Garland Science. With permission from W. W. Norton & Company, and courtesy of Mina J. Bissell, Karen Schmeichel, Hong Liu, and Tony Hansen.)

MOVIE 4.2 *Mitosis.* (Part 1: From Alberts B, Johnson A, Lewis J et al. [2014] Molecular Biology of the Cell, 6th ed. Garland Science. With permission from W. W. Norton & Company. Part 2: Courtesy of Patricia Wadsworth, University of Massachusetts at Amherst.)

MOVIE 4.3 *Transcription.* (With permission from Howard Hughes Medical Institute.)

MOVIE 4.4 *Apoptosis.* (From Alberts B, Johnson A, Lewis J et al. [2014] Molecular Biology of the Cell, 6th ed. Garland Science. With permission from W. W. Norton & Company. Part 1: Courtesy of Shigekazu Nagata and Sakura Motion Picture Company. Part 2: Courtesy of Joshua C. Goldstein, The Genomics Institute of the Novartis Research Foundation, and Douglas R. Green, St Jude Children's Research Institute.)

Chapter 6   Oncogenes and their roles in cancerous transformation

MOVIE 6.1   *EGFR signaling*. (From Weinberg R [2013] The Biology of Cancer, 2nd ed. Garland Science. With permission from W. W. Norton & Company. Storyboard and animation by Sumanas, Inc.)

MOVIE 6.2   *Responses of liver cells to EGF versus HGF*. (From Weinberg R [2013] The Biology of Cancer, 2nd ed. Garland Science. With permission from W. W. Norton & Company. Courtesy of Andrea Bertotti and Paolo M. Comiglio, Institute for Cancer Research and Treatment, University of Torino.)

MOVIE 6.3   *Angiogenesis*. (Part 1: Courtesy of Brant M. Weinstein, National Institutes of Health. Part 2: Courtesy of George E. Davis, University of Missouri School of Medicine. Storyboard and animation by Sumanas, Inc.)

Chapter 7   Tumor suppressors and their roles in resisting cancerous transformation

MOVIE 7.1   *CDK signalling*. (From Weinberg R [2013] The Biology of Cancer, 2nd ed. Garland Science. With permission from W. W. Norton & Company. Molecular modeling and animations in Chime courtesy of Timothy Driscoll/Molvisions. Chime conversion and QuickTime production courtesy of Sumanas, Inc.)

Chapter 9   The immune system and its relationship to cancer

MOVIE 9.1   *Phagocytosis*. (From Murphy K & Weaver C [2016] Janeway's Immunobiology, 9th ed. Garland Science. With permission from W. W. Norton & Company.)

MOVIE 9.2   *Antibodies*. (From Alberts B, Johnson A, Lewis J et al. [2014] Molecular Biology of the Cell, 6th ed. Garland Science. With permission from W. W. Norton & Company. Molecular modeling and animations in Chime courtesy of Timothy Driscoll/Molvisions. Chime conversion and QuickTime production courtesy of Sumanas, Inc.)

MOVIE 9.3   *T cell granule release*. (From Murphy K & Weaver C [2016] Janeway's Immunobiology, 9th ed. Garland Science. With permission from W. W. Norton & Company.)

MOVIE 9.4   *The immune response*. (From Murphy K & Weaver C [2016] Janeway's Immunobiology, 9th ed. Garland Science. With permission from W. W. Norton & Company.)

Acknowledgments

This book has profited significantly through input from many colleagues, friends, and family members. Joaquin Espinosa, Nancy Guild, and Joy Power (University of Colorado) read the text for scientific accuracy and the effectiveness of my scientific pedagogy. Dr. Kenneth McIntosh (Harvard Medical School) and Dr. John Austin (Columbia College of Physicians and Surgeons) read the medical sections of the book for accuracy and relevance. Hung Fan tried an early version of the manuscript on his classes about cancer at the University of California, Riverside. The publisher sought critical readings from ten reviewers, all of whom gave useful criticisms, and two of my siblings helped to improve my prose. I also give warm thanks to the people whose cancer stories are reported here. My wife Marjorie has been an invaluable source of support and constructive criticism from the outset, and I am grateful to my principal editors at Taylor & Francis, Monica Toledo, Denise Schanck, and Joanna Koster.

Cancer and the biology of human cells

CHAPTER 1

Cancer is a result of misbehaving cells. It is a disease that develops when a cell from one's own body fails to respond to controls that normally regulate cell behavior. When these controls are working properly, they assure that every cell acts in the best interests of the body as a whole. Cancer results from a rogue cell, one in which these controls have failed, so the cell grows and divides when it should not, forming more of its own kind at the expense of other cells. In this way, the rogue cells sap the body's energy and disrupt normal functions. If these cells are not eliminated, the body as a whole will become ill and may ultimately die. This book will help you understand the nature of cancer in its broadest sense. Since the disease arises from our own cells, an understanding of cancer must begin with an understanding of normal cells; from there we can build an understanding of the malfunctions that lead to cancer. This chapter therefore starts with an introduction to human cells, which is followed by an introduction to cancer as a disease.

WHAT IS A CELL? THE BASICS

WHEN NORMAL CELLS MISBEHAVE

CURRENT TREATMENTS OF CANCER AND PROSPECTS FOR THE FUTURE

LEARNING GOALS

1. List three structures and three functions that are shared by many different types of human cells.

2. Define the terms *trait* and *allele*, and give an example of a trait for which your two alleles could be different, given the traits of your parents.

3. Describe how cells that contain the same genetic information can perform different jobs.

4. Describe the changes in cell behavior that occur during cancerous transformation.

5. Identify the two behaviors of a cancer cell that are the most dangerous to a patient.

6. Give examples of how cancer is currently diagnosed and treated, stating one strength and one limitation of each diagnosis and treatment.

7. List the properties of cancerous cells that make this disease so hard to treat.

WHAT IS A CELL? THE BASICS

Cells are systems that can consume materials and energy from their environment and use them for their own growth and reproduction (MOVIE 1.1). In short, cells are the simplest living things. Every structure in your body is composed of cells and of the materials that cells make. However, most human cells are so small they cannot be seen with the naked eye; a cluster containing a million cells is not much bigger than a pinhead. Nonetheless, every cell is very complicated; one estimate by a physicist who cares about complex systems suggests that a single cell is more complex than a supercomputer or an Atlas rocket for launching satellites. Their small size and great complexity make cells hard to understand in detail, but quite a lot has been learned about what they are and how they work. Below is an introduction to cells; each topic relevant to cancer is treated with greater detail in Chapters 3 and 4.

Cells can have many shapes and sizes, depending on the jobs they do.

The egg and sperm that joined to start your life were each cells in their own right, albeit very different in size and shape (FIGURE 1.1A). Sperm, with their small elliptical heads and long tails, are specialized to swim and find an egg. Eggs, which are much bigger, are specialized to contain nutrients that will help the new generation start out well fed. When an egg and a sperm join, in the process called fertilization, they form a single cell that contains two internal compartments called nuclei, one nucleus from the sperm and one from the egg (FIGURE 1.1B). We will discuss the

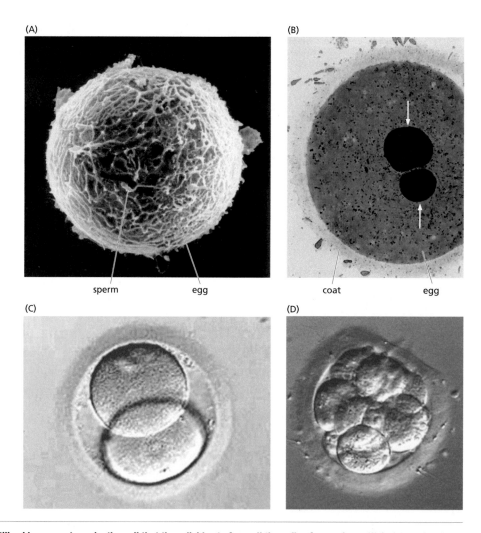

(A)

(B)

sperm   egg

coat   egg

(C)

(D)

FIGURE 1.1 **Human egg being fertilized by sperm to make the cell that then divides to form all the cells of an embryo.** (A) A picture showing many sperm (blue) trying to gain entry into an egg and become the one that fertilizes it. (B) A light micrograph of a slice through a fertilized egg showing the coat that surrounds the egg to protect it, the egg cell itself, and two interior spheres, which are the nuclei contributed by the egg and the sperm cells (arrows). The two nuclei will fuse to form one, whereupon this cell is called a zygote. It will divide many times to form all the cells of the body. (C) The results of the first division: this structure is now a two-cell embryo. (D) Three cell divisions have occurred since the egg became a zygote; now there are eight cells. (A and B, courtesy of Jonathan VanBlerkom, University of Colorado Boulder. C and D, courtesy of the Copenhagen Fertility Center.)

nucleus and its role in cell behavior later in this chapter. Shortly after fertilization, these nuclei fuse, and the result is a cell that is the parent of all the cells in your body. This single cell, which contains material from both egg and sperm, now divides to form two daughter cells, which divide again, making four cells, eight cells, etc. (FIGURE 1.1C,D). These cell divisions gradually build up all the cells needed to make first an embryo, then a fetus, then a child, and then an adult.

As those many cells form, they specialize, taking on different structures so they can accomplish different jobs for your body. Some become blood cells, some become nerve cells, some turn into muscle, and some help to make covering layers, like the skin that surrounds your body and the tissues that line the inside of your mouth, throat, stomach, and intestine. The task of understanding cells is in part learning what is common to them all and in part learning how and why cells become different.

All cells are built from materials with familiar names.

The most prevalent substance in a cell is water, but dissolved in that liquid are a number of materials you know about from the names of different kinds of food: proteins, fats, and sugars (whose general name is carbohydrates). Proteins and fats are important for cells because they are the materials from which much cellular structure is built; they also contribute to many internal cellular functions. Carbohydrates are important because cells use them to store energy as well as to accomplish some other jobs. In addition, cells contain minerals, like sodium, potassium, calcium, iron, and magnesium, as well as many small molecules, such as vitamins, sugars, and amino acids. Although most of these molecules are dissolved in water, some are attached to one another to build structures that are the cell's framework.

Finally, and most significant for this book, cells contain **DNA,** a long molecule made from two strings of smaller molecules hooked together end-to-end, like pop-it beads. DNA is important because it is the material cells use to store the information they need to be able to reproduce and specialize for making body tissues, such as muscle, skin, and bone. You will learn more about DNA in later chapters, because it is especially important for our understanding of cancer.

All our body's cells share some essential features.

All cells have a well-organized structure, though they can be remarkably different in appearance. The structure of each cell helps it to perform its many functions. One structure common to all cells is a very thin membrane, made of proteins and fat, that serves the cell as a boundary (FIGURE 1.2). This structure, called the **plasma membrane**, separates the inside of the cell from the rest of the world. Although it is thin, it is impermeable to most things, so the many molecules that a cell needs to survive don't leak out. However, anything that enters the cell, like molecules of food, must cross this boundary. Cells also make additional membranes that divide their interior into compartments that can accomplish specific functions. For example, membrane-bounded compartments called mitochondria (Figure 1.2B) convert the nourishment a cell consumes into a molecule that stores energy in a form that cells can use efficiently.

Many of a cell's components are positioned so they can contribute effectively to the well-being of the cell as a whole. For example, most of the cell's DNA is in another membrane-bounded compartment, the **nucleus**, where it is concentrated for active use (Figure 1.2). The nucleus of every human cell contains 46 long pieces of DNA, each of which is called a **chromosome**.

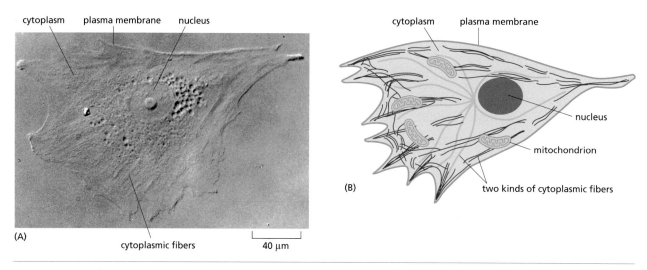

FIGURE 1.2 **The internal structures of a human cell.** (A) Image taken with a light microscope showing a human skin cell growing in a laboratory on a piece of glass. Optical tricks have been used to make cellular structures stand out. In this image you can see an overview of the cell and a few of the structures inside it, as labeled. (B) A diagram of a similar cell highlights multiple organelles. The plasma membrane surrounds the cell, defining its boundary. The nucleus contains the cell's DNA. The cytoplasm is the region inside the plasma membrane and outside the nucleus; it holds many organelles, such as mitochondria, the organelles responsible for producing the form of energy that cells use most efficiently. Two systems of cytoplasmic fibers are shown; these help to give the cell its shape. (A, courtesy of Casey Cunningham.)

SIDE QUESTION 1.1

If the plasma membrane is the boundary that holds in all the contents of the cell, why not just make it completely impermeable, so nothing can leak out?

One of the most important aspects of cells is their ability to store the instructions necessary to make a new cell. Much of this information is written into the structure of the cell's DNA. Since all cells need this kind of information, one of the processes required for cell growth and division is the duplication of its DNA. This process forms two identical copies of the original DNA, so there is a copy of the information for each of the daughter cells that will form when the parent cell divides.

Many other essential cellular processes occur in the space between the nucleus and the plasma membrane, a region called the **cytoplasm**. Here the cell processes food to obtain both the energy and the materials needed to make more DNA, proteins, carbohydrates, and fats. The cytoplasm is also the place where many cellular functions are carried out. For example, it is here that cells make multiple copies of a few specific proteins that assemble into fibers that give a cell its shape and allow it to move (Figure 1.2). The cytoplasm is also the place where cells generate and interpret signals that they send out to and receive from other cells in the body.

Our bodies are formed by a combination of cell reproduction and cell differentiation.

The fertilized egg from which you started life gave rise to every cell in your body. When you consider that there are trillions of cells in an adult human, this is quite amazing. Cell growth and division are the processes by which embryos are formed and organs, like heart and muscle, are made. Moreover, the adult body requires cell division to maintain itself in good order, including the repair of wounds. Cycles of growth and division are at the heart of both making and maintaining every living thing (FIGURE 1.3).

It is obvious, though, that not all the cells in our bodies are the same. Something in addition to cell reproduction is important for making an adult human. The processes by which cells become different, that is, make particular structures and take on specific functions, are called **differentiation**. FIGURE 1.4 shows an example of cells that are differentiating and

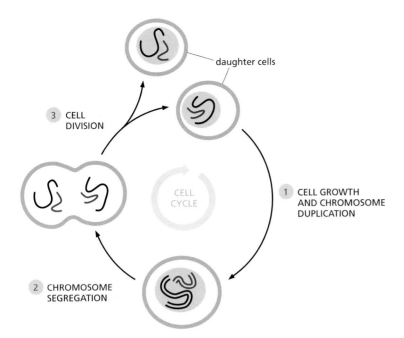

daughter cells

3 CELL DIVISION

CELL CYCLE

1 CELL GROWTH AND CHROMOSOME DUPLICATION

2 CHROMOSOME SEGREGATION

FIGURE 1.3 The cell growth and division cycle. During most of their lives, cells are taking in food so they can make the molecules and structures they need to do their jobs. For example, they use the energy and materials from food to duplicate their DNA and to make enough copies of all the other cellular structures to be sufficient for two complete cells (1); then they are ready to divide. During division, the cell first organizes and separates the two copies of its DNA (2), then it divides into two distinct objects called daughter cells (3). (Adapted from Alberts B et al. [2014] Essential Cell Biology, 4th ed. Garland Science.)

are grouped together to make skin, an example of what a biologist would call a tissue. The layer near the bottom of the image shows cells that are in the process of growth and division (blue arrows); their previous cycles of growth and division gave rise to all the cells that now lie above them. After having been born, as a product of a cell growth and division, the cells in the upper layers differentiated to become tough and water-resistant, so they can provide the protection against infection, dehydration, and potentially damaging blows that keeps our bodies healthy.

Cell differentiation is possible thanks to the ability of every cell to harbor a very large amount of information about how to make many different structures. The information in a cell is like a library that contains many books. In any given human cell, only a few of these books are being read; the rest are in storage. In reality, of course, this library is the DNA that makes up our chromosomes; the books are sets of **genes**. The word gene is familiar to us from the inheritance of **traits**, like brown eyes or red hair, but genes are actually specific stretches along a molecule of DNA that can be read by the cell as instructions, commonly for making a protein that will do a particular job, for example, helping to establish eye color (FIGURE 1.5).

Almost every human cell contains two copies of every gene, one contributed by the egg that came from mother and one by the sperm that came

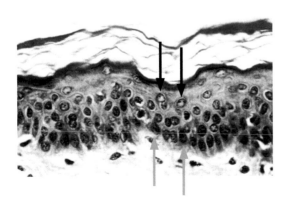

FIGURE 1.4 Cells in a tissue differentiate to do specific jobs. A light micrograph of a thin slice cut through the cells of skin. These cells have been stained with dyes that give different colors to their various parts. The top of the figure is the outer surface of the skin. It is covered by thin cells that are no longer alive. These cells have died because they filled themselves with protein fibers to protect the growing and dividing cells beneath. Under this top layer, the cells are in the process of differentiation. They still contain nuclei (black arrows) and cytoplasm (purple). Beneath these are other cells that are growing and dividing to make more cells that can differentiate (blue arrows). At the very bottom of the image, the clearer layer is mostly fibers that lie outside the cells that made them. These give our skin its physical properties, like strength and flexibility. (From Young B et al. [2003] Wheater's Functional Histology, 4th ed. Churchill Livingstone.)

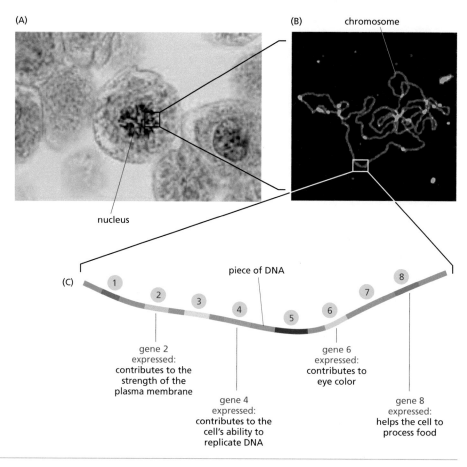

(A)

(B)   chromosome

nucleus

(C)   piece of DNA

1

2

3

4

5

6

7

8

gene 2
expressed:
contributes to the
strength of the
plasma membrane

gene 6
expressed:
contributes to
eye color

gene 4
expressed:
contributes to the
cell's ability to
replicate DNA

gene 8
expressed:
helps the cell to
process food

FIGURE 1.5 The DNA in a cell contains many genes, but only a few of them are expressed in any one cell at any one time. (A) A light micrograph of a cell stained to color its DNA dark blue. (B) An isolated chromosome is stained yellow to show its DNA, which forms a long thin molecule visible in a light microscope. (C) In this diagram of a short segment from a much longer DNA molecule, each number indicates a different gene positioned along this stretch of DNA. Some of the genes are expressed, while others are not. Genes 2, 4, and 8 are instructions for functions that are essential in every cell. Other genes are expressed only in certain cells; for example, gene 6 contributes to eye color. It is expressed in certain cells of the eye but not in cells that make up muscles, intestine, or any other tissue. Many other genes, such as the ones labeled 1, 3, 5, and 7, are not being expressed in this cell at this time. In other cells or at other times, they may be turned on when their gene product is needed. (A, courtesy of Ed Reschke/Getty Images; B, courtesy of JPK instruments.)

SIDE QUESTION 1.2

Think of three examples of traits in which your parents differ markedly from each other. Now consider those traits in yourself; do your traits resemble those of your father or your mother?

from father when the egg was fertilized. Our cells are said to be **diploid** because of this double content of DNA (FIGURE 1.6). Note, however, that although the two sets of genes include the same sets of instructions, the instructions are commonly not identical. Your mother and your father both have genes to define the color of their eyes, but they may have different traits, such as brown eyes or blue eyes. This is possible because they have different forms, or **alleles**, of the relevant genes. In the example of eye color, everyone has the genes that define the color of eyes, but there are different alleles of these genes that produce eyes that are blue, brown, or other colors.

A cell can become different from its parent cell through reading out different sets of genes.

Differences in cell structure and function are achieved by each cell opening one or more of its stored books, reading the information therein, and then following those instructions to build the objects described. Since each instruction is a gene, the process of implementing these instructions is called **gene expression**. For example, many of your white blood cells are protecting you from infection by reading the books that tell them how

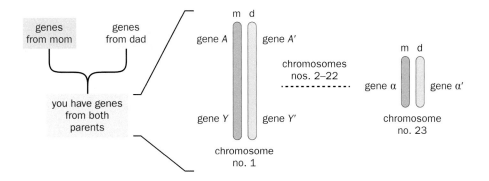

FIGURE 1.6 Both mother and father contribute genes to each individual in the next generation, so human cells have two copies of almost every gene. The pink and blue rectangles represent all the genes in the egg and sperm that made the zygote from which you grew; in the green rectangle, these genes have combined. All genes are on pieces of DNA called chromosomes, of which you have 46, half from mom (m) and half from dad (d). On each chromosome (numbers 1 and 23 are shown; the rest are not shown) there are multiple genes, represented here by genes *A*, *Y*, and *α* from mom and *A′*, *Y′*, and *α′* from dad. Gene *A* might code for a structure necessary to make cytoplasmic fibers. The two genes *A* and *A′* might or might not be identical, since they came from different parents; similarly for genes *Y* and *Y′*, *α* and *α′*, and all the other thousands of genes you have received from each parent.

to make molecules called antibodies, which help fight infections by binding to a bacterium or virus and inactivating it. Other white blood cells can inactivate pathogens by surrounding them with their plasma membrane, internalizing them, and digesting them like food (MOVIE 1.2). The books read by white blood cells are different from the books read by muscle or nerve cells, which don't make antibodies but do make other proteins necessary for them to do their own jobs. Almost every cell in any one person contains exactly the same library of information, but each cell reads only a fraction of the books it carries. Since many books of information are required to make the ~250 types of cells in our bodies, this library of instructions is very large. We now know that it contains about 20,000 distinct genes, each an instruction for how to build one or more structures that cells need to thrive, grow and divide, or differentiate.

The ability of cells to differentiate is the basis for the emergence of different kinds of cancer.

Each of the many cell types in our bodies is equipped to perform different functions that must be carried out to keep the body healthy. To perform these diverse functions, different cell types express different genes and therefore contain different structures (FIGURE 1.7). In this figure you can see a few examples of just how different cells from the same body can look: blood cells of two types, cells from the muscles that let you move your arms or legs, cells that make bone, and nerve cells, which a biologist calls neurons, that help you to perceive your environment, think, and respond. Even these few examples make the point that differentiation can lead cells to become very different in both structure and function.

The simplest way to think about this extensive diversity is to place the many cell types into categories. Doctors and scientists do this by thinking of each cell as part of a category of cells with similar or analogous functions. For example, the cells of skin, shown in Figure 1.4, are part of a tissue called an **epithelium**, a word that come from two Greek words, *epi*, which means on top of, and *thele*, which means nipple. In modern usage, epithelia (the plural) are tissues that cover parts of the body. Skin is covered by an epithelium, and so are many internal organs that are in some sense connected with the outside of the body, like your mouth, throat, stomach, and intestine. Many different cell types are classified as epithelial cells, and these are particularly important in our study of cancer, because more cancers arise from epithelia than from any other tissue.

(A)   (B)   (C)   (D)

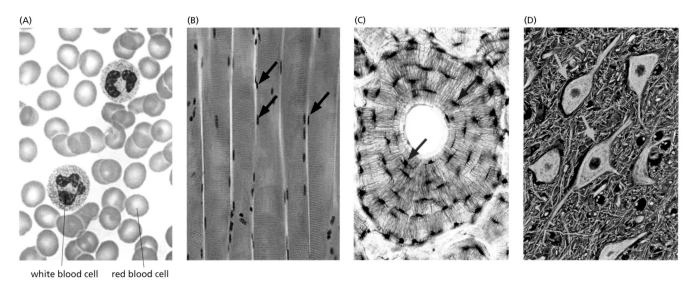

white blood cell   red blood cell

FIGURE 1.7 Light micrographs of stained, differentiated cells from different parts of the body. (A) Two kinds of blood cells are shown: red blood cells that carry oxygen from the lungs to cells in other parts of the body and white blood cells (stained purple in this picture) that help to fight infections. Note the irregularly shaped nuclei in the white cells, whereas the red cells have no nuclei. (B) Muscle cells are elongated tubes that contain many nuclei (black arrows). These cells form by the fusion of many single cells to make the big ones that allow our muscles to contract. (C) Bone is composed mostly of hard, precipitated minerals that lie outside of cells, but it also contains cells (red arrows) that make and secrete mineral-precipitating fibers. (D) Neurons (yellow arrows) in the human brain extend long processes that connect them with other neurons far away. The darker blue material in the background is largely composed of cells called glial, which support the neurons. (A, courtesy of Jarun Ontakrai/Shutterstock.com; B, courtesy of Glenda Stovall; C, courtesy of Tim Arnett; D, courtesy of Spike Walker.)

Another tissue that is important for cancer is blood. It is a little strange to think of blood as a tissue, since it is fluid, but it connects many important processes within your body and is, therefore, classified as a **connective tissue**. The most common cells in blood are the red blood cells that carry oxygen from your lungs to all other cells in your body and bring back the carbon dioxide that is produced by other cells as they process the nourishment you have eaten. The white blood cells, mentioned above, are another critically important cell type in blood, because of their role in protecting you from infection. Cancers can form from the cells that differentiate to make either kind of blood cell. Those that form from the precursors of white blood cells are called leukemias; they are the second most common kind of cancer in humans. Cancers derived from the precursors of red blood cells are comparatively uncommon.

The cells that make muscle or bone are clearly very different, both in their appearance in Figure 1.7 and in the character of the tissues they make, but they too can give rise to cancers. Finally, cancers can also arise from the cells of nervous tissues. They are rare, but they do occur.

Differentiation of cells is a big and important subject for biologists to understand.

The processes of differentiation are still under study by biologists who want to understand **development**, that is, how our bodies, and those of other organisms, form from a single cell through cycles of growth and division, accompanied by differentiation. Much has already been learned about the relevant pieces of DNA, about how a given cell chooses to express genes from one stretch of DNA rather than another, and about how specific instructions contained in DNA are implemented. Implementation is usually carried out by proteins that are made at the instruction of a given gene. Some of these proteins act inside the cell, and some act outside, for example, by forming a matrix of fibers that surrounds the cell

FIGURE 1.8 **Animals expressing a gene received from a different organism.** The sperm-forming cells that contributed to this litter of six mice were engineered to contain DNA taken from a jellyfish. This DNA included a gene that encodes a protein that makes the jellyfish green, so this trait has been passed on to half of the mice born from a female mouse fertilized by these sperm. This is an example of successful gene transfer and genetic engineering. (Courtesy of Ralph Brinster, University of Pennsylvania.)

and helps to define its environment. The processes of development are among the most remarkable of all aspects of life, and the study of development is one of the most vigorous and exciting fields of biology. It is truly awe-inspiring to see the pathways by which our bodies form and the ways in which the study of biology is revealing them.

Remarkably, the processes of development are shared by organisms with a wide range of appearances and complexities. For example, studies on development of bacteria, yeast cells, a kind of fruit fly, a little soil worm, a chicken, a simple fish, and a mouse have all been deeply informative about the ways in which the human body is formed. Perhaps even more remarkably, some specific pieces of DNA, that is, particular genes, can be transferred from one organism to another, and they work perfectly well in their new habitat (FIGURE 1.8). This is strong evidence for fundamental similarities that underlie the many different forms of life.

These similarities have been of tremendous importance in the progress scientists have made with understanding cancer. In part this is because cancer in many organisms is quite similar to cancer in humans, so one can learn a great deal about human cancers by studying similar diseases in animals one can keep in a laboratory. The great value of studying cancer in mice or chickens, for example, is that one can carry out experiments with these animals that could not morally be carried out with people. You will learn in later chapters how cancer has been induced in experimental animals through the action of radiation, toxic chemicals, and even viruses. The results from this work have led to discoveries about cancer that have subsequently saved many human lives.

In a healthy adult, most cells, but not all, are no longer growing and dividing.

Most cells in an adult's body have already differentiated to perform some function that is important for the health of the body as a whole. For example, red blood cells arise through the division and differentiation of cells in the marrow of our bones. Their differentiation includes the synthesis of large amounts of the protein hemoglobin, which carries oxygen from the lungs to other tissues. The final step in differentiation of red blood cells is the elimination of their nucleus, so all the space inside the plasma membrane can be devoted to hemoglobin and the carrying of oxygen. Mature red blood cells never grow and divide, and in fact they live for only a few months before our body kills and replaces them, thereby making sure that all the red blood cells in circulation are fresh enough to do their important jobs well. The muscle cells, bone cells, and neurons shown in Figure 1.7 are all differentiated for their tasks. In the body of an adult, they will not

How do you suppose all these different living things came to be so similar in the ways they develop?

SIDE QUESTION 1.3

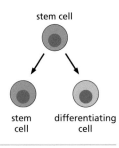

stem cell

stem
cell

differentiating
cell

FIGURE 1.9 **Stem cells are specialized to keep on dividing in an adult, producing both another stem cell and a cell that can differentiate to perform a particular job.** In this diagram of stem cell division, one daughter cell retains the character of a stem cell. It will grow and divide again. The other daughter cell can either differentiate immediately or divide several times to form many cells that will all differentiate.

SIDE QUESTION 1.4

The remarkable thing about a stem cell is that when it divides, its two daughter cells do different things: one continues to serve as a stem cell and will divide again, while the other differentiates to become a cell with special functions. How can two daughters of one cell become different? Note that this is a question scientists are still trying to answer with confidence. It's a hard question, but you can think of possible answers.

SIDE QUESTION 1.5

If you were looking at the cells of a neoplasm through a microscope, would you expect to see more or fewer cells in the process of division, compared to a normal tissue? Why?

grow and divide again. As depicted in Figure 1.5, these differentiated cells are expressing only the genes that are necessary for them to do their particular job. Thus, the control of gene expression is at the heart of differentiation.

In most tissues, however, a few cells remain largely undifferentiated and retain the ability to grow and divide. These are **stem cells**, which are specialized to continue producing new cells throughout the life of the organism. Stem cells undergo an unusual kind of cell division in which one of the two daughter cells continues as a stem cell, able to grow and divide again, but the other daughter will differentiate to make replacements for differentiated cells that have worn out or died (FIGURE 1.9). In this way, our bodies do preventive maintenance: they replace old cells with new ones before the old ones are failing to function properly. In epithelial tissues like the surface of skin or the lining of your intestine, this process goes on all the time and is important for keeping the body healthy. Skin cell replacement is also an important part of healing wounds. In another example, there are stem cells in your body's muscles. These are dormant much of the time, but if that tissue is injured or if persistent exercise creates a need for more muscle than was previously present, these stem cells enter the cell cycle, grow and divide, and produce more muscle cells as needed. Thus, cell division in the right place, at the right time, and in the right amount is an essential part of normal adult body function.

WHEN NORMAL CELLS MISBEHAVE

Cancer begins when the processes of normal tissue formation and maintenance go wrong. The cells of a given tissue should be growing and dividing just often enough to replace the differentiated cells that are dying or being lost. If the growing cells reproduce too often, they make more daughter cells than are necessary, and this can cause trouble. If the rate at which a given cell type is produced exceeds the rate at which that kind of cell is dying, there has been a loss of normal control, a cellular misbehavior that is characteristic of cancer. The medical name for this behavior is **hyperplasia**, and the resulting extra growth is called a **neoplasm**. Many neoplasms will cause no serious trouble (FIGURE 1.10), but when cells divide too often, the daughter cells often don't differentiate properly. If the process continues, the result is a mass of cells that cannot do any job very well and cannot contribute to the well-being of the body as a whole. Such a mass of cells is part of what is called a **tumor**, which is really just another word for a large neoplasm. It is a set of cells that have formed as a result of their uncontrolled behavior. However, for this cell mass to continue growing and to get larger than a fraction of an inch on a side (a few million cells), the tumor must induce nearby blood vessels to grow in its direction, so they will provide the kind of blood supply that keeps normal cells well nourished. If this happens, the cells in the tumor can continue to grow and divide, consuming food without achieving any useful purpose. This is wasteful, but it is unlikely to be lethal. If a tumor becomes big enough to be seen or felt, it commonly attracts medical attention and can be removed. If all the cells in the tumor are confined to that one region of the body, the growth is said to be **benign**, and it can usually be removed by surgery, which solves the problem.

What changes occur when a cell is transformed from a normal cell into a tumor-forming cell? This is a big subject. It will be treated with some detail in Chapters 6, 7, and 10, but the fundamentals of the changes can be described simply. A normal cell's growth and division cycle is controlled in part by information in one of those books mentioned above, that is, a particular set of genes. These genes must be expressed by every cell in the body, so all cells are properly instructed about how and when to reproduce. The information about when to grow and divide is conveyed to each cell by signals that come from nearby tissues or from the body as a whole. These signals are chemicals that either circulate in the blood or act locally in a given tissue. They define the rate of growth and division for each cell type. As a result of the instruction given by these signals, the number of cells formed in each tissue is just sufficient for its proper maintenance or repair. However, if the genes responsible for giving or receiving these signals are altered, then cell growth and division can start when it should not.

Changes that alter a gene are changes in the structure of the DNA. These are called **mutations**. If the DNA in a cell has been mutated so that cell divides too often, this division produces more cells of the same kind—that is, each daughter cell will also be mutated, so it too will divide more often than it should. The result is a group of identical cells, or a **clone**, that divides without obeying instructions; the consequence of all these cell divisions is a neoplasm or tumor.

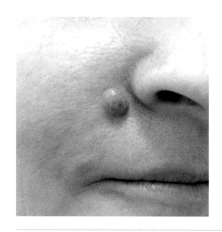

FIGURE 1.10 **A small, benign overgrowth of cells in the skin.** Small and localized growths, known as benign neoplasms, result from too many cell divisions in one region of the body. Benign neoplasms are common and not dangerous. [From skinsurgeryclinic.co.uk.]

A mass of misbehaving cells can progress to cancer by accumulated additional mutations.

A tumor, like the one described above, is not, however, cancer. Additional departures from normal behavior are required to change cells into ones that are likely to be life-threatening. The most significant of these is the loss of a cell's normal tendency to stay bound to its neighbors, serving as part of a particular tissue. This change produces cells that can wander away from the tumor without losing their ability to grow and divide (FIGURE 1.11). The result is cells that can migrate through the body and take up residence where they don't belong, starting again to grow and divide without control, and thereby forming new tumors in new places. This process is called **metastasis**, and tumors that behave like this, or have the potential to do so, are said to be **malignant**. They are life-threatening because it is hard for a doctor to find all such growths and eliminate them before they have become big enough both to make trouble in their own

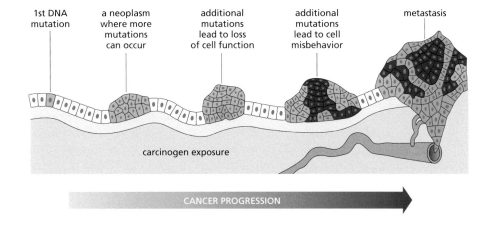

FIGURE 1.11 **The progression from a benign neoplasm to cancer.** From left to right, this figure diagrams the passage of time and the associated accumulation of mutations. The first mutation leads to a neoplasm, that is, a cluster of cells that all contain the first mutation. Additional mutations, indicated by cells of different colors, can then lead to the loss of normal cellular control, to further misbehaviors, and ultimately to the loss of local cell–cell attachments. Now the cells can wander away from their site of birth and enter a blood vessel to travel long distances and initiate growth elsewhere in the body.

SIDE QUESTION 1.6

Why is metastasis such a dangerous property in a cancer cell?

right and to start the whole process over again. Cells that behave in this uncontrolled manner form a malignancy and are real cancer cells.

The above description implies that there is a progression from normal cell behavior to malignant cancerous growth and that different tumors can be at different stages in this progression. This idea is correct, because **cancerous transformation**, that is, the process by which a cell acquires the properties of cancer, results from the accumulation of mutations in numerous genes that control multiple, specific aspects of cell reproduction and behavior. Such mutations are said to be **carcinogenic**, which means cancer-forming. Current knowledge about this transformation process suggests that around six genes must be altered before a normal cell is so changed that its daughter cells form a malignant tumor. According to scientists' understanding of mutation frequencies, this makes the emergence of cancer extremely unlikely. The fact that cancers do occur, and all too frequently at that, tells us that some carcinogenic mutations increase the likelihood that more mutations will occur; that is, they make the cell genetically unstable. Thus, cancerous transformation is the result of a series of unfortunate changes in a given cell's DNA. These changes include (1) mutations that increase its frequency of cell division, (2) mutations that reduce the fidelity with which cells pass information on to their daughters, and (3) mutations that loosen the strength of a cell's interactions with its neighbors, allowing metastasis.

Many characteristics of cancer cells are also found in normal cells of various types.

In spite of all these mutations, cancer cells are strikingly similar to some of the normal cells in our bodies. For example, stem cells are designed to keep on dividing and producing daughter cells, even in an adult. When a normal cell becomes cancerous and starts to divide in a tissue that would normally lack cell division, it is behaving a bit like a stem cell. In this way the behavior of cancerous cells is not really abnormal; it is the uncontrolled appearance of normal behaviors in the wrong place and at the wrong time.

SIDE QUESTION 1.7

Human sperm cells can migrate considerable distances as they seek an egg cell to fertilize. In what ways is this example of cell motion similar to or different from metastasis?

The same idea is relevant even for a deadly cancerous behavior like metastasis. One of the normal functions of white blood cells is to wander among other cells in the body (*see* Movie 1.2). White blood cells are designed *not* to stick to their neighbors; this property is essential for their ability to crawl through the walls of blood vessels (FIGURE 1.12A) and between the cells of other tissues, seeking out and destroying bacteria or other invading organisms that have gained access to the body's interior. These traits are necessary for white blood cells to fight infection effectively. White blood cells are not cancer cells because their cycles of growth and division are still under rigorous control. The fact that malignant cancer cells become capable of wandering in a way that resembles the normal behavior of white blood cells points out that cancer is not the *invention* of traits, it is the *emergence* of behaviors where they don't belong (FIGURE 1.12B). In short, carcinogenic transformation is a set of changes that lead to the loss of normal control on cell behavior.

Diagnosis of cancer requires the identification of inappropriate cell growth.

The determination of where the cells of a particular tumor lie along the pathway of cancerous transformation is one of the principal considerations of accurate cancer diagnosis. This assessment is usually accomplished through collaboration among several medical specialists. Commonly, a potential cancer is initially spotted by a general practitioner, who notices

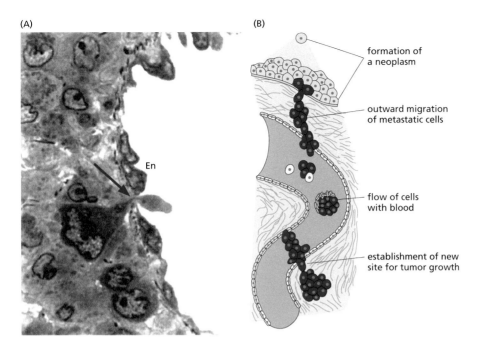

FIGURE 1.12 Picture and diagram of metastatic behavior. (A) A white blood cell (red arrow) is migrating through the wall of a blood vessel, labeled En. Once inside the vessel, the cell can move with blood to get around the body quickly. (B) Diagram of the whole process of metastasis, starting with the formation of a neoplasm, which is also diagramed in Figure 1.11. Metastatic cells leave their site of birth and initiate long-range migration by entering a vessel, where their movement to a distant part of the body can be quite fast. If they can then attach to the wall of a vessel and pass out through it, they can initiate a new site for growth. (A, from Song ML et al. [1986] *Tissue Cell* 18:817–825. With permission from Elsevier. B, adapted from Cotran RS, Kumar V & Robbins SL [1994] Pathological Basis of Cancer, 5th ed. W.B. Saunders Co.)

a medical problem. S/he will then refer the patient to a team that includes doctors with very specific training. This group commonly includes a **radiologist**, who is trained to take pictures of structures inside the body, using **X-rays** or analogous imaging tools. There is also a **pathologist**, a doctor trained to recognize abnormalities in cells and tissues, using microscopes to examine tissue samples taken from the patient. If the patient is found to contain a growth that appears to be cancerous, then an **oncologist** (a cancer specialist) joins the team to help characterize the problem and identify and administer the proper treatment.

Finding a tumor requires that it is big enough to show up in some kind of test. A skin tumor that grows on the surface of the body is comparatively easy to spot, so these tumors are usually caught early, well before their cells have left the original neoplasm and metastasized. Breast cancers, too, can often be identified early, through either careful self-examination, a doctor's scrutiny, or a mammogram, that is, an X-ray of the breast. Early identification of a tumor is important because it denies the tumor's cells the time to complete cancerous transformation and become malignant. If a growth is caught early, the abnormal cells can commonly be removed before they are far enough along the pathway of cancerous transformation to become metastatic.

However, tumors that arise deep in the body are generally harder to find early than those near the body's surface. Growths that form within an internal organ that is surrounded by enough space to let the tumor expand without causing pain, for example, an ovary or the pancreas, are very hard to spot. Such tumors often grow for quite a while without being noticed. This allows them to progress along the pathway to malignancy before they are detected.

There are, however, some comparatively gentle ways to look at parts of our insides and see if anything is growing where it should not. For example, a colonoscopy can examine the rectum and a significant fraction of the intestine by introducing light-carrying fibers that allow a video camera to image the inner surface of this body tube. Image quality with this method is now good enough to identify a pre-malignant condition that can be treated or removed, thereby preventing the formation of a malignancy. You will read more about these and other examples of important diagnostic tests in Chapter 2. Medical scientists are working hard to develop even better ways for finding tumors before they have developed cells that can metastasize.

Tumors are currently classified by their place of origin.

When a tumor is found, one knows where it has been growing, for example, in the liver or lung. Given the possibility of metastasis, however, this is not necessarily the place where the cancer began. Cancers of different origins can behave quite differently, and they are often best treated in ways that are specific for that tumor type, so identifying the site of cancer origin can be important. Commonly this is done by a close look at the structure of the tumor's cells. For example, if they look like intestinal cells, they probably came from that tissue. With modern microscopy, it is usually possible to pinpoint tumor origin with precision, and this can have a positive impact on the effectiveness of subsequent treatment.

Once a cancer's origin is established, a doctor knows an important aspect of cancer type. If the tissue of origin was an epithelium, the cancer is a **carcinoma**. Most cancers arising in the throat, stomach, and intestine are carcinomas. Many other internal organs are also made in part from epithelial tissues, so it is no surprise that carcinomas are the most common kind of human tumor. In addition, however, epithelial cells seem more inclined to cancerous transformation than other cell types (FIGURE 1.13). This tendency is probably a result of two things: (1) many epithelia are exposed to the outside world, so they are more easily affected by damaging conditions in the environment than cells confined to the body's interior; and (2) epithelia are continually being renewed by cell division,

SIDE QUESTION 1.8

All of the methods for identifying tumors described above are based on visualizing extra cells where they don't belong. Can you think of any properties of cancer cells that might identify a tumor even earlier, if we could figure out how to visualize it?

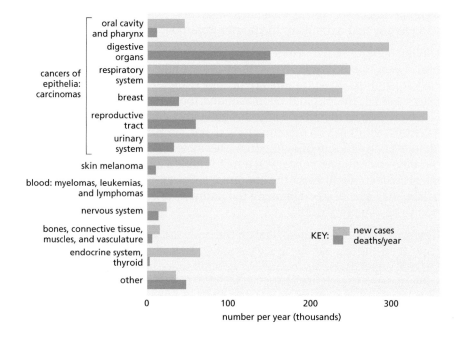

FIGURE 1.13 The frequency of cancer in various tissues of the human body. Green bars show the numbers of cancers detected in the United States per year in each human tissue. Light blue bars show the number of deaths per year from that kind of cancer. The differences between the lengths of the green and blue bars indicate the frequencies of cures. A small difference (low likelihood of cure) is often related to the chance that metastasis will have occurred prior to the time of cancer detection. (Data reported by American Cancer Society, Cancer Facts and Figures, 2014.)

so the cells that compose them have gone through more cell divisions than most other tissues. Their tendency to divide may make it easier for them to become cancerous, a subject that is considered more fully in Chapters 6, 7, and 11.

Cancer is also relatively common in cells of the reproductive systems and of breast tissues. The cells of some of these tissues reproduce comparatively often, and this reproduction is under the control of sex hormones, like estrogen in women and testosterone in men. This kind of control appears to make cells more likely to initiate reproduction when they should not.

Tumors also occur in connective tissues, that is, tissues that connect, support, or bind other tissues and organs. Cancers of solid connective tissues, such as bone, are called **sarcomas**, and they are comparatively rare. Cancers that arise from white blood cells are, however, much more common. These cancers are special because they almost never form solid tumor masses. **Leukemias**, myelomas, and lymphomas are three kinds of blood cancers that form when different blood-forming stem cells go through cancerous transformation.

Nervous tissue can also give rise to several kinds of tumors. Tumors sometimes arise from neurons when they are still very young, for example, in embryo. Cells that are going to become neurons are called neuroblasts. These can give rise to a particularly nasty tumor called a **neuroblastoma**. Such tumors are rare, but they are much more common in children than in adults. Nervous tissue also contains many additional cell types that are collectively called glia, which means glue. This name derives from the fact that when these cells were first seen, scientists presumed they were simply holding real neurons in place. We now know that glia perform numerous supporting functions that help neurons do their jobs. Glia are more likely to form tumors than neurons, probably because they, like epithelia, are more prone to growth and division than other cells in nervous tissues.

The dangers of cancers in different tissues are not, however, simply related to the frequency with which cancers arise in them; additional important factors are the ease with which tumors can be detected and treated. These differences are reflected in the different lengths of the green and blue bars in Figure 1.13. Chapter 2 describes the many methods by which cancers in different tissues can now be identified; cancer treatments are described in Chapters 8 and 10.

Cancer is far more common in older people than in the young.

Careful studies comparing the number of people diagnosed with cancer versus the patient's age at the time of diagnosis have found a steeply increasing curve (FIGURE 1.14). In the context of the fact that cells must accumulate multiple mutations that change them in several ways before they become truly cancerous, this makes sense; all those changes take time. However, the risk of cancer can also depend on environment, because factors outside the body can influence the rate at which mutations occur. The effects of these external agents are also likely to accumulate with age, so this aspect of carcinogenesis contributes to the greater frequency of cancer in old people. The large and important topic of environmental effects on cancer incidence is the subject of Chapter 5. However, the likelihood of getting cancer can also run in families, suggesting that some aspects of cancer prevalence are inherited; this kind of risk does not increase with age. Nonetheless, genetic inheritance of cancer risk can also

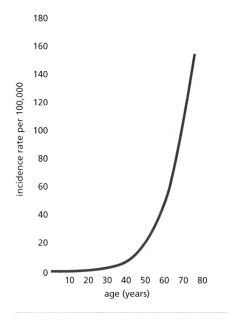

FIGURE 1.14 **The steep rise in the incidence of cancer with human age.** This curve indicates how much more likely it is to get cancer when you are older than when you are young. In this example, we see the age distribution of the number of newly diagnosed cases of colon cancer in women during one year in England and Wales. (Data from Muir C et al. [1987] Cancer Incidence in Five Continents, Vol. V. International Agency for Research on Cancer.)

be understood in terms of mutations, because some mutant alleles of genes relevant to cancer can be passed from parent to child, as described in Chapters 6 and 7.

CURRENT TREATMENTS OF CANCER AND PROSPECTS FOR THE FUTURE

The goal of most current cancer treatments is to remove or kill all the cancer cells in a patient's body. The biggest difficulty in achieving this goal results from the fact that the cancer cells developed from the cells of our own body. They are therefore sufficiently similar to normal cells that it is hard to kill them without harming normal cells too. A bacterial infection can be treated by giving the patient an antibiotic, a chemical that kills bacteria specifically. These drugs work by blocking some bacterium-specific process, such as the formation of an extracellular wall that bacteria build around their plasma membranes. Without their walls, bacteria burst open and die. Human cells don't build such walls, so this kind of antibiotic, such as penicillin, has very few bad effects on our health. It is, however, very hard to find chemicals that kill only cancer cells.

Current cancer treatment strategies include interventions with surgery, radiation, and toxic chemicals.

Surgical removal of a tumor is an excellent form of treatment, so long as the tumor is a single cell mass that can be identified with certainty and cut away without damaging the rest of the body. A kidney tumor that has not metastasized and is surrounded by an encasing layer of tissue is an example of a good subject for successful surgery. Analogous tumors can be found in other parts of the body: the thyroid gland, intestine, liver, and even lung. A tumor that lies deep within the brain, however, is an example of a tumor that is very hard to remove without causing serious damage. In cases where surgery is not an option, other treatments must be used. For example, tumor cells can be killed by exposure to high doses of certain kinds of radiation, such as X-rays. A skilled radiologist can shine beams of X-rays on a tumor and cause many of its cells to die, reducing tumor mass and sometimes even causing the tumor to disappear. These and many other cancer treatments are discussed more fully in Chapters 8 and 10.

If neither surgery nor radiation is effective or sufficient, for example, because metastasis has already occurred, there remains the option of **chemotherapy**, a treatment of the whole body with drugs that block the growth and division of cells and can even kill them. The difficulty with chemotherapies lies in the similarity between cancer cells and normal ones; most of these treatments are not truly cancer-cell-specific. The majority of cancer drugs in current use are inhibitors of some basic cellular process, like DNA replication or cell division. If they are good at blocking the reproduction of tumor cells, they commonly interfere with the reproduction of normal cells too, preventing essential tissue maintenance and repair. This lack of specificity accounts for many of the side effects that are common with cancer drugs. Hair may fall out, because the cells that make hair are blocked in their normal cycles of growth and division and are sometimes killed. Nausea and diarrhea are common, because the cells that normally divide to renew the lining of the intestine cannot do this job.

A significant achievement by medical cancer specialists has been the discovery of ways to administer chemotherapies so the damage done to

SIDE QUESTION 1.9

Describe how an ideal cancer treatment would work, even if you can't specify how to do it.

cancer cells is greater than to the rest of the body. The results are often very helpful for cancer treatment in the long run, even if their short-term effects are highly unpleasant. An example is the treatment for some acute leukemias in children; real cures can be achieved in >80% of the patients for reasons discussed in Chapters 8 and 10.

Some cancer treatments derive from ancient medical practices.

There is a group of cancer treatments that are not the fruits of medical science; these are often called alternative cancer therapies. Cancer has been a feature of the human condition for a long time; there is even evidence for cancer in mummies found in the tombs of ancient Egyptians. Medical traditions, like those developed in Asia and Africa, have long prescribed specific herbs, teas, and/or mushrooms for cancer prevention and treatment. Some of these are probably helpful in cancer medicine, but one must realize that if they were true cancer cures, this disease would have been eliminated many years ago. It follows that alternative medicine is not likely to provide a cure for cancer, though some of these treatments may reduce cancer risk. If you are dealing with a cancer diagnosis, some alternative treatments may be worth considering, but you can't expect them to provide a cure.

There may never be a true cure for cancer.

The above descriptions of cancer treatments pose the question: Isn't there some way really to cure cancer, that is, to turn cancer cells back into normal cells? At the moment, the answer to this question is, simply, no. Medical science does not know how to cause cancer reversal because we do not know how to undo the numerous mutations that are at the heart of cancerous transformation. There are some experimental treatments, now conceived and beginning development, that might ultimately reverse cancerous transformation, as discussed in Chapter 10. However, as you learn more about cancer biology, you will see why achieving such a goal will be very hard indeed.

Much scientifically based cancer research is currently focused on finding drugs or other treatments that are truly specific for cancer cells, inhibiting their growth and/or killing them while leaving normal cells unperturbed. This is not impossible because tumor cells take on a few properties that are not found in the normal cells of an adult body. Several treatments of this kind are already available for certain cancers, as described in Chapter 8. Such treatments are exciting advances, and they hold promise for even better things to come, as discussed in Chapter 10. Moreover, the methods of molecular biology are allowing oncologists to determine exactly which genes have been mutated in the course of a particular cancerous transformation. This advance is leading to a more personalized approach to cancer treatment, which is likely to improve both the treatments themselves and a patient's prospects for recovery, albeit at considerable financial cost.

There are at least four reasons why cancer will probably be a major medical problem throughout your lifetime and beyond. First, as described above, cancer is significantly more common in older people than in the young. Humans now live longer than in the past, thanks to improved nutrition, reduction of workplace hazard, and better medicines for the prevention and treatment of many diseases. All of us who live past middle age have an increased risk of developing a cancer at some time before

Are alternative cancer therapies more similar to surgery, radiation treatment, or chemotherapy?

SIDE QUESTION 1.10

we die. Second, cancer is intrinsically hard to treat. As described above, the germ that is causing this disease is derived by mutation from our own cells, so it is difficult to find treatments that are cancer-cell-specific. Third, cancers are highly variable. The specific mutations that lead to colon cancer in one person can be different from those causing the same disease in another, so the best treatment for a particular individual is not easy to find. Improvements in molecular biological tools for diagnosis are helping to solve this problem, but we are still a long way from truly effective, personalized treatments for most types of cancer.

Finally, the mutations that lead to cancerous transformation include changes that reduce the fidelity with which cellular instructions (molecules of DNA) are passed from one cell generation to the next. It follows that the emergence of cancer is accompanied by instability in cellular information management. As a result, the genes in cancer cells are changing all the time. A treatment that is effective in shrinking a tumor on one day will often lose its effectiveness as the tumor cells change.

Finding effective anti-cancer drugs is therefore an ongoing challenge, and it will probably take many innovations to give us treatments that approach being cures for most cancers. More practical approaches to the cancer problem currently lie in cancer prevention and early detection, as described in Chapter 11. Innovations in these fields are already improving cancer therapy and reducing the suffering that cancer causes. If the age of the American population is taken into account, the annual rates of death from cancer have recently been dropping by more than 1% per year, and even the rate of cancer incidence is going down. Thus, progress is being made; it is just not as fast as anyone would like.

This description of cancer should be enough to let us go forward and talk about some of the practicalities of cancer diagnosis. Our next subjects are the events that occur when an individual first becomes aware of their cancer. Subsequent chapters will provide a deeper understanding of the other issues raised above.

SUMMARY

Our bodies are made from cells and the materials that cells make. Cells are living systems that consume food and use it to make all the molecules necessary for their own growth and division into two cells that can repeat the process. The processes of cell growth and division allow us to grow from a fertilized egg into an adult, and they are essential for maintaining a healthy adult body. However, not all cells in our bodies are identical. Cells can differentiate to accomplish specific functions. Differentiated cells usually do not grow and divide, but they are replaced as they become worn out or die, thanks to the division of stem cells that continue to grow and divide throughout the life of an adult, making the cells necessary to maintain the body in a healthy condition.

Cell reproduction is under tight control. The number of cell divisions in each adult tissue matches the number of cell deaths and/or losses, thereby maintaining a healthy tissue. The transformation of a cell from normal to cancerous begins when that cell loses proper growth control, so it divides when it should not. Progression from this initial event into a dangerous cancer involves additional changes in cell behavior: first a loss in the ability to differentiate, then a loss in the fidelity of gene transmissions from one cell generation to the next, and finally a loss of the ability to stay attached to neighboring cells. These changes result in cells that can wander through the body, forming unwanted growths at sites distant from the original site of excess growth. This behavior, called metastasis, is at the heart of what can make cancer both lethal and difficult to treat.

Cancer diagnosis is commonly achieved by noticing a growth of cells where they don't belong, either through tactile detection of a lump or visualization of an inappropriate growth. Cancer can then be treated, either by cutting out the unwanted cells or by killing them. The latter can be achieved by exposing the cancer cells to radiation or to chemicals that interfere with their growth and induce them to die. However, none of these treatments is entirely satisfactory because they damage a body's healthy cells as well as the cancerous ones. Modern medical scientists are working to find treatments that are truly specific for cancer cells, but this task is difficult for at least two reasons: (1) cancer cells are derived from our own cells by mutation, so even though they have lost normal control, they are still quite similar to normal cells, making it difficult to find treatments that are cancer cell specific; and (2) cancer cells are genetically unstable, so they can change from one cell generation to the next. A treatment that kills unwanted growths one day may not work only a little while later. For these reasons, cancer is a hard disease to cure, and it will probably be many years before modern medicine is really in control of the cancer problem.

ESSENTIAL CONCEPTS

- A cell is the minimum unit of life. Each cell is able to take in food, which provides materials and energy for the synthesis of the molecules it needs to grow, divide, and form two daughter cells.

- All cells arise by the division of previously existing cells. The trillions of cells in a human adult are all descendants of the fertilized egg from which each of us started life.

- Our bodies are made from cells and the things cells make, like bone, the liquids of blood, and the fibers of connective tissues.

- Cells carry a large amount of information that tells them how to make many different molecules and structures. This information is stored in DNA. Each unit of information is a segment of DNA, called a gene; the information in DNA is called genetic information.

- Some genes are expressed in all cells, such as the genes that carry information for making cell membranes or for metabolizing the food a cell ingests.

- Cells can differentiate, that is, take on specific forms and functions, by expressing specific genes. Different genes are expressed in different differentiated cells to provide them with the specialized structures and functions they need to accomplish their jobs for the body as a whole.

- There are also genes that control cell growth. Cell growth and division are normally regulated very carefully. In an adult, the rate of forming new cells exactly balances the rate at with each cell type dies or is lost.

- Cancer is initiated when the normal controls on cell reproduction fail, so cells grow and divide when they should not, making a neoplasm. This failure is a result of mutations that inactivate one or more of the genes that control cell growth. Progression to a dangerous cancer occurs when cells acquire additional mutations that lead them not only to divide too often but also to lose their normal association with neighboring cells and wander through the body, setting up additional sites of growth, a process called metastasis.

- Cancerous progression is accompanied by the development of genetic instability, so when such cells divide, their daughter cells no longer contain exactly the same genetic information as the mother cell. Because cancer cells are prone to genetic change, they are hard to treat.

- Cancers are more common in older people, because their cells have had more time to accumulate mutations.

- Some tissues are more prone to cancer than others.

- Cancers are detected by identifying sites of inappropriate cell growth.

- Cancers are treated by trying either to cut out the unwanted cells or to kill them.

- Metastatic cancers are hard to treat because it is difficult to get rid of the cancer cells that have moved to other, not yet identified parts of the body. They can, however, be treated by administering chemicals that inhibit the growth of or induce death in cancer cells all over the body.

- Chemical treatment of cancer (chemotherapy) is usually not an ideal form of treatment because most drugs that damage or inhibit cancer cells have analogous effects on normal cells and make the body sick.

KEY TERMS

allele	diploid	nucleus
benign	DNA	oncologist
cancerous transformation	epithelium	pathologist
carcinogenic	gene expression	plasma membrane
carcinoma	gene	radiologist
chemotherapy	hyperplasia	sarcoma
chromosome	leukemia	stem cell
clone	malignant	trait
connective tissue	metastasis	tumor
cytoplasm	mutation	X-rays
development	neoplasm	
differentiation	neuroblastoma	

FURTHER READING

Mukherjee S (2011) The Emperor of All Maladies: A Biography of Cancer. Scribner. This is an interesting and moving account by a cancer doctor of both the care of patients and the history of cancer medicine.

The American Cancer Society (http://cancer.org) supports a very informative and reliable Website for people interested in cancer. It is designed first and foremost for people with cancer, so there are many items about diagnosis and treatment of specific cancers, as well as access to support groups. Interspersed among these items are accurate accounts of symptoms and treatments for many specific cancers.

The National Cancer Institute (http://cancer.gov) maintains a different but equally valuable Website from which people can learn about cancer. This site includes information that is useful for patients, the interested public, and cancer professionals.

Scitable by Nature Education (http://www.nature.com/scitable) is a Website designed to help the public learn more about several aspects of biological science, including what has been learned about cancer.

QUESTIONS FOR FURTHER THOUGHT

1. How would you describe what you now know about cancer to a friend who has never read about the disease?

2. Do you think cancer is an infectious disease? Why or why not? Is it inheritable? (You'll learn more about how to answer these questions later in the book.)

ANSWERS TO SIDE QUESTIONS

1.1   If the plasma membrane were perfectly impermeable, there would be no way to get food into the cell or waste products out. The cell would then quickly die.

1.2   There is no single answer to this question. Examples of traits that could be mentioned include eye color, hair color, skin color, height, colorblindness, pitch of voice, strength, and agility.

1.3   This is a very interesting question. For biologists, the answer is clear; the similarity in basic mechanisms across a wide range of organisms is a result of the fact that all organisms are related by evolution. As you go back in the history of life on earth, different groups of organisms, like the primates, all had a common ancestor. As you go back further, all mammals had a common ancestor. When you go back further still, all vertebrates had a common ancestor. This process of identifying common ancestors can continue on back to the appearance of modern cells, which occurred more than one billion years ago. Evolutionary biologists are confident that all living things have grown and evolved from a common ancestor. All organisms are similar in much of their basic biochemistry and molecular biology: for example, their plasma membranes, their DNA, and their ways of digesting food. This is because we all started from a common ancestor that possessed these properties and functions. These basic structures and functions are so fundamental to life that they have not changed much in the very long period of time since they first developed.

1.4   As mentioned, this is a hard question but fun to think about. The way biologists now understand it, there are two principal possibilities: (1) Something in the environment of a stem cell is stronger on one of its sides than on the other. When the stem cell divides, one daughter is more affected by that external environment than the other, so it takes on a different fate; for example, it retains its stem-cell character. Many biologists believe that the environmental effect on stem cells is provided by a cluster of nearby cells that secretes a signal responsible for making this local environment, which is called a niche. (2) The cell itself might contain some component that is asymmetrically distributed between the two daughter cells. For example, there could be a special region on the interior of the plasma membrane that binds key molecules. As a result of being tied down to one place in the cell, these molecules are inherited by only one daughter cell, and they affect the way that daughter develops, as opposed to the development of its sister cell.

1.5   Because the neoplasm forms as a result of too many cell divisions, there would probably be more cell divisions in the neoplasm than in a comparable sample of normal tissue.

1.6   Metastasis is dangerous because the property of wandering means that a cell that is dividing too often can now travel to other places in the body, where it can settle in and start dividing too often again. This property can lead to the formation of tumors in other parts of the body, a more dangerous situation than having all of the misbehaving cells in one place.

1.7   The migration of a sperm cell in its quest for an egg to fertilize is similar to the migration of a cancer cell in the sense that both are examples of cell movement. They are different, though, in many ways: (1) A sperm cell is built with a motor for rapid motion (the flagellum of the sperm), whereas metastatic cancer cells move slowly by crawling, or they simply flow along with blood. (2) A sperm cell was designed by nature to move, whereas most other cell types were originally designed to stay in one place. Their ability to move is a trait acquired through mutation. (3) A sperm cell has a particular cell it seeks, the egg, with which it will fuse to start a new generation, whereas a wandering cancer cell will not fuse with other cells. A metastatic cell will simply find a place in the body where it sticks well enough to settle down and start growing to form a new tumor.

1.8   This is another hard question with no single right answer. Examples of reasonable answers include the following: (1) We know cancer cells are growing and dividing more than most other cells. If we could find a way to stain cells that are synthesizing new DNA or going through cell division, that would give cancer cells a special color, making them easier to spot. (2) Cancer cells might make molecules that are not found in normal cells, for example, something important for DNA synthesis or mitosis. If we could stain that molecule, it would make cancer cells stand out. Note, however, that these two methods would also identify stem cells that were dividing normally for the sake of the host's body maintenance, and this feature would add complexity to cancer identification. Note also that the two ways mentioned here are actually approaches that are now in use for modern cancer diagnosis, as described in Chapter 2.

1.9   An ideal cancer treatment would be one that killed all cancer cells in the body and had no damaging effects on the body's normal cells. If only cancer cells put some unique molecule on their surface, a tag that could be used to recognize them uniquely and unambiguously, the problem of cancer treatment would become easy. The problem in cancer treatment is, of course, that cancer cells derived from our own normal cells and are sufficiently similar to normal cells that they are hard to single out.

1.10   Alternative cancer therapies are more like chemotherapy than any other conventional, scientific treatments. This is because they administer a natural product, like a tea or herb, which contains many chemicals. It is the action of these chemicals that has an effect on cancer, so the treatment resembles chemotherapy.

Cancer detection, diagnosis, and prognosis

2

CHAPTER 2

A person in whom a cancer is developing can occasionally sense that something is wrong. The inappropriate growth of cells can be visible or otherwise noticeable in either obvious or subtle ways. Sadly, this kind of detection is not very sensitive for early cancers, except those that form on skin. More commonly, a doctor detects cancer through the analysis of symptoms and specific tests. In recent years, there has been real progress in developing tools and techniques for early cancer detection. Some of these are simple and cheap, some are expensive and/or uncomfortable, and some carry health risks of their own. However, this group of innovations is now being used to screen people who are at risk of getting cancer, and this policy is helping to reduce the number of people in whom cancers grow undetected until they are difficult or even impossible to treat.

This chapter describes some of the methods now used to screen for and diagnose cancer. We start with the simpler screens, go next to the high-tech medical imaging methods that are used to locate and characterize cancer, and then move on to the ways in which a suspicious lump can be examined more thoroughly to see whether it contains cells that are benign or malignant. Finally, we will see how doctors use this information to decide on treatment and form a **prognosis**, a prediction for treatment outcome.

SIMPLE SCREENS TO IDENTIFY POTENTIALLY CANCEROUS GROWTHS

CHARACTERIZING A POTENTIALLY CANCEROUS LESION

MODERN INSTRUMENTATION TO IMAGE A PATIENT'S INSIDES

EVALUATING THE RESULTS FROM CANCER SCREENING TESTS AND MEDICAL IMAGING

CANCER PROGNOSIS: UNDERSTANDING WHAT A DIAGNOSIS MEANS

LEARNING GOALS

1. Name seven tests used to screen for cancer, and name the cancers for which each is effective.

2. Compare the reliability of various cancer screening tests that do not involve modern medical imaging. That is, how valid are the positive or negative results from each test?

3. Compare the information that results from the simple cancer screening tests with results from the five high-technology methods of medical imaging: endoscopies, sonograms, CT scans, MRI, and PET scans.

4. Identify the risks associated with each kind of low-tech cancer screening test and those associated with the five types of medical imaging used to detect and describe cancers.

5. Describe the strengths and limitations of light microscopy as a tool for characterizing biopsies of potentially cancerous tissues.

6. Explain what is meant by a cancer prognosis.

7. State the meaning of each number in the TMN system for staging cancer and describe the significance of each number for a cancer prognosis.

SIMPLE SCREENS TO IDENTIFY POTENTIALLY CANCEROUS GROWTHS

Some cancers can be identified through the use of simple screening procedures. Most of these are easy and cheap; they range from physical examinations to blood tests. Other tests may require special imaging equipment, but the results can sometimes be very useful in catching cancer early. In this section, we will review examples of cancers and the tests that can be used to catch them early, before they become malignant.

Cancer can sometimes be detected by a careful physical examination.

A general practitioner can spot several kinds of cancer on the basis of a simple physical examination. One of these examinations is a visual study of the patient's skin for abnormal growths, for example, moles or other patches of abnormal texture or color that indicate an inappropriate growth of skin cells. The most common of these abnormalities is a scaly, crusty patch of skin that results from the cellular damage done by ultraviolet (UV) rays in sunlight. This is a pre-cancerous condition called actinic keratosis (FIGURE 2.1A–C). Such patches of abnormal skin, known in general as lesions, are common on the heads and hands of fair-haired people who are of middle age or older, particularly those who live at high elevations or who spend lots of time outside. They can also be induced in younger people by frequent visits to a tanning salon. People with darker skin are less likely to acquire these lesions because their skin is better protected from the sun's damaging rays by pigments in their skin. Actinic keratosis essentially never develops in African, South Asian, or African-American people. If actinic keratosis is left untreated, the cells can acquire additional mutations and progress farther along a path of cancerous transformation, usually developing into a kind of skin cancer called **squamous cell carcinoma**. This cancer develops from the cells in skin that have already begun to differentiate by flattening out and expressing the protein, keratin, that makes normal skin scaly and tough (the word squamous comes from the Latin word for *scaly* or *layered*; see also Figure 1.4). Exposure to sunlight can also lead to **basal cell carcinoma** (FIGURE 2.1D). This is the most common kind of skin cancer because actinic keratosis takes a long time to develop into squamous cell carcinoma. Neither of these conditions is life-threatening, but both are sufficiently problematic that they should be avoided and/or treated.

For both conditions, early identification is important. This can be accomplished by regular physical examinations to spot such lesions early. Basal cell carcinoma is slow-growing and is readily treated either by surgery to remove the unwanted cells, commonly layer by layer, or by the local application of a very cold liquid, which will freeze the neoplastic cells and kill them. Actinic keratosis can be treated by scraping away the affected regions of the skin or with an ointment that contains chemicals that dam-

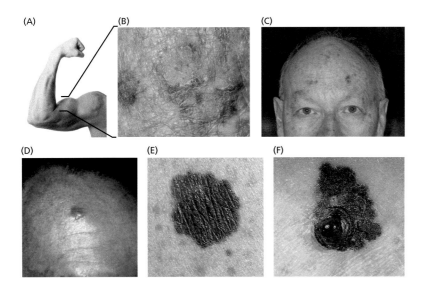

FIGURE 2.1 Examples of pre-cancerous and cancerous skin lesions that are easy to spot. (A) Arms are commonly exposed to sunlight and are therefore subject to cellular damage by UV rays. (B) A photo of actinic keratosis, a pre-cancerous condition characterized by a scaly, crusty patch of skin. This kind of growth can develop into a cancer, such as squamous cell carcinoma. (C) Head of an older man who has been exposed to strong sunlight for many years. Numerous skin lesions are visible. (D) A small basal cell carcinoma. (E) An early melanoma of the kind that forms during the initial stages of a cancerous transformation of melanocytes, cells that make the dark pigment melanin. (F) A more advanced melanoma that is already showing evidence of uncontrolled growth. Lesions of these kinds demand immediate attention because of their tendency to metastasize and form sites of malignant cancer elsewhere in the body. (A, courtesy of Vladimir Wrangel/Shutterstock.com; B, Diomedia/Diomedia/ Allan Harris; C, Diomedia/Diomedia/Hercules Robinson; D, courtesy of the American Academy of Dermatology; E and F, from Hodi FS & Wick MM [2002] Atlas of Diagnostic Oncology, 3rd ed [AT Skarin ed]. With permission from Elsevier.)

age dividing cells. This stops growth of the lesion. The treatment is easy, inexpensive, and highly effective. Left untreated, either of these neoplasms can grow large enough to disfigure the affected region. Squamous cell carcinomas occasionally become metastatic and are then life-threatening. More details about cancer treatments are presented in Chapters 8 and 10.

A less common but more dangerous kind of skin lesion, called **melanoma**, is also identifiable by inspection; if found, it deserves immediate and careful attention. The word **melanin** refers to a group of pigment molecules that darken the skin and protect the body from the damaging effects of sunlight. Darker skin is a result of more melanin because of either inheritance or exposure to the sun, which induces melanin production in many people. Melanins are made by a particular kind of cell called a melanocyte. During embryonic development, melanocytes form in one part of the embryo and then migrate to several other places, including the skin. Mature melanocytes can regain this ability to migrate through the body and are therefore more likely than most differentiated cells to become metastatic. Thus, if a melanocyte initiates cancerous transformation, it can become a dangerous cancer more quickly than cancers forming from most other cell types. Consequently, abnormal growths of melanocytes deserve particular attention.

The dark color of melanocytes makes their sites of proliferation easy to spot on the surface of the skin (FIGURE 2.1E). Fortunately, some melanocyte-containing growths are as benign as they are common, for example, moles. Moles are abnormal growths (neoplasms) where several types of skin cells, including melanocytes, have multiplied. Although moles are neoplasms, they are not dangerous. However, during the cell multiplications that lead to a mole, it is possible for a melanocyte to mutate; the wrong mutations can initiate cancerous transformation, leading to melanoma (FIGURE 2.1F). Such growths must be caught as early as possible because they have a nasty tendency to become malignant.

Keeping track of moles is something everyone should do. As a simple rule of thumb, a fast-growing mole with ragged edges is dangerous, whereas one with a hair growing from it is almost always benign. (The hair is an expression of normal skin cell differentiation.) A simple rule called ABCDE can help to identify a growth that needs attention: A is for asymmetry, as an asymmetrically shaped mole (as opposed to circular) is

FIGURE 2.2 Three palpation strategies for a breast self-examination. A neoplasm in breast tissue can sometimes be detected as a lump by careful palpation. This is commonly done with the pads of one's second, third, and forth fingers, moving systematically over the breast to identify any lump that feels different from the rest of the breast tissue. Shown here are three strategies of movement for lump detection.

breast self-examination:
manual inspection (reclining)

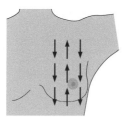  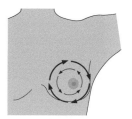

with fingertips close together, gently probe each breast in one of these three patterns

potentially dangerous; B is for border, as an irregular border is a bad sign; C is for color, where a mole with a mixture of colors, such as black and tan, is another bad sign; D is for diameter, because a mole that is more than one-quarter of an inch across is potentially dangerous; and E is for evolving, where a rapid change in the mole over time is bad. Careful self-examination, or examination by a partner, can greatly reduce the risk that a dark skin growth will progress so far as to become malignant. Identification of cancer-like changes in a mole at an early stage allows removal of the mole before its component cells accumulate the multiple mutations characteristic of carcinogenic transformation.

Breast cancer too can sometimes be detected early by a self-examination based on **palpation**, which is use of the hands to feel out trouble. All women should learn the skill of careful breast self-examination (FIGURE 2.2). A similar examination is sometimes done by a physician during a general examination, but self-examinations can be done at home at no cost. Studies by the American Cancer Society have shown that this type of screen is not very effective in catching breast cancer early, but if women know the shape and feel of their own breasts, they may be able to identify an abnormal growth and seek proper medical attention for it.

Another kind of cancer for which inexpensive screening is possible and advisable develops in the **prostate gland**, a part of the male reproductive system that makes the fluid that carries sperm during ejaculation. The prostate gland lies inside the body near the base of the penis, but it is most easily reached for palpation through the rectum in a test called a digital rectal examination (FIGURE 2.3). A doctor can don rubber gloves and use a finger to feel the patient's prostate gland. In this way, the doctor can tell whether the prostate has grown bigger than normal. Men in middle age or older have a comparatively high risk of getting prostate cancer, and although there are benign conditions in which the prostate gland enlarges, the digital rectal examination can sometimes identify a prostate tumor before it has grown big enough to likely include malignant cells. The digital rectal examination of a man's prostate is comparable to palpating a woman's breast in search of a lump that doesn't belong. Both of these tests are simple, cheap, and sometimes useful.

SIDE QUESTION 2.1

How sensitive is palpation? Test your own sensitivity by putting a row of objects with different sizes and shapes on a smooth table. Close your eyes and see which objects you can find and identify by feel. Now put the same objects on a rough surface and try again. Now put the objects under a piece of cloth, and see how well you do. The latter is a metaphor for the task a doctor faces in trying to palpate a small tumor inside your body.

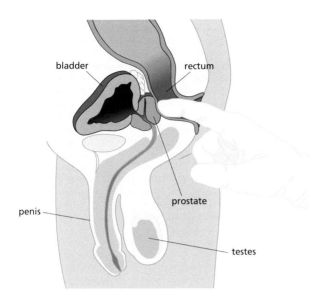

FIGURE 2.3 **Palpation of the prostate gland in a digital rectal exam.** A doctor can use palpation via the rectum to see whether a man's prostate gland is enlarged. In this figure, the man being examined is turned to the left, standing with his back to the examiner, and is bent over at the waist. An enlarged prostate identified by this method might be an indication of prostate cancer.

Sometimes it is wise to supplement palpation with additional screening tests.

Prostate tumors are sufficiently common in older men that many doctors use additional tests to supplement a digital rectal examination. One such test was developed from the fact that prostate cells produce a protein, called **prostate-specific antigen (PSA)**, which helps to keep semen fluid. A small amount of this protein is normally secreted into the blood. If the prostate gland is enlarged or damaged, more of this protein is commonly secreted. Therefore, a test for the amount of this protein in blood is an indirect test for the size and health of the prostate gland. For a while, it looked as if the blood levels of PSA might be an excellent indicator of prostate overgrowth, which might in turn be evidence of prostate cancer. More recently, patients with low PSA have been found to have prostate cancer and vice versa, so the value of the test has been questioned. However, the test is easy, cheap, and harmless, so it is frequently done on men of middle age or older, along with a digital rectal examination, as a screen for prostate cancer.

Even this combination of tests is still imperfect, so early identification of prostate cancer continues to be a problem, one that is compounded by two facts. First, benign enlargement of the prostate is common in older men, and this condition can be left untreated with no adverse effects. Second, screening for enlargement of the prostate gland by a method for medical imaging, like X-rays, is a bad idea. The location of the prostate gland means that X-rays have to pass near or through the testes. This treatment can damage the cells that produce sperm. Such a risk makes X-rays of this region a questionable medical practice, except for men who are old enough that they are not going to have more children.

Some kinds of cancer can be caught early by direct examination of the relevant cells.

Cervical cancer, a disease that affects a specific part of the uterus called the cervix (FIGURE 2.4), is readily treatable if caught early. Fortunately, early detection of this disease is relatively easy, using a simple and inexpensive method called a **Pap smear** (named after Georgios Papanicolaou, the doctor who invented the test). For a Pap smear, a doctor uses a device

FIGURE 2.4 Diagram of a tumor near the uterine cervix. The cervix of the uterus is a region near its opening to the vagina. Cancerous or pre-cancerous conditions in this region of the uterus are relatively common. Cancerous or pre-cancerous cells from this region can often be identified when a sample of the epithelium is taken for closer study by means of a Pap smear, as described in the text.

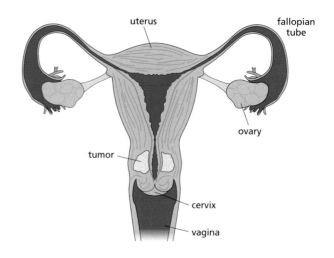

SIDE QUESTION 2.2

The small nuclei found in cells from a Pap smear of normal tissue are indications that these cells are properly differentiated. Do you find this statement reasonable? Start by looking at Figure 1.4A, which shows a light micrograph of a different kind of skin, and think about what happens to skin cells from the time they are born through cell division near the base of the skin until they are pushed up toward the skin's surface by the cell divisions occurring below them. How complicated are the patterns of gene expression needed to support cell growth and division versus simply making the fibers characteristic of a skin epithelium? Is this difference consistent with the smaller nuclei found in cells of a normal Pap smear? Why?

called a speculum to spread the labia and vagina, and a small brush at the tip of a thin rod is used to rub some cells from the surface of the uterine cervix. These cells are prepared for light microscopy, first by adding a chemical called a fixative to toughen them up and then by adding stains to give color to different cellular structures. A trained pathologist, or even a fancy computer program, can then examine the cervical cells through a light microscope, seeking cells that look abnormal (FIGURE 2.5). A Pap smear can be uncomfortable, given that the doctor must spread the labia and vagina, but the procedure does not require needles, knives, or anesthetic. Since the cells are examined after they have been removed from the body, this test poses no risk to the patient. Pap smears are commonly done during routine physical examinations and have greatly increased the chances of catching cervical cancer early.

Another simple screening test can sometimes identify cancer in the lower intestine. The damage caused by a cancerous growth can cause blood to leak into the intestine, allowing small amounts of blood to be passed from the body along with fecal material. A person of middle age or older, who is at some risk of cancer in the lower intestine, or **colon**, can collect a small sample of their feces on a sterile paper and send it to a lab, where a sensitive test for blood is applied. This test for occult fecal blood (occult comes from the Latin word for *hidden*) is cheap, noninvasive, and completely safe. It is quite effective as a test for colon cancer.

Additional cancers can now be caught early, thanks to endoscopies.

An **endoscope** is a flexible tube containing fibers that can carry light to look inside a patient. There are several kinds of endoscopes, all of them useful for diagnosis of disease; several are useful in screening for cancers. Endoscopies use your body's tubes that have one end open to the body's exterior, for example, the tubes you use for breathing, eating, urinating, and defecating. In all these procedures, the light-carrying fiber is slid into an interior region of the body to illuminate the area of interest. This fiber can also carry images from the region of interest out to a video camera, so the doctor can get a good look at internal tissues in a minimally invasive way.

A very useful form of endoscopy for cancer prevention is one that examines the colon. In a **colonoscopy**, the light-carrying fibers are passed through the anus and rectum and up into the colon. This test allows the doctor to look at the inner surface of part of a patient's intestine and see if there are abnormal growths, such as a **polyp**, on any of these tissues

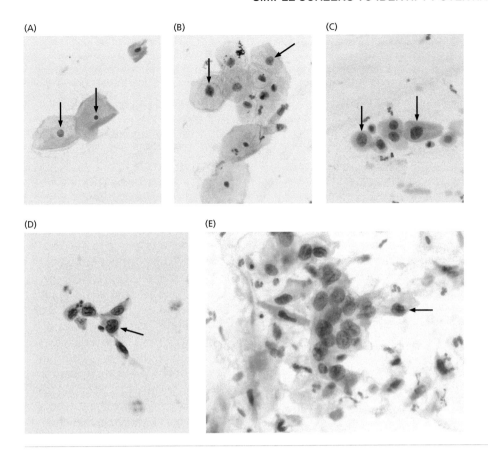

FIGURE 2.5 **Images of cells obtained by Pap smears, showing either normal cells or cells undergoing cancerous transformation.** (A) Normal cells of the kind that can be rubbed from the surface of a healthy cervix are shown. Note the small, darkly staining nuclei (arrows), which suggest these cells are not actively expressing many genes; their DNA is compacted. (B) Cells from a woman with a minor problem called a low-grade intraepithelial lesion. Some nuclei are now bigger, relative to the sizes of the cells (arrows). (C) These cells have even larger nuclei, indicating more gene expression, which implies a greater extent of cancerous transformation. (D) A similar enlargement of the nuclei, only more so. Now some cells are elongated, suggesting that they are capable of movement and could therefore metastasize. (E) A truly invasive carcinoma of the cervix. Many cells have enlarged nuclei, indicating a high rate of cell growth and division. Clearly, Pap smears can be very informative about possible cancer in the cervix. (From Matulonis U, Krag K & Genest DR [2002] Atlas of Diagnostic Oncology, 3rd ed [AT Skarin ed]. With permission from Elsevier.)

(FIGURE 2.6). During a colonoscopy, the same tool that carries light for imaging can also carry a wire loop with which to remove suspicious-looking growths for closer examination with a microscope (Figure 2.6D). A related method uses **fiber optics** to travel up the urethra (the tube that brings urine from the bladder to the place from which you urinate), allowing

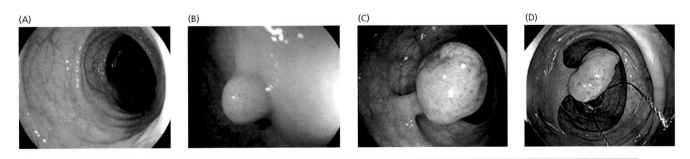

FIGURE 2.6 **Images of the inner wall of the colon, taken with an endoscope during a colonoscopy.** (A) Interior of a normal colon. The walls of this part of the intestine look quite smooth. (B) Close-up view of a region where a neoplasm has begun to grow, deforming the smooth surface of the intestine's wall. (C) A large polyp that has grown from the wall into the interior of the colon. (D) A different polyp, seen with a surgical loop that the doctor is about to use to remove the polyp for closer inspection. The loop can be passed over the polyp and pulled tight, so the abnormal growth is cut from the intestine's wall and brought out for examination with a light microscope. (A, courtesy of the NIH; B, from GastroMedicine & ENDOSCOPY, Melbourne, Australia.)

DIAGNOSIS CASE STUDY: Woods Patterson

Woods Patterson, a 55-year-old man, was down on his luck. Without either health insurance or the money to pay for medical treatment, he was unable to get an evaluation of a potentially dangerous situation. He described the problem like this: 'I'd known there was something wrong for several years, 'cause I'd been having trouble urinating. About 2012, I noticed that I was sometimes urinating blood. I talked about it with my doctor at the People's Clinic. She was worried on my behalf, but since I didn't have any medical insurance, there really wasn't a whole lot anyone could do. Finally, it got to the point where my doc said, 'You have GOT to get medical insurance.' So I made the rounds of all the social service organizations in town. About a week later, thanks to ObamaCare, I got my Medicaid card. That was good news!

So now I had medical insurance, but I wasn't able to find a urologist who would take Medicaid. Then I realized maybe I could get treatment at the Medical School in Denver, so I called them up around the middle of July. They told me they would take Medicaid, but it would probably be the end of the year before they could get to me. That made me worry; by that time I could be dead. So I spoke with a friend who knew some people at the medical school, and he talked with some of the doctors down there. Next week, when I called the office of the urology group they said they could see me in two weeks! That sure was good news.

When I got to the hospital, they took a urine sample, did a few other tests, then sent me home. In two days the hospital called to say that they had found some cancer cells in the urine, so they needed me back for another appointment. They didn't tell me that it would involve an endoscopy with a tube going up my penis; not so nice. But it was the right thing to do, and in fact I got kind of interested. They had a fancy display that showed what was at the tip of the fiber optic, so I could watch it on a big screen. When the tube got up to the bladder, I saw a big brown mountain. I asked the doctor, 'Is that my cancer,' and he said yes. I really was grateful for that, because the image gave me something to focus on. It was the first clear indication of what I was going to have to fight. Until then, cancer was pretty abstract. Seeing that big brown mass gave the whole thing a reality. I still use that mental picture today as a way to focus on what I am dealing with."

Identify four of your body's tubes that a doctor can examine by endoscopy. Via these tubes, what are the places inside your body that a doctor can examine?

a doctor to look into the bladder itself. While this operation is not used for routine cancer screening, it can be a valuable diagnostic tool if there are indications of trouble in the patient's urinary system (SIDEBAR 2.1).

Endoscopies are safer, simpler, and cheaper than most kinds of surgery, but they do carry some risk, and they have their own limitations. It is possible for the light-carrying fiber to pierce the wall of the body's tube used for entry. Although this is rare in the hands of a skillful doctor, it can have serious consequences. Moreover, some people resist the idea of having a doctor insert an instrument into their bodies, so they are reluctant to undertake a colonoscopy. A colonoscopy costs between $1000 and $5000, much more than an examination of occult fecal blood. A recent clinical study carried out in Spain showed that a modern method for identifying blood in feces, used every two years, was as good as a colonoscopy for identifying colon cancer and probably for reducing related cancer deaths. The combination of its economy and its greater acceptance by the interested public suggests that the United States might do well to alter its policy of frequent colonoscopies and rely more on tests for occult fecal blood.

CHARACTERIZING A POTENTIALLY CANCEROUS LESION

If a screening procedure identifies a growth or tissue that appears suspicious, a more thorough characterization can be carried out through a **biopsy**, meaning the surgical removal of a tissue sample for closer examination of its component cells. For example, if a Pap smear looks abnormal, a doctor can take a take a biopsy of the cervix by removing a small amount of tissue, commonly with a special pair of tweezers called biopsy forceps. Just how the biopsy is taken differs, depending on the location of the suspicious growth and the kind of access the doctor can get.

Biopsies are much less complicated and invasive than full-scale surgery, for which the patient must often be put to sleep with a general anesthetic and sometimes hospitalized for recovery after surgery. There are several tools for biopsies of different tissues. A core biopsy involves a needle, one or a few millimeters in diameter, with a very sharp cylindrical tip that can cut out a thin cylinder of tissue for more detailed examination, for example, with a light microscope. An aspiration biopsy involves a thinner needle that a doctor uses to suck cells from the suspicious lump, so long as the lump is not too tough. The thinner needle is easier on the patient, though structural information about the positional relationships among the cells is usually lost. Either of these procedures requires only a local anesthetic.

Biopsies allow doctors to study the cells from a suspicious lump.

If a sample has been taken from a potentially cancerous growth, the cells therein can be compared quite accurately with cells from normal tissue. The most common method for this kind of comparison is light micros-copy. A sample from the biopsy is first treated with chemicals called fixatives, which toughen the cells so they can be cut into thin slices. The slices are then treated with dyes that bring out structural features of the cells. A trained pathologist will have looked at many hundreds of cells of any given kind, so his/her eye can usually pick up abnormalities with high confidence. The resulting comparison can reveal that the cells in the suspicious lump look normal (good news), not quite right (less good news), or really quite abnormal (bad news). The degree of abnormality is a mea-sure of the extent to which the cells in the biopsy have changed, which is in turn a useful description of how far they have gone along the pathway that transforms normal cells into cancerous ones (FIGURE 2.7). As another example, light microscopy can be done on biopsies obtained during a colonoscopy. If a polyp is detected and removed, it too can be treated with fixatives, sliced, and stained for closer examination (FIGURE 2.8).

Pathologists have specific names for the ways cells in a biopsy may differ from normal cells. The term **dysplastic** means that the cells have lost some aspects of their normal structure. Dysplasia is commonly a result of only a short trip down the path of cancerous transformation, as shown in Figures 2.5B and 2.7B, where the abnormal cells are classified as a low-grade lesion or neoplasia. **Metaplastic** means that the cells have changed their appearance. In metaplasia, cells take on structural features of a different cell type. They still appear differentiated, but they don't look right for the tissue in which they were found (Figures 2.5C and 2.7C). Such cells are classified as high-grade problems, but the neoplasm is still in the place where it began to grow. **Anaplastic** means that the cells have lost essentially all the features that reflect normal differentiation for that tissue (Figures 2.5D,E and 2.7D). This implies that they are growing and dividing but not differentiating to perform their normal functions. Anaplasia suggests a long trip down the pathway of cancerous transfor-mation. Anaplastic cells are dangerous because they are likely to become metastatic.

Biopsies can also be used for analyses that look not simply at cell struc-ture but also at the molecules from cells in a potentially cancerous lump.

SIDE QUESTION 2.4

What are the strengths and weaknesses of needles as biopsy tools? In your answer, compare an aspiration biopsy with a core biopsy.

SIDE QUESTION 2.5

Why does the identification of anaplastic cells in a biopsy give a poorer prognosis than the detection of dysplastic cells?

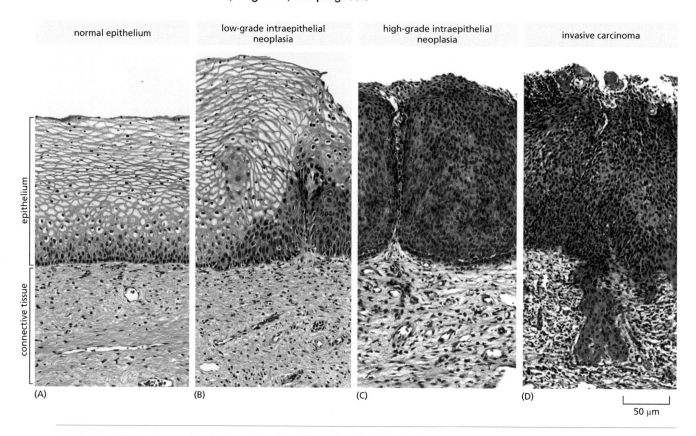

| normal epithelium | low-grade intraepithelial neoplasia | high-grade intraepithelial neoplasia | invasive carcinoma |

(A)   (B)   (C)   (D)

50 μm

FIGURE 2.7 Light micrographs of slices cut from four different cervical biopsies. The tissues in these four biopsies (A–D) range from normal to fully cancerous. Note that the term *epithelium* refers to the layer of cells near the surface of the cervix. The epithelium lies on top of *connective tissue*. The cells of the neoplasm are invasive when they can leave the epithelium and move into the underlying tissues. This property is characteristic of cancer. (Courtesy of Andrew J. Connolly.)

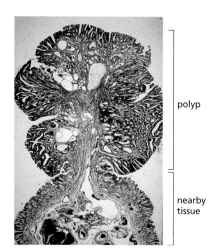

polyp

nearby tissue

FIGURE 2.8 Light micrograph of a slice cut through a polyp and nearby tissue. The structure of tissues in a polyp can be seen in this light micrograph of a stained slice of tissue obtained during a colonoscopy. None of the regions seen here shows indications of cancerous transformation. Finding a polyp like this motivates a doctor to caution the patient that close watch should be kept through another colonoscopy in a year or two. However, there is no indication that colon cancer has started. (Courtesy of Anne Campbell.)

For example, part of the biopsy can be used to see whether the cells react with an antibody that is known to bind to a cell-surface protein in a specific kind of tumor. With the tools of modern molecular biology, it is even possible to sequence the DNA of the cells from the lump, looking for specific mutations. These approaches have become increasingly informative as doctors have learned more about the ways in which different genes become altered as cells become cancerous. For the specifics of these changes to be meaningful, though, you need to know more about the details of the cellular changes that lead to cancer; these are covered in Chapters 6 and 7. In Chapters 9 and 10, you will learn more about the use of antibodies and DNA sequence information for refining a cancer diagnosis.

Microscopy of biopsies can show when cells are losing their normal associations with neighboring cells.

To become metastatic and malignant, a cancer cell must be able to wander (Figure 2.7D). Changes in cell–cell contact are therefore something that a pathologist will look for when examining the cells in a biopsy. The most stringent way to assess the presence of wandering cells is to find cells that are already in the wrong place. Occasionally, a biopsy from one tissue will contain cells that look as if they came from an entirely different tissue. For example, a biopsy from the liver might contain cells that originated in the colon. This is clear evidence for metastasis. More commonly, metastasis is detected because cells that belong to the biopsied tissue are found outside that tissue in a vessel or organ nearby. Blood vessels are thought to be the major route by which metastatic cells travel through the body, but the volume of blood in our veins and arteries is so large that wandering cells are hard to find there. A more practical way to detect metastasis is to take a close look in a system of vessels and glands called the **lymphatic system**.

In the lymphatic system, the **lymphatic vessels** (often called simply lymphatics) form a branching array of tubes that run through much of the human body (FIGURE 2.9A). Lymphatics contain **lymph**, a transparent, yellowish fluid that bathes all the cells in our body, and they drain it into a major blood vessel near the heart. Lymphatics also serve as conduits for the movement of white blood cells. There is no pump analogous to the heart to move lymph around. It is pushed slowly along by both the contraction of nearby muscles and gravity. Lymphatic vessels can also serve as pathways for the slow movement of metastatic cancer cells, so they are of great interest to cancer specialists.

Along the lymphatic vessels there are swellings with well-defined structures, called **lymph nodes**. These are sites where many specialized cells accumulate (FIGURE 2.9B). Lymph nodes are important parts of the body's **immune system**, that is, the cells and tissues that make antibodies and several kinds of white blood cells that can identify and kill foreign organisms that might harm the body. Our bodies contain hundreds of lymph nodes where white blood cells congregate and interact with other cells, checking to see if those cells really belong. (The immune system in general and white blood cells in particular are discussed more fully in Chapter 9.) A cancer cell that has wandered away from a tumor mass will often enter a lymphatic vessel and travel with the slow flow of lymph. Commonly, these cells stop for a while in a lymph node, as the cells of the immune

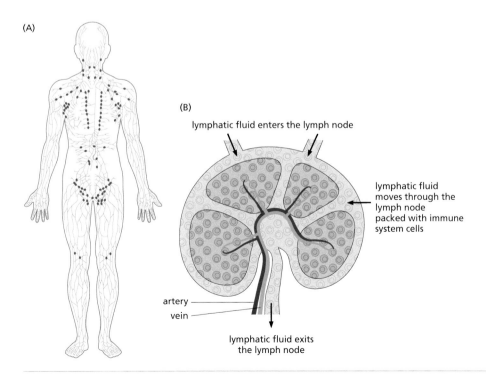

FIGURE 2.9 **Diagrams showing lymphatic vessels and lymph nodes.** (A) Lymphatic vessels and lymph nodes are distributed throughout the human body. Indeed, this diagram shows only a fraction of the many nodes and lymphatic vessels that are present throughout the body. In combination, they are well-positioned to drain extracellular fluid from all over the body back into the bloodstream. (B) A single lymph node shows the lymphatic vessels that bring lymph into this organ and take it out again, so the lymph can flow back into the bloodstream. Also shown are blood vessels that bring blood to nourish the cells of the lymph node and allow those cells to interact with materials and cells in the blood.

system examine the molecules on their surfaces to determine whether they come from your body or from an invader. Cancer cells are, of course, products of our bodies, so they are often sufficiently similar to our normal cells that they are accepted as part of self and allowed to move on, wandering through the body until they find a place where they can stick and settle, initiating metastatic growth. Despite being accepted by the lymph node as belonging to the body, cancer cells usually look sufficiently different from white blood cells and the other normal cells of a lymph node that they can be recognized in a light microscope by a good pathologist. Biopsy of lymph nodes is a simple and not very destructive way for doctors to find out if cells have already left their site of origin in the **primary tumor** and are now on their way to set up sites of metastatic growth.

Biopsies of lymph nodes can provide important evidence about the metastatic behavior of a potentially cancerous lesion.

The detection of cancer cells in a lymph node is a bad sign because it means that metastatic behavior has begun. At this stage, simply removing the primary tumor is not going to remove all the cancer cells that already exist. Conversely, the lack of cancer cells in a node is a good sign, but it is not proof that metastasis has not occurred. Perhaps metastatic cells had not yet arrived at that lymph node, or maybe there were not enough wandering cells for the pathologist to notice. The resulting uncertainty is an example of the difficulties of interpreting results from the tests now available for detecting and characterizing cancers.

In relatively recent times, doctors would commonly biopsy quite a few lymph nodes (4–18), hoping that, by looking at multiple samples, they would get an accurate assessment of metastasis. Indeed, the fraction of lymph nodes that contain cancerous cells is still used as a score to grade a tumor's cancerous progression. A downside of removing this many nodes, however, is that after the operation, the lymph in that region of the patient's body doesn't drain properly back into the blood, so the patient often experiences painful swelling. More recently, it has been recognized that metastatic cells from tumors in particular organs are most likely to migrate to specific lymph nodes. These are now regarded as **sentinel nodes**, because the presence or absence of cancer cells in them is a good indication of metastatic behavior. If the sentinel nodes are removed and studied, it is usually not necessary to remove other lymph nodes, and this makes the patient's life, post-biopsy, considerably easier.

MODERN INSTRUMENTATION TO IMAGE A PATIENT'S INSIDES

Medical imaging is a set of high-technology methods that provide doctors with several ways to look inside our bodies. The endoscopies mentioned above are one example, but there are more. The best established of these is the use of X-rays. Many of us have had X-ray pictures taken by a dentist in search of tooth decay or by a physician checking to see if a bone is broken. X-rays are also used in cancer diagnosis to make a shadow picture of the body in the region of concern (**SIDEBAR 2.2**). An X-ray of breast tissues, called a **mammogram**, is an important example of the use of X-rays to improve the evaluation of a potentially cancerous growth. A lump, such as one first identified by palpation, might be a tumor, or it might be just a bag of fluid encased in layer of tissue, which is called a **cyst**. Distinguishing between these possibilities requires a closer look. Any lump situated in normal tissue will deflect and/or absorb some

DIAGNOSIS CASE STUDY: Rosemary

Rosemary is a 68-year-old professor in an American research university. Her experience with cancer began about 15 years ago when she went for a routine mammogram. The hospital where her X-rays were done was very efficient, but the downside of this efficiency was that she heard within minutes that her breast contained a lump that looked 'undoubtedly cancerous.' She was completely surprised and totally horrified.

Given the result from the mammogram, a physician applied a local anesthetic and inserted a needle to take a sample of the cells in the suspicious lump for further study; in other words, a biopsy. Within an hour and a half, a pathologist had examined the extracted tissue, and although his conclusion was tentative, the indication was clear: the lump looked like cancer. There would have to have to be an operation, and it should be done soon.

Rosemary reported, 'As soon as I heard this diagnosis, I raced to a phone and called a doctor friend to ask what this all meant and what I should do. I had never had an experience like this before. It was literally hard to breathe. Oddly, I have asked several friends who have been in the same situation, and they all experienced an overwhelming sensation that it was hard to breathe. This physical sensation was even more powerful than emotion, although it was almost certainly a manifestation of fear. That night I was supposed to give a lecture at a branch of my university. Although I felt shaken, I decided that I should go ahead with the lecture, because I didn't want this thing to get the better of me. My husband was about to go on a business trip, which was important to him, so I really didn't want to involve him more than I had to. It felt as if this was only the first of a whole string of tasks I would have to undertake while living with a cancer diagnosis. So I drove myself about 15 miles to the off-campus site and gave my lecture. The talk went fine, but driving home I got horribly lost, which I think showed how I really felt.'

X-rays, so the portion of the X-ray film that is just behind the lump gets a lower exposure and therefore doesn't get as dark (FIGURE 2.10A,B). The resulting image contrast can usually be interpreted by an experienced doctor. If the presence of a lump is confirmed in an X-ray, and its position is specified with accuracy, a doctor can then take a biopsy of the tissue for closer examination, for example, by light microscopy. The information from microscopy can include the detection of dysplastic or even anaplastic cells. This information is usually definitive for the identification of cancerous growth (FIGURE 2.10C–E).

Mammograms are sufficiently sensitive that they sometimes reveal growths that are either too small to feel or positioned so they are hard to detect by palpation. Indeed, mammograms are sufficiently successful in the early detection of lumps in breast tissue that many doctors recommend them on a regular basis for people who are at some risk of breast cancer. For example, the American Cancer Society suggests a mammogram every year for women ages 45–54. However, if no suspicious lump is detected during this period, then the procedure should be used only every other year thereafter. A policy like this is designed to take advantage of medical imaging to screen for cancer when the likelihood of finding one is sufficiently high. When the risk of cancer is low, for example, in young women, the test is inappropriate. Mammograms cost about $100, and although they are good for identifying lumps, they are not perfectly accurate. Moreover, X-rays can damage cells, so they should only be used where a real medical advantage can be achieved.

The combination of X-rays and biopsy is also effective for detecting and characterizing tumors in lung, liver, and many other tissues. Conventional X-ray images, like a chest X-ray or mammogram, are two-dimensional (2-D) pictures of all the structures in one part of the body, superimposed on film or an electronic detector. Essentially, the X-rays pass through the body and then encounter the imaging device. An X-ray image therefore displays what is called a **projection** of the body's structure onto the

If your mother or aunt had a regular mammogram and was told she had a suspicious lump in one breast, what would you advise her to do?

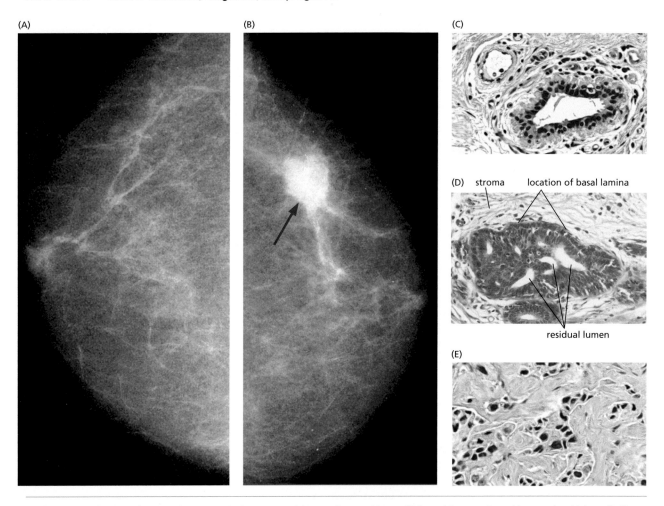

FIGURE 2.10 Images of a normal breast and a breast containing an abnormal lump. (A) Normal tissue as it would appear in a high-quality X-ray picture (mammogram). (B) A comparable image of a breast containing an abnormal growth (arrow) that would immediately attract a doctor's attention. (C) The kind of image a pathologist would see by light microscopy in a stained slice cut from a biopsy of normal breast tissue. The open spaces are ducts that can carry milk from the tissue that makes it to the nipple. Surrounding the ducts is normal connective tissue. (D) A similar micrograph from a biopsy containing an abnormal growth that is not yet a true cancer. The lumen of the duct is being closed off by unwanted growth of the cells that line it. The surrounding connective tissue (also called stroma) is still separated from the cells that line the duct by a basal lamina, a layer of fibers that separates all normal epithelia from the underlying tissue. (E) A similar micrograph of truly cancerous tissue. These cells should be forming a duct, but no such structure is present because the cells have lost differentiation. The stromal tissue is still essentially normal. The differences between panels C, D, and E represent the kind of information that pathologists use to categorize a biopsy. (A and B, from Campos SM & Hayes DF [2002] Atlas of Diagnostic Oncology, 3rd ed [AT Skarin ed]. With permission from Elsevier. C and E, courtesy of A Orimo. D, adapted from AT Skarin [ed] [2002] Atlas of Diagnostic Oncology, 3rd ed. With permission from Elsevier.)

plane of the detector (FIGURE 2.11). As such, X-ray images are useful for cancer screening and for some aspects of evaluating the extent of a tumor's growth, but the information they provide is limited. Nonetheless, they are useful and are commonly used (SIDEBAR 2.3).

A more informative X-ray imaging procedure generates a three-dimensional (3-D) view of a selected part of the body. A **computed tomography** scan (**CT scan**) uses multiple 2-D X-ray pictures of the same region, taken over a wide range of angles, to provide a computer with sufficient information to calculate an accurate, 3-D model of the structures in that region of the body (SIDEBAR 2.4). The radiologist in charge can then use a computer to display many different computer-generated slices cut from this 3-D model, so CT scans are almost invariably more informative than a simple X-ray photograph. For example, they are usually more sensitive for finding a small lump of tissue that doesn't belong. Tumors as small as one-third of an inch in diameter show up quite clearly. Moreover, when

(A)

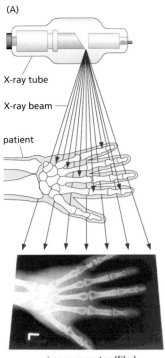

X-ray tube

X-ray beam

patient

image receptor (film)

(B)

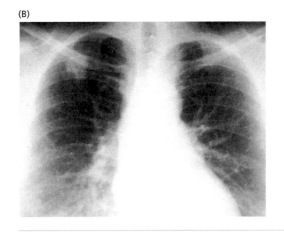

FIGURE 2.11 Formation of an X-ray image and an example of a chest X-ray, showing an area of abnormal growth. (A) An X-ray-generating tube is analogous to a flashlight, but it emits X-rays rather than visible light. X-rays can penetrate human tissues, but they pass better through soft tissues than through hard tissues, like bone. A detector placed behind the object being studied is therefore unevenly illuminated. Bone and other hard tissues absorb and scatter more X-rays than soft tissues, so behind them, the detector receives less illumination. The resulting shadow image is shown here as a negative, so less exposure shows up as bright. (B) In this frontal view of a patient's chest, the ribs and spine are obvious. The red arrows indicate an object that is not expected and might be a cancerous growth. A biopsy would then be an effective way to learn whether or not this growth is cancerous. (B, Hodi FS & Wick MM [2002] Atlas of Diagnostic Oncology, 3rd ed [AT Skarin ed]. With permission from Elsevier.)

dyes that scatter X-rays are used to fill specific spaces, such as the interiors of blood vessels, the contrast and detail make a CT scan a powerful tool for cancer detection and characterization, as well as for other kinds of medical diagnosis (SIDEBAR 2.5). CT scans can also be used to follow the progression of a cancer and assess both its rate of growth and some aspects of the extent of metastasis (FIGURE 2.12).

There are, however, three problems with CT scans. First, they cost 3–10 times more than a traditional X-ray picture. The typical current cost for a

DIAGNOSIS CASE STUDY: Rob

At the age of 31, my son Rob was a healthy, athletic, and energetic young businessman. He had moved to Budapest, Hungary, where he worked for a company doing industrial real estate management and development. He was an accomplished squash and tennis player, he jogged regularly, and he took pleasure in his good health. That summer he and I had gone for a run together; both his energy and stamina were much better than mine, so he went on ahead when I ran out of gas. That evening, though, he mentioned that when he ran nowadays, he often had pains in his knees, and he was thinking of seeing a doctor about it.

A little later, he saw a Hungarian orthopedist, and for the next month or so, he was treated for possible arthritis. At about this time, his company transferred him and his wife, Suzy, to their New York City office. Again, Rob saw an orthopedist to seek relief from aching knee joints, but other than that, he felt well

and was excited about his new job, new house, and new life. Over the next months, the joint pains got worse, and the treatments didn't help. Moreover, he just didn't feel as if he had his normal level of energy. Finally, a friend advised him to see a good general practitioner who would examine his condition from a broader point of view. After a physical examination, this doctor clearly smelled trouble. He ordered a chest X-ray and some other tests, which Rob dutifully had done.

The first news I heard was a call from Suzy saying that Rob had speckles in his lungs. This was so far from my expectations for a problem related to painful knee joints that I didn't know what to make of it. Shortly thereafter, Rob and Suzy saw his physician again and were told gently but plainly that he had advanced lung cancer. It had already metastasized to his bones and his central nervous system. At first, none of us could believe it.

SIDEBAR 2.3

SIDEBAR 2.4

ADVANCES IN MEDICAL IMAGING: Tomography

Tomography is a method for getting 3-D information about objects that can only be viewed with 2-D pictures, like the projection of body structure one gets from a single X-ray picture. A CT scan uses a computer to assemble many 2-D X-ray images, taken over a wide range of angles, into a reliable 3-D model. Indeed, CT computations can be used to generate 3-D models from pictures taken by many imaging methods, producing a variety of useful views of our inner tissues.

The idea behind tomography is related to the way our brain uses the images formed by our two eyes to achieve stereo vision and give us a sense of three dimensions in the world around us. The small differences between the images in our two eyes can be interpreted by our brain as depth; for example, the distance between us and specific features of the objects under view. Tomography differs from stereo vision in using lots of images, taken over a wide range of tilts, to generate a 3-D view. The large number of tilted views means that tomographic images are more reliable and quantitative than a stereo image; indeed, computed tomography can be used to make 3-D models of very complex structures.

When an object like a human's abdomen is imaged from many angles, each image looks slightly different. A computer can be taught how to use these small differences to assemble the multiple views into a single 3-D structure. When about 100 different views of the same structure are collected over a range of views covering 180°, these images can be used to compute the structure imaged, resulting in a unique 3-D object. It is an accurate determination of the structure, because the solu-

tion found by the computer is the only 3-D object that could have given rise to all those multiple 2-D views (Figure 1).

Once the 3-D model is in computer memory, it can be sampled by slicing it at any position and orientation that is informative for seeing the structures of interest. These slices look quite a bit like regular X-ray pictures, but they are not projections. They are displays of information from only a single layer cut from the 3-D image. As seen in Figure 2.12, one

can generate tomograms from CT scans made at different times to see how a tumor is progressing. The instruments now used to obtain CT scans are efficient and comparatively easy to use. Indeed, the many devices now available for computed tomography have provided doctors with powerful ways of looking inside people, taking advantage of the different imaging methods that scientists and engineers have brought to the field of medical imaging.

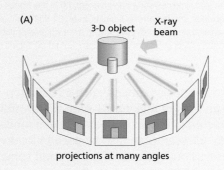

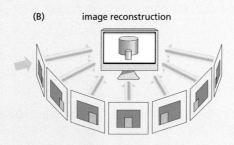

Figure 1 The principles of tomography. (A) The sample to be studied is a 3-D object. The broad green arrow represents an X-ray beam coming at the object from one angle. To collect the data for a tomogram, the orientation of the X-ray beam is rotated relative to the object under study, so pictures can be taken from many angles. The narrower green arrows with a dark green core, coming out the other side of the object, show the continuation of each of these beams after its X-rays have been scattered and absorbed by the object. The small pictures with black borders at the ends of those arrows show what the 3-D object looks like from each of the orientations used. The different projections do not look the same, because they have been taken from different angles. These images are like the X-ray images a doctor would collect while making a CT scan, but the same principles apply for 3-D imaging with MRI or PET scans as well. (B) This set of images can be used by a computer to generate a 3-D model for analysis. Now the green arrows are geometrical ideas, generated in the computer. They represent projection of the previously recorded images into a single volume in computer memory. Each of the rays contributes information into the central region, allowing a computer to calculate a reconstruction. When all these rays are projected into one volume in the computer's memory, and when their intensities are added up correctly, one gets a quite accurate version of the starting 3-D object. This is computed tomography; its principles apply, regardless of the kind of radiation used to generate the projections.

DIAGNOSIS CASE STUDY: John

John was a construction worker who ran his own roofing company. At the age of 54, he was diagnosed with a kidney cancer that was described as terminal, and he was given 5–10 months to live. Remarkably, he lived on for about 10 years, thanks to a combination of his strong constitution, good treatment by several doctors, and his amazingly positive attitude toward life and medical problems.

The first symptoms of his condition emerged when he went for a regular physical exam, which he had been doing every two years since he turned 40. As is common for such an exam, the doctor took a sample of urine for a urine analysis. There was some blood in that urine sample, so the doctor repeated the test. The second test was OK, suggesting that the first test had given a false positive. Two years later, exactly the same thing happened. After another two years he went back for a physical, and his doctor's assistant described John as a perfect example of a really healthy man in his early 50s. The fact was, though, he just didn't feel right. There was nothing obvious, but he wasn't as strong as usual, and he got tired. He had a little bulge on his abdomen, but he was in such good shape, he thought maybe it was just hard abs. At his next physical, though, there was blood in his urine again, so his doctor was able to convince the insurance company to pay for a CT scan. The result was dramatic.

The doctor reached John by phone as he was driving home and said that the news wasn't good. On a whim John asked, 'What, do I have cancer?' said the doctor answered, 'Yes, call me when you get home.' The CT scan had revealed a very large tumor on one of John's kidneys; it was so big and in such a bad place that it was going to be hard to remove by surgery. John's response was that he still felt good, and he was on his feet, so the problem couldn't be that bad. He did, however, go to see an oncologist.

chest X-ray ranges from $200 to $300, but CT scans range from $700 to $3000. Thus, a CT scan is warranted only when high-quality, 3-D images are needed for proper description of a medical condition.

The second problem with CT scans is the amount of X-irradiation they require, which is 10–100-fold greater than that needed for a single X-ray picture. X-rays can damage cells, as we will discuss in some detail in Chapter 5. At low levels of irradiation, the cell can commonly repair this damage, even if it affects a crucial molecule like DNA. At higher levels of irradiation, DNA damage can become a serious problem because it introduces mutations, and mutations can be carcinogenic. The subjects of DNA

(A)

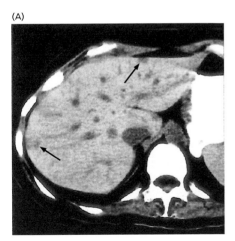

(B)

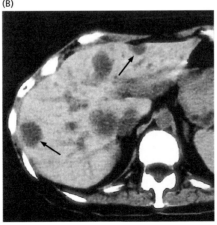

FIGURE 2.12 **Computed slices from CT scans of the liver of a patient with liver cancer.** (A) Results from a CT scan taken at the time of diagnosis are shown. Arrows point to small lesions that are probably sites of cancerous growth. (B) A similar image of a slice from a CT scan of the same patient taken several months later. Again, the arrows point to sites of suspicious growth. These are now clearly bigger than in the picture taken earlier. Moreover, additional sites of similar appearance are now evident. This is strong evidence for a diagnosis of advanced liver cancer. (From Shaffer K & Van den Abbeele AD [2002] Atlas of Diagnostic Oncology, 3rd ed [AT Skarin ed]. With permission from Elsevier.)

damage and repair are very important in the development and progression of cancer, so we will deal with them in Chapters 3–7. Here, it is important simply to know that minimizing one's exposure to X-rays is good medical practice.

The third difficulty with CT scans comes from the chemicals that are often used to improve image contrast. Iodine is commonly injected into a patient's vein before a CT scan, because this heavy atom scatters X-rays very well and greatly improves image contrast for blood vessels. Iodine can be coupled chemically with a benign molecule, like a sugar, making a substance that dissolves readily in water, yet it still does a good job of scattering X-rays and has essentially no harmful effects on the human body. Moreover, the iodine-containing chemical is eliminated from the body by urination after a few hours. This means there is a convenient window of time, while the iodine is still in the body, in which to take high-contrast X-ray images, first of blood vessels and then of the urinary tract as the stain is excreted. However, a few people (1–3%) have a negative reaction to such contrast-enhancing agents, for example, nausea and vomiting or skin irritation. A very few people have a really severe negative reaction to these chemicals, but in spite of these potential problems, the enhancement of X-ray image quality that contrast agents offer means that they are commonly used.

Occasional X-ray pictures, like one mammogram every few years, expose a patient to levels of radiation that are no different from the radiation we all experience as a result of our environment. For example, we are exposed to radiation from the cosmic rays that come from outer space (particularly if we fly long distances in an airplane) and from radioactivity in soil (see Chapter 5). A patient will commonly get the same dose of radiation from this background during two years between mammograms as she does during the mammogram itself. Frequent CT scans, on the other hand, are a bad idea, unless there is no alternative for solving an immediate medical problem. So we are faced with an irony: one of the best methods for detecting cancer uses radiation that can be a factor in causing cancer in the future. A recent study has found, however, that although a CT scan does increase the risk of cancer, the effect is small. The lifetime risk of getting some kind of cancer is about 40%; after having had one CT scan, the risk goes up to 40.05%. However, this very small increase is a bit misleading, because it describes the increase in risk for all kinds of cancer, not cancer in the organ that was irradiated during the scan. Nonetheless, the data show that the ratio of benefit to risk for a CT scan is favorable when a serious medical issue calls for it.

Given that X-rays can cause damage to cells and tissues, other medical imaging technologies that do not use this form of radiation have become important in cancer medicine. Each of these methods is described and evaluated in the text that follows.

Ultrasound can produce images of structures inside our bodies.

Sound whose frequency is too high for the human ear to detect is called **ultrasound**. Ultrasound penetrates the soft tissues of the body quite well, but it is partially reflected from every surface where the texture of the body changes, for example, where a tissue ends and fluid or bone begins or vice versa. The reflected sound intensities can be collected to make a picture of structures inside the body, forming an image called a **sonogram**. There are several kinds of devices that generate images from ultrasound, but they all work in essentially the same way: the instrument has a probe, often hand-held, that emits a very short pulse of ultrasound.

SIDE QUESTION 2.7

Why does an X-ray scattering stain in blood vessels improve the image of a tumor in a CT scan?

SIDE QUESTION 2.8

Given the information you have just learned about the strengths, costs, and dangers of X-rays, would you favor using CT scans over mammograms for regular breast exams? Why or why not?

(A)

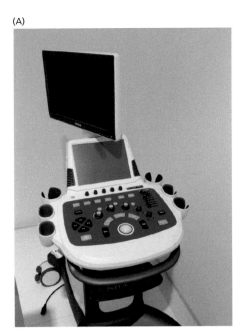

(B)

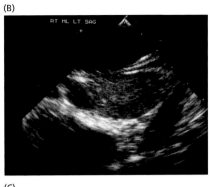

(C)

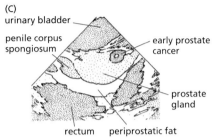

urinary bladder

penile corpus spongiosum

early prostate cancer

prostate gland

rectum   periprostatic fat

FIGURE 2.13 **Sonography of tissue to identify a tumor.** (A) A typical device for making a sonogram is small enough to be used conveniently at a patient's bedside. (B) A sonogram of the prostate, where a tumor was suspected, and surrounding regions. (C) An interpretive drawing, done by an expert. It is clear from these images that sonograms are not as easy to interpret as high-quality X-ray images or CT scans. Nonetheless, their economy and lack of damaging side effects make them valuable tools in cancer diagnosis. (A, from Wikimedia Commons, CC BY-SA 2.5. B and C, from Oh WK & Kantoff P [2002] Atlas of Diagnostic Oncology, 3rd ed [AT Skarin ed]. With permission from Elsevier.)

The same probe then measures the time it takes to get echoes from surfaces inside the body; the quicker the return, the closer the object; the stronger the echo, the greater the difference in texture at that distance from the probe. When the probe is moved around, the instrument records four things: where the probe is placed, the direction it points, the time till the echo returns, and the strength of the echo. This information is converted into an image, which is displayed on a computer screen for the doctor to examine (FIGURE 2.13). Two great advantages of sonography are its comparatively low cost (only $200–300 for a 3-D image) and the fact that there is no evidence that ultrasound increases the frequency of mutations or does any other damage to the body. Sonograms are even used during prenatal care to examine a fetus in its mother's womb.

Sonograms can sometimes detect tumors, but their principal limitation lies in their comparatively poor display of structural detail. This means they are not very sensitive to subtle differences between normal and abnormal tissues (see Figure 2.13B). For example, if a region of the liver has just begun cancerous progression, a sonogram will commonly not be able to reveal structural differences between the cancerous and normal tissue. Moreover, high-frequency sound doesn't penetrate fat very well, so for some patients, the pitch of the sound must be lowered. The lower frequencies are better at penetrating fat, but they reduce the detail that can be seen.

However, clever doctors and scientists are always looking for ways to enhance the quality of images from existing instruments. One way to improve sonograms is to enhance the contrast they show between blood vessels and the surrounding tissues, just as iodine enhances contrast for X-rays. Currently, this is being done by making a solution that contains a large number of tiny air bubbles and injecting this solution into a patient's vein. As the blood containing the solution circulates, nearby blood vessels

Some bats use ultrasound to detect their prey, for example, flying insects. Compare what a bat must do to catch a moth with what a doctor must do to identify a tumor in a sonogram.

SIDE QUESTION 2.9

DIAGNOSIS CASE STUDY: Scott

Mary is the mother of Scott, a healthy young man recently graduated from college, who at age 3 was diagnosed with a serious kidney tumor. Mary described the events leading up to the diagnosis like this: 'Early on there were no signs of trouble. Scott was fine, running around and playing with his friends. I had noticed for couple of weeks that his belly was a bit swollen, but that's sort of what kids look like, so I didn't pay much attention. After a few more days, though, my husband and I decided we should get him looked at. The minute the pediatrician felt Scott's abdomen, he called a radiologist and made an appointment for an ultrasound, even though it was late in the day. The doctor was really nice about it and called his family to say he was going to be home late. That was our first indication that this might be something serious. I was kind of stunned, but I still wasn't thinking about any kind of cancer.

When the report came back from the ultrasound, the doctor said that Scott had a mass on his kidney, and it could be one of three things: a neuroblastoma, which is a horrible cancer that would be really bad news, a Wilms tumor (a tumor of certain kidney cells that is relatively common in children), or simply a bag of fluid. Our pediatrician said he would make an appointment for further diagnostic imaging at the Children's Hospital in the nearby big city, and he'd let us know what we should do next. As we went home, we were practically speechless.

That evening our doctor called to say we had an appointment at Children's Hospital for the next morning. That night was really the worst part of the whole experience. I couldn't sleep, and my husband tossed and turned beside me. I was most worried about Emily, Scott's older sister, because she and Scott were so close. If anything happened to him, she was going to be devastated.

The next morning, we went to the hospital. That was hard, because even though they do their very best to make a children's oncology clinic a happy place, it just plain isn't. Scott had a CT scan, which meant that he had to have a needle poked into his arm to inject a dye into his blood vessels. The doctors were very honest with him about it. They told him it would hurt, but he was pretty brave about it. After the scan, the radiologist came back and she was beaming. She announced that Scott had a Wilms tumor, and she made it sound as if we had just had a new baby. Apparently, the diagnosis was so clear that they never even took a biopsy of the growth for further study. This really was good news, because a neuroblastoma would have been much more serious; if it was on the kidney, it would already have metastasized. So we felt happy too."

show up more clearly. This approach has recently been given a new twist by constructing bubbles that have a special affinity for a protein found on the inner surface of recently formed blood vessels. New blood vessels are common in growing tumors, so when the bubbles stick to the walls of these vessels, they add contrast to tumor tissue and make it easier to see. Sonograms can be very useful, but given their limited contrast and resolution of detail, they are not a complete answer for identifying cancers. They are often used in concert with other medical imaging methods that can provide images with greater detail (SIDEBAR 2.6).

MRI scanners provide remarkably sensitive cancer detection by combining a strong magnetic field with radio waves.

Magnetic resonance imaging (**MRI**) uses a powerful magnet and radio waves to generate informative 3-D images of structures within a patient's body. First, the patient is placed inside a very strong magnet (FIGURE 2.14A). The field from the magnet forces the nuclei of atoms in the patient's body to turn toward a common orientation. The protons in atomic nuclei are positively charged, and they spin. Any moving electric charge generates a magnetic field, so atoms with the right nuclear properties contain tiny electromagnets. Normally, the magnetic fields of individual atomic

(A)   (B)   (C)

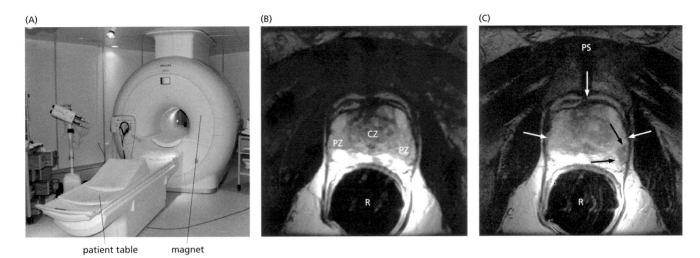

patient table   magnet

FIGURE 2.14 **Magnetic resonance imaging (MRI) of a prostate tumor.** (A) In preparation for imaging by MRI, the patient lies on a moveable table, and the part of his/her body to be imaged is slid into the hole in the donut-shaped magnet. (B) A slice from an MRI-generated tomographic reconstruction of a normal prostate gland. The dark circle marked R is the rectum, which passes very near the prostate gland. CZ and PZ mark the central and peripheral zones of the prostate gland, respectively. (C) A similar image of a prostate that contains a tumor. White arrows mark the edges of the prostate. PS refers to nearby connective tissue. The growth that does not belong is marked by black arrows. It is quite easy to see, thanks to the quality of this imaging method. (A, from Jan Ainali, CC-BY-3.0, via Wikimedia Commons. B and C, from Oh WK & Kantoff P [2002] Atlas of Diagnostic Oncology, 3rd ed [AT Skarin ed]. With permission from Elsevier.)

nuclei tip every which way; however, when atomic nuclei are placed in the strong magnet, they are biased to tip in the direction of the applied magnetic field (SIDEBAR 2.7).

The magnetic field is especially effective in orienting the nuclear magnets of certain atoms, like the hydrogens in water. Thanks to this orientation phenomenon, MRI scanners can use a radio wave of the right frequency to measure the number of rotatable hydrogen atoms per unit volume, which is proportional to the amount of water in any given place. An MRI scanner generates contrast from the differences in water content in different tissues, for example, between blood, bone, muscle, and fat. MRI can produce even more contrast following the addition of certain elements whose atomic nuclei give stronger signals than hydrogen. **Gadolinium** works well for this purpose, so a solution of this element is sometimes injected intravenously to stain blood vessels and produce MRI images of particular clarity. Many doctors think that MRI with gadolinium provides the best diagnostic test for characterizing cancer in the brain. A limitation of MRI, however, is that if there is any air in the way of the radio waves, as in the lungs or gut, image quality is degraded. Nonetheless, MRI is now well-established as a way to evaluate cancer of the prostate gland (FIGURE 2.14B,C), and it looks promising for the detection of breast cancers.

A problem with using MRI for cancer screening and diagnosis is that the instrument is expensive (several million dollars). Each MRI scan costs $1000–3500, and the fee for professional interpretation of the images adds significantly to this number. Moreover, the instrument is commonly very noisy when collecting data, and some people don't like getting inside the strong magnet. There is only very limited evidence that it does you harm, but it feels claustrophobic. Note, however, that if you have a piece of steel or other magnetic metal implanted somewhere in your body, this method is not for you! In spite of these limitations, MRI is being used more and more frequently to help doctors see cancerous growths.

ADVANCES IN MEDICAL IMAGING: Magnetic Resonance Imaging (MRI)

MRI takes advantage of the fact that many atomic nuclei are magnets, due to the fact that positively charged protons in the nucleus are spinning. A spinning charge generates a magnetic field, symbolized by the black arrow in Figure 1A. When this tiny magnet is exposed to a strong magnetic field (Figure 1B), the nuclear magnet tends to align with the external field. When the partially aligned atomic magnets are exposed to radio waves of the right frequency, energy is added to the spinning nuclear magnet, so the nuclear magnet can flip into a higher-energy orientation (Figure 1C). When the radio wave generator is shut off, the atomic magnets relax back to their normal orientation in the magnetic field, and as they do so, they emit a radio wave of their own, albeit a weak one (Figure 1D). This energy can be detected by an appropriate antenna, which measures the time, strength, and duration of the emitted radio wave. These properties of the emitted radio wave are the information used to generate an MRI image.

The signals detected in MRI can be made informative by adding one more feature to the setup just described. Before sending the pulse of radio waves, the MRI instrument turns on a second, much weaker magnet that makes the magnetic field stronger in one region of the body than another; it establishes a magnetic field gradient, meaning there is a smooth and well-characterized variation in magnetic field strength with position across the patient's body. The frequency of the radio wave emitted by a relaxing atomic magnet depends on the strength of the magnetic field that surrounds it. The antenna that detects the weak radio waves emitted by the relaxing atomic magnets measures both the frequency and duration of each wave. A computer then combines the strengths of the signals measured at each frequency with knowledge about the magnetic field gradient, and it makes a map. If there are more nuclei emitting radio waves in one region of the body than another, then there will be more signal from that region, so it will show up as brighter in the computer-generated image. Said another way, by using the two signals (the frequency and duration of each emitted radio wave), the MRI machine and its associated computer can make an image that displays the amount of water per unit volume in different parts of the body. By collecting this information from a wide range of angles, the computer gets enough information to make a 3-D model, just as it does in making a tomogram from a CT scan.

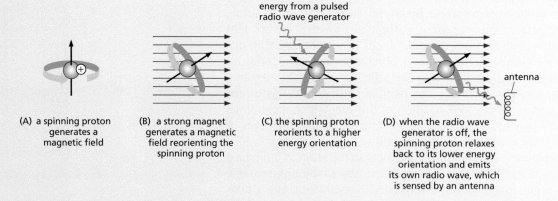

energy from a pulsed radio wave generator

antenna

(A) a spinning proton generates a magnetic field

(B) a strong magnet generates a magnetic field reorienting the spinning proton

(C) the spinning proton reorients to a higher energy orientation

(D) when the radio wave generator is off, the spinning proton relaxes back to its lower energy orientation and emits its own radio wave, which is sensed by an antenna

Figure 1 The principle of signal generation in magnetic resonance imaging (MRI). (A) Each proton in an atomic nucleus is spinning. Since the proton is positively charged, it generates a weak magnetic field (black arrow). (B) A strong magnet can generate a powerful magnetic field (many red arrows). When this field is applied to spinning protons, it tends to orient them. (C) A pulsed radio wave of the right frequency can add energy to the spinning nuclear magnets (wavy blue arrow), forcing them into a different orientation relative to the strong magnetic field. (D) When the external radio wave is turned off, the nuclear magnets can relax back to their preferred orientation in the strong magnetic field. As they do so, they emit a weak radio wave (wavy blue arrow), which is detected by an antenna. When the strength, frequency, and timing of the emitted waves are measured, the information can be used by a computer to generate an image that displays the number of protons per unit volume that were reoriented by the applied fields and waves. This shows the distribution of water or stain in the sample being viewed.

PET scanning is a powerful way to identify tumors, particularly if metastasis has occurred.

Positron emission tomography scanning is an imaging method that reveals how rapidly the cells of a given tissue are using sugar for energy. Cancer cells consume a significant amount of sugar to fuel their growth, so PET scanning is a way to locate cancerous growths, wherever they might be. In preparation for a **PET scan**, the patient is injected with a chemical that is similar to table sugar but is labeled with a highly radioactive atom. The cells of the body take up this special sugar and try to use it for energy. However, this sugar has been built in a way that blocks a cell's normal chemical reactions; it gets stuck in the cells that consume it. It persists in those cells for hours, so the radioactive atoms are temporarily localized to the cells a doctor wants to locate.

The radioactive atom is fluorine-18, a form of fluorine that has an unstable atomic nucleus. Half of these nuclei fall apart every ~2 hours, releasing subatomic particles called positrons. These particles indirectly generate two **gamma rays**, which travel in opposite directions. Gamma rays are much like X-rays, only more energetic. They are sufficiently energetic to pass right out of the body's tissues and escape into the space around the body, where the PET scanner has detectors that sense the gamma rays and determine the direction from which they came. In short, places where the special sugar was concentrated act as sources of gamma rays. By having detectors all around the body, the PET scanner gets the information necessary to make a tomogram that identifies the 3-D positions of the sites of gamma-ray emission (FIGURE 2.15).

Some normal cells in skin, the intestine, and the testes show up in a PET scan, because they too contain cells that grow and divide. The scan also detects normal cells in the brain because they require a lot of sugar to keep working hard as they send and receive messages, even though they are not growing and dividing. Kidneys and the bladder will also appear in a PET scan, because the special sugar that was not taken up by cells accumulates in urine. These extra signals mean that the image from a PET scan shows a faint outline of the body as a whole in addition to the strong signals that come from fast-growing, cancerous tissues. To obtain an even better image of the body, some modern PET scanners include a source of X-rays to make a CT scan as well as a PET scan, called a PET-CT instrument (FIGURE 2.16A). Any tumor that contains growing and dividing cells

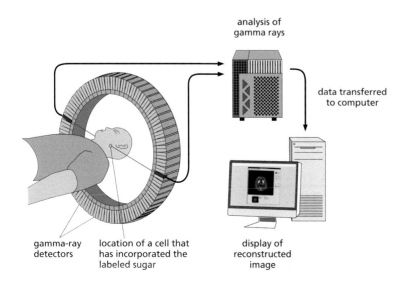

analysis of gamma rays

data transferred to computer

gamma-ray detectors

location of a cell that has incorporated the labeled sugar

display of reconstructed image

FIGURE 2.15 Instrumentation for PET scanning. A patient has previously been injected with a solution of a special sugar labeled with radioactive fluorine. Now gamma rays (red lines) are emitted from the cells that incorporated that sugar. These rays are emitted in all directions. Some of the rays are sensed by a set of gamma-ray detectors, which encircle the patient's head. Signals from these detectors are fed to an analysis system, and then a computer uses the information to calculate where the gamma-ray-emitting cells must be in three dimensions, hence the term tomography.

(A)

(B)

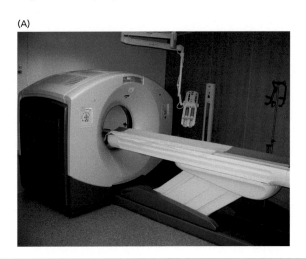
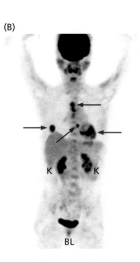

FIGURE 2.16 **Instrumentation for PET scanning and results from this kind of imaging.** (A) A modern PET-CT scanner is capable of both detecting gamma rays and performing a CT scan, thereby providing general structural information into which to place the results from gamma-ray detection. (B) PET scan of a patient with metastatic lung cancer. Red arrows mark sites of metastatic growth; K = kidneys and BL = bladder. (A, GE Discovery D600 PET/CT system, with 16 slice CT system, gantry diameter 70 cm. B, from Fighting Cancer with Nanotechnology, The Kavli Foundation, www.kavlifoundation.org.)

SIDE QUESTION 2.10

Compare the strengths and weaknesses of CT scans, MRI, and PET scans. Which imaging method would you expect to be most effective at detecting metastatic growth?

can then be placed within the context of the body as a whole, regardless of its position (FIGURE 2.16B).

The great advantage of a PET scan is that tumors are revealed simply from the rate at which their cells take up sugar. Indeed, many cancer cells contain mutations that have modified the way they handle the sugars they ingest, leading them to use the food they consume inefficiently. This means that cancer cells take in more sugar than normal cells for a given rate of growth and division. Thus, they are particularly likely to give a good signal in a PET scan. As a result, PET scanning works to detect many kinds of tumors. Moreover, the method is sufficiently sensitive that it can pick up relatively small growths, as little as about one-fourth of an inch (6 mm) in diameter.

Both CT scans and MRI have sufficiently good resolution that they too can see tumors this small, but the doctor needs to know where to focus these instruments in order to find a tumor. If one doesn't know where a metastatic tumor might be, a whole-body CT scan would be required to find it. This is not a good medical choice because it would involve more X-irradiation than is wise for the patient's long-term health. A PET scan, on the other hand, can form a single image of the entire body and identify tumor cells, wherever they are. The images are low in structural detail, but the contrast between tumor and normal tissue is good, making tumor detection easy. However, PET scanning is costly ($3000–6000) and not entirely benign, given that gamma rays, like X-rays, can cause mutations and other kinds of damage to cells. Fortunately, the amount of radiation that the patient experiences is low to moderate, so if there is a serious risk that cancer is present, a PET scan can be the best way to find it.

EVALUATING THE RESULTS FROM CANCER SCREENING TESTS AND MEDICAL IMAGING

In this chapter, you have seen several ways of looking at a human body to get information that helps to identify a cancerous growth while it is still small. Some of these methods are simple, such as examining the skin or palpating parts of the body. Other methods use fiber optics to look for cells that show signs of cancerous behavior. Then there are the methods that rely on high-tech imaging devices. All these methods, from touch to the detection of gamma rays in a PET scan, have certain things in common that are worth thinking about to understand the strengths and limitations of cancer detection and diagnosis.

All methods for cancer detection rely on a few basic ideas.

One way to evaluate a screening or diagnostic method is to look at the amount of abnormal structure that is required to give a detectable signal. How big, how hard, or how dense must a lump be, relative to its surroundings, in order to be seen or felt? The ability of the instrument in use (our hands or something more complex) to detect and characterize small differences is a fundamental aspect of the detection process. In short, the method must give a **signal**. This can be the presence of a lump that our fingers feel or a mass that imaging methods detect as different from surrounding structures. For there to be a signal, there must be **contrast**, that is, a difference between the structure of interest and its background. However, if a method is very sensitive, it may give contrast from lots of irrelevant things around the lump of interest, adding **noise** to the image and making valid detection difficult. Thus, contrast is not the only issue in obtaining a good signal. There must be a good **signal-to-noise ratio**, that is, the level of signal strength over the background noise, before we can get reliable information (FIGURE 2.17).

Probing with our fingers (palpating) or imaging with any form of radiation will produce a signal-to-noise ratio based on the quality of the information generated. Good image quality means sharp definitions of boundaries and a clear view of what might be a lump of cancerous tissue, as opposed to a blob of noise. Examples of such noise are clusters of blood vessels, chance variations in nearby tissues, or a cyst. PET produces images through locating a signal that is concentrated particularly well in cancer cells, which take in a lot of the radioactive sugar. This signal can be detected

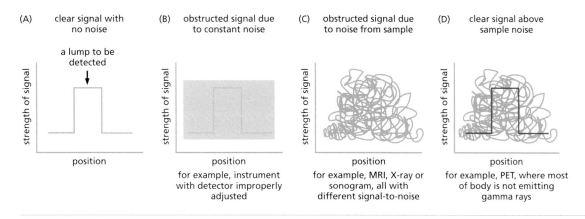

(A) clear signal with no noise
(B) obstructed signal due to constant noise
(C) obstructed signal due to noise from sample
(D) clear signal above sample noise

a lump to be detected

strength of signal — position

strength of signal — position
for example, instrument with detector improperly adjusted

strength of signal — position
for example, MRI, X-ray or sonogram, all with different signal-to-noise

strength of signal — position
for example, PET, where most of body is not emitting gamma rays

FIGURE 2.17 **The idea of signal-to-noise ratios in detection of a possibly cancerous lump.** (A) The gray line with a big bump in the middle represents a lump to be detected, for example, by a medical imaging device. In this figure, there is no noise, so the signal is obvious. (B) The signal has not changed, but it is now surrounded by meaningless signal that acts as noise, making it much harder to see the signal. This sort of noise can occur with any imaging method. It is usually due to some feature of the instrument and is comparatively easy to remove with some sort of filter. (C) Again, the signal is unchanged, but now the noise is not a uniform signal with no meaning; it is many signals, each rather similar to the one we want to detect. This kind of noise might occur with any imaging device, but a good example is a sonogram of a lump that formed in a part of the body that is reflecting a lot of sound, making the signal from the lump hard to detect. (D) Exactly the same noise as in panel C, but the imaging device has put a special kind of contrast into the lump. It is now red and shows up rather clearly in spite of the noise; the signal-to-noise ratio has been improved, thanks to the new contrast. Examples of this effect are found in a CT scan with iodine injected to add contrast, in MRI with a contrast agent like gadolinium, and in a PET scan, where the cells in the lump are concentrating the special labeled sugar and therefore emitting more gamma rays than any of the surrounding tissues.

against a comparatively blank background, giving good contrast and an excellent signal-to-noise ratio (Figure 2.17D).

An example will help to illustrate the importance of a high signal-to-noise ratio when searching the body for cancer. Imagine that you are trying to see the weeds in a lawn, catching them early to dig them out before they have gone to seed. Distinguishing the green leaves of the weeds you want to dig out from the green blades of grass that you want to keep requires sharp images in which the leaves and blades can be distinguished. If, on the other hand, the weeds were red, they would stand out against the green blades of grass and detection would be easy; no carefully formed image of leaf shape would be required. A PET scan delivers images that are a bit like red weeds against a green background, so sharp imaging is not necessary. Improving image signal-to-noise ratio is at the heart of many innovations in imaging technology, be it for cancer diagnosis or any other characterization of a diseased state.

Doctors do, however, have concerns beyond signal-to-noise ratios in deciding whether to prescribe a test that uses an advanced imaging method. As mentioned previously, some methods include exposure to X-rays, which can increase cancer risk when used in high doses. Some methods are far more costly than others, and their frequent use drives up the cost of medical care. However, if doctors refrain from using one of these expensive instruments, they run the risk of being accused of medical malpractice at a later time; they failed to use an established procedure that might have detected a cancer at an early stage. Hospitals, likewise, have a good reason to encourage expensive imaging; they bought the necessary equipment at considerable cost. When it is not in use, they are failing to bring in money to pay off their investment. Thus, only insurance companies (who will usually have to pay) and patients (who may have to pay and will certainly be receiving whatever radiation is involved) have a motivation *not* to do a scan. Given the opposing issues of not wanting the nuisance and exposure of a scan but wanting to get an accurate diagnosis, the patient is likely to be ambivalent. This situation can lead to the overuse of medical imaging. On the other hand, advanced medical images can provide essential information that might lower long-term medical costs by reducing the extent of subsequent treatment and/or hospital stays. For all these reasons, modern methods of medical imaging are highly regarded and widely used, even though they do carry costs and some risk. TABLE 2.1

SIDE QUESTION 2.11

Consider two people with equally good eyesight, one of whom has color blindness while the other does not. Which individual would you expect to have a better signal-to-noise ratio in spotting unknown features in their environment? Why?

Table 2.1 Common Methods for Cancer Detection and Diagnosis

Method	Instrument	Result	Use	Sensitivity	Cost	Danger
Palpation	Hand	Detect lumps	Breast, prostate	Moderately sensitive	Low	Zero
Visualization	Eyes	See abnormalities	Skin, melanoma	Sensitive	Low	Zero
Occult fecal blood	Chemical tests	Detect blood, stool	Colon cancer	Moderately sensitive	Low	Zero
Pap smear	Swab, microscope	Find abnormalities	Cervical cancer	Sensitive	Modest	Zero
Sonography	Ultrasound detector	Detect growths	Anywhere	Moderately sensitive	Modest	Zero
Colonoscopy	Endoscope	Find polyps	Colorectal	Very sensitive	Modest	Very low
Mammogram	X-ray photograph	Detect lumps	Breast	Sensitive	Modest	Low
Chest X-ray	X-ray photograph	Detect lumps	Lung, abdomen	Sensitive	Modest	Low
CT scan	X-ray and computer	Detect lumps	Anywhere	Very sensitive	High	Moderate
MRI	Magnet, computer	Detect lumps	Anywhere	Very sensitive	Very high	Very low
PET scan	PET scanner	Detect fast growth	Anywhere	Very sensitive	Very high	Moderate

compares the most common methods of cancer screening and diagnosis, including the relative value and cost of each.

The quality of a cancer screen is measured by the frequency with which it gives the correct answer.

Given the strengths and limitations of the various cancer screening tests and procedures, which are best and how often should they be done? The answers to these very reasonable questions are not as obvious as one might think, although some general statements can serve as guides. For example, remember the very strong dependence of cancer frequency on age (see Figure 1.14). Before the age of 40, the incidence of cancer is so low that one needs a very good screen to make it likely one will find a tumor that merits medical attention. Given that many of the screens do include some risk, it is clear they should not be used on young people who have no special risk of cancer and are otherwise in good health.

The quality of any screening test or procedure is measured by the frequency with which it gives a correct diagnosis. If it reports a cancerous growth when none is there, this is called a **false positive** or a **type 1 error**. If a test fails to detect a cancerous growth that is actually there, this is called a **false negative** or a **type 2 error**. The quality of a cancer screening process is assessed by the frequencies of false positives and false negatives it delivers: the lower the number of times for each of these errors, the better. Unfortunately, none of the screening methods now available performs as well as anyone would like. The best statistical information available about the quality of a screening procedure is for the use of mammograms to detect breast cancer. Sadly, the data suggest that this apparently reliable test is not as good as you might expect. Sometimes a breast lump is detected and identified as cancer when there is only a small chance that it would grow into a dangerous cancer (a false positive). Sometimes a precancerous lump is missed (a false negative). The latter is more common in mammography of young women, making the idea of screening early in life even less appropriate. This issue is discussed in the context of available data in SIDEBAR 2.8.

Limitations in the diagnostic tools available have raised concerns about how much cancer screening is good for the patient.

As discussed in Sidebar 2.8, existing screens and procedures for identifying breast cancer have limitations. The same is true for all cancer screens, with the possible exception of skin cancer, where direct observation is easy and cheap. In addition to the concerns about false positives and false negatives, there is the worry that some small tumors, which were identified by a sensitive screen and therefore treated, might have **regressed**, that is, gone away on their own, or just never progressed into a dangerous cancer. The treatment resulting from these detections was an unnecessary expenditure of time and money. In addition, there is certainly anxiety associated with being told that one has a lump where it doesn't belong. Is that anxiety warranted, even when no cancer is ultimately discovered? Oncologists are now collecting data on the frequency with which screens identify lumps that turn out to be benign, but it will be impossible with current methods to know what might have happened if a given lump had not been treated. As an example of the importance of screening, look back at Rosemary's story in Sidebar 2.2, where a routine mammogram identified a lump, and subsequent biopsy showed that the lump deserved immediate attention. Screening to achieve early detection is certainly important for reducing fatal cancers, but better screening

SIDE QUESTION 2.12

If it were illegal to sue doctors for malpractice, do you think the cost of medicine would go up or down? Do you think the quality of health care would go up or down? Why?

SIDE QUESTION 2.13

How would you describe the difficulty of determining the frequency with which a cancer screening procedure, like mammography, identifies a growth that would have regressed on its own?

SIDEBAR 2.8

ADVANCES IN MEDICAL IMAGING: The Accuracy of Mammography

The medical profession has become concerned about the value of mammograms, specifically, the ability of this imaging technology to spot breast tumors early and help women avoid serious cancer. Mammograms have the potential to relay false positives that lead to both unnecessary treatment and considerable anxiety for the patient. They can also fail to develop sufficient signal-to-noise in the critical region of the image and therefore miss a lump that should have been identified. That is a false negative.

The problem is how to know whether a given mammogram has helped or hindered. One study of this issue, published in 2009 in the journal *Annals of Internal Medicine* (Vol. 151, pp 727–737), found that regular mammograms for women ages 40–74 reduced their subsequent incidence of death from breast cancer by 15–20%. The estimates of false positives ranged from 1 to 10%, as identified by a second round of imaging (a biopsy was thought unnecessary). The higher number of false positives came from the younger patients, who had a lower risk of breast cancer in the first place. The false negatives were women whose mammograms were described as clear of medical problems but who were subsequently diagnosed with breast cancer. Taken together, these data suggest that mammograms for middle aged women are doing a pretty good job of improving public health.

This same study evaluated both clinical examination and self-examination of breasts by palpation.

This form of screening led to an increase in the number of lumps detected and therefore of biopsies that were subsequently done. However, the results of those biopsies commonly led to the conclusion that the lump was benign. Palpation did not lead to a decrease in the numbers of deaths from breast cancer.

One must realize, though, that these kinds of studies on cancer diagnostics are difficult and can lead to conflicting results. For example, a 2014 publication by the Canadian National Breast Screening Study in the *British Medical Journal* reported no reduction in breast-cancer-related deaths from mammography for women 40–59 years old. This study compared data about cancer diagnosis in women who received an annual mammogram with a control group of women who received an annual physical exam but no mammogram.

There is now increasing evidence that mammography can lead to overdiagnosis, meaning the identification of small tumors that would never have grown into a health threat if they had simply been left alone. Overdiagnosis is not good for the patient because the treatment of a tumor always carries some health risk: possible deformity from surgical removal of the lump, toxicity to normal cells from X-radiation or some other method used to treat the tumor, plus the anxiety that accompanies a diagnosis of cancer. Some studies suggest that at least 20% of the mammogram-detected breast lumps could have

been left alone with no bad effects on patient health.

The American Council of Radiologists has reinvestigated some of the studies that suggest mammography does not help. Using data from Sweden, they estimated that, statistically, 7.5–10 deaths were prevented among 2500 women invited to be screened by mammography every 2–2.5 years for 10 years. Likewise, data from the mammography screening program in the United Kingdom suggest that, statistically, 2–2.5 lives are saved for every over-diagnosed case. These studies suggest that regular mammograms for middle-aged women are wise.

Mammography does, however, produce some false negatives. In younger women and those with particularly dense breast tissues, there is a significant chance that a tumor will be missed, particularly if it is small but fast-growing at the time of examination. Estimates of this kind of mistake range from 6 to 46%, with the higher numbers associated with younger patients.

It follows that we do not currently have an infallible screening test for breast cancer, and this is an issue that merits new research. For all these reasons, it is unwise for women younger than 40 to have a mammogram, unless they are at special risk of breast cancer as a result of trends in their family or an unusual exposure to cancer-causing conditions. After the age of 40, regular mammograms are probably a good idea, in spite of the current controversy.

procedures would clearly help to make sure that time and money are not wasted.

Developing better screens for cancer would improve early diagnosis of the disease.

Catching a tumor before metastasis has begun is certainly desirable. Thus, many scientists have been motivated to seek novel approaches to cancer screening tests and procedures. Tests for specific substances in the blood are one potentially valuable approach to cancer identification. As was hoped with the PSA (prostate-specific antigen) test for prostate cancer, there may be specific proteins that could be recognized in blood and used as cancer **markers** to detect numerous manifestations of this disease. A quantitative measure of the amounts of such proteins might even assess how far the cancer had advanced. At the moment, this idea is still a dream, but it may develop into a way to look for cancers quickly, comparatively cheaply, and noninvasively. A test for a cancer marker in the blood would not, of course, tell you where the tumor was located, but if you knew a tumor was present, thanks to this kind of test, it would probably be worth a PET scan to find it. Identifying through simple blood tests a subpopulation of people who really are at risk of cancer would make the expense and risk of advanced medical imaging worthwhile. Early diagnosis through cancer markers not only might save lives but also would improve public health at a reasonable cost.

Existing methods of medical imaging could also be brought to bear in early diagnosis of populations that are at risk of a particular cancer. For example, as discussed in Chapters 5 and 11, cigarette smoking causes a marked increase in a person's risk of getting lung cancer, particularly after 20 or more years of steady smoking. Performing CT scans every few years on the chests of people in this group might identify tumors early and make a significant difference in their likelihood of their dying from lung cancer. When caught early, lung cancer can be treated more effectively than is possible after the patient is complaining about shortness of breath and chest or joint pains (see Sidebar 2.3). However, this is the kind of proposal that would lead to a significant increase in the cost of health care. One could argue that the money would be better spent on research to identify methods that reduced the number of people who smoke.

CANCER PROGNOSIS: UNDERSTANDING WHAT A DIAGNOSIS MEANS

If a diagnosis of cancer has come from a responsible physician, there will certainly be a recommendation about treatment. Unless time is very short, for example, as a result of a late diagnosis and a desperate situation, it can make good sense to seek a second opinion. If the two opinions disagree, the patient can try to learn why. This usually means requesting that the doctors talk through the factors that have led to their opinions. Indeed, many hospitals have a policy of discussing each cancer patient's case among several doctors, so different points of view can be aired and compared. Taking the time to be reasonably confident about any major step in personal healthcare is well worth the extra time, and it is usually worth any extra expense.

A diagnosis of cancer commonly leads to a description of the treatment options that medical science can offer and the likely outcome, or prognosis, for each possible treatment. This sounds simple enough, but the outcomes may be described in ways that are not meaningful to the patient. They are commonly based on **cancer stages**, meaning how far the cells of the tumor have progressed through cancerous transformation, and the extent of metastasis, if any, as we will describe. Moreover, a prognosis is often given in terms of **probabilities** or chances that the patient will be alive after some number of years. For example, the prognosis may be the likelihood of five-year survival, given a specific kind of treatment. This kind of information is both scary and unsatisfying; one usually assumes a long life, and no one likes to be considered a statistic. To make sense of this kind of prediction, we will first take up how cancer is staged and then discuss the nature of probabilities.

The stages in cancer progression are quite clearly defined.

The stages in cancer progression range from 0 to 4: the lower the number, the better the prognosis. These staging numbers are based on three criteria. First and most important is the condition of the tumor itself: its size, the degree to which the tumor cells look abnormal (dysplastic, metaplastic, or anaplastic), the shape of the tumor's edges, and the presence or absence of secondary tumors nearby. These tumor properties are represented by the number assigned to T for tumor.

The second number used for staging cancers is N for lymph node. Remember that the lymph nodes are sites where migrating tumor cells can often be found. N tells the number of nearby lymph nodes in which cancer cells were detected: the higher the number, the worse the prognosis, because it implies a more extensive spread of the cancer cells. Note, however, that with the increasing use of sentinel nodes as sites to look for wandering tumor cells, this number may be less meaningful, because not so many nodes were sampled at biopsy.

The third factor in cancer staging is represented by M (for metastasis): either 0, for no distant growths detected outside of lymph nodes, or 1, if such a growth has been found. TABLE 2.2 summarizes these aspects of tumor staging and adds a little detail.

These TNM rankings can be lumped into a general stage number, where stage 1 is the condition in which things are highly localized; stage 2

Table 2.2 **The TNM Cancer Staging System**
Primary tumor (T): refers to the size and extent of the main or primary tumor.
TX: Main tumor cannot be measured.
T0: Main tumor cannot be found.
T1, T2, T3, T4: Refers to the size and/or extent of the main tumor. The higher the number, the larger the tumor or the more it has grown into nearby tissues. These stages may be further divided to provide more detail, such as T3a and T3b.
Regional lymph nodes (N): refers to the number of nearby lymph nodes that contain cancer.
NX: Cancer in nearby lymph nodes cannot be measured.
N0: There is no cancer in nearby lymph nodes.
N1, N2, N3: Refers to the number and location of lymph nodes that contain cancer. The higher the number after the N, the more lymph nodes that contain cancer cells.
Distant metastasis (M): refers to whether the cancer has metastasized.
MX: Metastasis cannot be measured.
M0: Cancer has not spread to other parts of the body.
M1: Cancer has spread to other parts of the body.

involves local and nearby growths, including nearby lymph nodes; stage 3 represents a bigger tumor and tumor cells detected in more lymph nodes; and stage 4 is an aggressively growing tumor accompanied by sites of metastatic growth in a distant organ, for example, the spread of colon cancer to the liver. Clearly, the higher the number, the worse the prognosis.

Even with this information, though, it is hard for a patient to know what stage or TMN numbers mean. Medical science will usually express the implications of a given stage as a likelihood of survival for some period of time. For example, the diagnosis of stage 2 breast cancer is associated with 80–90% chance of survival for at least five years, assuming the patient gets proper treatment. The chance of survival is commonly stated as a range because there are ranges in the severity of the cancers that are called stage 2 and because people differ in the way their bodies respond, both to the disease and to its treatment. Proper treatment usually means that the primary tumor and nearby lymph nodes are removed by surgery and the patient receives an appropriate regimen of radiation and/or drugs to kill malignant cells, as described in Chapter 8. With stage 3 breast cancer, the chances of five-year survival drop to 55–70%. For stage 4, five-year survival is down to about 20%.

Probabilities are a part of cancer prognoses, so it is important to understand them.

We are all familiar with statements about chance in our normal lives. Weather reports talk about the chances of rain, sportscasters predict the chances of a win, and people who bet on cards or any other game of chance are usually well aware of the probability that a particular card will turn up when they need it. Statements about chances or probabilities (two words that mean the same thing in this context) can be based on two quite different sets of factors. In cards, there are 52 cards in a pack, and 4 of them are twos. This means that if a deck is well shuffled, so that all the cards are in random order, there are 4 chances out of 52 of drawing a two; the probability of a two is $4/52 = 1/13 = 0.077$. A probability like this is based on the structure of the situation, and it is very reliable.

The statements of chance or probability in weather prediction or in cancer prognosis are different from the structurally defined probabilities of cards and dice. They come from historical records. Cancer prognoses are based on the medical histories of large numbers of people, their condition at diagnosis, and what happened to them in the following years. Over the last 50 years and more, oncologists (doctors specializing in cancer) have kept good records about the effectiveness of various treatments. An example of such data for patients with bladder cancer, treated in two different ways, is shown in FIGURE 2.18. In this graph, the vertical axis displays the fraction of people diagnosed with bladder cancer who are still alive after increasing amounts of time, shown on the horizontal axis. The two curves in the graph compare these facts for groups of patients who received different treatments. These data show that cancer patients who received both surgery and a treatment called Bacillus Calmette-Guerin (BCG) are more likely to survive than those who underwent only surgery. Data like these encourage both doctors and patients to use the more effective treatment and thereby get a better prognosis. Since cancer treatments are improving all the time, the historical data on prognosis may be somewhat pessimistic, but they are not off by a lot.

Data collected over many years indicate that women with stage 2 breast cancer who undergo a **lumpectomy** (removal of only the cancer mass

SIDE QUESTION 2.14

What is the probability that a die (singular of the familiar word dice) will land on a flat surface with three dots on top? How do you know?

FIGURE 2.18 Historical information comparing survival after different treatments for bladder cancer. The vertical axis shows the fraction of people still alive and with no recurrence of their cancer after treatment. The horizontal axis shows time since treatment. The graphs display the fraction of people still alive at each time after treatment, either surgery alone or surgery plus a stimulation of the patient's immune system with a treatment called Bacillus Calmette-Guerin (BCG). Clearly BCG improves patient prognosis compared with surgery alone. (From Patard J et al. [2001] *Urology* 58:551–556. With permission from Elsevier.)

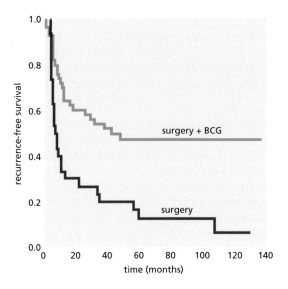

and the immediate surrounding area, not the entire breast) experience no measureable difference in prognosis relative to those who undergo a **mastectomy** (removal of all tissue from the affected breast). Said another way, the less invasive lumpectomy is as good for the patient's prognosis as the more invasive treatment. This kind of finding is good news for all because it shows that the more drastic and expensive treatment provides no added value.

Similar studies have shown that when a malignant breast tumor is small and/or when the tumor cells are relatively normal, a variety of chemotherapies are effective in blocking cancer growth; no significant improvement in outcome is achieved by employing radiation treatment along with surgery and chemotherapy. These kinds of data are very useful for identifying the minimal treatment that is effective for a given cancer at a given stage.

Most doctors giving a prognosis won't go farther than stating survival percentages. These historical records are the only facts they have to go on. However, there are also personal factors that affect outcomes. For example, a study on the fraction of people with a given condition who die within a given period may not have taken account of the patient's age or weight. Thus, a particular patient's prognosis may be better than the quoted figures, or it could also be worse. Prognoses describe likelihoods, not certainties.

As of the early twenty-first century, we don't yet know how to get the information that would turn vague cancer prognoses into some kind of certainty. There is currently a big effort in the biomedical community to increase the amount of information about each patient that is available for assessing his/her medical situation. A topic of great current interest is the condition of the patient's genes, which can be characterized using methods based on molecular biology that will be described in Chapter 10. A wide-ranging characterization of the patient's genotype, that is, the structure of all his/her genes, may allow a more accurate prediction of the way her/his body will respond both to the tumor and to the prescribed treatment. This kind of evidence may change the accuracy of prognosis, but in terms of a patient's feelings, it is not clear that this will be an improvement. Is it better to know that you *will* die within five years or to know that you *might* die? This question is one of many that make living with cancer a very personal matter, which is the subject of Chapter 12.

SUMMARY

There are now many ways to identify cancers. A physical exam and a few simple, inexpensive tests are good first steps, but modern methods for imaging structures inside the body are now able to improve upon what a doctor can learn from looking at a patient from the outside only. Some cancers are hard to detect, even with the best methods now available, but others, like breast, prostate, and colon cancer, have become sufficiently visible that screening for them with methods for medical imaging is common, even when no symptoms of cancer are present. Screening tests and procedures have medical value, given that catching a growing cancer early makes it easier to treat effectively. However, none of the screens so far available is perfect. Sometimes a growing cancer is missed, and sometimes a lump that was thought to be cancer is benign. Moreover, some kinds of medical imaging used in cancer diagnosis may increase the risk of cancer in the future; for example, the considerable exposure to X-rays that would come from multiple CT scans. Nonetheless, scientific data empower doctors with the knowledge they need to reduce the risk of cancer going undetected until it has advanced so far as to be untreatable.

Some methods used for cancer screening can also help to characterize cancer if it does arise. The study of biopsies by light microscopy or with molecular methods can tell a trained pathologist a great deal about the extent to which cancerous progression has already occurred, thereby permitting a more accurate diagnosis. Identification of cancer cells in tissues other than the primary tumor, such as nearby lymph nodes, can provide direct evidence about metastatic behavior. Some of the methods for advanced medical imaging, such as PET scans and MRI, can also be informative about metastases.

From a good characterization of a cancer, both doctor and patient have the information necessary to compare the case at hand with many similar ones that have previously been studied and documented. Historical information allows a doctor to give a prognosis, that is, a prediction about the outcomes, based on different kinds of treatment. This information can be valuable in choosing which treatment(s) to undergo.

ESSENTIAL CONCEPTS

- There are several simple but valuable ways of detecting cancer early. Some of these are inexpensive, noninvasive, and quite effective. As you reach middle age, they are important tests to do on a regular basis.

- Doctors and scientists have developed multiple ways to look inside the human body. These methods can often identify a cancerous growth that was otherwise invisible. Some of these methods, like mammography and colonoscopy, are easy and cheap enough that they are now used for regular cancer screening in people of middle age or older, who have a significant risk of developing cancer.

- Modern technology has allowed the development of additional tools for medical imaging. Some of these are sensitive ways to look for cancers, but they can be expensive. Some of them carry cancer risk on their own, so they are not appropriate for everyone.

- If there is good reason to believe that cancer is present, the more expensive methods, like CT scans, MRI, and PET scans can be valuable ways to find the cancer, image it, and determine its extent.

- When cancer is found, detailed information about its stage of progression can be obtained through the study of both cancer location and biopsy.

- A trained pathologist can use a light microscope to learn a lot about a cancer's type and stage, but a fuller characterization involves biopsies of nearby tissues, like lymph nodes, seeking evidence about metastasis.

- Biopsies can also be used for molecular analysis of cancerous tissue. Currently this information can help with prognosis, but as methods improve, molecular study will probably become a key way to obtain detailed information about cancer stage and appropriate treatment.

- Cancer can be staged, on the basis of the shape and distribution of the tumor, the lack of differentiation in its component cells, and the extent of its spread to other parts of the body.

- Pathologists and oncologists have large amounts of data about the likely outcome of a cancer identified at a given stage, following various treatments. These data allow a doctor to give a prognosis.

KEY TERMS

anaplastic	immune system	polyp
basal cell carcinoma	lumpectomy	positron emission tomography (PET scan)
biopsy	lymph	
cancer stage	lymph nodes	primary tumor
colon	lymphatic system	probability
colonoscopy	lymphatic vessel	prognosis
computed tomography (CT scan)	magnetic resonance imaging (MRI)	prostate gland
		prostate-specific antigen (PSA)
contrast	mammogram	regress
dysplastic	mastectomy	sentinel nodes
endoscope	melanin	signal-to-noise ratio
false negative	melanoma	sonogram
false positive	metaplastic	squamous cell carcinoma
fiber optics	palpation	type 1 error
gamma rays	Pap smear	type 2 error

FURTHER READING

Most books on cancer screening and diagnosis are written for cancer specialists, and they are neither easy nor pleasant reading. There are, however, two reliable Websites that contain vast amounts of information on every known form of cancer. The two sites listed below contain detailed information on all the common cancers and on the methods for screening and diagnosis mentioned here. They also give access to information about some rare cancers. If you are curious about a particular cancer, these Websites are excellent places to visit.

The American Cancer Society (http://cancer.org) supports a very informative and accurate Website for people interesting in cancer. It is designed first and foremost for people with cancer, so there are many items about diagnosis and treatment of specific cancers, as well as access to support groups. Interspersed among these items are reliable accounts of symptoms and treatments for many specific cancers.

The National Cancer Institute (http://cancer.gov) maintains a different but equally valuable website from which people can learn about cancer. This site includes information that is useful for patients, the interested public, and cancer professionals.

QUESTIONS FOR FURTHER THOUGHT

1. Why is it better to catch a cancer early, rather than late?

2. Why bother with all this expensive and possibly harmful medical imaging? Why not simply take a biopsy in the first place?

3. Given the power of PET scans and MRI for describing cancers, why not just perform one of these on all middle-aged people?

4. If you were a doctor whose patient had a palpable lump in her breast, what would you do next? Why? How would you argue with the insurance company to pay for the treatment?

5. Why does distant metastasis carry a poorer prognosis than a nearby secondary tumor?

ANSWERS TO SIDE QUESTIONS

2.1   This is really an experiment; there is no correct answer.

2.2   Yes, the small, compact nuclei are reasonable. When a cell is growing and dividing, many genes must be expressed. Simply to make the fibers that make a skin cell tough requires the expression of only a few genes, so the nucleus is not as active in these differentiated cells. Thus, the logic based on thinking about levels of gene expression is consistent with what a doctor sees in a Pap smear.

2.3   As mentioned in the text, a doctor can pass an endoscope, containing fiber optics for viewing, through the anus to look at both the rectum and lower intestine. An endoscope in the urethra can look all the way up into the bladder. Endoscopes can also go down the throat from the nose or mouth to look all the way into the stomach. These devices can also go into the breathing tube to look at the topmost parts of lungs.

2.4   Strengths of a needle biopsy include the fact that a sharp needle can penetrate the skin and go to the place of interest in the body without a doctor having to make a large cut. This minimizes pain and disruption of the skin. The biopsy needle can withdraw enough tissue for many kinds of cancer screening tests, for example, the study of the cells by microscopy and some molecular tests. Two limitations of biopsy needles are the difficulty a doctor faces in getting the needle tip to exactly the right place in the body and withdrawing enough tissue for all relevant studies. The advantage of an aspiration biopsy is that the thin needle is less damaging to the patient's tissues. The disadvantage of an aspiration biopsy, relative to a core biopsy, is that sucking cells out through the thin needle can disrupt their relative positions, destroying some of the structural features a pathologist might want to see.

2.5   If the cells in a biopsy are anaplastic, rather than dysplastic, it implies that they have already changed a great deal from normal. To have changed that much, they must have experienced multiple mutations. These mutations can take the cells a long way down the path of cancerous transformation, whereas if the cells still looked rather normal, fewer mutations would have occurred.

2.6   This is a question for discussion; it doesn't have a right or wrong answer. It is designed to provoke conversation about the information presented regarding the strengths and weaknesses of information from a mammogram. On balance, most physicians would probably favor having a biopsy of the suspicious lump to get more information about it. Another possibility would be to order additional medical imaging, such as an MRI scan, but this method is expensive, and it would not reveal the cells of the lump with sufficient detail to see if they are dysplastic or even anaplastic. This kind of detail requires a biopsy.

2.7   An X-ray scattering chemical in blood vessels will provide an X-ray image in which these vessels show up more clearly. Because actively growing tumors need blood for nourishment, adding a stain that improves the visibility of blood vessels will reveal a tumor more clearly than a simple X-ray would do.

2.8   This is again a question for discussion. Some mammography services are now advertising their use of CT scans for cancer screening; they think it is worthwhile, and perhaps it is. On the other hand, the added cost will be paid by either the patient or an insurance company, driving up health care costs. At this time, it is hard to know whether the added sensitivity for detection of small lumps that comes with a CT scan is of sufficient medical value to merit the additional cost and exposure to radiation.

2.9   A bat sends out pulses of ultrasound and uses the echo to find a flying insect. If there is nothing other than flying insects in the air around the bat, any reflection of sound is a signal to be followed, because it is leading to food. If the bat is flying through the branches of a tree, there are many reflections of its ultrasound that have nothing to do with insects. This is an example of noise in the information received. A sonogram includes reflections of sound from many internal organs, so the doctor's problem is using the structural information received from the echoed ultrasound to see if there is something structurally wrong. Thus, the doctor faces a significant problem of noise in the signal received. The quality of this signal must be very good to give an informative ratio of signal to noise.

2.10   Although the quality of images produced by CT scans and MRI is significantly better than that from PET, these methods see tumors only by the disruption of the structures normally seen. The contrast in PET is derived directly from the rate of sugar consumption by different regions of the body. Although some tissues, like brain, will always show up, a tumor in most regions of the body will be seen quite clearly against a comparatively blank background. Thus, the signal-to-noise ratio of PET is better for tumor detection than that obtained with the other two imaging methods. For this reason, PET scanning is the preferred method for seeking metastatic growths.

2.11   People with color blindness are unable to distinguish some colors from others, for example, red from green. As a result, their vision does not provide as much information as in the case of people with normal vision. Because people with color blindness get less visual information, they are receiving less signal, so their signal-to-noise ratio is lower than that of people with normal vision.

2.12   These are again hard questions with no right or wrong answer. Some would argue that the absence of the right to sue doctors for malpractice would drop the cost of health care but lower its quality. Other positions can be defended. These are interesting and important questions to think about.

2.13   The problem here is one of action versus inaction. If a small lump has been detected by mammography, a doctor could note that fact but suggest that the patient do nothing about it. If health records were good, the practice of medicine could then keep track of what happened to that patient and whether a cancer developed at a later time. However, most patients would not be willing to take that chance. If the lump is biopsied, the doctor would know almost immediately more about whether treatment was advisable. If the lump was removed, it has been treated, so one cannot know what would have happened if it had not been treated. All these considerations make it hard to collect information about what would have happened if no treatment had been done.

2.14   The probability is 1/6 because the die is a cube and therefore has six equivalent faces. There is an equal probability that it will land on any face, so on average the face with three dots will show up in one out of every six throws.

The organization of normal human cells

CHAPTER 3

Cancer results from the misbehavior of cells. Mutations (changes in DNA) alter the structure of genes and thereby change cellular functions, such as the way cells respond to signals that tell them when to grow and divide. To develop a deeper understanding of cancer, however, you will need to know the answers to many questions: How do mutations change DNA? How does a change in DNA alter cell function? Which functions are changed to make cells lose their normal controls and become cancerous? How does cell malfunction cause a neoplasm? What changes allow metastatic growth? The answers to these and many other questions about cancer require an understanding of the structures and functions of normal cells.

In this chapter and the next one, you will experience a brief journey into **cell biology**, the study of cell structure and function. You will see that cells not only take in food to grow and divide, they also respond to their environment, interpret the signals they receive, and make choices that are analogous to rational decisions. They react to multiple external and internal factors in ways that maximize their contributions to the well-being of the body as a whole. Moreover, cells can make particular molecules, change their shape, and move toward or away from chemical stimuli. They can emit signals for other cells to sense, and they can even destroy themselves. From the information presented, you will gain sufficient background to understand all these complex behaviors, so you will be able to see how mutations and related changes in a cell's DNA lead to cancerous transformation. To achieve this goal, we will start with the parts from which a cell is made, and then we'll look at how the information stored in DNA is used by cells to make these various parts. Finally, we can look at how these parts fit together to make a cell.

HOW SCIENTISTS STUDY CELL STRUCTURE

THE STRUCTURES OF CELLULAR MACROMOLECULES

MACROMOLECULES IN THE CONTEXT OF CELL STRUCTURE

PROTEIN FIBERS FOR CYTOPLASMIC STRUCTURE AND FUNCTION

REGULATING A CELL'S MANY PROCESSES

LEARNING GOALS

1. Diagram a typical cell as it appears in a light microscope. Label its structures, and describe the functions of the cell's most important parts.

2. Compare light and electron microscopy as tools for the study of cells; identify the strengths and limitations of each.

3. Compare the three kinds of cellular macromolecules. Specify the building blocks from which they are made, and give examples of the structures and/or functions they achieve for a cell.

4. Diagram how cells store genetic information, and read it out to make essential cellular molecules.

5. Distinguish the roles that DNA plays in the synthesis of proteins versus lipids.

6. Describe how a mitochondrion participates in a cell's use of energy from food.

7. Contrast the behaviors and functions of the three fiber systems of the cytoskeleton.

8. Define an enzyme, and describe how cells regulate the activity of enzymes.

HOW SCIENTISTS STUDY CELL STRUCTURE

The information now available about cells reveals much about what happens when cells are behaving as they should and also what happens when they go awry. Acquiring this knowledge has taken a great deal of time and work, largely because cells are so small and complicated (FIGURE 3.1). As mentioned in Chapter 1, most of the cells in our bodies are so tiny that a lump containing a million cells is not much bigger than a pinhead. Thus, visualizing a single cell requires a microscope. A good light microscope will let you see the overall shapes of cells, focus in on the multiple structures within them, and watch cells as they perform their various tasks (SIDEBAR 3.1). Light microscopes have been developing over several centuries, and they are now essential tools in cell biology and other cancer-related sciences.

Chemistry and molecular biology have provided many facts that are essential for understanding cell structure.

The science of biochemistry, initiated more than a hundred years ago, has produced methods and knowledge that have allowed scientists to isolate and characterize many of the molecules from which cells are built. One of these molecules is DNA, whose structure was discovered in the early 1950s. Since then, a special field has developed to study the biology of DNA and how it functions in cells. Work in this field, which is called

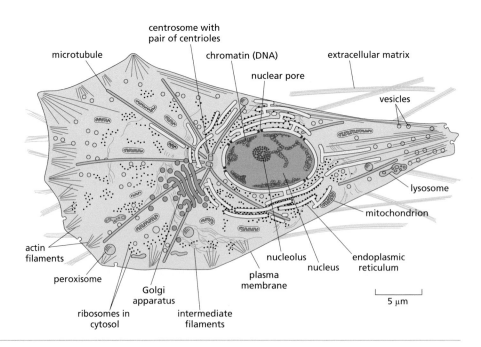

FIGURE 3.1 Diagram of a cell with labels on many of its parts that are visible with a light microscope. When human cells are taken from the body, for example, in a biopsy, and transferred to a sterile vessel that contains a suitable growth medium (water, salts, sugars, amino acids, vitamins, etc.), they will attach to the bottom of the vessel and spread out. They then look much like this diagram. Each cell contains many different structures, as labeled here and discussed further in the chapter. (Adapted from Alberts B, Johnson AD, Lewis J et al. [2015] Molecular Biology of the Cell, 6th ed. Garland Science.)

A CLOSER LOOK: Microscopy

If you use your eyes to look at an object very closely, you can see details that are about one-hundredth of an inch across. In the metric system, this size is the same as 0.2 millimeter (mm) or 200 micrometers (μm) (1 mm = 1000 μm). (Scientists prefer meters and millimeters to inches, because they are so much easier to think about, once you get used to them.) Most human cells are 10–50 μm in diameter, so even if you could see one with your naked eye, it would only look like a speck. A microscope is an instrument that forms an enlarged image of an object, magnifying it so our eyes can see relevant detail (Figure 1).

Viewing cells with a light microscope.

A light microscope contains pieces of glass shaped to form lenses, which bend rays of light, so all the light that emerged from one point on an object will converge on a single point in an image (Figure 2A). Whether the image is bigger or smaller than the object depends on the object's position relative to the lens (Figure 2B,C). If you need to make the image much bigger than the object, you can use two lenses in tandem; the second gives the image an additional stage of magnification (Figure 2D). Such a device is called a compound light microscope (Figure 3). This kind of microscope was invented in the late sixteenth century and has been used by biologists ever since.

In a compound microscope, the first lens can magnify by as much as 100× and the second lens by another 10×, so in combination the lenses magnify an object by 1000×. At this magnification, a 10 μm cell appears 10 mm across, about the size of a dime. This size lets you make out details within the cell, which is very useful for studying cell structure and function (see Figure 3.2).

When you use a microscopic to look at a cell, what are you actually seeing? Think about a picture you have taken with a camera, and call to mind the colors the camera recorded, perhaps those in your friend's shirt or the sky. What color is a cell? Most animal cells have no color; they are transparent. To reveal aspects of

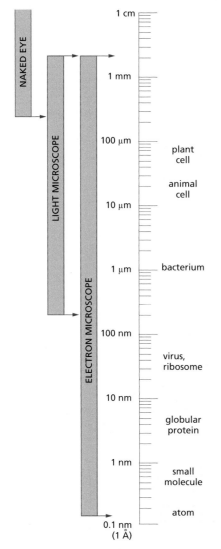

Figure 1 **The relative sizes of various biological structures that can be visualized with the human eye, with a light microscope, or with an electron microscope.** Each kind of microscope extends the range of objects our eyes can see by a considerable amount. (Adapted from Alberts B, Johnson AD, Lewis J et al. [2015] Molecular Biology of the Cell, 6th ed. Garland Science.)

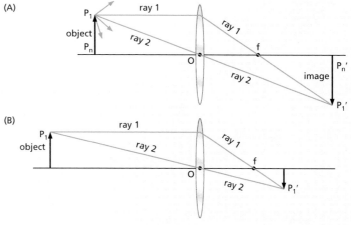

Figure 2 **Forming an image with a lens.** The size of the image formed by a lens depends on its position relative to the lens. (A) In a light microscope, a lens bends the paths of the light that pass through it. Ray 1 goes horizontally, passes through the lens, and is bent by the lens so it passes through a point called the focal point (f). Ray 2 passes through the center of the lens (O), so it is not deflected. The short blue arrows coming out of P_1 represent other rays of light emerging from this point. Any of those rays that enters the lens will be bent by its curved surfaces so they too converge on the point P_1'. For this reason, the point P_1' will get all the light that came out of P_1 and passed through the lens. P_1' is the image of P_1. Similarly, light coming from any other point on the object, such as P_n, will be deflected so it arrives at P_n'. This is how a lens makes an image. (B) The same lens is forming an image of the same object, which is now placed farther from the lens; the resulting image is smaller, as it would be in a camera. (*Continued on page 62.*)

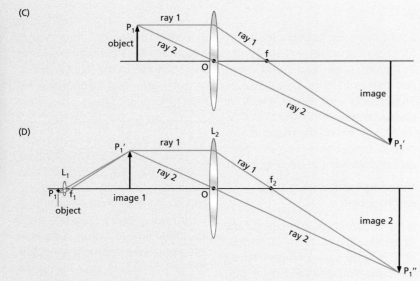

(C)

(D)

Figure 2 Forming an image with a lens (*continued*). (C) The same lens imaging the same object, now closer to the lens; the image is bigger than the object, as with a magnifying glass. (D) In a compound microscope, two lenses are used in tandem so their magnifications can multiply one another. The object viewed can now be very small. A first lens (L_1), which is small and curved so its focal point (f_1) is close to the lens, forms a first image of the object. This is significantly magnified relative to the object. A second lens (L_2) uses the first image as its object and bends the rays of light that come from P_1' to form a second image. Here, P_1'' is the image of P_1', which is the image of P_1.

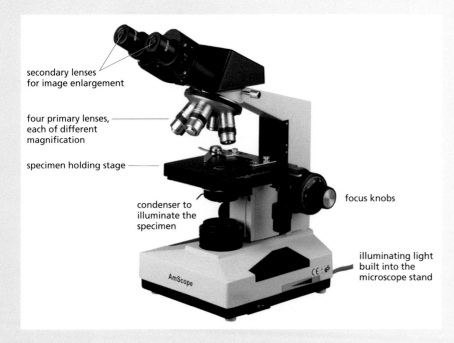

Figure 3 A typical compound light microscope. A typical compound light microscope produces an image that can be seen either by eye or with a camera (not shown). The specimen, that is, the object being viewed, sits on a stage where it can be illuminated from below. Images are formed by one of the primary lenses, each called an objective lens, and then by the secondary lenses, called oculars, one for each eye. (Courtesy of AmScope.)

their structure, a scientist must either color the cell in some way, as in Figure 3.2, or use an optical trick to provide contrast; that is, meaningful differences in the lightness and darkness of different features in the cell.

To color a cell, one commonly stains it with dyes that give different colors to different parts of the cell. Unfortunately, this treatment can damage the very structures you want to observe. To get contrast from a live, unstained cell, scientists use a special instrument called a phase microscope, which uses interference between the light that has interacted with the object and light that has not. The interference generates contrast, showing the cell in shades of gray (Figure 4).

In recent years, microscopes have been developed to visualize a particular kind of stain that is said to be fluorescent. A fluorescent molecule is one that absorbs light of one color and emits light of a different color. For example, some fluorescent molecules absorb blue light and emit green. If you shine a very bright blue light on such a dye and then look at the sample through a filter that will transmit only green light, everything looks black except the object that is fluorescent. This property of fluorescence makes such dyes an excellent way of looking for the position of specific molecules in the context of the cell's structure, so fluorescence microscopy has become an exceptionally valuable technique for light microscopy of cells (Figure 5).

Beyond contrasting the structures of a cell, there is the question of how much detail can be seen in a given microscopic image. This depends on the microscope, but even a very good light microscope can reveal details at only about 0.2 μm. This means that the image of a cell that is 10 μm in diameter will show details that

(A)   (B)

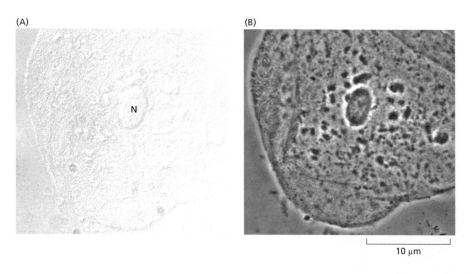

10 μm

Figure 4 **Images of a human cheek cell taken with transmitted light, showing the improvement in contrast that is achieved with a phase microscope.** (A) A simple image of a single cell; the nucleus (N) is just visible in this low-contrast view. (B) The same cell imaged with phase optics in which the interference between light scattered by the specimen and unscattered light gives contrast that appears to the eye as dark and light. Both the nucleus and many cytoplasmic granules are now clear. (Courtesy of ED Salmon, University of North Carolina.)

are about one-fiftieth the size of the cell itself. This aspect of an image is called its resolution; the better the resolution of an image, the more detail you can see. A good light microscope can resolve numerous structures within the cell, like the nucleus and some other substructures, but the individual molecules from which these structures are built are much smaller—about 0.002 μm in diameter. These are so small that they simply blend together to form a continuous feature in the light microscopic image. The limitation on resolution is not a result of faulty manufacture or microscope design; it is a result of the nature of light and the ways light interacts with the materials that make up a cell.

Seeing more cellular detail with an electron microscope.

In the early twentieth century, German physicists invented a microscope that used electrons instead of light to form images. In this instrument, the lenses are electromagnets that bend the paths of the electrons in much the same way that glass lenses bend the paths of light. Once electron microscopes became easy to operate, biologists used them to get about 100-fold better resolution than they could with a light microscope (Figure 6). The new level of detail, however, came at a cost. Live cells contain lots of water. Water does not interfere with light microscopy, but it poses a serious problem for electron microscopy. The electron beam requires a very good vacuum. If you put a live cell into such a vacuum, its water boils, and the structure of the cell is destroyed. Electron microscopy brings to biology a big improvement in resolution, but cells must be prepared in special ways to withstand the vacuum of the instrument.

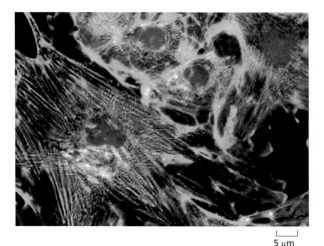

5 μm

Figure 5 **A fluorescence micrograph of mammalian cells.** These cells from the lining of a cow's biggest artery were grown on a thin piece of glass, called a coverslip, to which they attached tightly, so they spread out over a comparatively large area and became quite thin (the dimension projecting toward you, which you can't see). The nuclear DNA is stained with a blue fluorescent dye, the mitochondria are stained with an orange fluorescent dye, and the fiber bundles in the cytoplasm are stained with a green fluorescent dye. (Courtesy of Nikon Instruments, Inc.)

SIDEBAR 3.1

(A)

(B)

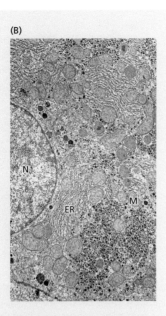

The preparation of cells for microscopy is a serious issue that scientists argue over frequently. Two labs can work on the same cell type, using different methods to prepare their specimens, and get quite different results. Who is right? Sometimes it takes years to figure that out. Usually the confirmation of one view or the other comes from a different but related method that produces results which agree with one of the views but not the other. This situation makes science interesting but slow for arriving at a valid understanding of a hard problem.

Figure 6 Electron microscopes provide better resolution than a light microscope, revealing many details of intracellular structure. (A) An electron microscope in which modern electronics and computer controls make an instrument that is comparatively easy to use for imaging biological material at high resolution. (B) An electron micrograph of a thin slice cut from a mammalian liver cell. The nucleus (N), a mitochondrion (M), and some endoplasmic reticulum (ER) are marked. At this level of resolution, one can even see detail inside a mitochondrion. (A, Alberts B, Johnson AD, Lewis J et al. [2008] Molecular Biology of the Cell, 5th ed. Garland Science. (B) Courtesy of Daniel S. Friend.)

molecular biology, has achieved many insights into how cells process and use the information stored in DNA to control cell growth and reproduction. Although molecular biology is a comparatively new science, its contributions to our understanding of cellular biology have been extremely important.

Through these and other scientific disciplines, scientists have learned how cells carry out the tremendous variety of functions they perform within the human body. Cells from different tissues accomplish a wide range of tasks. To do so, they display a variety of shapes, as seen in FIGURE 3.2. Some cells are roughly cubical, some are spherical, some are elongated, and some have long arms that can make contact with other cells many cell diameters away. These different shapes help each kind of cell to do its allotted jobs effectively; the function of a cell is dependent upon its structure, including the presence of specific proteins that regulate the chemical reactions the cell can perform. Many aspects of cell biology follow from one basic idea: the structure and function of a cell's parts contribute to the structure and function of the cell as a whole. This idea means that each cell type has distinctive features and roles, but there are some structures and functions that are common to essentially all cells. These widely used structures and processes are the focus of this chapter and the next.

Cells are built as a hierarchy of structural components.

The complexity of cell structure is easiest to think about as a hierarchy of substructures. The smallest parts of this hierarchy are atoms, which

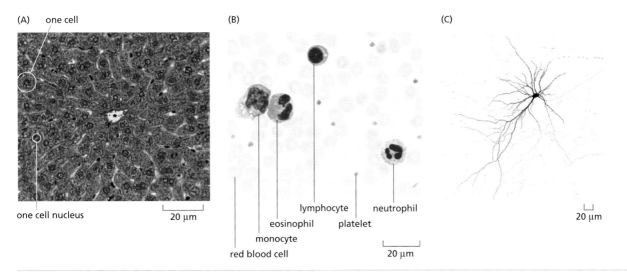

FIGURE 3.2 **Light microscopy of diverse cells.** (A) Light micrograph of a slice cut from a mammalian liver. Each cell is roughly a cube, and the blue circle within each cell is its nucleus. The red color comes from a stain called eosin, which binds to protein; the dark blue is hematoxylin, a dye that binds to DNA in the nucleus and RNA in both nucleus and cytoplasm, making the area outside the nucleus purple. (B) Light micrograph of a smear of blood. The pale, pinkish circles are red blood cells; the small blue circles are platelets, which help blood to clot; the bigger blue cells with internal structure (dark blue-stained nuclei) are different kinds of white blood cells. Each kind of white blood cell performs a different function in our body's defense against disease, as we will discuss in Chapter 9. (C) Part of a nerve cell grown in the laboratory. The great length of the thin arms extended by this kind of cell allows nerve cells to communicate over long distances. (A, © 2015 The Regents of the University of Michigan. B, from Alberts B, Johnson AD, Lewis J et al. [2002] Molecular Biology of the Cell, 4th ed. Garland Science. C, from Hiester BG, Galati DF, Salinas PC & Jones KR [2013] *Mol Cell Neurosci* 56:115–127. With permission from Elsevier.)

assemble into small biological molecules, like sugars. Small molecules, in turn, join together to form **macromolecules** (large molecules). All cellular macromolecules are formed as chains of small molecules, linked together end-to-end to form **polymers**. While some cellular macromolecules function on their own, many of them assemble with others to form macromolecular complexes that can carry out intricate cellular processes: for example, the replication of DNA and the synthesis of proteins. You can see this hierarchy of cellular structures in FIGURE 3.3, either by starting at the bottom and working your way up, or starting at the top and working down.

Many macromolecular complexes can combine to make an **organelle**. Cells contain numerous organelles, each of which includes hundreds to thousands of macromolecules organized to carry out multiple chemical reactions or other functions for the cell. Organelles are large enough to be visible in a light microscope. Commonly, organelles are positioned relative to one another in systematic ways. For example, the cell nucleus is commonly near the cell's center; the centrosome and the Golgi apparatus lie next to the nucleus (see Figure 3.1). The concept of structural hierarchy continues to even larger sizes, because cells themselves are commonly organized in groups to make a tissue. Several tissues often cooperate to make an organ, like a heart, liver, lung, or intestine; these are the structures that serve as components of your body, providing the functions necessary for you to live.

To reveal the cellular parts of this hierarchy with more clarity, we will now describe several biological structures in more detail, starting from the smallest category that is unique to cells: the macromolecules. Then we'll look at the functional relationships among these structures and initiate a description of how these relationships go wrong when cancer develops.

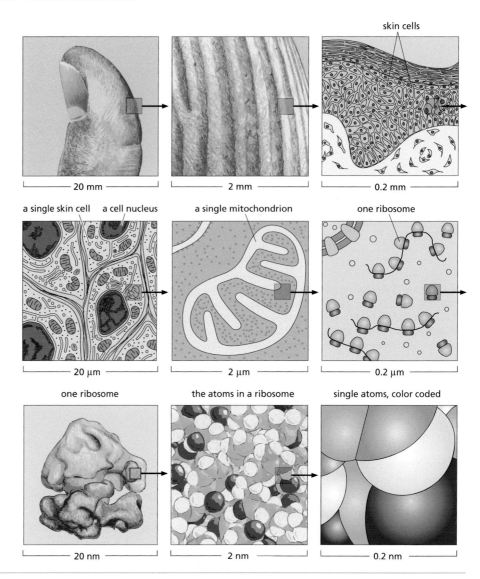

FIGURE 3.3 **The relative sizes of things.** Starting at the upper left, with a thumb and a scale in millimeters, these drawings show steps in size down to the level of the atoms from which all molecules are made. Each drawing is 10× higher in magnification than the previous one. These images give a feeling for the relative sizes of a cell's components, which is important for understanding the hierarchical character of biological structure. (Adapted from Alberts B, Johnson AD, Lewis J et al. [2015] Molecular Biology of the Cell, 6th ed. Garland Science.)

THE STRUCTURES OF CELLULAR MACROMOLECULES

Many macromolecules contain tens of thousands of atoms, so in comparison with other molecules, they are very large. Cellular macromolecules are of three kinds: nucleic acids (a general word that includes both DNA and RNA), proteins, and polysaccharides. Together with molecules of fat, which biologists call lipids, these molecules are the building blocks for all cellular structures. Each kind of macromolecule is made from a different kind of small molecule; as a result, each macromolecule has unique characteristics.

One important aspect of the relationships among different kinds of macromolecules is a result of the ways cells store and use information. The structure of DNA contains information that is used to make the macromolecules called RNAs, which in turn contain information on how to make proteins. The proteins are a very diverse group of macromolecules.

Some are used to build cellular structures, like membranes and fibers; others are important in controlling the cell's chemical reactions. The latter proteins are **catalysts**, which means they speed up a particular chemical reaction. The rates of chemical reactions are important to cells because they define many aspects of cell structure and function, including growth and differentiation. Proteins that are catalysts are called **enzymes**. Enzymes accomplish a large range of jobs for every cell. They control the ways in which cells use the materials and energy in food, and they are the cell's tools for making DNA, RNA, polysaccharides, and lipids.

DNA is the macromolecule that stores most of a cell's information.

The idea that a molecule can store information seems strange. The way DNA does this job was obscure until scientists learned about its structure. The structure of DNA allows it to serve as the books of information mentioned in Chapter 1. One aspect of DNA's structure important for information storage is that each molecule of DNA is very long. The DNA in a human cell is arranged in 46 pieces, each about 100 million times longer than it is wide. If you made a model of DNA in which each piece was a cylinder 1 inch in diameter, each cylinder would be about 1600 miles long!

DNA is built from two strands, so it is often referred to as double-stranded DNA or a DNA duplex. These strands twist around one another to make a double helix (FIGURE 3.4). Each strand of the double helix is a polymer

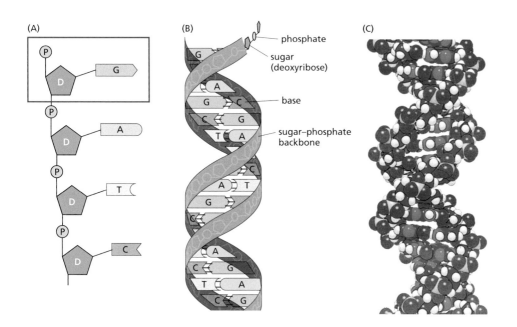

FIGURE 3.4 **The structure of DNA.** (A) A diagram of nucleotides and the way they join to form a chain (polymer) that can be part of DNA. The red box outlines one nucleotide, which is composed of a phosphate group (P), a sugar (deoxyribose, D), and a base (either G, A, T, or C). The phosphate of each nucleotide links to the sugar of the next nucleotide to form a chain called the sugar–phosphate backbone. (B) A diagrammatic view of a DNA double helix. Two strands of nucleotides, like the one shown in part A, twist around one another to form a double helix in which the backbones (the ribbons containing sugars and phosphates) are on the outside and the bases face in. Because of this arrangement, the bases on one strand can interact with bases on the other strand. Given the shape of each base, C fits well with G and A fits well with T. The bonds in the sugar–phosphate backbone are very strong. The bonds between complementary bases (red lines that connect C with G and A with T) are much weaker, but taken all together, they are strong enough to hold the two backbones together, explaining the stability of double-stranded DNA, which is also called a DNA duplex. (C) A realistic image of DNA in which each atom is represented by a colored sphere: carbon is blue, hydrogen is white, oxygen is red, and phosphorus is orange. This representation shows the dense packing of atoms in a DNA double helix. Tight packing is part of the reason that DNA is so stable. (Adapted from Alberts B, Johnson AD, Lewis J et al. [2015] Molecular Biology of the Cell, 6th ed. Garland Science.)

made from many copies of only four small molecules called **nucleotides**. Each nucleotide contains one unique part that chemists call a base (Figure 3.4A). The names of these bases are adenine, cytosine, guanine, and thymine, but we will represent them simply by the first letters of their names: A, C, G, and T. In addition to a base, each nucleotide contains a sugar molecule, called deoxyribose (labeled D in Figure 3.4A), and an arrangement of phosphorus and oxygen atoms called a phosphate group (labeled P in Figure 3.4A). The sugar and phosphate parts of all these nucleotides are identical, so only the bases distinguish one nucleotide from another. Thus, the nucleotides too are referred to as simply A, C, G, and T. Each strand of DNA in a human chromosome is many millions of nucleotides long, long enough to contain a very large number of bases and therefore a large amount of information. The key question then is, How is that information stored?

Each set of three bases in a sequence along one strand of DNA spells a specific genetic word, called a **codon.** The codon AAA means one thing, AAC means something else, and so forth. You can think of a strand of nucleotides in DNA as a string of letters that spells many words. For example, AAA AAC CCC CCG GGG GGA is a set of six genetic words that one might find along one strand in a molecule of DNA. When there are enough words in a row to have biological meaning, they make a gene. Each piece of human DNA is long enough to spell thousands of genes.

The fact that DNA is a made from two strands of nucleotides twisted around one another to form a double helix is of critical importance to the way this macromolecule works in a cell. One aspect of the relationship between DNA's structure and its function is that the shape of the double helix is independent of its nucleotide sequence; any sequence of As, Cs, Gs, and Ts will produce the same double-helical polymer. Thus, DNA sequences can include many different genetic instructions, but they will always form a double helix. A second important aspect of DNA's structure is that the sequences of nucleotides on its two strands have a fixed relationship. For the strands to fit together and form a double helix, the base A on one strand is always opposite the base T on the other strand, and the base C on one strand is always opposite the base G on the other strand (see Figure 3.4B). Thus, if you know the sequence of bases on one strand of DNA, you know the sequence on the other strand. These sequences are said to be complementary. This structural property means that if a gene is written in the sequence of bases on one strand of DNA, it is not written on the other strand; that sequence is complementary, not identical.

The double-helical structure of DNA is very stable for two reasons. (1) Its bases fit snugly together across the double helix, making many weak chemical bonds. These bonds, in sum, are strong enough to hold the two strands together firmly. (2) The sugar and phosphate parts, which are identical in all nucleotides, are linked by strong chemical bonds. These bonds form strands called the sugar–phosphate backbone of DNA. The sugar–phosphate bonds are so strong that they don't break, even when a solution containing DNA is heated in water to near boiling. Given the strong bonds along the sugar–phosphate backbone and the weaker but numerous bonds between complementary bases on the two DNA strands, a DNA double helix is sufficiently stable to last for many years. Studies on DNA taken from ancient Egyptian tombs or from caves occupied by early humans show that these molecules can last for thousands of years.

The structure of DNA was discovered in the 1950s by a young American biologist, James Watson, and a British physicist, Francis Crick. For a brief description of how they made their important discovery, see **SIDEBAR 3.2.**

A CLOSER LOOK: Discovering the structure of DNA.

SIDEBAR 3.2

James Watson and Francis Crick were an unusual team because they were personally and intellectually so different. However, each brought to the problem they were studying—the structure of DNA—a keen scientific insight and a passion for understanding the relationship between molecular structure and function. Crick's understanding of mathematics allowed him to recognize, from information collected by others, that DNA was built as a helix with a certain kind of symmetry. He was also a strong believer in the idea that nature's solutions to important problems, like the structure of genetic information, would be simple. Watson came with a deep understanding of genetics and a brash confidence in the then-unproven idea that DNA was the place where genes were stored. These points of view allowed the two of them to think about how information might be written into the structure of a macromolecule.

Watson and Crick knew from chemical studies of DNA, carried out elsewhere, that DNA always contained equal numbers of Gs and Cs, as well as equal numbers

of As and Ts; however, the number of As was not equal to the number of Gs, nor Cs to Ts. This led them to the concept of pairing G with C and A with T to build the final structure. They played with atomic models of the bases and found ways in which they could fit snugly together (**Figure 1**). They also realized that the linkages between sugars and phosphates would build sugar—phosphate strands that could form a stable helix. Most important, they realized that if the two strands of nucleotides were built with complementary bases, the sequences of bases on the two strands would offer a simple way to solve the problem of DNA replication: uncoil the double helix and let each half serve as a template for the assembly of complementary nucleotides, building a new strand that was exactly like the old one from which it had just separated. This insight meant that the molecular structure they proposed for DNA solved a fundamental problem in biology: how information could be passed from one generation to the next. Their paper describing this model is only one page long, but it is one of the most

famous and important papers in all of biological research. Watson and Crick were awarded the 1962 Nobel Prize in Physiology or Medicine for their work.

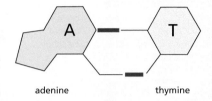

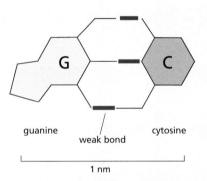

Figure 1 Watson found a way in which some of the bases fit nicely together. Using simple molecular models for the bases, Watson played with their relative positions and found orientations in which the bases from DNA, known to be present in equal amounts, fit together quite nicely.

The structures of DNA and RNA are similar but distinct.

Both DNA and RNA are found in the cell's nucleus, and both molecules are chemically acidic, hence their collective name, nucleic acids. Like DNA, RNA is built from nucleotides, linked together by sugar–phosphate bonds to form a polymer (FIGURE 3.5). Like DNA, the linear sequence of nucleotides along the sugar–phosphate backbone is an important feature of RNA's structure. However, the four bases used to make RNA are not quite the same as those in DNA. They include adenine, cytosine, and guanine (A, C, and G), but RNA does not contain thymine (T). Instead, it uses the base uracil (U), which is structurally similar to T. RNA also uses a slightly different sugar in its sugar–phosphate backbone (ribose rather than deoxyribose).

Unlike DNA, RNA is usually single-stranded. It can, however, fold to form loops that twist into short stretches of helix when two sequences of bases along the RNA polymer happen to be complementary (Figure 3.5B). The structures that RNA molecules will take on are therefore quite variable; this variability shows up in the functional variability of different kinds of

FIGURE 3.5 **RNA is formed from a chain of nucleotides, analogous to the chains that form DNA.** (A) As in DNA, nucleotides (boxed in blue) are connected by a backbone of sugars (ribose, R) linked together by phosphates (P). The sugar in RNA is slightly different from the one in DNA, which accounts, in part, for why RNA rarely forms long double helices. As in DNA, four different bases are attached to successive sugars, but in RNA the bases are A, C, G, and U (uracil) in place of T. (B) Although RNA doesn't form long double helices, it can fold in such a way that complementary bases form bonds analogous to the ones that hold double-stranded DNA together: G pairs with C, and A pairs with U. The curved blue line in this figure represents a sugar–phosphate backbone. Along this chain, some complementary bases form bonds that help to hold this RNA in a specific shape. Such shapes are important for the function of some RNAs. (Adapted from Alberts B, Johnson AD, Lewis J et al. [2015] Molecular Biology of the Cell, 6th ed. Garland Science.)

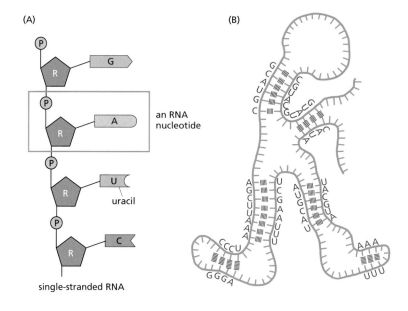

RNA. Moreover, most RNA is significantly less stable than DNA. Below you will see how this is actually an advantage for the cell.

A key feature of RNA structure is that when cells make it, they do so by assembling nucleotides into a polymer whose base sequence is exactly complementary to the base sequence along one strand of a DNA duplex (FIGURE 3.6). The process of RNA synthesis is called **transcription**, because the information along a DNA strand is simply written into a slightly different form. The sequence of the RNA made in this way is always the same as one strand of DNA in the region transcribed (except for the substitution of U for T), and it is complementary to the other strand. The DNA strand whose sequence is like the RNA transcript is called the sense strand, but it is the complementary DNA strand that is used as a template to define the RNA sequence by the pairing of complementary nucleotides.

Cells transcribe a given stretch of DNA into a piece of RNA through the action of an enzyme called **RNA polymerase** (FIGURE 3.7A). (Note that enzymes are commonly named by the reaction they catalyze with the suffix -ase added to the end of the name.) RNA polymerase speeds up the reactions required to make a chain of RNA that is complementary to the template strand of DNA. First it matches a nucleotide on the template DNA strand with a complementary nucleotide selected from the many nucleotides floating around in solution. Then it links that nucleotide with other nucleotides already in the RNA polymer being made, forming a sugar–phosphate bond. RNA polymerase is like a machine that puts the nucleotides together to make the RNA polymer. However, the sequence of bases in any given piece of RNA is defined by the sequence already present along one strand of DNA. In this way an RNA transcript carries information (nucleotide sequence) from a particular stretch of DNA and can take it to other parts of the cell. For this reason, many of the transcripts made from DNA are called **messenger RNAs (mRNAs)**.

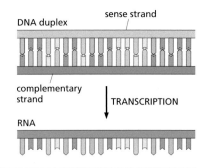

FIGURE 3.6 **RNA is made from DNA by transcription.** During transcription, one strand of the DNA duplex serves as a template. It defines which nucleotide will be added as the RNA grows in ength, using the same complementarity of bases that is seen between the two strands of a DNA duplex. Note that the nucleotides on the RNA molecule are similar to the nucleotides on the upper strand of the DNA duplex, which is called the sense strand; the complementary strand of DNA is called the template. (Adapted from Alberts B, Johnson AD, Lewis J et al. [2015] Molecular Biology of the Cell, 6th ed. Garland Science.)

For RNA polymerase to read a meaningful sequence of nucleotides from DNA, it must know where to start the transcription processes. A region of DNA that can bind RNA polymerase and initiate transcription is called a **promoter**. Transcription continues until RNA polymerase reaches a series of DNA nucleotides called a **transcription terminator**. At this point, the RNA is released and RNA polymerase detaches from the DNA. A single promoter can bind RNA polymerase over and over, meaning that

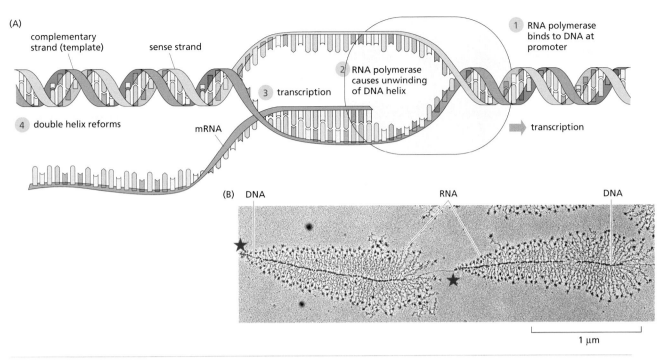

FIGURE 3.7 How a cell reads information from a DNA duplex. (A) A particular sequence of nucleotides in DNA, such as TATA, will bind proteins that initiate the transcription of RNA from one of the two strands in a DNA duplex. Transcription is accomplished by the protein RNA polymerase, which uses the information on one of the DNA strands to define exactly which nucleotide to add at each place as the RNA is made. (B) An electron micrograph of a single, long DNA duplex that has bound many copies of RNA polymerase. Each polymerase bound initially at a promoter (one of the two regions marked by red stars) and then moved to the right along one of the two DNA strands, making RNA as it went. The farther it has moved, the longer the RNA transcript it has made. When RNA polymerase reaches a transcription terminator, it falls off, as does the now-complete RNA transcript. (B, courtesy of Ulrich Scheer.)

RNA can be transcribed many times from a given segment of DNA (FIGURE 3.7B). This process is at the heart of how different genes can be expressed at different levels in any given cell. You can think of DNA as a technology for information storage and RNA as a set of devices for information retrieval, as discussed in SIDEBAR 3.3.

The many copies of mRNA that result from multiple transcriptions of a given gene carry the information that was stored in that particular gene (its sequence). These mRNAs can travel to any part of the cell where they can be of use. The cell keeps its DNA safe in the nucleus, while it makes copies via RNA polymerase for use elsewhere in the cell. Now, the instability of RNA becomes important. Cells degrade the RNAs they make, so a given mRNA will convey information only for a limited period. For DNA to keep expressing a given gene, it must either stabilize that RNA transcript or continue to transcribe that region of DNA.

Many genes act by directing the synthesis of specific proteins.

mRNA carries information from the nucleotide sequence of a particular stretch of DNA to other places in the cell, where it can be used to direct the construction of a specific protein. The cellular sites for protein synthesis are called **ribosomes**. You saw ribosomes in Figure 3.3 as structures that look in a microscope like small particles scattered around in both the cytoplasm and the interior of mitochondria. Each ribosome is a complex of proteins and pieces of special RNA that are important for ribosomal structure, not for carrying information. Ribosomes are made in a specialized region of the nucleus called the nucleolus (see Figure 3.1) and then

Compare the process of getting information from books with the process of transcribing DNA into RNA.

SIDE QUESTION 3.1

Why is the stability of a messenger RNA an important issue for the cell?

SIDE QUESTION 3.2

A CLOSER LOOK: The properties of nucleic acids.

DNA is unusually stable for such a large molecule. Many criminals are now in jail as a result of the stability of DNA they left behind while committing a crime. Indeed, pieces of functional DNA have been recovered from ancient Egyptian mummies and even much older animals that were frozen during an ice age. The sequences of these DNA molecules appear to have remained unchanged over thousands of years. Stability like this is highly desirable for a material used to store information. Indeed, it would be great if videotapes, DVDs, and even books were that stable! The closest that mankind has come to this kind of stability is stone tablets.

Cells keep their genetic information in this highly secure material, but they also have to get at the information it contains. The strategy of reading the information in DNA by transcribing it into RNA gives the cell a working copy of the information, but it leaves the master copy safe in the nucleus. RNA is not as stable as DNA, but it is stable enough to get the word out to the cytoplasm, where the message can be interpreted by translation on a ribosome. In fact, the lesser stability of RNA is an advantage. Once the message has been used, it can be degraded, so the information coming out of the nucleus doesn't always say the same thing; old news is replaced with new.

Another remarkable thing about a cell's DNA is the large amount of information it can carry. The best current estimates tell us that the nucleus of each cell in the body contains about 25,000 genes, each with a part that encodes RNA and parts that regulate RNA transcription. Many of these genes can be expressed in several ways by altering the processes that modify the RNA transcript as it becomes a message, so in fact, our DNA hosts the information to make more than 100,000 different proteins. In addition, biologists have recently discovered a wealth of small RNA molecules that are transcribed from short stretches of DNA but not translated into protein. These molecules help to regulate the expression of other genes. Taken all in all, there are probably more than 200,000 useful molecules encoded in our DNA, ready to be expressed when the cell needs them. Each of these molecules carries out some important function for the cell, as if it were a tool or even a machine.

Imagine being able to carry all this information in an object that is only a fraction the size of a cell, which we have already described as small. This masterpiece of microminiaturization is at the heart of why cells can be both small and complex. Although this is hard to relate to computer memory with any precision, the number of bytes in a cell's useful DNA corresponds to about 1 gigabyte of computer data (10^9 bytes in scientific notation). If you scaled the volume of a cell's nucleus up to an object the size of a USB flash drive, it would give you a device that stored about 10 billion gigabytes. Not bad!

transported into the cytoplasm. Most cells contain thousands of ribosomes, so they can make proteins quickly when needed.

Proteins are polymers of amino acids, not nucleotides. The process of protein synthesis **translates** the sequence of codons in mRNA into a sequence of amino acids that builds a specific protein. Stated another way, the process of protein synthesis turns a gene sequence into an amino acid sequence (FIGURE 3.8). The three-letter code mentioned previously for DNA now becomes critically important. Many of the codons specify a particular amino acid in a language called the genetic code. For example, the sequence of nucleotides cited earlier, AAA AAC CCC CCG GGG GGA, codes for six specific amino acids, called lysine, asparagine, proline, proline, glycine, and glycine. (Note that a given amino acid can be specified by more than one codon.) The key idea here is that the sequence of nucleotides in the strand of DNA defines the sequence of nucleotides in mRNA, and the mRNA sequence defines the sequence of amino acids for a particular protein. Francis Crick, who was so important in developing an understanding of DNA structure, was equally important in proposing a way by which cells might interpret the genetic code, as described in SIDEBAR 3.4. He and others made huge contributions to what has been called the central dogma of biology: DNA controls the synthesis of RNA, which controls the synthesis of protein.

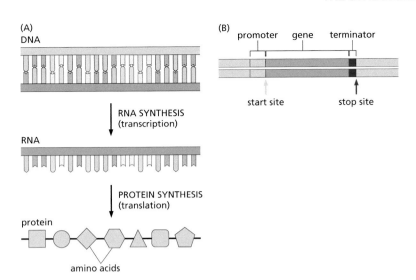

FIGURE 3.8 The flow of information from DNA to RNA to protein and its regulation by a promoter. (A) One strand of DNA serves as a template for the synthesis of a messenger RNA. This molecule of mRNA can then pass out of the nucleus, moving into the cytoplasm where it will encounter ribosomes. Ribosomes translate the codons in an mRNA into a sequence of amino acids that will make a particular protein. (B) Promoters are sites where RNA polymerase can bind to DNA and begin transcription; terminators are places where RNA polymerase falls off the DNA and stops transcription. Between these two sites lies the structural gene, the DNA that encodes the gene product. The amount of expression of a particular gene is defined in part by the activity of its promoter. (Adapted from Alberts B, Johnson AD, Lewis J et al. [2015] Molecular Biology of the Cell, 6th ed. Garland Science.)

A CLOSER LOOK: Working out how a nucleotide sequence can define a sequence of amino acids.

SIDEBAR 3.4

Francis Crick made many significant contributions to our understanding of molecular biology, including an idea called the adaptor hypothesis. Not long after his work with Watson on the structure of DNA, Crick realized that cells had to have a way to convert the genetic information written in the sequence of DNA bases into the sequence of amino acids in the proteins that would be the gene products. There had to be some sort of adaptor that would have nucleotides at one end, able to read the sequence of bases in a codon, and an amino acid at the other end, so the information in the codon could be translated into the process of adding a specific amino acid to a growing amino acid polymer. This idea was first published in 1955, but it took quite a few years to collect the data to test it. Once that information was in hand, the idea turned out to be strikingly valid and to lie at the heart of how translation works.

The adaptors turned out to be a set of special RNA molecules, now called transfer RNAs or simply tRNAs. Each tRNA has a loop at one end, called the anticodon loop, which has a sequence that is complementary to one specific codon. At the other end of the tRNA is an amino acid that is the meaning of that codon (Figure 1).

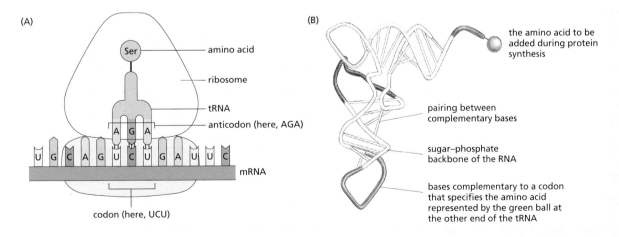

Figure 1 The structure of Crick's "adaptor". (A) Cartoon representation of mRNA associated with a ribosome (large green structure) and with a kind of RNA molecule now known as transfer RNA or simply tRNA. At one end of the tRNA is an amino acid. At the other end is a sequence of bases that is complementary to one of the codons that dictates the addition of this amino acid in the language of the genetic code. (B) Diagram of the structure of a tRNA molecule, showing the anticodon loop in blue and the amino acid attached to this tRNA as a green sphere. There is at least one tRNA for every amino acid and one for almost every codon. Before protein synthesis, each tRNA becomes attached to the right amino acid, thanks to a specific set of enzymes that charges the tRNA, making it ready to help with protein synthesis. During protein synthesis, the tRNA adapts the language of the genetic code (a specific codon) to a language appropriate for protein synthesis (a specific amino acid). (Adapted from Alberts B et al. [2015] Molecular Biology of the Cell, 6th ed. Garland Science.)

Ribosomes in the process of translating a particular mRNA into protein will pause briefly at each codon. This allows time for the right tRNA to bind that codon through base-pairing at its anticodon loop (**Figure 2**). Now the right amino acid is in place to be added to a growing amino acid chain. Once this amino acid has been transferred from the tRNA to the elongating protein, the tRNA is released, and the ribosome uses energy to move the mRNA along by one codon. Now the process is repeated, bringing in whichever tRNA is appropriate for the next codon on the message. The ribosome, working with one mRNA and the whole set of tRNAs, is a little like an assembly line. Functioning as a unit, it churns out protein at an impressive rate. The process continues until it reaches a codon like UAG, which is a termination codon. There is no tRNA for the termination codons, so the process of amino acid addition stops, and the completed amino acid polymer falls off the ribosome.

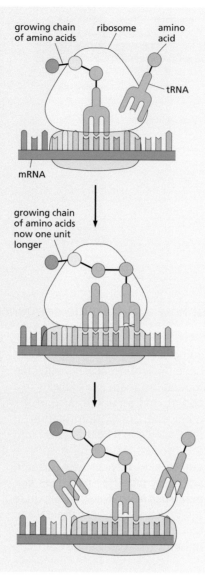

Figure 2 Transfer RNAs at work in protein synthesis. A ribosome is bound to an mRNA and is in the process of translating its nucleotide sequence into a sequence of amino acids. Two tRNAs are in action. Both come to sit on the mRNA, base-pairing with complementary bases of a codon on the message (that's how this particular tRNA with its attached amino acid was selected over other possible tRNAs that might have bound to the ribosome). The transitions from the top figure to the lower ones show the stages in shifting the growing amino acid chain over, so the new amino acid can be added, and the already used tRNA can be released. This represents one step in the synthesis of a protein.

Some nucleotide sequences act like punctuation in a sentence. During transcription, a gene is read along one strand of the DNA double helix, beginning at the promoter and ending with a transcription terminator. Likewise, the nucleotide sequence in an mRNA includes punctuation that tells the ribosome where to begin translation and where to stop. AUG is the sequence that initiates protein synthesis. Translation of the nucleotide sequence into an amino acid sequence then continues, three bases at a time, until the ribosome encounters any one of the sequences UAA, UAG, or UGA; these codons stop translation and induce the ribosome to fall off the mRNA (**FIGURE 3.9**). Working out the meaning of the codons was a major breakthrough in our understanding of how cells work. The necessary research was carried out in several labs, located all over the world. For a brief description of this research, see **SIDEBAR 3.5**.

There are 20 different amino acids commonly used to make proteins, as opposed to the four nucleotides used to make DNA or RNA. With this variety of building blocks, the range of protein structures that could be made is truly immense, given that any of the 20 different amino acids could in principle be inserted at each place along a given amino acid chain. Most proteins are between 100 and 1000 amino acids long. Taking 500 as a representative number, 20 possibilities for each of the 500 positions means

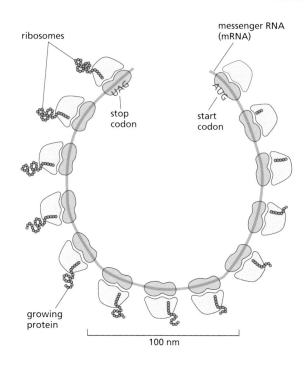

ribosomes

messenger RNA (mRNA)

stop codon

start codon

growing protein

100 nm

FIGURE 3.9 The information carried from DNA by messenger RNA is translated into a specific amino acid sequence by ribosomes. At a place where the message contains the sequence AUG, known as a start codon, the ribosome begins to translate the base sequence along the mRNA into an amino acid sequence. Near the end of the message there is a stop sequence that tells the ribosome to stop translating and fall off. Thus, the sequence of DNA that was transcribed into mRNA defines the sequence of amino acids in the resulting protein. (Adapted from Alberts B, Johnson AD, Lewis J et al. [2015] Molecular Biology of the Cell, 6th ed. Garland Science.)

that the number of possible proteins is represented by the number 1, followed by 650 zeros. This number is so big it doesn't even have a name! Out of all these possibilities, we make only a few thousand proteins in any one cell at any given time. During our entire lifetime, our entire body makes no more than about 100,000 different proteins. The number of

A CLOSER LOOK: Working out the genetic code

Even after the structure of DNA was known and the idea of an adaptor was in the air, it was still unclear exactly how a sequence of nucleotides could specify a sequence of amino acids in a protein. A number of physicists and chemists were interested in this problem, and their thinking helped to define its solution. If each nucleotide specified one amino acid, then the four nucleotides could specify only four amino acids. It was already known that proteins were made from 20 amino acids, so a one-letter code was clearly insufficient. If two nucleotides defined an amino acid, then the code could specify 16 amino acids; this is better but not good enough. (With two nucleotides serving as a genetic word, the possible words would be AA, AC, AG, AT, CA, CC, CG, CT, GA, GC, GG, GT, TA, TC, TG, and TT.) With a three-letter

code, the number of possible genetic words goes up to 64 (you can write out all those words for yourself), which is plenty, and it implied that there would be several genetic words, now called codons, that corresponded to the same amino acid.

Several scientists tried to crack this code in the sense of discovering which codon(s) corresponded to which amino acid. A few codons were easy, because it was straightforward at the time to make a polymer of only one kind of nucleotide. Marshall Nirenberg was the first to learn that poly(U) was a genetic messenger. (Remember, it is the nucleotides in RNA that actually get translated into amino acids to make a protein, so the relevant nucleotides include U, not T.) This synthetic RNA could be added to a complex cell extract, which con-

tained ribosomes together with virtually all the small molecules from a cell, including all the amino acids, plus ATP added as a source of energy. This mixture produced a polypeptide constructed from multiple copies of the amino acid called phenylalanine (abbreviated Phe or F). His lab used the same trick to learn the meanings of the other simple codons and different tricks to get most of the other possible three-letter codons. He and a few other scientists are credited with having cracked the genetic code, for which three of them were awarded a Nobel Prize.

A remarkable feature of the genetic code is that it is almost universal. Bacteria and humans use the same set of codons to spell each amino acid. This is a powerful statement about the fundamental similarity of all life.

SIDEBAR 3.5

(A)

side chain

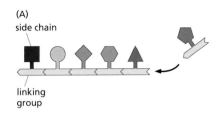

linking
group

(B)

amino acid abbreviated name	character of the amino acid side chain
D, E	negative charge, water-soluble
R, K, H	positive charge, water-soluble
N, Q, S, T, Y	water-soluble, but no charge
A, G, V, L, I, P, F, M, W, C	water-insoluble, like a fat

FIGURE 3.10 Diagram of a short amino acid polymer and a table of amino acid properties.
(A) Symbolic representation of six amino acids, five of which are already polymerized while one is still free in solution. All amino acids share an identical region referred to as the linking group. The linking groups are used to assemble amino acids into a polymer. Each amino acid also has a specialized region called its side chain, represented by the colored shapes. There are 20 different side chains, which define the 20 amino acids used to make proteins. (B) A table of those 20 amino acids, each represented by its one-letter name, analogous to the one-letter names for the nucleotides and bases. To the right are some chemical properties of each amino acid, indicating its charge and water solubility or lack thereof.

What are some of the possible consequences for a protein of a mutation that changes the sequence of the DNA that encodes it? What might happen if the gene's promoter were mutated?

proteins we make is constrained by the information in our genes. Our DNA includes the information to make sequences of amino acids that will fold up to make proteins that can accomplish useful tasks for the cell.

How do the different amino acids differ, and what is the importance of these differences? One part of each amino acid is identical to the comparable part in all other amino acids; this part, which can be called a linking group, defines the way these small molecules polymerize to form a chain during protein synthesis (FIGURE 3.10A). This chain forms a backbone for the protein, analogous to the sugar–phosphate backbone in nucleic acids. Polymers of amino acids are often called **polypeptides**, because the chemical name for the bond between linking groups is a peptide bond. In addition to the linking group, each amino acid contains a second part called a side chain. The side chains give each amino acid specific chemical properties (FIGURE 3.10B). Some side chains contain positive or negative charges; these amino acids are said to be polar, and they are extremely soluble in water, just like the sodium and chloride ions that form when table salt dissolves in water. Other side chains are not electrically charged, but they are chemically similar to a sugar. Like table sugar, they dissolve readily in water. Still others, called nonpolar side chains, are fatty and consequently insoluble in water, reacting to an aqueous solution much as oil separates from water.

When side chains with these different chemical properties are linked together in the same polymer, they define the way the polymer will fold up when placed in an aqueous environment, like cytoplasm. The shape of that fold is defined by the chemistry of all the amino acid side chains in the polymer and their positions along the polypeptide chain (FIGURE 3.11A). The fold will be most stable when the water-soluble side chains are placed on the outside and the fatty side chains are on the inside. The protein's shape will also depend on the facts that positive and negative charges attract one another, but charges of the same kind repel. When the fold of a polypeptide places water-soluble side chains on the protein's outside, the whole polypeptide becomes water-soluble. The relative positions of amino acid side chains along a polypeptide, that is, the sequence of its amino acids, has a big impact on the way a given chain of amino acids will fold and thus on the structure of the protein.

FIGURE 3.11B illustrates four different representations of a single polypeptide that has been folded in one way; the pictures provide four different views of the same structure. They display different amounts of detail about the positions of all the atoms in the multiple amino acids that have been linked together in this polypeptide. The upper left diagram shows only the backbone of the polypeptide, that is, the identical linking groups of each amino acid. The ribbon diagram in the upper right also shows only the backbone, but it is a particularly useful way of displaying a protein's fold. It shows the path taken by the polypeptide in space, with graphic features that emphasize different aspects of local structure, such as a helix or a sheet formed at different places along the backbone. Both of these displays leave out enough atoms (the side chains are not shown) that you can see the polymer's fold very clearly. Proteins are often represented by ribbon diagrams. The two lower models show all the atoms in all the amino acids. On the left, sticks are used to show the bonds between atoms; on the right, each atom is represented with a sphere that approximates the atom's size. Now you can see how compact a protein really is, but the fold of the polypeptide chain is obscured by all the atomic detail.

A protein's structure defines its function.

The structure of each biological component defines the functions it can accomplish, and proteins are no exception. The amino acid sequence dic-

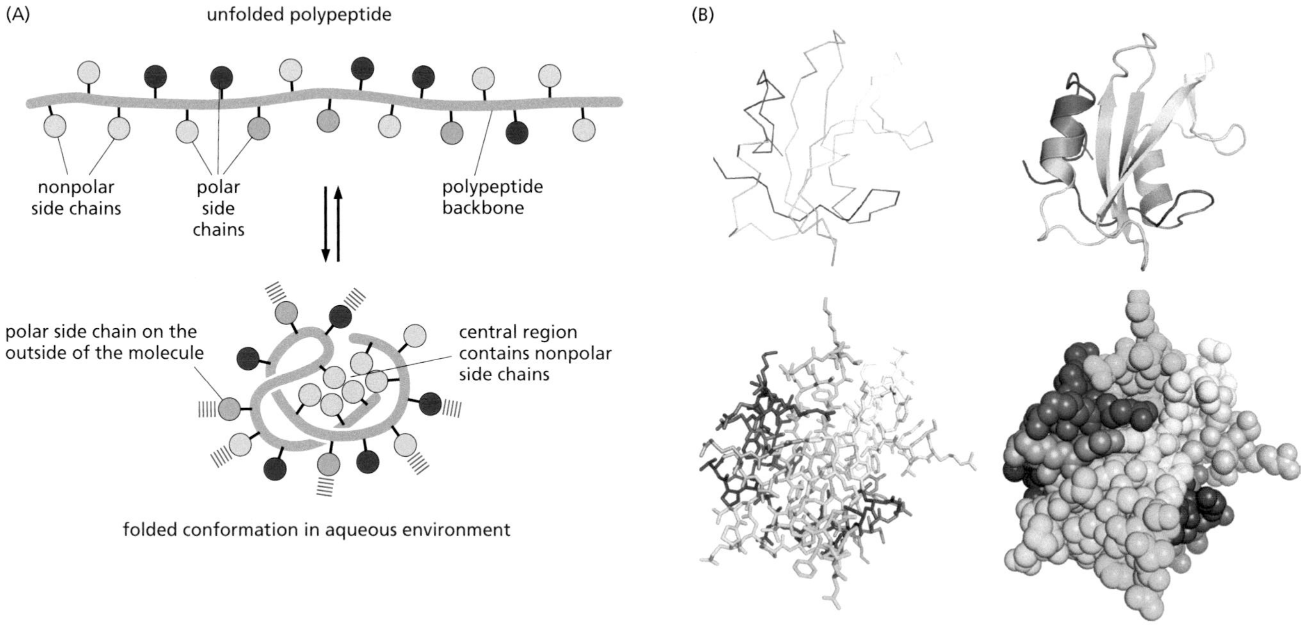

FIGURE 3.11 **The folding of a polypeptide to make a protein.** (A) The backbone of a polypeptide (gray-green) is formed by a chain of linking groups. This chain can fold to place the nonpolar amino acid side chains in the middle, so they are surrounded by amino acids whose polar side chains are soluble in water. As a polypeptide begins to emerge from a ribosome, it is initially unfolded, but as more of the mRNA is translated and the polypeptide gets longer, folding begins. Once folding is complete, the polypeptide has adopted a structure that is stable in an aqueous solution, for example, the cytosol. After folding, the surface of the protein consists almost entirely of amino acids with water-soluble side chains; most of the water-insoluble, nonpolar side chains will have been buried. (B) Four ways to display the structure of a protein. The upper left diagram shows only the polypeptide backbone. This display reveals the way the amino acid chain has folded. The upper right image again displays the protein backbone, but now ribbons are used to make the fold in each subregion of the protein more obvious. Dark blue and orange show places where the backbone has formed a helix; green fading into yellow shows regions where the chain has doubled back on itself to make a sheet, albeit one that is not flat. The other regions of the diagram indicate loops with no regularity of structure. On the lower left, the same protein is shown with stick models. All the atoms are represented by short colored bars that display the bonds between adjacent atoms. On the lower right, all of the protein's atoms are shown as spheres or parts of spheres. This model displays a protein's structure in a way that is analogous to the display of DNA in Figure 3.4C. The colors are the same in all four representations. (A, adapted from Alberts B, Johnson AD, Lewis J et al. [2015] Molecular Biology of the Cell, 6th ed. Garland Science. B, PDB code 1SHA.)

tates the fold that a protein will adopt, leading to the formation of a structure that can accomplish a given task for the cell. For example, many proteins fold in a way that leaves an indentation on their surface, forming a pocket. Such pockets serve as binding sites by which the protein can interact with other molecules (FIGURE 3.12A). One important example of proteins interacting with other molecules is the way cells receive information from signals sent by other cells. Cells communicate by way of signaling molecules; how one cell behaves is often directed by information that comes from another cell by way of a molecular signal. Such signals have specific meanings, and they get this specificity by binding strongly to only the one particular protein that serves as a signal receptor. The receptor, in complex with the signaling molecule, will then initiate a particular process or job in the receiving cell. Specificity is achieved by the strength of the binding between the signal and its receptor. This strength depends on the way the signal and its receptor fit together (Figure 3.12A). Protein shape and the bonds that a protein's surface can form with other molecules specify the binding interactions that allow proteins to be very choosy in the molecules with which they interact (FIGURE 3.12B).

As mentioned previously, certain proteins, called enzymes, are catalysts that control the rates of specific chemical reactions that occur in a cell. For example, when sugar molecules enter a cell, they are immediately altered by an enzyme that connects the sugar with a phosphate group, like the phosphates in the sugar–phosphate backbone of nucleic acids.

Why is tight binding between a signaling molecule and its receptor important for signal specificity?

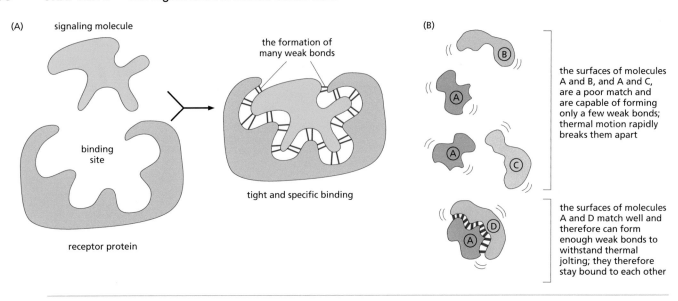

(A) signaling molecule

binding site

receptor protein

the formation of many weak bonds

tight and specific binding

(B)

the surfaces of molecules A and B, and A and C, are a poor match and are capable of forming only a few weak bonds; thermal motion rapidly breaks them apart

the surfaces of molecules A and D match well and therefore can form enough weak bonds to withstand thermal jolting; they therefore stay bound to each other

FIGURE 3.12 Schematic representation of why the shape of a protein's surface is important for its tight binding to other molecules, a property that is often important for protein function. (A) Part of a receptor protein (green) includes a pocket with a complex shape. If a signaling molecule (yellow) with the right shape comes along (through random thermal motions) and bumps into this protein, it will stick tightly in the pocket, thanks to all the bonds it can form there. This kind of tight binding is what allows protein receptors to be specific for particular cellular signals. Signaling molecules that have the wrong shape won't fit in the receptor's pocket, and therefore they do not bind and activate the receptor. (B) Molecules A, B, and C have no part of their surfaces that will fit with one another. D and A, on the other hand, have complementary shapes, so they fit well and can bind each other tightly. This is the idea underlying all cases of proteins binding other molecules, such as small molecules or other proteins. These binding reactions build the structures found in cells. (Adapted from Alberts B, Johnson AD, Lewis J et al. [2015] Molecular Biology of the Cell, 6th ed. Garland Science.)

This enzymatic reaction changes the shape and electric charge of the sugar so it cannot leak back out of the cell. The enzyme that adds the phosphate to the sugar controls the amount of sugar that will stay in the cell for nourishment. Another example of an enzyme is lysozyme, which serves as molecular scissors to cut polymers in the wall of a bacterium. The chain of amino acids that makes lysozyme folds to form a groove or cleft that runs down one side of a protein (FIGURE 3.13); the sugar polymer fits right into that groove. In the groove there are amino acids whose structures allow them to weaken the bonds between the sugars in the bacterial polymer, so it breaks. Lysozyme is found in tears and other body fluids. It helps to defend our bodies against bacterial infection.

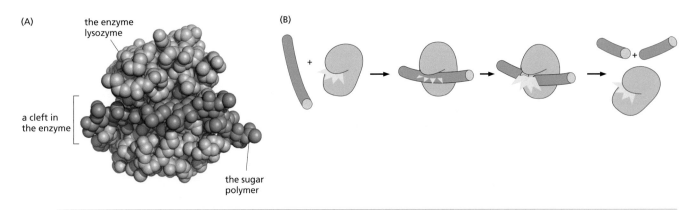

(A) the enzyme lysozyme

a cleft in the enzyme

the sugar polymer

(B)

FIGURE 3.13 Enzymes are proteins whose polypeptide chain has folded in such a way that they are catalysts. (A) Lysozyme (green) has a long pocket that can bind a sugar polymer (blue) that normally resides in a bacterial cell wall. (B) After this sugar polymer binds to the cleft in lysozyme, the enzyme catalyzes a reaction that cuts the polymer. This action weakens the cell wall, so the affected bacterium will break open and die. (A, courtesy of Richard J. Feldmann; PDB code 3AB6.)

Some enzymes bind to other enzymes, assembling into an enzyme complex that contains several different catalytic activities and is therefore capable of complex functions. Some proteins are not enzymes but bind to other copies of themselves, so they build up cellular structures, such as the cytoplasmic fibers mentioned in Chapter 1. The diversity of protein structure and function is a clear manifestation of the large amount of information stored in a cell's DNA.

Transcription factors are proteins that alter the frequency with which a gene is transcribed.

Only a subset of a cell's genes is transcribed and translated at any one time. It follows that there must be factors that regulate gene expression in ways that are valuable for the behavior of the cell. For example, for a cell to divide, its DNA must first be replicated. DNA replication is accomplished by an enzyme called DNA polymerase, which must be made by the expression of specific genes. During cell differentiation, on the other hand, DNA polymerase is not needed. The genes that encode this enzyme are turned off, but the genes needed for differentiation are turned on. This regulation of gene expression is accomplished in part by the ability of promoters and nearby sequences of DNA to bind a special class of proteins called **transcription factors**. The binding of a transcription factor to the DNA near a particular promoter alters the stability and activity of RNA polymerase when it is bound to that promoter and therefore alters the frequency with which that gene is transcribed (FIGURE 3.14). The frequency of transcription from a particular promoter is one factor that governs the amount of a given mRNA in the cell; another factor is messenger stability. The regulation of gene expression will come up again in Chapter 4, and particularly in Chapters 6 and 7, when we discuss the genes whose transcription is altered during cancerous transformation.

Polysaccharides are polymers of sugars.

Polysaccharides are macromolecules formed by the polymerization of sugars. They serve diverse purposes in cells. Some polysaccharides are

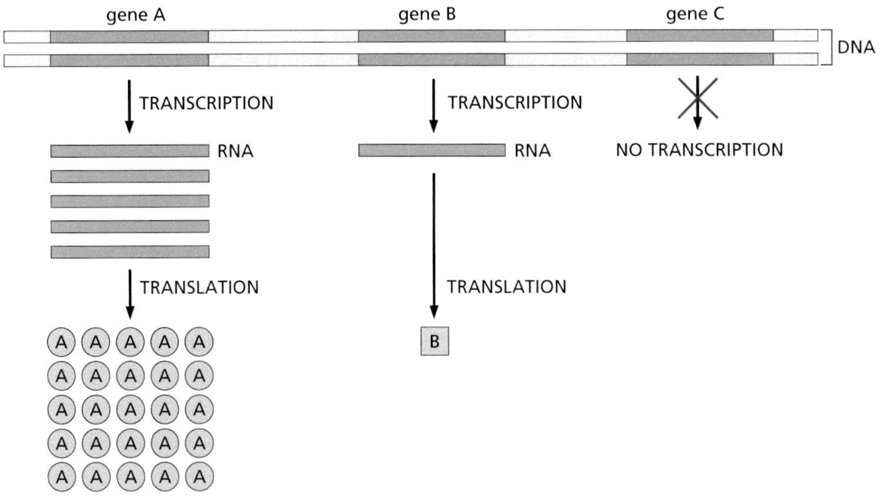

Why is the speed of a chemical reaction important for what goes on in a cell? Wouldn't it be better to have all of a cell's reactions go at top speed all the time?

SIDE QUESTION 3.5

FIGURE 3.14 **Transcriptional regulation of gene expression.** The diagram shows a segment of a DNA duplex that includes three genes. Gene A is being expressed at a high level as a result of a strong promoter that has been activated by transcription factors. Gene B is being transcribed but at a much lower level, resulting in less mRNA and protein product. Gene C is not being transcribed at all, so no protein product can be made. In a healthy cell, each protein is made at a level that corresponds to the need for that protein's function. (Adapted from Alberts B, Johnson AD, Lewis J et al. [2015] Molecular Biology of the Cell, 6th ed. Garland Science.)

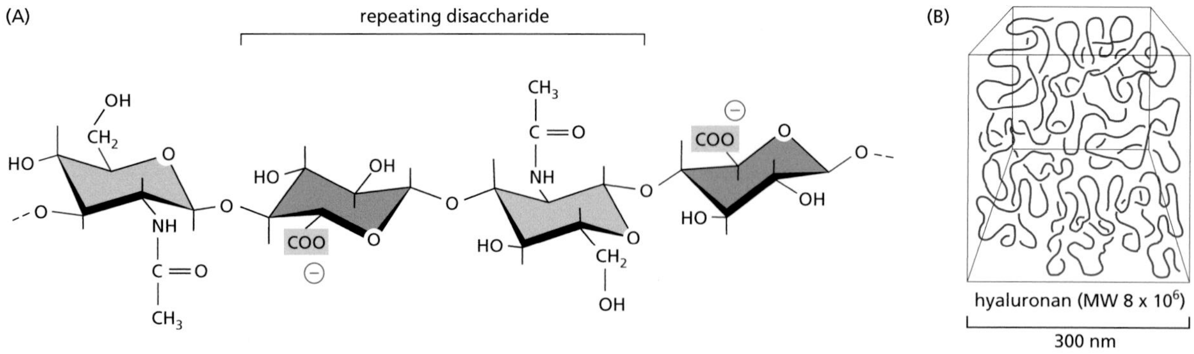

FIGURE 3.15 Polysaccharides are polymers of sugars. (A) Sugars are composed of carbon (C), oxygen (O), hydrogen (H), and sometimes nitrogen (N) arranged in particular patterns. In this figure, the bent polygons represent rings of carbon atoms linked together by bonds. The C at each vertex is not written, but its presence is understood. A vertex contains a letter only when it is occupied by an atom other than carbon, for example, O indicates oxygen. The atoms attached to these rings give each sugar its particular properties. Here two different sugars, colored orange and green, are linked to form a disaccharide. Many disaccharides can be linked end-to-end in a single chain to form a polysaccharide. Sugars can also be linked in ways that make branches. (B) The squiggly red line represents a very long polysaccharide chain. Polymers like this will fold into disordered structures, but the chemical character of the sugars along the backbone of such a huge macromolecule gives the structure as a whole its size and general shape. Molecules like this can be a part of mucus, or they can be tough enough to be part of a stiff plate, like the cartilage that pads the places where two bones rub over one another, as in a knee. (Adapted from Alberts B, Johnson AD, Lewis J et al. [2015] Molecular Biology of the Cell, 6th ed. Garland Science.)

structural, coating the outer surface of the plasma membrane; some are *chemical*, storing energy derived from food; and some are *informational*, lying on the cell's surface to help identify it as part of a certain cell type. Polysaccharides are highly variable because they can contain many different sugars, and those sugars can be linked to one another in multiple ways. In FIGURE 3.15A, four sugar molecules are displayed as clusters of atoms, showing their component carbon (C), hydrogen (H), oxygen (O), and nitrogen (N) atoms. The short lines between the atoms represent the chemical bonds that hold these atoms together. The folded polygons filled with different colors are a shorthand way of displaying additional atoms and their arrangement. At the vertices (where two lines meet in a corner), there is a carbon atom. Many biomolecules contain so many carbon atoms that molecular diagrams omit the Cs; the presence of a C is assumed at any unlabeled vertex.

The atoms attached to these polygons are part of what gives each sugar its special properties; some groups of atoms, like COO, are shown with a negative sign (COO$^-$) because this group of atoms makes an acid, which becomes negatively charged when dissolved in water. Other groups, like CH-CH$_2$ have no charge; they are nonpolar and are poorly soluble in water. In short, the molecular properties of each sugar are defined by the atoms that comprise it and the ways those atoms are arranged. (Note that in this figure two of the polygons representing these sugars are colored green and two are orange. The two units with the same color have the same structure, although at first sight they look different. One is just like the other, turned upside down.)

Two sugars linked side by side make a disaccharide, which means simply two sugars. Many polysaccharides are formed from the polymerization (attaching together, end-to-end) of disaccharides. In FIGURE 3.15B, the squiggly red line represents a very long chain of disaccharides, many of which are negatively charged because of acid groups, like the ones shown in Figure 3.15A. This polysaccharide is so big that it weighs the same as about 8 million hydrogen atoms. At such a large size, the polysaccharide is easily seen in an electron microscope, and it can even be detected in a light microscope (see Sidebar 3.1).

The linking together of molecules to produce a polysaccharide is regulated by enzymes. A simple sugar by itself is usually stable and unreactive, but it can be activated when an enzyme causes it to combine with a different molecule that stores some chemical energy. Once a sugar is activated, it can go on to form bonds with other sugars, as shown in Figure 3.15A. The formation of sugar–sugar bonds is controlled by yet another enzyme. Thus, there is a hierarchy in the flow of information that controls how polysaccharides are assembled: certain genes contain the information necessary to make the mRNAs that can be translated into the proteins that serve as enzymes that assemble sugars into polysaccharides. When polysaccharides are needed, for example, to store the energy from food, the expression of the right genes is turned on, and the relevant enzymes are made. These enzymes catalyze the reactions that assemble the poly-saccharides and store the extra energy.

SIDE QUESTION 3.6

Why are polysaccharides a good way to store energy?

Lipids are molecules used by cells to make membranes, store energy, and convey signals.

Lipids are essential building blocks for cellular membranes. They are also used to store energy and to convey chemical signals. For example, several hormones, like estrogen and testosterone, are lipids. Although lipids vary in shape (FIGURE 3.16), a feature common to them all is their poor solubility in water. The near-insolubility of lipids is the result of their many nonpolar carbon–hydrogen bonds and their lack of polar atoms, such as oxygen or phosphorus. However, certain lipids, like those shown in Figure 3.16, contain one part that is sparingly soluble in water, even though the rest of the molecule is not. As a result, these lipids can gather together into the thin sheets that form a cell's membranes. Most membranes are just two lipid molecules thick, so they are often called lipid bilayers. These bilayers are commonly built from the molecules shown in Figure 3.16B. These molecules organize into two layers, arranged so their water-insoluble parts face each other, letting the water-soluble parts face out. Now the coats on the two surfaces of the membrane are water-soluble, and the membrane is stable in an aqueous solution.

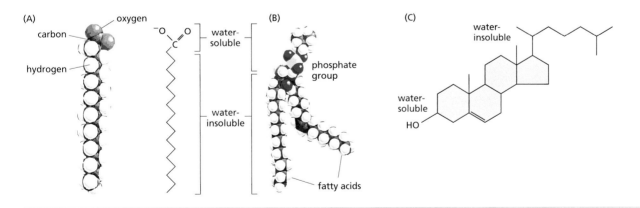

FIGURE 3.16 **Structures of some important lipids.** (A) A lipid, called a fatty acid, is represented here in two ways: by an atomic model (left) and by a shorthand diagram (right) in which the bonds between carbon atoms are shown as lines. A carbon atom is at the vertex between any two lines, and the hydrogens that bind to that carbon are not shown. This kind of short hand is often used to diagram biomolecules. Fatty acids are poorly soluble in water because of their long chains of carbon and hydrogen atoms, which will separate out from water the way oil separates from water in an Italian salad dressing. The two oxygens (red) at one end of the molecule give this lipid its acidic character and some water solubility. (B) A phospholipid is made by combining two fatty acids with two small, carbon-containing molecules, a phosphorus atom (yellow), and some oxygens (red). All biological membranes are made from two layers of phospholipids like this, combined with other lipids and proteins. (C) Diagram of cholesterol, another important component of biological membranes. This kind of lipid is called a steroid; several hormones, like estrogen and testosterone, are steroids. Steroids are an important kind of molecule, not only for cell structure but also for conveying signals from one cell to another. (Adapted from Alberts B, Johnson AD, Lewis J et al. [2015] Molecular Biology of the Cell, 6th ed. Garland Science.)

SIDE QUESTION 3.7

Both polysaccharides and lipids are made by enzymes, not by DNA or RNA. Does this mean that there is no genetic control on the synthesis of these biomolecules?

Lipids, like the cholesterol shown in Figure 3.16C, also contribute to membrane structure. Their water-insoluble parts dissolve in the water-insoluble parts of a lipid bilayer. The amount of cholesterol in a membrane helps to define its fluidity, which is important for the behavior of the proteins that mix with lipids to form the biologically important membranes that surround many cell structures. Note that estrogen and testosterone, which resemble cholesterol, can enter a cell because their water-insoluble structure lets them pass right through the lipid bilayer that surrounds a cell.

Like polysaccharides, lipids are produced by the action of enzymes. The right enzymes can use molecules that come into the cell as food, rearranging their atoms to form structures like those shown in Figure 3.16. Once again, we see genes at the top of an informational hierarchy, where gene expression makes the appropriate mRNA. This message is translated on ribosomes into a polypeptide that folds to form a protein with a structure that allows it to serve as an enzyme to catalyze a chemical reaction and perform a needed cellular function. In this case, that function is either the synthesis or the breakdown of specific lipids.

MACROMOLECULES IN THE CONTEXT OF CELL STRUCTURE

You are now armed with enough knowledge about the principal molecules from which cells are made that we can look at the ways these molecules are organized to make a self-replicating structure, like a cell. We start this tour at the outer edge of the cell and work our way in to a description of intracellular organelles.

The boundary of every cell is its plasma membrane.

Regardless of its shape, every cell is surrounded by a very thin lipid bilayer called the plasma membrane. Even though the plasma membrane is only two lipid molecules thick, it is sufficiently impermeable that it prevents most molecules from passing through it (FIGURE 3.17A). The importance of this property is easy to see when you think of all the molecules that reside inside a cell. If they could leak out, the cell would die. The plasma membrane achieves its barrier function as a result of its lipid layers (FIGURE 3.17B). The two fatty molecules are enough to keep most water-soluble molecules from crossing the membrane. In fact, the plasma membrane is generally so impermeable that to allow entry of the water-soluble molecules necessary for life, for example, the sugars and amino acids needed as food, the membrane is spanned by specialized proteins that serve as channels (Figure 3.17B, shown as green blobs). Some of these channels even work like pumps that drive particularly valuable molecules into the cell. For example, amino acids, which are needed for protein synthesis, are often pumped across the plasma membrane, assuring that the cellular concentration of nutrients is high enough for cell survival and growth.

SIDE QUESTION 3.8

What kinds of molecules would you expect to get across the plasma membrane without having to go through a protein channel?

Where cells abut one another, as in the sheet of cells that forms an epithelium, the plasma membranes of adjacent cells are very close. This proximity allows the cells to stick to each other, which, in turn, prevents cells from wandering, a metastatic behavior that normal cells avoid. To make these cell–cell adhesions strong, the surface of interaction between cells is sometimes increased by the extension of fingers from one cell into the space occupied by its neighbor (FIGURE 3.17C).

SIDE QUESTION 3.9

What kind of amino acids would you expect to find on the surface of a membrane protein in the places where it spans the thin layer of lipid?

Cell adhesion in some epithelia is so important that the plasma membranes of their component cells contain special structures that link neighboring

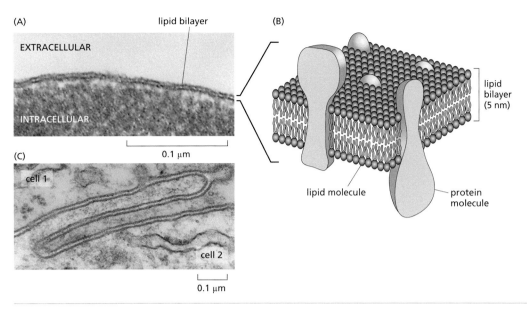

FIGURE 3.17 Structure of the plasma membrane. (A) The edge of a red blood cell, as seen in an electron microscope. The plasma membrane defines the boundary of the cell. (B) Diagram of the molecular structure of this membrane. Red molecules are phospholipids (as diagrammed in Figure 3.16B), each of which has two fatty tails and one charged head (red ellipsoids on the membrane's two surfaces). This double layer of lipid makes a structure so thin that single proteins (green) can span the membrane's thickness. The notation 5 nm means 5 billionths of a meter. In spite of being so thin, the membrane is quite impermeable to water-soluble molecules. Molecules like sugars and amino acids can cross the membrane only where there is a specific mechanism, such as a membrane-spanning protein that can serve as a channel, to let them through. (C) Electron micrograph of two cells in close contact. You can trace the membranes of cells 1 and 2 though this little maze and see how close two neighboring cells can be. Note that the cytoplasms of the two cells, marked with different colors, are separated by the two closely apposed plasma membranes and a thin layer of extracellular space. The material in the extracellular space contributes to the adhesion of these cells to one another. At the upper right, one membrane is bending in to form a "bud" that can pinch off and bring some extracellular fluid into the cytoplasm of that cell. (A, courtesy of Daniel S. Friend. B, adapted from Alberts B, Johnson AD, Lewis J et al. [2015] Molecular Biology of the Cell, 6th ed. Garland Science. C, from Staehelin LA (1974) *Int. Rev. Cytol.* 39:191–283. With permission from Elsevier.)

cells (FIGURE 3.18A). For example, the epithelium that lines the intestine is the boundary between the interior of that tube, its lumen, and the outside wall of that tube, which contains blood vessels that convey nutrients to the rest of the body. The intestine's lumen contains both bacteria and digestive enzymes that break down food into small molecules that can pass through the plasma membrane. These bacteria and enzymes must not be allowed to cross the wall of the intestine and get into the blood, so the cells of intestinal epithelia are well bonded to one another. Membrane specializations called a tight junction, a zone of adherence, and a desmosome all contribute to this special cell adhesion (FIGURE 3.18B).

Other membrane proteins are designed to bind molecules that float by in the extracellular fluid (FIGURE 3.19). These membrane proteins are not involved in cell adhesion, but they work like antennas to pick up signals from the environment. They are **receptors** for molecules that can convey information from one part of the body to another. For example, some of the hormones that regulate cell growth work by binding to a specific membrane protein and setting off a series of chemical reactions inside the cell. Such receptors allow cells to cooperate in doing what the body needs to accomplish. Membrane proteins that receive signals are particularly important in cancer biology because soluble signals include information about whether or not a cell should divide. We shall see in Chapters 6–8 that receptors like this are key molecules in both the emergence and treatment of cancer.

Many cells are surrounded by an **extracellular matrix**, often referred to simply as the **ECM**. The ECM is composed of fibers built from proteins and polysaccharides that are made by cells and secreted into the extracellular

SIDE QUESTION 3.10

Proteins that work as channels through membranes have a hole down their middle that is hydrophilic, or water-soluble. How is that structure different from the organization of most proteins?

SIDE QUESTION 3.11

As you have seen, several membrane proteins help to hold cells together in tissues. What would happen to a tissue if it was treated with an enzyme, like the ones produced in your stomach, that digests proteins down to their component amino acids?

FIGURE 3.18 **Cells in epithelial tissues, like the wall of the intestine, are held together by bonds that make the tissue strong.** (A) A drawing of cells lining the intestine. The lumen is the interior of the intestine, where the processes of digestion are breaking food down into small molecular components. The cells that line the intestine bind to one another with specialized junctions that give the tissue its strength and keep the contents of the lumen from leaking into the space where blood would carry its digestive enzymes off to other parts of the body. (B) An electron micrograph of the place where two intestinal cells are stuck to one another. The tight junction (TJ) serves as a boundary that keeps the contents of the lumen in place. The zone of adherence (ZA) helps the adjacent cells to bond tightly with one another. The desmosome (D) is a cell surface specialization that works like a spot weld between adjacent cells, again strengthening their adhesion. (B, from Staehelin LA (1974) *Int. Rev. Cytol.* 39:191–283. With permission from Elsevier.)

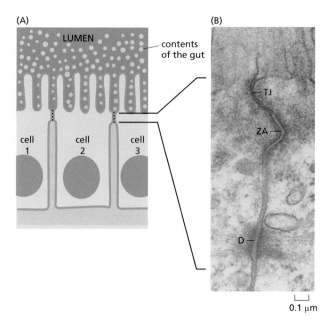

space. The ECM is important to the cells that reside in it because it forms part of their external environment (FIGURE 3.20). The physical character of the ECM depends on its structure and composition. This figure shows an unusually well-ordered region of ECM, composed of fibers made largely from a single protein called collagen. Between the darkly stained, extracellular fibers, there are many polysaccharides that help to give the ECM both its texture and chemical properties, like its electric charge. These fibers are not well stained by the methods used to prepare this electron micrograph, so they are invisible in this picture. The electric charges on polymers of the ECM bind many small molecules that would otherwise float freely in the aqueous solution between the fibers. Some of the small molecules bound by ECM are the signaling molecules to which cells make receptors. Thus, if the ECM is altered, the change may lead to a release of signaling molecules, which can then bind to receptors in cells and initiate

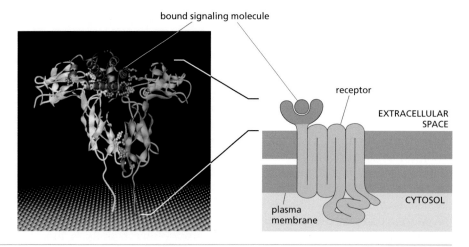

FIGURE 3.19 **Some membrane proteins are like antennas that receive signals from outside the cell.** (Left) A ribbon diagram showing the fold of the part of a receptor protein that lies just outside the cell in extracellular space. The receptor protein is formed from two identical parts (green and blue) that bind to a signaling molecule (red), giving the cell a message. (Right) A schematic of a membrane-bound receptor protein, showing not only the extracellular portion of the protein, which binds the signaling molecule, but also the parts of the protein that weave back and forth across the membrane. Some parts of this long polypeptide chain are exposed on the outside surface of the membrane and some on the inside, where they can transmit the fact of signal molecule binding to the cell's interior. (Left, from deVos AM, Ultsch M & Kossiakoff AA [1992] *Science* 255:306–312. With permission from AAAS.)

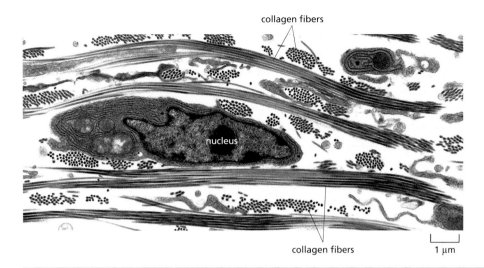

collagen fibers

nucleus

collagen fibers

1 μm

FIGURE 3.20 **An example of extracellular matrix.** Near the center of this electron micrograph is a cell that is specialized for making proteins of the extracellular matrix (ECM). This cell has an obvious nucleus, which is making mRNA that encodes the protein collagen, an important component of ECM in many parts of the body. Outside the cell, there are many bundles of the fibers that collagen forms. Collagen fibers are one of the materials that make skin strong. The bundles of collagen fibers are surrounded by polysaccharides that do not stain well for electron microscopy, so they do not show up in this image. (From Ploetz C, Zycband EI & Birk DE [1991] *J Struct Biol* 106:73–81. With permission from Elsevier.)

cellular responses. A striking example of the importance of the ECM is that when a cell becomes cancerous, it alters the ECM that surrounds it. These changes will often affect the behavior of nearby cells. Thus, although everything outside the plasma membrane is literally outside the cell, the ECM is important for cell behavior.

Inside the plasma membrane, cells are divided into multiple compartments. Go back and look carefully at the cytoplasm in Figure 3.1, and you will see the extent and variety of this compartmentation. Each intracellular compartment is important for a cell's ability to grow, divide, and perform functions essential for the well-being of the body as a whole. Below, each of these cellular compartments is discussed separately, starting with the biggest, which is one of the most important.

The nucleus contains most of the cell's DNA and serves as an information-processing center.

The largest structure in most cells is the nucleus, wherein DNA is stored (FIGURE 3.21A) and transcribed into RNA. The nucleus is the primary place for regulation of gene expression. The nucleus is also where DNA is replicated, a process that must precede cell division, so each daughter cell will have a complete copy of the genes it needs to grow and divide again. Most human cells are diploid: their nuclei contain one set of chromosomes from father and one from mother. A diploid cell, therefore, provides two copies of every gene. Commonly, when the expression of a given gene is called for, both the maternal and paternal copies of that gene are expressed.

All the genes in an individual cell comprise its **genome**. The principal functions of the nucleus are the maintenance, propagation, and controlled expression of the genome. In humans, the amount of DNA in the nucleus is astounding. There are literally billions of nucleotides in the DNA of every nucleus. The total length of all the chromosomes in one cell is almost 6 ft. In contrast, the number of protein-encoding genes in the human genome is only about 25,000. This means that the amount of DNA in a human cell far exceeds the length required to encode our proteins.

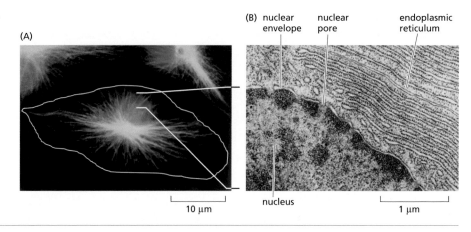

FIGURE 3.21 **Images of nuclei.** (A) A fluorescence light micrograph in which the cell's DNA is stained red-orange and cytoplasmic fibers outside the nucleus are green. The position of the plasma membrane has been traced in white. (B) An electron micrograph showing a portion of a cell's nucleus and a region of the surrounding cytoplasm. In this image the DNA appears as darkly staining material, some of which lies up against the inner surface of the nuclear envelope. At some places, the continuity of the nuclear envelope is interrupted by nuclear pores, which allow an exchange of material between nucleus and cytoplasm. In the cytoplasm, the membranes of the endoplasmic reticulum are clear. Some of these membranes are studded with small, darkly staining particles. These are ribosomes, which carry out protein synthesis. (A, courtesy of Mark Ladinsky. B, courtesy of Keith Porter.)

SIDE QUESTION 3.12

How could a change in the packing of DNA influence gene expression?

We carry around a lot of DNA whose function is not accounted for in the above descriptions. Scientists who study genomes have more recently discovered that DNA encodes a large number of RNAs that are not translated into proteins. Many of these are now known to help regulate the expression of protein-encoding genes, but there are probably still additional functions of RNAs to be discovered. Nevertheless, all our DNA is packed into a nucleus whose diameter is only about 1/400,000 of the DNA's total length. Human DNA is therefore compacted by several levels of folding.

Electron microscopy of a nucleus shows DNA-containing material packed quite densely in regions near the edges of the nucleus (FIGURE 3.21B). The folding of DNA becomes even more evident when a nucleus is broken, so the nuclear material can spill out (FIGURE 3.22). The DNA that seeps from a broken nucleus is not simply a bundle of double helices. It is wrapped around small, disc-shaped clusters of proteins called histones. All the histones are positively charged, so they attract the negative charges of DNA's sugar–phosphate backbone, making a stably folded material called **chromatin** (Figure 3.22B–D). Each cluster of histones and the DNA wrapped around it is called a nucleosome. However, the chromatin in the nucleus is not the comparatively straight structure seen in Figure 3.22B,C. It twists, loops, and folds on itself, making the complex material seen leaking from the nucleus in Figure 3.22A. Within an undisrupted nucleus, this folding is even tighter, as indicated by the dense staining near the edges of the nucleus in Figure 3.21B. The turns and loops of chromatin accomplish two functions: they compact the length of the DNA, so it will fit into a nucleus, and they pack the DNA so tightly that transcription factors are sometimes prevented from binding to promoter sequences. This kind of packing interferes with the ability of RNA polymerase to initiate transcription. By hiding some parts of the DNA in this way, the tight packing of chromatin helps to down-regulate the expression of genes whose products are not currently needed. Tight packing of chromatin is thus a negative regulator of gene expression. It is interesting that when cells go through a cancerous transformation, their chromatin sometimes becomes disorganized. As a result, genes that should be silent can be expressed. This leads both to the production of proteins that do not belong in a normal cell of that type and to the overproduction of pro-

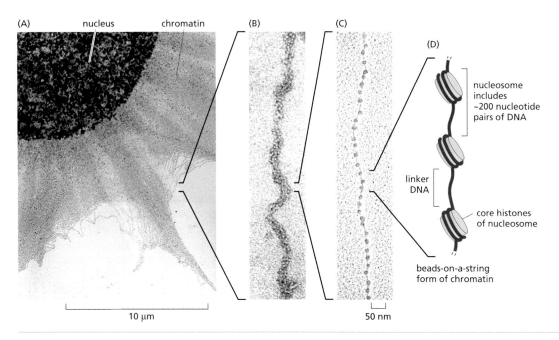

(A) nucleus   chromatin   (B)   (C)   (D)

nucleosome includes ~200 nucleotide pairs of DNA

linker DNA

core histones of nucleosome

beads-on-a-string form of chromatin

10 μm   50 nm

FIGURE 3.22 **A cell's DNA is wrapped around protein particles to keep it compact.** (A) An electron micrograph showing the nucleus of a cell that has been broken open, so its DNA is leaking out. This DNA looks like a network of strands. (B) An electron micrograph showing a highly magnified view of one of these strands. This material doesn't look like a simple double helix because the DNA is wrapped around particles made from clusters of proteins called histone, which are in turn twisted together into an ill-defined fiber. This DNA–protein complex is the material called chromatin. (C) When chromatin is stretched, by changing the chemical conditions in the medium, it looks like beads on a string. Each bead is a nucleosome. (D) A diagram of the interactions between DNA (red) and histone protein complexes (yellow). These form strings of particles, the nucleosomes, from which chromatin is made. (A and C, courtesy of Victoria Foe. B, courtesy of Barbara Hamkalo.)

teins that should be expressed only very slightly. Such events are one of the ways in which cancer cells misbehave.

The boundary of the nucleus is a semi-permeable envelope.

The boundary around the nucleus is commonly referred to as the **nuclear envelope** (see Figure 3.21B). It is more complex than the plasma membrane because it is two membranes thick. The nuclear envelope is actually a system of flattened, membranous sacs that fold to form both an envelope around the nucleus and also a set of cytoplasmic compartments called the **endoplasmic reticulum**, more commonly called simply the **ER** (FIGURE 3.23). The portion of this membrane system that surrounds the nucleus is penetrated by pores, each of which is filled with a complex of proteins that serves as a filter. This filter defines what can get into and out from the nucleus. For example, the nuclear pores allow mRNA to move out of the nucleus and into the cytoplasm, where it is translated into proteins. Nuclear pores also allow some proteins to get from the cytoplasm, where they are made, into the nucleus, where they can perform their functions.

The cellular space outside the nuclear envelope but inside the plasma membrane is the **cytoplasm**. It is a complex, watery fluid that contains thousands of different molecules, each with a specific role to play in cell function and behavior. The liquid part of the cytoplasm is called the **cytosol**. Some of the molecules in cytosol, such as sugars, contain only about 35 atoms and are comparatively small and simple. Additional small molecules in the cytosol are formed during the breakdown of food. Also present are atoms derived from minerals, such as sodium, potassium, chloride, magnesium, calcium, iron, and zinc. All these atoms and molecules play roles in nutrition or other aspects of cellular life.

Identify three proteins with distinct functions that must all be able to get into the nucleus for that organelle to do its jobs properly.

SIDE QUESTION 3.13

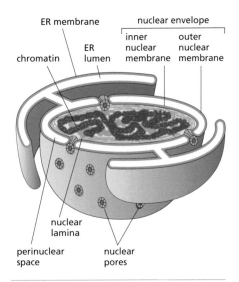

perinuclear space

nuclear pores

FIGURE 3.23 **Diagram of the nuclear envelope, its pores, and its connections with the endoplasmic reticulum (ER).** The nuclear envelope is made from two layers of membrane. This double-layered structure is penetrated by nuclear pores that allow materials made in the nucleus, such as mRNA, to get out to the cytoplasm and materials made in the cytoplasm, including proteins like RNA polymerase, to get into the nucleus. The outer membrane of the nuclear envelope is continuous with the membranes of the ER, so the perinuclear space between the two membranes of the nuclear envelope is continuous with the interior (lumen) of the ER.

Proteins are made on ribosomes located in the cytoplasm.

Cytoplasm contains many ribosomes (see Figure 3.3), each of which is a complex of about 80 proteins and three pieces of special RNA. Ribosomes, like mRNA, are made in the nucleus. They pass out of the nucleus through the filter that plugs each nuclear pore. In the cytoplasm, ribosomes work like molecular machines, using the information in each mRNA molecule to synthesize a particular protein. Some of the proteins made on ribosomes are enzymes that control the chemical reactions called **metabolism**—that is, the breakdown of food and the conversion of its atoms and energy into the molecules needed for cell growth, division, and differentiation. The ribosomes that make these proteins, and many others proteins as well, float free in the cytosol, so as they make a protein, the resulting polypeptide chain folds up in the cytosol, where it will subsequently function (FIGURE 3.24, top).

Some proteins must move to places other than the cytosol to do their jobs, for example, the proteins that penetrate the plasma membrane and function as channels. These proteins are commonly made on ribosomes bound to the ER membranes, not on ribosomes free in the cytosol. Still other proteins are made to be secreted, so they can function in the extracellular space. These proteins too are made on membrane-bound ribosomes. The ribosomes that will translate mRNAs encoding either of these kinds of proteins become bound to the ER just as soon as they begin to translate either membrane-associated or secreted protein (Figure 3.24, bottom). When these ER-bound ribosomes make a protein, the growing chain of amino acids can either pass through the membrane and fold up in the interior of the ER or it can pass right into the membrane itself, becoming part of an enlarging membrane area. Which place the growing protein will go depends on the chemistry of the amino acids from which

FIGURE 3.24 **Ribosomes bound to the ER can place the proteins they make either in the lumen or in the ER membrane itself.** When ribosomes are free in the cytosol, they can bind mRNA and make proteins that stay in the cytosol, where they fold to form the right structure and execute their functions. When ribosomes are bound to the ER, the polypeptide chains they make go into the ER and can be released either into the interior of that compartment or into the membrane itself, depending on the solubility of the side chains on the amino acids from which the new protein is made. (Adapted from Alberts B, Johnson AD, Lewis J et al. [2015] Molecular Biology of the Cell, 6th ed. Garland Science.)

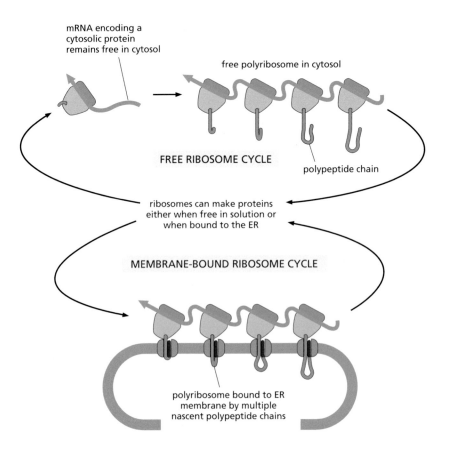

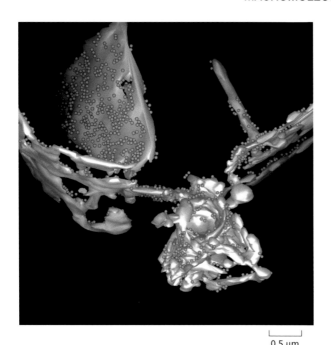

0.5 μm

FIGURE 3.25 **The endoplasmic reticulum (ER) is a large and complex cytoplasmic membrane system that can take on various shapes.** A 3-D model of ER membranes generated by electron tomography. Blue dots are ribosomes, which are bound to ER membranes and are synthesizing either membrane proteins or proteins for export from the cell. The multiple shapes of the ER in a single cell are clearly seen. (Courtesy of Matthew West and Gia Voeltz, University of Colorado Boulder.)

it is made. If many of the protein's amino acids have nonpolar side chains, that protein will naturally tend to flow into the fatty part of the ER membrane. If the protein has short stretches of nonpolar amino acids, these can serve to span the lipid layer of a membrane, allowing the protein to have one part in the ER lumen and one part in the cytosol. Thus, the chemistry of the amino acids, selected on the basis of information in a given mRNA, defines many aspects of a protein's structure, localization, and function.

ER membranes are also involved in several kinds of chemical reactions. For example, the ER harbors proteins that serve as enzymes for making many of the cell's lipids. Thus, as membrane lipids are produced, they can slip right into an ER membrane, increasing its area. The ER membranes also include enzymes that can modify the chemistry of foreign molecules that happen to get into the cell, along with food. These molecules can include not only toxins from nature, such as those made by some plants, but also man-made chemicals, like drugs. The ability of the ER to alter foreign molecules is important in cancer biology, as you will see in future chapters, because the chemical versatility of this membrane system acts for both good and bad in modifying cancer-causing chemicals that we may happen to ingest.

Given all the different chemical reactions going on in the ER, it is not surprising that this membrane system can take on more than one form. Much of the ER from a very simple cell is shown in FIGURE 3.25. Some of it is shaped as flattened sacs and some is tubular. Ribosomes can be seen scattered over much of its area but not all. This diversity of ER form is probably associated with specific differences in the chemical actions of different parts of the organelle, but this issue is one that scientists are still sorting out.

Membranes made in the ER can travel elsewhere in the cell after being processed in the Golgi apparatus.

Another membranous organelle in cytoplasm is the **Golgi apparatus** (often referred to simply as the *Golgi*, named after Camillo Golgi, who discovered the structure in the early twentieth century as a result of its

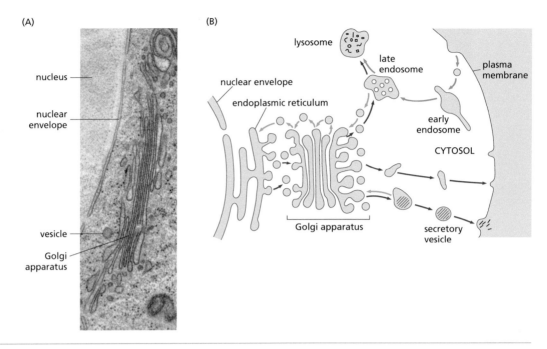

FIGURE 3.26 Traffic of proteins made by ribosomes attached to the ER is accomplished by vesicles that move from the ER to the Golgi apparatus and then on to other places in the cell. (A) An electron micrograph of a Golgi apparatus situated near the cell's nucleus. Some vesicles lie between the nucleus and the Golgi. Other vesicles are visible near the flattened sacks of membrane that are characteristic of Golgi structure in most cells. (B) A diagram that shows the complexity of membrane traffic. Material made in the ER is carried to the Golgi by small vesicles (short red arrows). These fuse to form a flattened sack on one side of the Golgi. This sack contains enzymes that modify proteins, both inside the sack and in the membrane of the sack. As new vesicles continue to come from the ER, new sacks form and push the older ones toward the opposite side of the Golgi. There, new vesicles bud off to carry the modified membrane and its content to other sites in the cell (long red arrows). Some of these vesicles go to the plasma membrane, fuse with it, and thereby augment the area of the plasma membrane. The interior of these vesicles carries proteins that the cell is going to secrete. Other vesicles find different cytoplasmic targets, here shown as late endosomes and lysosomes. Note that there is also a backward flow of membrane (blue arrows). This includes vesicles that bud in from the plasma membrane, a process called endocytosis (green arrows), and a flow of vesicles that return enzymes and membrane components to the ER, keeping the system working. (A, courtesy of Mark Ladinsky. B, adapted from Alberts B, Johnson AD, Lewis J et al. [2015] Molecular Biology of the Cell, 6th ed. Garland Science.)

unusual staining properties, as seen in the light microscope). Like the ER, the Golgi is made from sacs and tubes of membrane (FIGURE 3.26A). The Golgi grows by receiving membrane vesicles that bud off from the ER. The Golgi contains enzymes that modify those membranes and then send them to other cellular locations where that particular kind of membrane is needed. Vesicles from the ER arrive on one side of the Golgi apparatus, where they fuse to form a flattened sac (FIGURE 3.26B, short red arrows). As more vesicles arrive, they form a new sac on top of the old one, causing the sacs to progress slowly from one side of the Golgi to the other. As the sacs move through the Golgi, enzymes in that organelle modify the proteins in the membranes as well as those within the sacs. Most of these modifications attach sugars to proteins, tailoring them for their future functions at different places in the cell.

Once a membrane sac has reached the side of the Golgi away from the ER, its membrane buds off as vesicles. These newly formed vesicles moved through the cytoplasm to whatever destination is appropriate for the proteins and lipids they now contain (Figure 3.26B, long red arrows). Some vesicles go from the Golgi to the plasma membrane where they fuse with it, adding their membrane to the surface of the cell so it can increase in area as the cell grows. During this fusion process, the interior of each vesicle is released to the outside of the cell, a process termed **exocytosis**. This is how proteins made by ribosomes bound to the ER can be secreted from the cell. Still other vesicles leave the Golgi and fuse with vesicles that are formed by an inward bending of the plasma membrane, a process called **endocytosis**, as depicted in Figure 3.26B (green arrows near the plasma membrane). Endocytosis allows cells to take material from outside the cell and mix it with proteins made in the ER. One product of this fusion is yet another membrane-bound compartment called a **lysosome**. These structures contain digestive enzymes, like the ones in your stomach and intestine, that can break down proteins, lipids, polysaccharides, and nucleic acids, releasing their small-molecule subunits into the cytosol for the enzymes of metabolism to use in building new macromolecules for the cell. Cells use lysosomes both to degrade extracellular material that has been brought in by endocytosis and to combine with old membrane-bound compartments of the cell and degrade them.

The migration of membranes from one cellular site to another is called **membrane traffic**. Membrane traffic takes proteins and lipids that were made in the ER and processed in the Golgi, then moves them to wherever they are needed, thanks to the mechanical activity of the cellular machinery for moving vesicles, described later in this chapter.

Mitochondria are yet another kind of membrane-bound, cytoplasmic organelle.

Mitochondria are cytoplasmic compartments that use some of the molecules brought into the cell as food to make other molecules that cells need. For example, pieces broken from sugar molecules are used by mitochondria to make a molecule called adenosine triphosphate (**ATP**). ATP carries a form of chemical energy that cells can use for almost all the jobs they do, such as synthesizing macromolecules, pumping chemicals across membranes and moving the cell from one place to another. (Note that the base in ATP is the same "A" found in both DNA and RNA.)

Most cells contain numerous mitochondria (FIGURE 3.27A). Commonly they are distributed throughout the cytoplasm, though in cells that have a high and localized demand for energy, such as cells that pump digested

Consider an analogy between a lysosome and a company's policy of doing preventive maintenance. In what ways is the analogy apt and how does it fail?

SIDE QUESTION 3.14

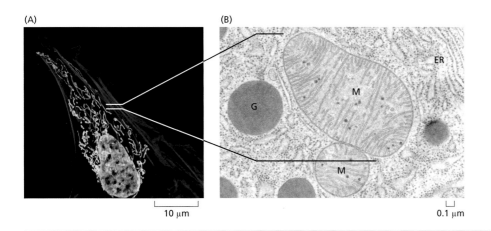

(A)

(B)

10 μm

0.1 μm

FIGURE 3.27 Images of mitochondria taken with different microscopes. (A) Light micrograph that uses fluorescence optics to image a cultured mammalian cell. DNA in the nucleus is stained light blue, microfilaments in the cytoplasm are purple, and mitochondria are yellow. (B) Electron micrograph of a thin slice cut from a liver cell. The image shows two mitochondria (M). These organelles are built from two membranes: an outer membrane that surrounds the organelle and defines its interior, and an inner membrane that folds to form tubes and platelike structures. Other membrane-bound structures in this micrograph are endoplasmic reticulum (ER), which is studded with ribosomes, and granules (G) that contain lipid and protein. The granules and ER do not appear in the light micrograph, because they have not been stained with a fluorescent dye. (A, courtesy of Dylan Burnette and Jennifer Lippincott-Schwartz, National Institutes of Health. B, from Porter KR & Bonneville MA [1973] Fine Structure of Cells and Tissues, 4th ed. With permission from Lea & Febiger.)

food molecules from the intestine into the blood, the mitochondria are concentrated where energy is most needed. Each mitochondrion has two membranes: an outer one that defines the perimeter of the organelle and an inner one that is folded to make many flattened sacs within the mitochondrion (FIGURE 3.27B). Mitochondria carry out two essential functions for the cell: the production of ATP and the chemical reactions of some important metabolic processes. ATP synthesis is carried out on the inner mitochondrial membrane; the metabolic functions are accomplished mostly between the two membranes. This is yet another example of a cell using compartments defined by membranes to carry out multiple, quite different chemical reactions in a comparatively small space.

The synthesis and use of ATP are very important to the cell, so they deserve a few more words. Mitochondria are like a power plant, because they take energy in one form and convert it into a form that will be more useful. The energy they use is in food; the energy they make is in the form of ATP. ATP can store energy because it contains three phosphate groups, analogous to the ones in the sugar–phosphate backbone of DNA (FIGURE 3.28). The phosphates in these molecules are all negatively charged. Recall that negative charges repel one another. If ATP is dissolved in water (H_2O), the two molecules react, and the bond between the last two phosphates will break, allowing one negatively charged phosphate to separate from the rest of the molecule and diffuse away, leaving behind adenosine diphosphate, or ADP. When ATP is by itself in aqueous solution, this falling apart is very slow, because even though the molecule's negative charges are pushing one another apart, the bonds between the phosphates are strong enough to hold ATP together. However, enzymes called ATPases speed up the breaking of those bonds; ATPases are built so they can capture some of the energy released when the negative charges on ADP and phosphate fly apart. Almost all of a cell's pumps, motors, and synthetic enzymes are ATPases, capable of splitting ATP into ADP and phosphate to get the energy they need to do their appointed jobs.

Mitochondria make ATP by combining ADP with a phosphate group, using energy taken from sugar to push the two repelling molecules together,

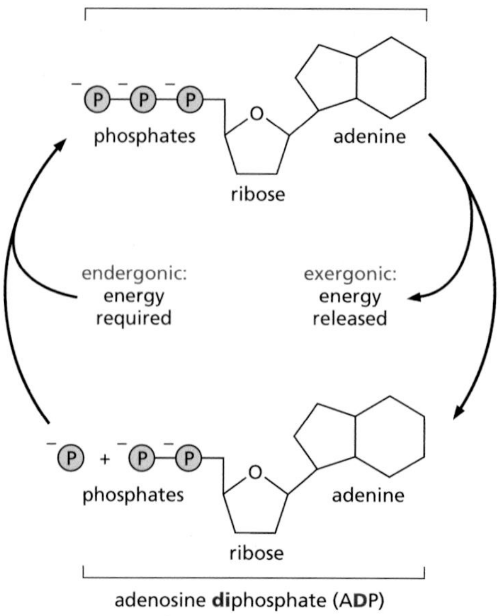

SIDE QUESTION 3.15

ATP has been called the energy currency of the cell. Consider this analogy carefully. In what ways is it accurate and in what ways limited?

FIGURE 3.28 The structure of ATP and its high-energy bonds. The components of ATP are adenine (a base), ribose (a sugar), and three phosphate groups. Adenine is made from two rings of atoms, hence its shape. Ribose is a sugar, like the ones shown in Figure 3.15, but it has only five atoms in its ring, hence its shape. Adenine and ribose together make adenosine, and adenosine triphosphate (ATP) is adenosine with three phosphate groups attached. Each phosphate is negatively charged. These charges repel one another, which makes the phosphates tend to fly apart. Under normal circumstances, they are held together by chemical bonds. There are, however, enzymes that speed up the breaking of the bonds between phosphates, allowing one phosphate to split off, releasing energy and leaving adenosine with only two phosphate groups attached. Now it is called adenosine diphosphate (ADP). The repulsion between the phosphates means that when the phosphate flies away from ATP, there is a release of energy that can be harnessed to do work for the cell. On the other hand, making ATP by combining ADP and phosphate requires energy; that is the job of mitochondria (see Figure 3.29).

adenosine **triphosphate (ATP)**

endergonic: energy required

exergonic: energy released

adenosine **diphosphate (ADP)**

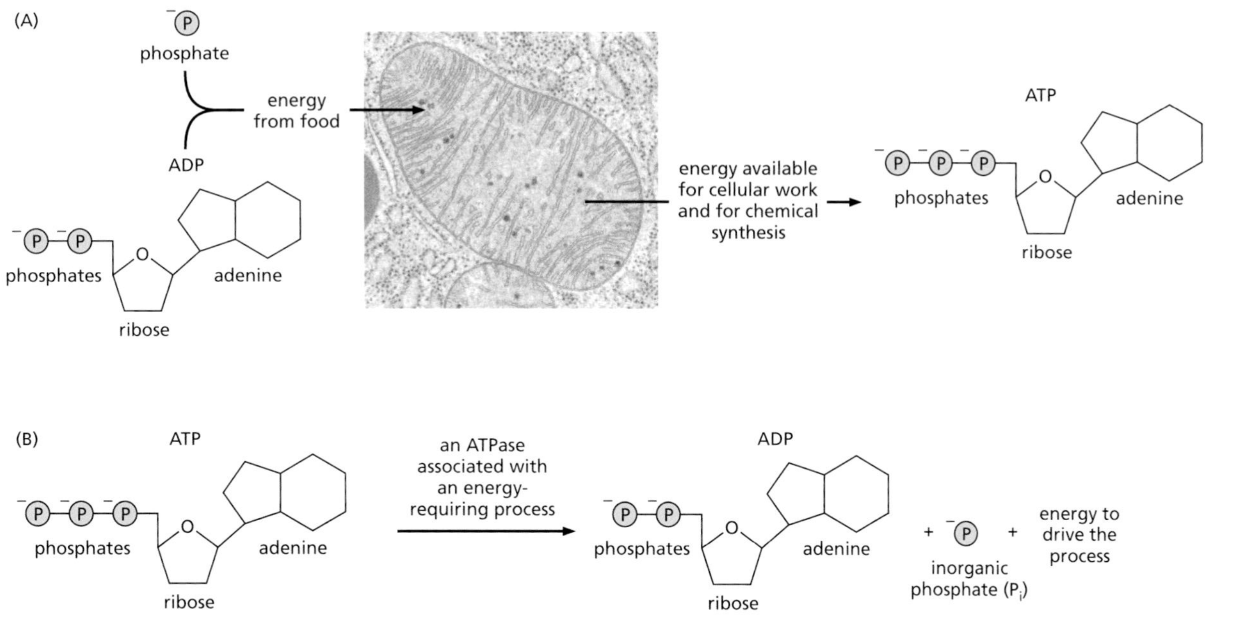

(A)

phosphate

energy
from food

ADP

phosphates

ribose

adenine

energy available
for cellular work
and for chemical
synthesis

ATP

phosphates

adenine

ribose

(B)   ATP

phosphates

adenine

ribose

an ATPase
associated with
an energy-
requiring process

ADP

phosphates

adenine

ribose

+ ⁻(P) +

inorganic
phosphate (Pᵢ)

energy to
drive the
process

FIGURE 3.29 The cellular cycle in which mitochondria capture energy from food and use it to make ATP, which can then go elsewhere in the cell and be used as a source of energy to drive energy-requiring processes. One important role of mitochondria is the synthesis of ATP. The bonds between phosphates in ATP store chemical energy, so this molecule can be used anywhere in the cell to help drive processes that require energy. (A) Phosphate and ADP enter a mitochondrion. This organelle uses energy from food to put them together and form a molecule of ATP, which has an additional energy-rich bond. This energy can be used anywhere in the cell. (B) When a phosphate is split off from the rest of ATP, chemical energy is released. This energy can be used by ATPases to run pumps, drive molecular motors, or support complex biosynthesis. Thus, a mitochondrion uses up food energy to convert ADP and phosphate into ATP, which can then be used as a source of energy anywhere in the cell. (From Porter KR & Bonneville MA [1973] Fine Structure of Cells and Tissues, 4th ed. With permission from Lea & Febiger.)

forming a phosphate–phosphate bond (FIGURE 3.29). The energy-rich ATP can then diffuse anywhere in the cell for use by various different ATPases. ATP synthesis is essential for keeping a cell healthy.

The preceding account of cytoplasmic membrane systems has included descriptions of several different structures and numerous kinds of membrane proteins. To help you keep track of them and get a sense of the relationships among the various membrane types, FIGURE 3.30 is a light micrograph of a cell in which multiple membranous structures have been stained with different colors. The various kinds of membrane proteins are summarized in SIDEBAR 3.6.

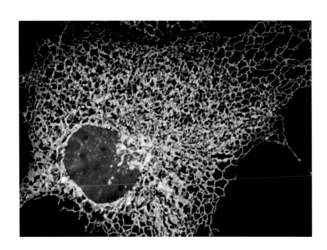

FIGURE 3.30 Fluorescence light micrograph of a cultured mammalian cell, stained to show several cellular membrane systems. Multiple organelles are stained with different colors to make them visible. The DNA of the nucleus is stained magenta. The ER (cyan) appears as a network of tubules and sheets that spreads through much of the cytoplasm. Mitochondria are dark purple. The Golgi (yellow) lies next to the nucleus. (It is not really inside the nucleus, but part of it lies above or below the nucleus, so in this planar image the two are superposed.) The cytoplasm also contains small vesicles (red) that move through the cytoplasm as part of membrane traffic. This kind of preparation reveals the complexity of a cell's cytoplasmic membrane systems. (From English AR, Zurek N, & Voeltz GK [2009] Curr Opin Cell Biol 21:596–602. With permission from Elsevier.)

SIDEBAR 3.6

A CLOSER LOOK: Classes of membrane proteins.

The paragraphs on membrane proteins and membrane traffic list enough different functions that it may be confusing. Here is that same list, organized in a different way to help you grasp the full diversity and importance of membrane proteins and their functions, putting them into the context of the ways they are important in cancer.

Structural: Some membrane proteins support the adhesion of a cell to its neighbors. These help to make tissues, which in turn make organs, so they are essential for normal body structure. When cells lose their ability to adhere to neighbors, they can then move

about in abnormal ways. That is part of what goes wrong when a cell becomes metastatic.

Transport: Some membrane proteins form channels for the transport of specific, water-soluble molecules in and out of the cell. These are important in achieving the right balance of ions in cytoplasm. They are also important in nutrition (taking in food molecules) and in some kinds of secretion (when cells need to send signals to one another).

Biosynthesis: Some membrane proteins, especially those in the ER, take advantage of lying in the nonpolar part of the lipid bilayer from which the membrane is

made to carry out chemical reactions that won't go forward in an aqueous solution.

Signaling: Some membrane proteins are receptors for signals that either float by in solution or are bound to the surface of neighboring cells. These proteins are like antennas, in the sense that they receive information from the environment, but they are also like switches, because once they receive a signal, they turn on a set of processes inside the cell. Such signals are part of the normal regulation of cell behavior in a healthy body; they are among the things that go wrong during cancerous transformation.

PROTEIN FIBERS FOR CYTOPLASMIC STRUCTURE AND FUNCTION

Cytoplasm contains three kinds of protein fibers that are collectively called the **cytoskeleton**. They form a framework for cell structure, just as our skeletons help to define the shape of our bodies. When we consider these fibers, though, the term *skeleton* is something of a misnomer. Some of the fibers can work like muscles as well as like bones, and other fibers are much more dynamic than a skeleton. They can grow and shrink within seconds, allowing the cell to respond quickly to a changing environment.

The cytoskeleton is required for cell growth, division, structural differentiation, and motility. During normal cell growth, the fibers of the cytoskeleton define cell shape, but during cell division, these same fibers rearrange to make two specialized machines, one that organizes and segregates the already duplicated chromosomes into two identical sets, and another that helps the parent cell divide into two. During differentiation, when a cell becomes specialized, fibers of the cytoskeleton take on different arrangements to reorganize the cell and give it a shape that helps it carry out its designated function. The cytoskeleton also makes a cell motile, so it can move to wherever it is needed. The cytoskeleton of a white blood cell, for example, helps it travel to a site of infection. When cells become dysplastic during cancerous transformation, the cytoskeleton becomes disordered. Moreover, the process of metastasis, in which a cancer cell wanders away from its primary location to establish a tumor elsewhere, is an example of cell motility made possible in part by the cytoskeleton. Learning how such motions are controlled is an important part of cancer research, and preventing unwanted cell motions is one avenue for cancer treatment.

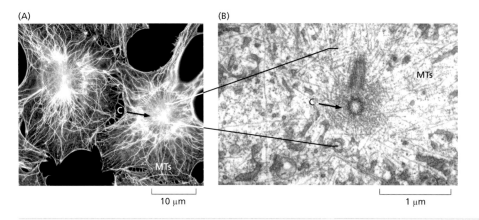

FIGURE 3.31 Micrographs of centrosomes and the microtubules that grow from them. (A) Two mammalian cells, imaged in a fluorescence microscope, with their nuclei stained blue-green and their microtubules (MTs) stained yellow. The black arrow points to a centrosome (C), which is lying below the nucleus, not in it. (B) An electron micrograph of a centrosome. It contains two centrioles, which are cylindrical structures that can initiate the growth of a cilium or flagellum. One centriole is surrounded by ill-structured material that can nucleate microtubule growth, which is the function of the centrosome. Microtubules are made by polymerization of the protein tubulin; they help to control the cell's shape and motility. (A, from Torsten Wittmann, The Scripps Research Institute, The Nikon Small World Competition, 2003. B, from Kent McDonald, University of California, Berkeley.)

Microtubules are thin but stiff cytoskeletal fibers.

Microtubules are one type of cytoskeletal fiber. They are made by assembling many copies of a single protein, called tubulin, into a hollow, cylindrical fiber. The tubular structure of a microtubule gives it rigidity, which allows these fibers to provide the cell with a robust framework. Microtubules often extend radially from the centrosome, a cluster of protein complexes situated beside the nucleus (FIGURE 3.31). The primary functions of the centrosome are to initiate the growth and regulate the number of microtubules. Microtubules extending from a centrosome can push or pull on other cellular structures, like mitochondria, the ER, vesicles derived from the ER or Golgi, and even the plasma membrane. Thus, they can adjust the position of organelles within the cytoplasm. One way they do this is by serving as tracks for a remarkable kind of ATPase that uses the energy stored in ATP to walk along a microtubule, pulling a cargo, such as a vesicle or a piece of ER, as they go. These enzymes are called microtubule-dependent motors, and human cells make about 50 different kinds of them.

Another important cargo of microtubules and their associated motors is the cell's DNA. When a cell is getting ready to divide, the centrosomes duplicate. The pair of centrosomes then reorganizes the cell's microtubules to make the cellular machine that will organize and segregate the chromosomes, as we will describe in more detail in Chapter 4.

How do you suppose the growth of microtubules can alter cell shape?

SIDE QUESTION 3.16

Microfilaments are thin and flexible cytoskeletal fibers.

Microfilaments form a thin, gel-like network immediately inside the plasma membrane. This gel helps to define the membrane's shape (FIGURE 3.32A, arrow). The growth of this gel can push the plasma membrane outward, helping a cell to advance in a given direction. Microfilaments can also form tight bundles just inside the plasma membrane. These force the membrane to project outward from the rest of the cytoplasm, forming a filopodium (Latin for fibrous foot) (FIGURE 3.32B). Microfilaments also occur in bundles that run through the cytoplasm. These clusters of fibers can bind motor enzymes, analogous to the motors bound by microtubules. The motors in microfilament bundles can push individual microfilaments

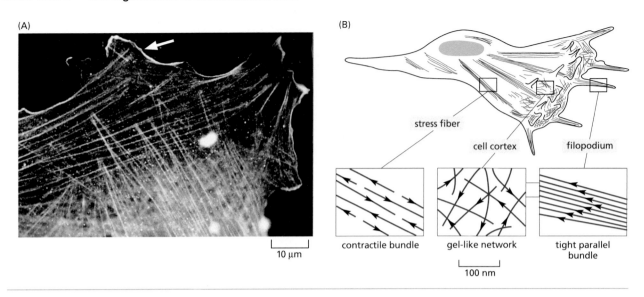

FIGURE 3.32 Microfilaments are important for cell shape and motion. (A) Fluorescence light micrograph of a cell whose microfilaments have been labeled by binding of a fluorescent dye to the protein actin, from which microfilaments are made. There are microfilaments just under the plasma membrane, the region called the cell cortex (arrow). Other microfilaments form fiber bundles deeper in the cytoplasm. (B) Diagram of a cell's microfilaments. The fiber bundles deep in the cytoplasm are called stress fibers. The microfilaments in the cell cortex form a gel-like network that helps to define cell shape. In some cells, there are slender projections from the cell surface, called filopodia, that contain tightly bundled microfilaments. (A, courtesy of Grenham Ireland. B, adapted from Alberts B, Johnson AD, Lewis J et al. [2015] Molecular Biology of the Cell, 6th ed. Garland Science.)

over one another, forcing the bundle to contract. This action builds up tension in the cell and exerts forces on any extracellular materials bound to the outer surface of the plasma membrane. This action is part of the machinery by which microfilaments help cells move; they pull the body of the cell along, following its advancing membrane. After microtubules have separated the duplicated chromosomes, microfilaments form a belt around the cell's equator and pull it tight. This is a major part of the machinery that helps a cell to divide. Just as normal cell movements use microfilaments and microtubules, the initial movements of metastatic cancer cells depend on these same cytoskeletal fibers. Cancer biologists are therefore seeking ways to block the activity of the cytoskeleton to interfere with metastasis.

Intermediate filaments are stable and tough cytoskeletal fibers.

Intermediate filaments are thicker than microfilaments but thinner than microtubules, which is the origin of their name. They are comparatively stable and tough. Commonly, they support a particular cell shape that allows some cellular structures to remain in place as other structures move around (FIGURE 3.33). Intermediate filaments are prevalent in skin cells, where they provide the tough framework needed to protect the rest of the body from injury and to prevent water loss.

To help you relate the many cellular structures just described, FIGURE 3.34 presents an electron micrograph of a slice cut through a mammalian cell grown in the laboratory. Some of the many cytoplasmic structures mentioned earlier are labeled.

REGULATING A CELL'S MANY PROCESSES

Most cellular processes must be controlled—that is, accomplished at the right time, in the right place, and to the right extent. We know that cellular

SIDE QUESTION 3.17

Some human genetic diseases involve mutations in the genes that encode the proteins that form intermediate filaments in skin. What you would expect to be the symptoms of a patient who carries a mutation that makes his intermediate filaments floppy and weak?

(A)   (B)

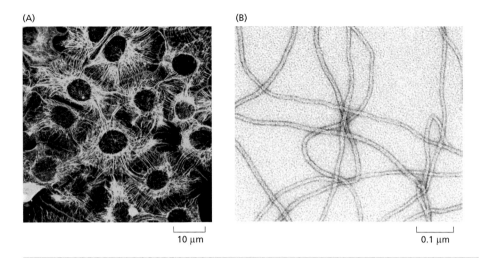

|——| |——|
10 µm   0.1 µm

FIGURE 3.33 **Intermediate filaments are the third component of the cytoskeleton; they are important for cell strength.** (A) A light micrograph in which a fluorescent dye has been attached to a protein subunit of an intermediate filament protein. The organization of intermediate filaments is established by motors pulling them along microtubules. (B) An electron micrograph of some purified intermediate filaments. They are of constant diameter (~10 nm) and considerable length. They are flexible but not easily stretched. (A, courtesy of Kathleen Green and Evangeline Amargo. B, courtesy of Roy Quinlan.)

control is necessary because its loss leads to cancerous transformation. This poses the question: How are cellular processes regulated so they contribute to a cell's proper function and to its ability to work toward the health of the body as a whole? As you can imagine, this is a big subject. Here we look at two important aspects of it: the way activities of proteins are turned up or down and the way cells receive signals from elsewhere in the body.

Many proteins can be activated or inactivated by adding a phosphate group to their surface.

In describing proteins, we emphasized that the structure of a protein defines its function. The most important part of this definition is the sequence of amino acids in the polypeptide chain, which in turn defines

(A)   (B)

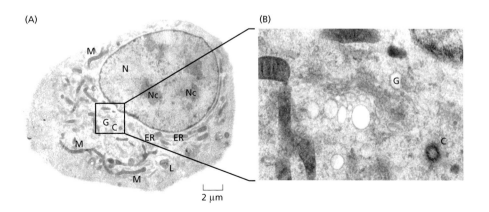

|——|
2 µm

FIGURE 3.34 **Overview of mammalian cell structure as seen by electron microscopy.** (A) Slice cut from a mammalian cell grown in the laboratory, showing its prominent nucleus (N) with condensed chromatin lying near the nuclear periphery. The nuclear envelope separates the cytoplasm from the nucleoplasm, which contains four darkly staining nucleoli (Nc), where ribosomes are being made. This slice of cytoplasm contains many mitochondria (M), a Golgi apparatus (G), a lysosome (L), and many components of the endoplasmic reticulum (ER), which is studded with ribosomes. Near the nucleus is the centrosome (C), though at this low magnification, the microtubules that radiate from it are not visible. (B) The region boxed in part A is shown at higher magnification. Now the membranes of the Golgi are clear, though the microtubules radiating from the centrosome are still not visible. To view these fibers, see Figure 3.31B, which is shown at still higher magnification. (Courtesy of Mark Ladinsky, California Institute of Technology.)

(A)

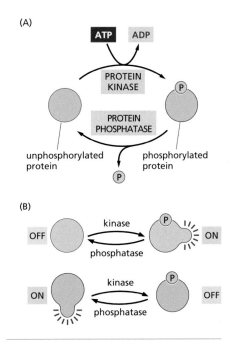

(B)

FIGURE 3.35 **Reactions that put a phosphate onto a protein or take it off.** (A) A protein (green circle at left) is about to have a phosphate added, via the process called phosphorylation. An enzyme called a protein kinase catalyzes this reaction, converting ATP to ADP and leaving the protein with a phosphate added to its surface. The phosphate is negatively charged. These charges, added to the protein's surface, alter its structure and therefore its function. A protein phosphatase is an enzyme that cuts a phosphate off from a protein, returning the protein to its unphosphorylated state. (B) Two distinct proteins are portrayed, one pale blue and the other green. Each can be phosphorylated and dephosphorylated, in each case causing a change in the protein's structure (and therefore its function). In one case, the phosphorylation turns the protein on; in the other case, it turns the protein off. Cells contain many proteins that display these kinds of regulation.

SIDE QUESTION 3.18

Describe in your own words why it is important to have a phosphatase to counter the action of a protein kinase.

the way that protein folds. Once a protein has been made and folded, however, its structure can be modified by reactions called **post-translational modifications** (**PTMs**). These are chemical reactions in which the structure of a protein is altered by the addition of some new molecular piece. You have already heard about one PTM: the addition of sugars to ER-derived proteins as they pass through the Golgi. These PTMs help determine how membrane traffic will direct each membrane protein to its site of action. Another PTM, in fact the most common one, is the addition of a phosphate group, the same cluster of atoms we saw in the structure of ATP and in the sugar–phosphate backbone of nucleic acids.

Phosphates are added to proteins by enzymes called **protein kinases**. These enzymes bind both a molecule of ATP and another protein, the one to which the phosphate will be added. The bound protein is called the substrate of the protein kinase. The kinase catalyzes a reaction that cuts the terminal phosphate from ATP and attaches it to an amino acid of the substrate protein (FIGURE 3.35). The importance of this reaction lies in the fact that phosphate carries negative electric charges. When these charges are added to an amino acid, they change the distribution of electric charge in that part of the protein's surface. This alters protein shape. A change in shape means a change in function. A protein kinase can therefore alter the action of any protein it **phosphorylates**—that is, attaches a phosphate group to. Interestingly, some phosphorylations make proteins function faster, while others slow a protein's function down (Figure 3.35B). Thus, protein kinases are regulators of protein action, but learning whether a protein kinase activates or suppresses a given substrate protein's function requires detailed knowledge of both the protein kinase itself and its protein substrate. About 30% of all human proteins are phosphorylated at some time in the course of their action, and there are about 500 different protein kinases to do these many control jobs. In Chapter 6, you will learn that genes encoding protein kinases are often altered by the mutations that change a cell's behavior, making it cancerous.

An equally important aspect of cellular control is that for every reaction that turns a process on, one or more processes turn it off. This makes sense, because if activation of an enzyme were permanent, the cell wouldn't be able to stop the reaction catalyzed by that enzyme. For every protein kinase that changes the activity of a protein by adding a phosphate, there is another enzyme, known as a **protein phosphatase**, that cuts the added phosphate off, returning the protein to its initial state (Figure 3.35). Thus, a balance of protein activation and inactivation modulates many cellular processes.

Many cellular processes are regulated by signals from outside the cell.

As described earlier, some proteins in the plasma membrane, and even some that reside in cytoplasm, are designed to work as receptors for information-carrying molecules secreted by other cells (see Figure 3.19). Once a signaling molecule is bound, its receptor causes significant changes in the cell by initiating a chain of chemical reactions that lead to the modification of multiple proteins, often by PTMs (FIGURE 3.36A). Depending on just what signal is received, a cell can respond in any of several ways. Even more remarkably, many human cells contain multiple types of receptors, each specialized to bind and respond to a different signaling molecule. These many receptors can act on the cell in combination (FIGURE 3.36B). For example, multiple types of signals and their receptors can influence a cell's decision whether to grow and divide or to differentiate. Another kind of signal can induce a cell to destroy itself,

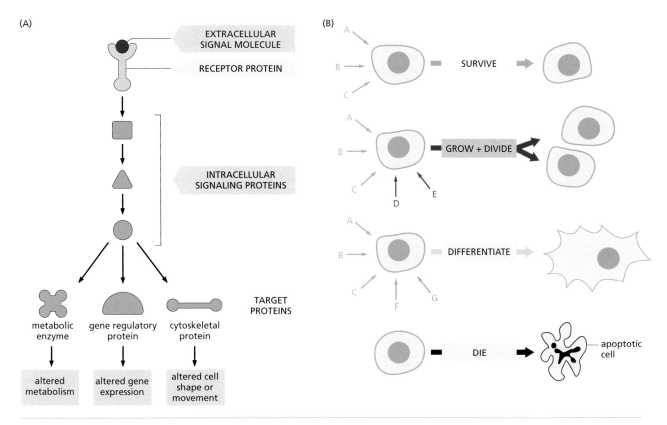

FIGURE 3.36 Some membrane proteins are receptors for signals carried by molecules floating in the extracellular space. (A) A signaling molecule (red circle) has bound its receptor (green) and initiated a cascade of steps to cause changes in cell behavior. This kind of cascade is a common feature of the pathways by which cells receive and convey information. Such cascades will come up many times as we look at the details of cancer, because pathways like this are often changed when a cell is transformed to display cancerous behavior. (B) Each cell in the human body has multiple receptors so it can receive multiple different signals at the same time. Diagrammed here is a set of seven signals (lettered A–G) that can be interpreted by the same cell. Depending on the combination of signals received, the cell will behave in drastically different ways. (Adapted from Alberts B, Johnson AD, Lewis J et al. [2015] Molecular Biology of the Cell, 6th ed. Garland Science.)

a process called apoptosis (Greek for falling off). In fact, many human cells need to receive certain signals simply to stay alive. Understanding the cell's range of signals and the interpretations of those signals is an important part of cancer biology. For example, carcinogenic transformation blocks the signal for apoptosis, preventing cancerous cells from self-destruction, even when they should commit suicide for the good of the body as a whole.

Most cellular processes occur in cancer cells much as they do in normal cells

The structures from which cancer cells are built and the mechanisms by which they function are all strikingly similar to the analogous structures and processes in normal cells. This statement applies to DNA replication, gene expression, chromosome segregation, membrane traffic, and even many aspects of metabolism. The differences between normal cells and cancerous cells are largely in the controls of these basic cellular processes. For example, cancerous cells will initiate DNA replication even when they have not been instructed to do so; this is why they divide more frequently than they should for the well-being of the body as a whole.

Cancer cells also make more mistakes in replicating and segregating their genes than do normal cells. This happens because cancer cells are deficient in some of the controls that keep both DNA replication and chromosome segregation accurate. By and large, however, the mechanisms by which cancer cells grow and divide are fundamentally the same as those of normal cells. This is one reason why cancer is so hard to treat effectively. Treatments that target cancer cell growth and division, which are designed to stop the disease from progressing, have essentially the same destructive effect on the normal cells needed for a person to survive.

SUMMARY

The complexity of a cell's structure is built up in a hierarchy. Atoms are organized to make small biological molecules, such as sugars, amino acids, nucleotides, and fats. These small molecules can then be assembled into polymers: polysaccharides, proteins, DNA or RNA, and the fatty parts of membranes. These polymers are, in turn, organized into macromolecular complexes that can perform complex functions. For example, a ribosome is a macromolecular complex that can read the nucleotide sequence of an mRNA and assemble amino acids in the right order to make a protein. Macromolecular complexes assemble to make organelles, like a mitochondrion, which can do even more complex tasks, such as metabolizing food and turning food energy into a form that is useful for the cell (ATP).

You know from previous chapters that cells are the subunits of the tissues that build organs; thus, all life is organized in a hierarchical fashion. Constructing this ordered hierarchy requires information, just as human-initiated building projects require plans. Much of the information for building a cell is stored in DNA. This information can be retrieved by transcription of RNA from specific segments of DNA that correspond to specific genes. Messenger RNA is translated on ribosomes into a sequence of amino acids that defines the structure of a particular protein. Because human DNA is long and contains billions of nucleotides, it contains enough information to make many thousands of different RNA molecules, many of which are translated into proteins.

Each cell makes several thousand different proteins in order to accomplish the many jobs that must be done for a cell to perform its functions for the body in which it resides. Some proteins are enzymes that act as catalysts to speed up specific chemical reactions. Other proteins bind to other macromolecules to make cellular structures, like the components of the cytoskeleton. Proteins also serve as receptors for signals that come from other cells. They serve as channels and pumps in membranes. The activity of many proteins can be regulated, for example, by phosphorylation, so a cell can balance the activity of its many processes for the good of the cell as a whole.

Cells are organized into compartments, so the many chemical reactions a cell must carry out are kept separate and can be controlled independently. The cell's genome is localized in the nucleus, which is responsible for genome maintenance, controlled gene expression, and DNA replication. Cytoplasm, the space outside the nucleus and inside the cell's plasma membrane, is divided into multiple subcompartments, such as the ER, the Golgi apparatus, lysosomes, and mitochondria. Each of these compartments is the site of specific and essential cellular functions.

Cytoplasmic organization is controlled by the cytoskeleton, which is made from three kinds of protein fibers with quite different properties. The cell uses the cytoskeleton to position its parts correctly. The cytoskeleton can also change both the shape and the position of a cell, helping it to function in the most useful way for the body as a whole.

Mutations in the DNA that encodes cellular proteins can lead to aberrant protein functions. For example, changing the action of a protein kinase that phosphorylates multiple proteins can lead to inappropriate activation or inactivation of many enzymes, making significant trouble for the cell. This is why mutations in certain genes, such as those that encode protein kinases, can cause dramatic changes in cell behavior, as is seen during cancerous transformation.

ESSENTIAL CONCEPTS

- Cells take in food to grow and divide. Thus, a cell is the basic unit of life.

- The components of cells are made by assembling small molecules into large ones: nucleotides assemble to form DNA and RNA, amino acids assemble to form proteins, lipids assemble to form membranes, and sugars assemble to form polysaccharides. These steps of assembly are one part of a structural hierarchy needed to build a cell and to assemble cells into tissues and organs.

- DNA is the cell's principal site for storing the information needed to build another cell. This information is written into the sequence of nucleotides along one strand of a DNA double helix.

- A gene is a unit of biological information written into the structure of DNA. Each gene is an instruction for how to make a particular RNA molecule. Many RNAs are, in turn, instructions for how to make particular proteins.

- Information stored in a messenger RNA molecule instructs ribosomes to assemble a specific sequence of amino acids into a polymer called a polypeptide.

- Each polypeptide folds in a pattern defined by its sequence, making a protein.

- Each protein performs some function for the cell that made it. Some proteins work as enzymes that control the speed of a chemical reaction; some build cellular structures, such as fibers that will help the cell to move.

- Cells are surrounded by a plasma membrane, which makes a boundary between the inside and outside of the cell.

- Membranes are lipid bilayers containing proteins that work as channels or pumps for the transport of ions, sugars, and other molecules of food, such as amino acids. Some membrane proteins are receptors for signals from outside the cell.

- The activity of proteins can be modulated by post-translational modifications, such as phosphorylation. For each such reaction, there is also a reaction that reverses the modification and thereby reverses the change in activity.

- Normal cells and cancer cells use fundamentally the same mechanisms to accomplish their basic functions. The differences between them are largely a matter of control over these basic functions.

KEY TERMS

ATP	cytosol	intermediate filament
base	endocytosis	lysosome
catalyst	endoplasmic reticulum (ER)	macromolecule
cell biology	enzyme	membrane traffic
chromatin	extracellular matrix (ECM)	messenger RNA (mRNA)
codon	exocytosis	metabolism
cytoplasm	genome	microfilament
cytoskeleton	Golgi apparatus	microtubule

mitochondrion	polypeptide	ribosome
nuclear envelope	post-translational modification (PTM)	RNA polymerase
nucleotide		transcription
organelle	promoter	transcription factor
phosphorylate	protein kinase	transcription terminator
polymer	protein phosphatase	translation
	receptors	

FURTHER READINGS

Watson JD & Crick FHC (1953) Molecular Structure of Nucleic Acids. *Nature* 171:737–738. The famous paper by Watson and Crick that describes a model for DNA structure is in a scientific journal called *Nature*. It's only about 1 page long!

There are several Websites that include quite clear and accurate descriptions of cellular processes. Their graphics and videos may help you to learn some of the cellular processes important for cancer. Two good ones are Vision Learning (http://www.visionlearning.com/) and Scitable (http://www.nature.com/scitable).

iBiology (http://www.ibiology.org/) is a Website containing many videos on various aspects of cell biology.

Useful texts on cell biology that will tell you much more about each subject covered here:

Alberts B, Johnson AD, Lewis J et al. (2015) Molecular Biology of the Cell, 6th ed. Garland Science.

Lodish H, Berk A, Kaiser CA et al. (2016) Molecular Cell Biology, 8th ed. W.H. Freeman.

Pollard T & Earnshaw WC (2007) Cell Biology, 2nd ed. Saunders.

QUESTIONS FOR FURTHER THOUGHT

1. You have seen chromatin, the DNA-containing structure in a cell's nucleus. Think about chromatin as part of a structural hierarchy, and sketch out its components from atoms to the cell nucleus.

2. Trace the flow of information from a gene to the action of the protein it encodes. Now try this for a protein in a mitochondrion that contributes to ATP synthesis.

3. Identify five cellular processes that require ATP as a source of energy.

4. Give three examples of the membranes a cell uses to separate its functions. Do you see any advantages that the cell achieves by expending the energy and macromolecules necessary to make those membranes?

5. Cells use polymers of proteins, that is, chains of many copies of the same protein added end-to-end, to make the fibers of the cytoskeleton. What are two advantages of making fibers in this way?

6. The addition of a phosphate to a protein's surface changes the protein's structure and function. What are some other ways in which a protein's function is altered?

7. To get images that show the inner parts of a mitochondrion, why can't you just magnify pictures taken with a light microscope?

ANSWERS TO SIDE QUESTIONS

3.1   When you are reading a book, the information is immediately available for your use. You could act on it right away, yet the book itself was unchanged by your having read it. This is a little like RNA polymerase reading a nucleotide sequence on DNA and transcribing it into RNA for use by the cell. The analogy is better, though, if you think of DNA as a book in a library that doesn't let its books circulate. If the information you want is complicated, you would probably want to take the book to a copy machine and make a copy of the sections that are important for you. Now you can take the copy with you to help you carry out the instructions faithfully. The copy you made in the library is quite like mRNA, made in the nucleus, and you are now the ribosome, able to translate the instruction into action.

3.2   The stability of mRNA is important because messages of any kind must be used and then set aside in order to be meaningful. If the message you get is always the same, it is both boring and uninformative. You got the message, and now you are ready to go on to other things. Because mRNA is usually NOT stable, its message naturally dies away, unless more of the same message is always being made and sent. Its disappearance allows the cell to go on with other business.

3.3   If a sequence of nucleotides in a gene is changed, there will commonly be a change in the amino acid that gets put into the protein during translation. This might make no difference, for example, if the substituted amino acid is chemically similar to the one that it replaces. But if the mutation changes the codon so a chemically different amino acid is inserted, it would probably have a serious effect on the structure of the protein and thus on its function. It is well known by geneticists that not all mutations have bad effects on the gene product. Still, some mutations do. A mutation in a promoter would not affect the structure of the gene product, but it might have a big effect on how much of the protein was made at any particular time. It might even inactivate the promoter, so the gene was no longer expressed.

3.4   If tight binding is required for a signal to be received and interpreted, the cell is assured that it won't be responding to the wrong signal. Given that there are likely to be multiple signals around, a cell wants to respond ONLY to the one that was made to activate that particular receptor. Other signals might bind to the same receptor weakly, but if tight binding is required, they won't convey information to the cell; only the right signal for a given receptor can convey a message.

3.5   Chemical reactions are the means by which cells make the molecules they need. All the molecules in a cell were either inherited from the mother cell, or transported into the cell like a food molecule, or made by the cell itself, using the chemical reactions that go on within its plasma membrane. The rates of all these reactions are controlled by enzymes. The rate of a reaction depends in part on how much of the starting materials are available, but it also depends on the activity of the enzyme that catalyzes the reaction. Often the same starting molecule (for example, a sugar from the diet) is broken down in different ways, depending on which enzyme is active at the moment. Thus, the molecules a cell makes from sugar depend heavily on the activities of the relevant enzymes. If all enzymes worked at the same rate, the cell would not be able to grow, divide, and respond to its environment.

3.6   Polysaccharides are, by definition, polymers of sugar. They store energy well because the many sugars they contain can be obtained by cutting the polymer up into simple sugars, which can then be broken down to release energy-rich molecules that help with ATP synthesis, providing the cell with the energy it needs. Breaking down a polymer is generally quicker and easier than making it.

3.7   It does *not* mean there is no genetic control on lipid and polysaccharide synthesis. Both these kinds of molecules are made by enzymes, and enzymes are proteins. Proteins are encoded by genes, so there is genetic control of the synthesis (and breakdown) of lipids and polysaccharides, although the control is indirect.

3.8   To cross the plasma membrane without a channel, a molecule must be sufficiently soluble in the lipids of the membrane bilayer. Thus, we could expect many lipids to be able to enter without a channel. In fact, both testosterone and estrogen (lipids like the one shown in Figure 3.16C) are sufficiently lipophilic that they can cross the membrane without a channel to help.

3.9   Most membrane proteins exist in a complicated environment. Part of their sequence runs through the hydrophobic (nonpolar) part of the membrane, the lipid bilayer. If the protein spans the membrane, then it also has two parts in an aqueous environment: one part in the cytoplasm and the other part in the extracellular fluid. One expects that their membrane-spanning part(s) will have hydrophobic (nonpolar) amino acids on their surfaces, so they can interact easily with the hydrophobic (nonpolar) parts of the lipids. The portions of the protein that are in either the cytosol or the extracellular medium will be much like proteins that inhabit those regions: polar (hydrophilic) amino acids will be on the protein's surface.

3.10   The central core of most cytosolic proteins is nonpolar, or hydrophobic. This is one of the factors that helps to stabilize their fold (Figure 3.11). Membrane-associated channel proteins are different from this general pattern as a result of the environment in which they live and work: the lipid layer of a membrane. As mentioned in the text, they have hydrophobic (nonpolar) amino acids on their surfaces, so they can interact well with the lipids in the membrane. Their hydrophilic (polar) core gives them an aqueous channel through which ions, sugars, etc., can pass. When compared with cytosolic proteins, it is as if they were made inside-out.

3.11   The enzyme would digest the proteins exposed on the surface of the cell. (Note that the enzyme can't get through the lipid bilayer, so it shouldn't affect proteins inside the cell.) Thus, the proteins that bind adjacent cells will be degraded, lose function, and allow the cells to fall apart. Interestingly, this is one of the ways that cells can become metastatic. They secrete proteases that do exactly this job, breaking the intercellular bonds and setting tissue cells loose.

3.12   You have seen that gene expression requires the synthesis of mRNA, which requires RNA polymerase to bind the right part of the cell's DNA. DNA binding by RNA polymerase is activated by promoters and the proteins that bind to them. If the DNA is folded so tightly that neither the promoter-binding proteins nor RNA polymerase can get at the relevant region of DNA, then no gene expression will occur.

3.13   Three important nuclear functions that are carried out by proteins are RNA synthesis (by RNA polymerase), regulation of gene expression (by transcription factors), and DNA replication (by DNA polymerase).

3.14   Lysosomes degrade organelles that the cell has identified as worn out. (Note that lysosomes do other things in association with endocytosis, as suggested in Figure 3.26B. We have not yet talked about these other roles of the lysosome.) So long as organelles are degraded before they fail to function properly, the analogy with preventive maintenance is very good.

3.15   The word currency comes from the Latin word *currens*, which means moving or traveling. Currency means money that is traveling around, being used in exchange. In that sense, ATP is a currency with which the cell pays for energy-requiring services. It is not a bartering system, because a membrane pump or a synthesizing enzyme doesn't really trade back anything directly for the energy it uses. It does, however, perform a job that is good for the society of the cell as a whole. In this way ATP is like a currency in a big economy, where one earns money doing a job (equivalent to digesting sugar and using it to power the conversion of ADP plus phosphate into ATP) and spends the money for things one needs (the right cytoplasmic environment, which is achieved by ion pumps and the syntheses necessary for making needed parts, etc.).

3.16   A microtubule is a polymer of the protein tubulin. The microtubule elongates by adding tubulin to one of its ends. If the polymer gets as long as the diameter of the cell, then it must either stop growing or push on the inner surface of the plasma membrane. In fact, the bonds that hold the tubulin molecules together in the microtubule are strong enough to keep on pulling in additional subunits and elongating the polymer, even when it must deform the plasma membrane to do so. This behavior of growing microtubules has been shown by several kinds of experiments.

3.17 The skin of these patients doesn't have the mechanical strength of normal skin, so even gentle rubbing can cause one layer of skin to peel from another, leading to a wound. Lymph flows in, and the outer surface of the skin forms blisters.

3.18 If a given protein kinase adds a phosphate to a particular protein and turns it on, the protein will stay turned on until something turns it off. If nothing turns it off, the enzyme will stay up-regulated and continue to do its job, even if that action is no longer needed. Clearly this could lead to trouble in the same way that a furnace that turned on and never turned off would make a house VERY uncomfortable. It is the balance of turning on and turning off that gives a cell the control it needs to behave correctly.

How normal cells reproduce and differentiate

4

CHAPTER 4

Now that you have a sense of cells as structures, we can look at the mechanisms that underlie a cell's abilities to grow, divide, and differentiate. It is the control of these mechanisms that goes wrong when cancer develops, so understanding the processes that will be perturbed by carcinogenic mutation is at the heart of understanding cancer itself. We start by taking a closer look at the cell cycle, with a focus on DNA replication and cell division. Then we examine the processes by which cells differentiate; that is, how they regulate the expression of genes whose products are essential for the functional specializations that occur when cells undertake specific jobs. Next, we look at control of cell behavior: how cells receive and interpret signals, such as an instruction to grow and divide versus an instruction to differentiate. Finally, we examine cellular processes for quality control, such as sensing and repairing damage to DNA. Cells are skillful not only at repair but also with processes that monitor the fidelity of key events, like DNA replication and chromosome segregation. If mistakes are made, cell cycle progression is slowed, allowing time to get things right. Remarkably, if adequate quality cannot be achieved, a cell can opt to kill itself, an extreme pathway of differentiation and behavior.

CELL GROWTH AND DIVISION CYCLES

CELL DIFFERENTIATION AS AN ALTERNATIVE TO CELL GROWTH AND DIVISION

SIGNALING BETWEEN CELLS

HOW CELLS COPE WITH MISTAKES

CHECKPOINTS ARREST CELL CYCLE PROGRESSION TO IMPROVE THE FIDELITY OF CELL REPRODUCTION

LEARNING GOALS

1. Diagram the process of DNA synthesis, showing how the idea of complementarity between bases that face each other across a double helix leads to the formation of two identical double helices.

2. Diagram and label the stages of a cell's growth and division cycle, including a description of what happens to the cell's DNA at each stage of the cycle.

3. Describe what happens at the level of gene expression when a cell produces specific proteins as it differentiates.

4. Outline the steps of cell reproduction and differentiation that occur in the wall of the intestine in order to keep that tissue in good working order.

5. Define the roles of stem cells in the maintenance of a tissue and in the production of blood.

6.   Specify two functions of the extracellular matrix.

7.   Diagram the steps in a cell's responses to a growth factor and relate these events to a cell's control over the onset of S phase

8.   Define a cell cycle checkpoint. Describe two checkpoints and explain how they help to maintain the integrity of the genome.

9.   Compare what happens to a normal chromosome during cell division with what happens to a chromosome that has suffered an unrepaired double-strand break.

10. Define apoptosis and state three reasons why this process is essential for normal human health.

CELL GROWTH AND DIVISION CYCLES

In Chapter 1, you learned that our bodies grow from a fertilized egg into the trillions of cells that comprise an adult human body by means of many cycles of growth and division. Now that you know something about the components of cells, we can look at the cell cycle in more detail. Since cell proliferation is a result of the growth and division cycle, the control of this cycle is an essential aspect of normal cell behavior. Changes in cell cycle control lead to the overproduction of cells characteristic of cancer.

The cell cycle includes distinct times of growth and division.

The cell cycle is divisible into segments, based on the biochemical processes that occur at those different times. Cells are the smallest they will ever be immediately following division, the time when two cells are born. Their most important jobs now are to recover from division and grow in size. During the time between divisions, called **interphase**, each cell must make enough parts to supply both of the cells that will form at the next division (FIGURE 4.1). The most important synthetic process a cell will undertake is the replication of its DNA, because that process makes a complete copy of the instructions that each of the daughter cells will need to thrive. (In this book, we will use the term DNA replication to refer

FIGURE 4.1 **The cell growth and division cycle.** The cell cycle is sometimes described in two parts, as shown on the inner circle of this diagram: interphase, which is the time of synthesis and growth, and mitosis, when duplicated chromosomes are segregated and the cell divides. A typical time for the whole cycle, for human cells growing in culture, is about 1 day. Interphase is divided into three phases: G1, S, and G2. In G1, a cell grows and samples its environment, looking for instructions. If it receives instructions to divide, the cell will "start", or pass its restriction point, and initiate DNA synthesis. DNA replication occurs during S phase (S stands for synthesis), taking 4–6 hours for completion in most human cells. After DNA replication is complete, the G2 phase begins, during which the cell checks its DNA for errors and synthesizes more of the RNAs and proteins it needs to make all the materials necessary for two complete cells. Once it has completed these tasks, the cell will initiate mitosis and segregate its chromosomes to achieve nuclear division. Then the cell divides in two by a process called cytokinesis.

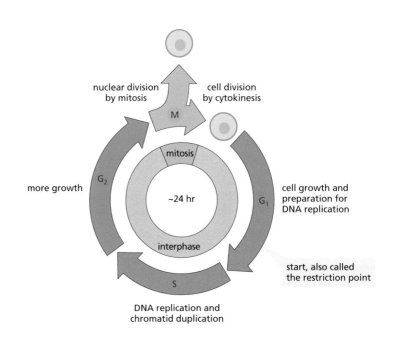

to the complete doubling of a cell's DNA. DNA synthesis is the biochemical process by which DNA is replicated.) The period of DNA synthesis, called the **S phase**, does not begin immediately after division; there is usually a gap, called **G1**, before S phase starts. During that gap, the cell uses its DNA to make many different mRNAs, which are then translated into proteins. Some of these proteins are enzymes that make more lipids, so the cell's membranes can grow. The cell also receives and interprets signals that are instructions about whether it should grow and divide again or differentiate into some specific cell type. Late in G1, the cell either leaves the cell cycle to specialize into a particular cell type or commits to another round of division. If the signals received by the cell promote division, the synthetic events necessary for cell duplication begin, so that decision time is often called start or sometimes the restriction point (Figure 4.1).

Much of our understanding of the cell cycle and its control has come from the study of cells in isolation from the body as a whole. This situation results simply from the practical difficulties of looking at living cells in detailed ways when they are part of an organ or tissue. Our understanding of S phase, G1, and start has come largely from observations and experiments on either mammalian cells grown in the laboratory, using methods called **cell culture**, or organisms that grow as single cells, like many yeasts. For more information about cell culture and what can be learned from it, see SIDEBAR 4.1.

SIDE QUESTION 4.1

The serum from a fetal calf works better for promoting the growth of cultured cells than serum from an adult cow. Why?

A CLOSER LOOK: Cell Culture

Ideally, one would always study a cell in its natural environment, but close examination of human cells within the human body is often impossible. Recently, some fancy methods have been developed that use powerful lasers to look inside an animal's body and follow the behavior of cells that have been labeled with fluorescent dyes. These methods are now used on mice, but most people wouldn't put up with that kind of treatment. Because of the difficulties of visualizing events inside animals, it has been common for many years to examine the behavior of live human cells by taking a biopsy and dispersing its cells for close examination while they grow in a fluid that mimics the nurturing environment of blood. However, this procedure poses four questions: how do you get the cells in the first place, how do you keep them healthy, how do you look at them, and how relevant are the events you see in culture to the corresponding events in an animal's body?

Obtaining cells for study in culture. A biopsy yields a sample of tissue in which the cells are still attached to one another. These links can be broken by brief treatment with a **proteinase**, that is, an enzyme that degrades proteins; this activity damages the intercellular links. If the treatment is brief, the plasma membrane protects the proteins inside the cell from damage, and the cells of the tissue simply disperse.

Keeping cultured cells healthy. The suspension of dispersed cells can then be transferred to a dish or bottle that contains a liquid with many of the components found in blood. This is called a cell **culture medium** (Figure 1). It is an aqueous solution containing all the minerals that cells need: sodium, potassium, chloride, magnesium, calcium, phosphorous, iron, zinc, etc. The medium must also be neither too acidic nor too basic, so it contains chemicals called buffers that help to regulate its acidity. It must also be

kept at the body temperature and contain sources of food energy, like sugars, and all the amino acids that the cells cannot make for themselves (the essential amino acids), plus all the vitamins. In addition, the medium contains growth factors that will bind to receptors on the cells' surfaces and induce them to grow and divide. The simplest and most economical way to do this is to add a derivative of blood, called **serum**, which is blood minus the blood cells and the proteins that allow blood to clot. The most effective serum for cell culture comes from fetal animals; serum from fetal calves is the most commonly used.

These conditions are good for growing human cells, but they are so nourishing that they will also grow many other organisms, like bacteria, fungi, or almost anything that happens to come along. Thus, one must handle the medium and the cells with **sterile technique**, which means using both extreme

SIDEBAR 4.1

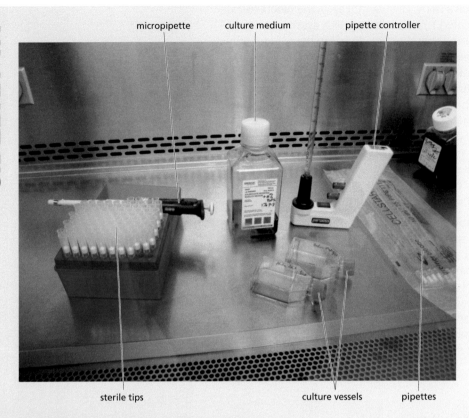

micropipette   culture medium   pipette controller

sterile tips   culture vessels   pipettes

Figure 1 Tools for cell culture. The bottle (center) contains culture medium for nourishing the cells. The two small flasks with orange tops are culture vessels for growing cells and keeping them sterile. The micropipette is designed to measure and transfer small amounts of liquid by use of sterile tips. The pipettes are sterile plastic tubes for transferring larger amounts of liquid, using the pipette controller that has a motor-driven pump to help with accurate liquid transfers. The culture vessels have been loaded with culture medium and are ready to receive an injection of cells, which will be measured with the micropipette. These tools are all in a cabinet that is both sterile and contained. There is a glass window (not visible) that lies between these tools and the scientist doing the work, so no bacteria, viruses, or other pathogens can be transferred either to or from the culture.

cleanliness and special chambers to make sure that unwanted organisms are kept away from the cells of interest (Figure 1).

In a medium like the one described, several kinds of human cells will grow and divide quite well, reproducing once in every 12–56 hours. This is not fast compared with a bacterium, which under the same conditions will reproduce 3 times per hour, but it means that, in a week, 100 originally explanted cells will have grown to a population of about half a million. Thus, cell culture is a convenient way to make many cells for detailed study.

This kind of culture is called a primary culture, because the cells have just come from the host organism. Primary cells will duplicate as many as 70 times, but after

a while, they stop their cell cycles in G1 and won't divide anymore. Cancer cells, on the other hand, will keep on growing in culture indefinitely, so long as you provide them with a suitable medium. There is an irony here: cancer cells, which are lethal to their host, are themselves immortal, meaning that they can keep on growing forever. A great deal of cell culture work has gone into efforts to understand the various ways in which cells change when they become immortal, and this subject will come up again later, after you have learned more about the details of cancerous transformation. Now we can say simply that culturing cancer cells is comparatively easy and is often done. Indeed, more research on cultured human cells has been done with the cancer cell line called HeLa

(short for Henrietta Lacks) than on any other single cell type. This is because HeLa cells, derived from an aggressive cervical carcinoma, are robust and easy to grow in large numbers, even though the woman from whom they came is now long dead!

Studying cultured cells. Cultured cells can be detached from the surface on which they are growing by the same treatment that dispersed them from their original tissue: a brief treatment with proteinase. Now they can be washed and transferred to any chamber that is convenient for further study, such as the dish in **Figure 2**, which has a window on its bottom made of thin, flat glass. The medium in this dish will keep the cells growing while you study them with a microscope. Several

(A)
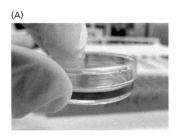
(B)
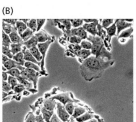
(C)
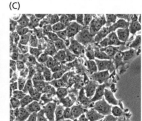

Figure 2 **Growing cells in culture.** Many human cells can be grown in dishes or other flat-bottomed vessels, where they attach to the surface and flatten out. (A) A small culture dish filled with cell culture medium (red). Scientists wear gloves to protect themselves from any viruses or other pathogens that might be growing with the cells of interest and to protect the culture from any pathogens on the scientist's hand. (B) A micrograph of epithelial cells growing in culture. Red arrow points to the circular nucleus of one cell. (C) A micrograph of the same culture, 24 hours later. Now there are enough cells to fill most of the surface of the dish, so most cells are in immediate contact with their neighbors; this is known as confluence. Under these conditions, normal cells stop their cycles of growth and division. Cancer cells, on the other hand, keep on growing, even after they have become confluent.

SIDEBAR 4.1

pictures of cells shown in this book were taken that way (for example, Movies 1.1 and 1.2; Figures 3.21A, 3.27A, 3.30, 3.31A, 3.32A, and 3.33A; and Figures 4 and 5 in Sidebar 3.1). Similar chambers can be used to keep cells alive and growing for days, so it is possible to follow long-term cell behaviors, like differentiation. One important use of this method is to compare the behavior of normal and cancerous cells, as shown in Movie 4.1.

One must remember, though, that cells from an organism did not start their lives attached to glass; they will therefore grow in a more natural way if you put them on a support that is more like their normal environment. For this purpose, scientists are now using samples of connective tissue or even just protein polymers from connective tissue added to their culture medium; this mixture provides a more natural environment for the cells to continue growing and migrating in three dimensions, perhaps differentiating into the form and function they had when they were in their host.

Many biochemical studies have been done on cultured cells by growing a few million cells, breaking them open to make a homogeneous mixture, and then using chemical methods to purify one or a few proteins, nucleic acids, or small molecules from this cell homogenate. Proteins that are abundant, like some membrane proteins, histones from chromatin, and some cytoskeletal proteins, can all be studied this way. Even enzymes can be purified from cultured cells, if the methods are good.

DNA synthesis requires the concerted action of many proteins.

The essential feature of DNA synthesis is that each of the two existing DNA strands serves as a template that defines the nucleotide sequence of a new strand (FIGURE 4.2). The result is two identical double-helical structures, called DNA duplexes, one for each of the daughter cells that will form at the end of interphase. Synthesis begins when proteins designed for the job achieve a local separation of DNA's two strands at specific sites called **replication origins** (FIGURE 4.3A). Special enzymes called DNA polymerases then bind to the single-stranded DNA to initiate the replication process (the enzymes are not shown in this figure). The two ends of a single site of strand separation are both suitable for DNA synthesis. As DNA polymerases bind to these places, they become known as replication forks. The DNA polymerases at the two forks now move in opposite directions, unwinding the double helix and synthesizing new DNA on each of the old strands that are present at each fork. At every nucleotide along an old strand, **DNA polymerase** adds to the end of a new strand the one and only nucleotide that is complementary to the nucleotide at that point on the old strand, assuring that the sequence of the new DNA strand is exactly complementary to the sequence of the old DNA strand (FIGURE 4.3B).

FIGURE 4.2 DNA replication uses the pairing of complementary nucleotides to build two identical double helices. The upper part of this figure shows a typical DNA duplex. The orange bands represent the sugar–phosphate backbones, and the complementary bases are paired in the space between the backbones. Lower down, the diagram shows this duplex unwound, as occurs during DNA synthesis, separating the two old strands and making space for DNA polymerase to get to work. Each old strand serves as a template to define the choice of nucleotide to be added at every position as two new strands are made. This duplication process assures that each old strand is now coupled with a perfectly complementary new strand, making two identical DNA duplexes.

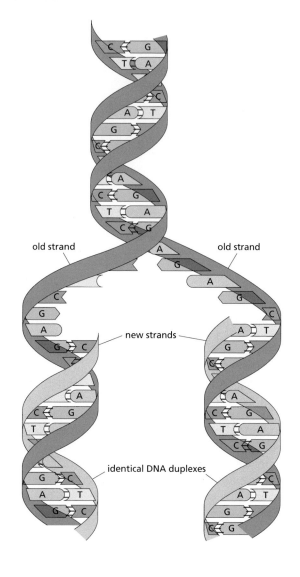

FIGURE 4.3 Each origin of replication produces two replication forks that propagate in opposite directions. (A) Replication begins by a local unwinding of the DNA double helix, producing a loop where DNA polymerases can bind (these enzymes are not shown). (B) Replication continues on each old strand as DNA polymerase selects the free nucleotide from solution that is complementary to the nucleotide at that position on the old strand. The enzyme attaches the new nucleotide to the growing new strand by forming a link at the sugar–phosphate backbone; then the enzyme moves on to repeat the process. In short, the old strand acts as a template that defines the sequence of the new strand made by DNA polymerase.

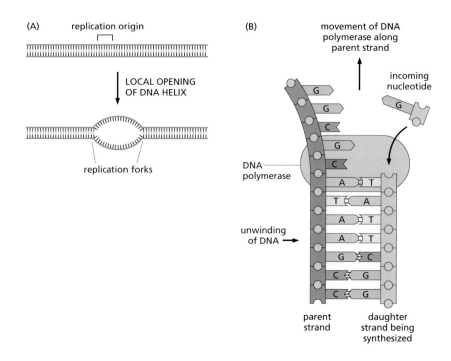

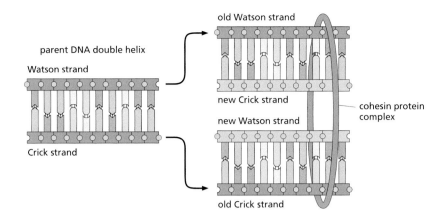

FIGURE 4.4 **The essential ideas of DNA replication.** The two strands of a DNA double helix (here designated as Watson and Crick) are complementary. DNA replication is achieved by splitting the parent strands apart and synthesizing two new strands, each complementary to the old strand with which it is paired. As DNA is replicated, the identical DNA duplexes, called sisters, become tied together by loops of a large protein complex called cohesin. The cohesin complexes keep the sister DNA duplexes together until the time of chromosome separation, as the cell prepares for division.

To make this easier to describe, people often give names to the two strands of a DNA duplex: Watson and Crick, in honor of the scientists who figured out DNA structure. The old Watson strand templates a new Crick strand, and the old Crick strand templates a new Watson strand, leading to the formation of two, identical Watson–Crick DNA duplexes (FIGURE 4.4). These identical DNA duplexes are called sisters. As DNA synthesis proceeds, the cell couples the sisters to one another by a protein complex called cohesin. This process makes a single mechanical unit out of the two identical copies of the cell's genetic information, a situation that becomes important both at the end of interphase, when the cell gets ready to divide, and if either duplex becomes damaged and must be repaired.

A single origin of DNA synthesis could initiate the duplication of an entire DNA double helix, but human chromosomes are so long that, under those conditions, S phase would take a very long time. It is therefore common for each chromosome to have multiples origins of replication. This strategy works well to shorten S phase, because when the forks that are propagating from neighboring origins bump into one another, the DNA polymerases simply fall off, an enzyme called ligase connects the adjacent ends of sugar–phosphate backbones, and the job is done. Given the many origins of replication that participate in DNA synthesis, S phase in a human cell takes only 4–6 hours, during which billions of bases are faithfully replicated. Now the cell contains two complete copies of all 46 chromosomes, one copy for each daughter cell that will form at the time of division.

Cells must duplicate many things to be ready for division.

The replication of DNA is of paramount importance in a cell's preparation for division, but many other macromolecules and organelles must also be synthesized to provide all that is necessary for the formation of two viable cells. The cell doubles its number of mitochondria by adding proteins and lipids to existing mitochondria, making them bigger. Each mitochondrion then pinches in two, making enough copies of this organelle for two daughter cells. The amount of ER doubles, and the number of ribosomes does likewise. Most cells keep only one Golgi apparatus throughout interphase, but just before cell division, this organelle fragments into many little vesicles that distribute approximately equally to the daughter cells. Each newly formed cell then forms its own Golgi apparatus by reassembling these vesicles during G1. The many synthetic events required to make two complete cells take longer than the replication of DNA, so after S phase is complete, there is another gap in time, called **G2**, before cell division begins (Figure 4.1).

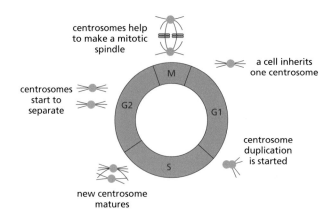

FIGURE 4.5 Duplication of the centrosome, another essential event in the cell's growth and division cycle. Each cell inherits one centrosome when it is born by cell division. Mitogenic signals stimulate not only the replication of DNA but also the duplication of the centrosome. When the new centrosome first forms, it cannot yet initiate microtubules, but later in the cell cycle it acquires this ability. Later in G2, the two centrosomes move slightly apart. As the cell enters M phase, each centrosome can initiate many microtubules, helping to form the mitotic spindle for chromosome segregation.

SIDE QUESTION 4.2

From what you have learned so far, what are all the things a cell must synthesize to make a sufficient dowry for its daughter cells?

Another essential activity in preparing for division is to make the proteins that will form two cellular machines, one that can organize and segregate the duplicated chromosomes and another that can divide the cell into two. To accomplish these jobs, the organelle called the **centrosome** gets special attention. This microtubule organizer is initially present in only one copy, but about the beginning of S phase, the centrosome starts to double (FIGURE 4.5). The centrosome is not surrounded by a membrane, and it contains no nucleic acid. Its duplication is accomplished simply by assembling the proteins that make up its structure into a new and functionally equivalent unit. Exactly how centrosomes duplicate is not yet known, but by the end of S phase, the cell contains two of these organelles. Initially, they lie close to one another, continuing to function as a single organizer of the cell's microtubules. At the start of cell division, the centrosomes separate (Figure 4.5). This separation is important because the centrosomes play key roles in organizing the microtubule-based machine that will segregate the chromosomes.

Separation of the already duplicated chromosomes and centrosomes is accomplished by a microtubule-based machine called the mitotic spindle.

The mitotic spindle is the cellular machine responsible for DNA segregation in a process called **mitosis**, a word that comes from the Greek word for thread. This word was chosen when the phenomenon was first seen in microscopes, during the latter part of the nineteenth century. It is appropriate because when the cell gets ready to divide, each piece of DNA becomes compacted into a threadlike structure that is just visible in a light microscope. The word spindle comes from the overall shape of the collection of microtubules that the cell builds to separate the duplicated chromosomes; at one stage during its formation and function, this structure looks quite like the spindles that were used many years ago for twisting strands of wool into yarn.

Mitosis involves cooperation between structures from both nucleus (the chromosomes) and cytoplasm (the centrosomes and microtubules, plus many proteins that bind to them). These components can work together because at the time of spindle formation, the nuclear envelope disperses, allowing the microtubules that grow from the cytoplasmic centrosomes to extend into the region that used to be the nucleus (FIGURE 4.6A–C). Some of these growing microtubules bind to specialized sites on each chromosome and exert forces on them, pushing and pulling to align the chromosome at the plane that lies midway between the two centrosomes (Figure 4.6D). For a dynamic view of this process, see MOVIE 4.2.

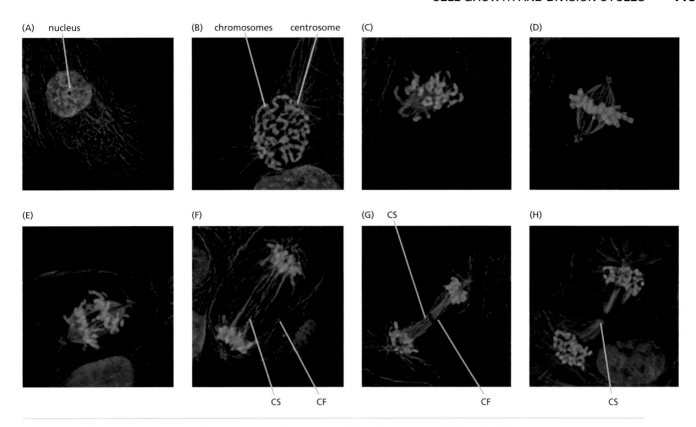

FIGURE 4.6 **Structural changes in a cell as it goes through mitosis.** In all these images, nuclear DNA is stained blue and microtubules are stained red. (A) This cell is in the very early stages of mitosis. The chromosomes are condensing into visible fibers, but the nuclear envelope is still present, separating the nucleus from the cytoplasm. (B) This cell is in the late prophase stage of mitosis. Its chromosomes are distinct objects, and the microtubules are rearranging. The nuclear envelope has not yet dispersed, but since it is not stained, it is invisible in this image. (C) The nuclear envelope has broken down, so the microtubules can interact with the chromosomes and begin to organize them for segregation. (D) By this time, the organization of the chromosomes is complete: all chromosomes are attached to microtubules, and they have been moved to the equator of the spindle, which acts like the starting line of a race. This is the stage at which the chromosome segregation machinery looks like a spindle. (E) Early in the segregation process, the chromosomes separate and move toward the ends of the spindle. (F) Later in the segregation process, the central spindle (CS) elongates to increase the separation between the two sets of chromosome before the new nuclei form. Meanwhile, the cleavage furrow (CF, which doesn't show because it is not stained) has begun to pinch in at the cell equator. (G) With the two sets of chromosomes far apart, the nuclear envelopes are beginning to re-form. The microtubules of the central spindle have been bundled together by the cleavage furrow. (H) Finally, the nuclear envelope is re-forming, the chromosomes can soon return to their interphase condition, and cytokinesis has pinched the cell into two distinct objects. (Courtesy of Bill Earnshaw.)

The microtubule attachment sites on chromosomes are called kinetochores, because they control the motions of the chromosomes as they are aligned and segregated (the word kinetic comes from the Greek word for motion). There is one kinetochore on each of the two sister chromosomes (the ones made by DNA replication and bound together by cohesin). The kinetochores on a pair of sisters attach to microtubules that grow from the centrosomes at opposite ends of the spindle, so once the spindle is fully formed and the chromosomes are organized, sister chromosomes are attached to sister centrosomes. Now, a very specific proteinase becomes active and cuts the cohesin molecules into pieces. This allows the two copies of each chromosome to separate. The microtubules of the spindle pull each chromosome toward the centrosome to which it is attached, segregating the two identical sets of genes into two distinct pools, one at each end of the spindle. The spindle then elongates, moving the two sets of chromosomes as far apart as possible within the parent cell (FIGURE 4.6D–F).

Once the two sets of chromosomes are sufficiently separated, the cell initiates a division process called **cytokinesis**. This action is accomplished by another fiber-based machine, a ring of microfilaments that forms just

SIDE QUESTION 4.3

What do you suppose would be the effect of activating the above-mentioned proteinase too early, so the identical copies of the chromosomes separate before the spindle is fully organized?

under the plasma membrane at the region where the chromosomes were placed when the spindle was fully formed (Figure 4.6D). This ring of microfilaments is attached to the inner surface of the plasma membrane. The microfilaments bind a motor enzyme that makes the filaments slide over one another, pulling in on the plasma membrane and pinching the cell in two (FIGURE 4.6G,H). Again, for a dynamic view of this process in a cultured cell, watch Movie 4.2.

CELL DIFFERENTIATION AS AN ALTERNATIVE TO CELL GROWTH AND DIVISION

Most of the cells in an adult human body are not involved in proliferation and tissue replacement; they have specific jobs to do. For a cell to do its job, it must contain special gene products, usually proteins, that will either build the structures the job requires or serve as enzymes to increase the rates of chemical reactions necessary for that function. You can pick any human tissue you find interesting, and there you will find cells that have become specialized to perform one or more functions important for the body as a whole.

When a cell in the G1 phase of the cell cycle receives signals that tell it to differentiate, it leaves the cell cycle and begins to express the proteins necessary for its new set of functions (FIGURE 4.7). The expression of particular genes whose products are needed for cell differentiation requires that RNA polymerase bind to and become active at exactly the right places on the cell's DNA; that is, the parts that contains the nucleotide sequences that encode the needed proteins.

The regulatory part of a gene binds multiple proteins, each with a role in the control of gene expression.

Each gene can be thought of in two distinct parts: a structural part that contains the sequences that will be transcribed to make the necessary RNA, and a regulatory part that controls both when that gene will be expressed and how much expression will occur (FIGURE 4.8). The regula-

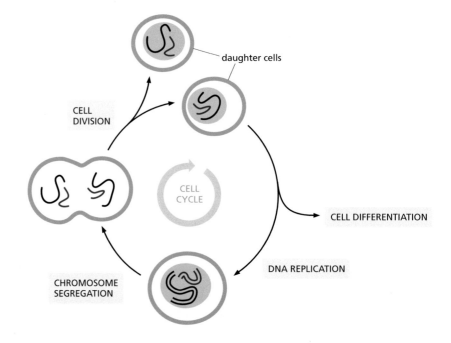

FIGURE 4.7 **A cell must leave the growth and division cycle to allow differentiation.** A cell that has been going through cycles of growth and division can leave the cycle and differentiate. Commonly, differentiation is an alternative to growth and division. Usually, a cell that has begun differentiation will not return to the cell cycle. Cancer cells, on the other hand, revert from the differentiated state when they become dysplastic. Commonly, they then begin to divide too frequently. (Adapted from Alberts B, Johnson AD, Lewis J et al [2015] Molecular Biology of the Cell, 6th ed. Garland Science.)

daughter cells

CELL DIVISION

CELL CYCLE

CELL DIFFERENTIATION

DNA REPLICATION

CHROMOSOME SEGREGATION

tory part of each gene contains a **promoter**, which is a sequence of DNA that can bind RNA polymerase. In human cells, the promoter commonly contains repeats of the nucleotides T and A, so it is called the **TATA box** (Figure 4.8). Because promoters are sequences in DNA, they are present throughout the life of the cell. This poses the question, How does a cell regulate the level of expression of specific genes? The answer lies in the fact that RNA polymerase will bind a TATA box in an effective way only when it has formed a complex with several other proteins, some called the TATA binding protein (TBP) and others called **transcription factors**, abbreviated TFs (Figure 4.8). Transcription factors are proteins that help to assemble an active RNA polymerase on the right region of DNA; that is, at the promoter for a gene that should now be expressed.

To help RNA polymerase recognize and read out the right genes at the right times, the cell contains a large number of different TFs. Multiple transcription factors (TFIIA, TFIIB, etc., in Figure 4.8) must all assemble with the TATA binding protein to recruit RNA polymerase and initiate transcription. Indeed, as this figure implies, there are other factors that contribute to the regulation of gene expression. For a cartoon description of the transcription process, see MOVIE 4.3. Some TFs have an affinity for a particular sequence of nucleotides near the promoter that occurs only once or perhaps a few times in the cell's entire genome. These are the TFs that help to turn on specific genes. They bind these particular DNA sequences so tightly that they force the DNA duplex to bend at the point of binding (FIGURE 4.9). This change in DNA structure increases the affinity of that stretch of DNA for the general transcription factors (TFIIA, TFIIB, etc.), promoting formation of the complex seen in Figure 4.8. These sequence-specific TFs help to control where RNA polymerase will act and thus which genes are expressed. For example, to start S phase, cells must express the genes that encode DNA polymerase and other proteins necessary for successful DNA replication, such as the cohesins. There is a specific set of TFs that regulates the expression of genes needed for DNA synthesis. You will see below that an important aspect of cancerous transformation is the mutation of genes that control these transcription factors, leading to a loss in proper regulation of the genes they control. The result is the onset of S phase, even when the cell has not received instructions to divide.

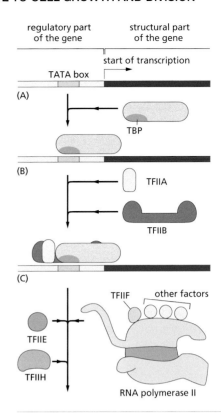

FIGURE 4.8 Each gene has a structural part that can be transcribed to make RNA and a regulatory part that controls both when and how often the gene is expressed. (A) The regulatory part of a gene contains the promoter. Part of the promoter is a TATA box (green), which is near the starting point for transcription of the structural part of the gene (red). (B) A TATA binding protein (TBP) binds the TATA box. Now multiple transcription factors can bind to the TATA binding protein and make it ready to bind RNA polymerase. Here, TFIIA and TFIIB (described in speech as tee eff 2 A or B) join the complex. (C) Now additional transcription factors bind the subunits of RNA polymerase, guided by other factors, which ultimately make a functional complex that can initiate transcription. (Adapted from Alberts B, Johnson AD, Lewis J et al [2015] Molecular Biology of the Cell, 6th ed. Garland Science.)

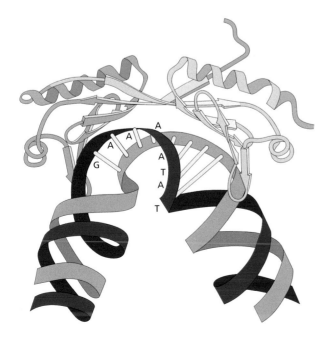

FIGURE 4.9 Diagram of a way transcription factors can interact with DNA. Some transcription factors have the ability to bind to only a particular sequence of DNA. With this ability they can guide a full transcription-initiating complex to the regulatory region of a specific gene, starting RNA synthesis at this place but nowhere else. These transcription factors (green and blue) bind DNA of the right sequence (red strand) very tightly, so tightly that they can bend or deform the double helix. Structural change is one of the ways that DNA is modified to become able to bind RNA polymerase and initiate transcription. (Adapted from Alberts B, Johnson AD, Lewis J et al [2015] Molecular Biology of the Cell, 6th ed. Garland Science.)

SIDE QUESTION 4.4

Transcription factors are proteins, which means there must be genes that encode the mRNAs that lead ribosomes to make them. Would you expect that the genes for transcription factors are always expressed, or is their expression likely to be regulated? How would you regulate the expression of a gene expression regulator?

SIDE QUESTION 4.5

Use Figure 4.7 to make an argument in words that excessive cell growth is likely to be associated with a dysplastic state.

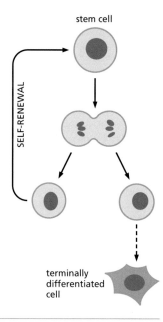

stem cell

SELF-RENEWAL

terminally differentiated cell

FIGURE 4.10 The division process characteristic of a stem cell. The parent cell goes through mitosis and cytokinesis, producing daughter cells that often look identical. However, these cells are different in ways important for their behavior. One daughter cell continues to serve as a stem cell and divide again; the other daughter cell will differentiate. As part of differentiation, it can serve temporarily as a transit amplifying cell, dividing several times to produce many identical copies of itself. (Adapted from Alberts B, Johnson AD, Lewis J et al [2015] Molecular Biology of the Cell, 6th ed. Garland Science.)

Cell differentiation is usually an alternative to cycles of cell growth and division.

During the G1 phase of the growth and division cycle, a cell makes a decision about whether to divide or to differentiate. If a cell commits to growth and division, it will stop that cycle only if it gets into serious trouble; for example, if it runs out of food or becomes damaged. If a cell commits to differentiation, on the other hand, it will continue down that path. As indicated in Figure 4.7, the decision of whether to cycle or differentiate occurs at a control point just before start. That point is important for cancer, so we will return to this issue as we get into more details about a cell's decision-making processes.

The dichotomy between differentiation and cell cycling is one of the key criteria by which doctors characterize tumors. If tumor cells are well differentiated and still look like the normal cells of that tissue, then the tumor is likely to be benign. If the cells differ slightly from those of the original tissue, the dysplastic state you read about in Chapter 2, then the tumor poses a greater risk to the patient. If the tumor cells have lost most of the features typical for their tissue of origin, that is, they are anaplastic, then the tumor is very likely to be dangerous. These cells can pass through growth and division cycles more quickly than dysplastic or normally differentiated cells, so the tumor they form will probably grow faster. Moreover, their loss of differentiation probably means that they are no longer making the membrane proteins important for cell–cell adhesion, increasing the chances that they could become metastatic. Thus, the basic biology of cell differentiation is important for cancer at the levels of both carcinogenesis and cancer diagnosis.

The stem cell strategy.

Since most tissues are made of differentiated cells that have left the cell cycle, we are faced with the question, How do tissues make the cells they need to replace old cells that are no longer working properly, the preventive maintenance described earlier? In many cases, specialized cells form through the asymmetric division of stem cells. Such divisions produce one daughter cell that retains its stem cell character and can divide again, while the other daughter cell differentiates (FIGURE 4.10). When many differentiated cells are required, the differentiating cells will often divide several times before they begin to express the proteins needed for their differentiated functions. While dividing to produce many identical daughter cells, these cells are called **transit amplifying cells**. To make this important issue more concrete, we will now look at two examples of tissues where cancer is relatively common and where stem cells normally give rise to transit amplifying cells that go on to differentiate and perform useful functions.

The intestine uses the division of stem cells to stay healthy.

The intestine, sometimes called the gut, is the tube that connects your stomach with your anus. As food moves down the intestine, the nourishment is taken from it and passes out through the wall of the intestine into surrounding blood vessels. At its bottom, the intestine contains only things your body can't use; these are the feces that you pass with a bowel movement. The wall of the intestine is built from cells that have differentiated to do two important jobs: (1) transport the products of digestion into your blood, so the resulting nourishment can be carried to all parts of your body, and (2) survive in the face of all the digestive enzymes that are breaking down the food you ate into small molecules that can be transported into your body. To achieve these goals, the intestine wall includes stem cells that are continuously growing and dividing asymmetrically, producing more stem cells and other cells that will differentiate. To make the many cells that are needed, the wall also includes transit amplifying cells whose symmetric cell divisions help the few cells produced by stem cell division to become sufficiently numerous that the whole system keeps on working. It is probably because of all these cell divisions that cancers of the intestine are relatively common. The more complete description given below shows how cell growth and differentiation work together to form a tissue that performs an essential adult body function. It also provides a context within which you will be able to understand the comparatively frequent emergence of intestinal cancers.

The differentiated cells of the intestine wall are epithelial cells. Most of them are specialized to be pumping machines that can transport nutrients (amino acids, sugars, etc.) from the inside of the intestine, called its **lumen**, to its outside, where blood vessels are nearby (FIGURE 4.11). The nuclei of these cells express genes that encode the membrane proteins to

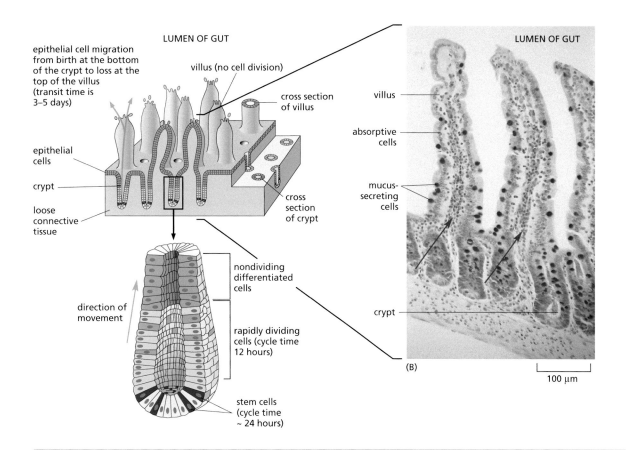

FIGURE 4.11 **Organization of cells in the intestine wall.** (A) Diagram of the intestine wall showing epithelial cell birth, differentiation, and migration. The surface of the intestine wall is lined with finger-shaped clusters of cells, each called a villus. These project into the lumen of the intestine, greatly increasing the surface area of the gut wall and thereby the rate at which digested food can be transported into the blood. The transporting cells are called absorptive cells, because their pumping action helps the body to absorb food. Crypts are places where stem cells are dividing to produce new stem cells and transit amplifying cells, which divide several times and then differentiate. The divisions of the transit amplifying cells push existing cells out along the villus, sloughing them off from the tip. These cells are then digested, and their parts are recycled into the body. The lower diagram displays a crypt in more detail, showing the positions of stem cells (red), transit amplifying cells (yellow), and differentiated cells (dark blue), which are pushed out of the crypt onto a villus. (B) Low-magnification light micrograph of a slice through three villi of the intestinal wall that separates the lumen of the gut from the space where there are blood vessels (red arrows). In one villus, an absorptive cell and a crypt are labeled. The cells stained red are making and secreting mucus, which helps to protect the absorptive cells from the digestive enzymes in the gut lumen. (Adapted from Alberts B et al (2015) Molecular Biology of the Cell, 6th Edition. Garland Science: New York.)

do these pumping jobs. However, pumping is hard work, consuming lots of ATP, so these cells also make many mitochondria. In addition, the job of these pumping cells is dangerous. The gut lumen contains digestive enzymes that break down the already chewed-up food into its component small molecules: amino acid from proteins, sugars from polysaccharides, nucleotides from nucleic acids, etc. These molecules can be pumped across the plasma membranes of the cells that line the intestine and get into the bloodstream. However, the enzymes that digest food pose a threat to the integrity of the intestine; they can also degrade the macromolecules of the epithelial cells that line the intestine's wall. Were this to happen, there would be a leak in this important tube, allowing the contents of the gut lumen into the body: bad news. One defense against this dangerous situation is that some of the cells in the gut wall differentiate not as pumps but to make mucus, an aqueous solution of charged polysaccharides. These cells secrete mucus onto the surface of the gut epithelium. The mucus helps to protect the pumping cells from the digestive enzymes.

The front-line position of intestinal cells makes them a bit like soldiers under fire: short tours of duty make sense. Otherwise they might lose their ability to do their job well. To achieve this goal, the wall of the gut is organized so each epithelial cell is born by cell division in specialized in pits within the intestine wall (Figure 4.11). These pits, which are called crypts, contains stem cells that can grow and divide, producing one daughter cell that will differentiate and another that will continue as a stem cell, ready to grow and divide again. To provide enough cells to line the intestine's wall, the differentiating cells divide a few more times, serving as transit amplifying cells. After those divisions, the resulting cells stop dividing and differentiate completely by expressing the genes that encode all the proteins needed for their pumping or mucus-secreting functions.

The continued divisions of the stem and transit amplifying cells push the differentiated cells away from the crypts where they were born. Gradually, the differentiated cells move outward along fingerlike projections of cells called villi (the singular is *villus*, the Latin word for shaggy hair) that project into the intestine's lumen. After 3–7 days, depending on the part of the intestine being considered, the epithelial cells reach the tip of the villus, where they die, are sloughed off, and join the food that is still in the intestine. These cells are digested like food, and their parts are recycled into your body as small molecules. Just as with the epithelium of skin, mentioned in Chapter 1, Figure 1.4, the cells of the intestinal epithelium differentiate to perform a function, do their job, and then die for the good of the body as a whole. This is another example of cell division occurring at a rate that just balances the rate of cell death. Such balances help to keep the body in good running order. Biologists call this kind of balance **homeostasis**, which means a set of processes that maintains an appropriate state.

An aspect of cancer mentioned in Chapter 1 is that the cells of most tumors lose at least part of their differentiated structure as they grow and divide in an uncontrolled way. Many of the cells in a tumor behave like transit amplifying cells; they are not sufficiently well differentiated to do the jobs of the tissues from which they grew. Tumors also contain some cells that behave like stem cells. These were first identified experimentally as cells important for the ability of pieces cut from a tumor to initiate a new tumor after their transplantation to a different experimental animal. The role of these cancer stem cells is still under investigation, but they may be important for the development of metastases.

SIDE QUESTION 4.6

There is a disease that runs in families called familial polyposis. Its symptoms are that many members of these families develop large numbers of polyps on the lining of their intestines. From what you know about cells lining the intestine, what cellular events are likely to lead to this condition? What are likely properties of the protein product of the gene mutated in this familial disease?

It is clear, though, that if the rate of division of either the intestinal stem cells or the transit amplifying cells were too high, too many cells would be produced and a neoplasm would form. It is this kind of overgrowth that makes the polyps discussed in the section on screening for intestinal cancers by colonoscopy. Moreover, if the resulting cells really lose control of gene expression, they can lose their ability to differentiate and to stick to their neighbors. The resulting cells can become cancerous and even metastatic, a hallmark of serious cancerous transformation.

Blood contains many kinds of cells that grow and differentiate from a single type of stem cell.

The most common cells in blood are the **erythrocytes**, which is Latin for red cells. These cells are filled with many copies of the protein hemoglobin, which binds oxygen. This property enables erythrocytes to carry oxygen from your lungs to the tissues of your body. Erythrocytes also help to carry the waste product of sugar breakdown (carbon dioxide) from the tissues of your body back to the lungs, so you can exhale it. In addition to erythrocytes, blood also contains diverse cell types that are collectively called **leukocytes** (Latin for white cells). All these cells arise through distinct processes of differentiation from a single kind of multipotent stem cell that lives in bone marrow. The asymmetric division of these stem cells produces cells committed to differentiate into various kinds of blood cells, so they are called multipotent progenitors (FIGURE 4.12). Daughters of these cells become progenitors either of lymphocytes or of a more diverse group of cells, called the myeloid line. This figure shows many cell types, each with its own lineage of formation and its own set of functions in the body. For example, the platelets that break off from megakaryocytes are important for blood clotting. The cells colored blue are involved in the body's immune response, helping to recognize and kill invading organisms that could make you sick. These cells are also important in the body's defenses against cancer, so in Chapter 9 there is a much fuller description of the immune system. In Chapters 8 and 10 you will learn how doctors are trying to harness the immune system for several kinds of cancer therapy. For our current purposes, use this diagram simply as an illustration of the ways in which one or a few stem cell types can form a variety of cells with distinct functions.

Each type of blood cell results from a distinct pathway of differentiation in which certain genes are expressed through the regulated synthesis of the mRNAs they encode and of the proteins that are the products of those genes. In short, the differentiation of these cells is brought about by controlled expression of the right genes. We will come back to a few of the white blood cell types when we talk about different kinds of leukemia.

Blood-forming stem cells and the progenitor cells they produce are unusual, because they can give rise to so many distinct cell types. Moreover, they continue to perform this function throughout your life, providing the cells necessary to keep your blood healthy. Most of the necessary cell divisions are supplied by transit amplifying cells in each of these differentiation pathways. For example, it requires about a million cell divisions every second, just to provide the new erythrocytes that are needed to replace the old ones that die through natural processes. The production of leukocytes, one the other hand, is more dependent on conditions in your body. When you have an infection, the production of some kinds of leukocytes increases markedly. Both of these cases are examples of the body's ability to produce the right numbers of each cell type to maintain tissue homeostasis and keep the body in good working order.

Cancers of blood cells are called leukemias. There are several different leukemias, depending on which kind of progenitor has lost control of its growth. However, all leukemias are, in a sense, metastatic. Why?

SIDE QUESTION 4.7

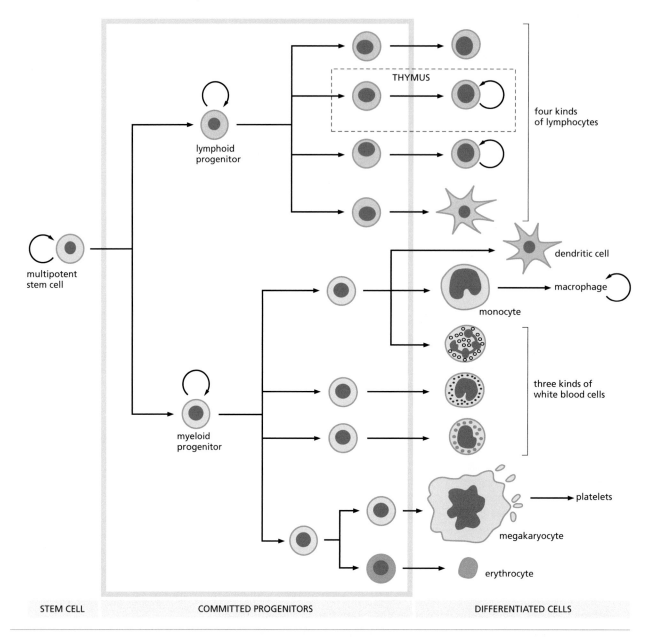

THYMUS

four kinds
of lymphocytes

lymphoid
progenitor

multipotent
stem cell

dendritic cell

macrophage

monocyte

three kinds of
white blood cells

myeloid
progenitor

platelets

megakaryocyte

erythrocyte

STEM CELL   COMMITTED PROGENITORS   DIFFERENTIATED CELLS

FIGURE 4.12 Many types of blood cells are produced by a single, multipotent stem cell. Blood contains many kinds of cells in addition to the familiar red blood cells (erythrocytes) that carry oxygen. Erythrocytes and the many kinds of white blood cells, also called leukocytes, all form from a single kind of stem cell that lives in bone marrow. There are four kinds of lymphocytes (blue), which are the major cell types of the immune system, described in Chapter 9. Macrophages and other kinds of white blood cells that develop from the myeloid progenitor are important in the phagocytosis of pathogens. The other cell types shown are not discussed here, but this diagram gives you a sense of the complexity of cell differentiation in blood formation. (Adapted from Alberts B, Johnson AD, Lewis J et al [2015] Molecular Biology of the Cell, 6th ed. Garland Science.)

The lifespan of differentiated cells varies from one tissue to the next.

Not all cell differentiation leads quickly to cell death. Many of the body's cells begin to differentiate as a particular organ forms during embryonic life. Thereafter, these cells can survive in a differentiated state for a long time. Red blood cells live about 4 months, the cells that line blood vessels last for months to years, bone cells live as much as 30 years, and many nerve cells, called neurons, survive for most of your adult life. In long-lived cells, the expression of genes for differentiated function is perpetually turned on, while the genes for DNA synthesis and cell division are kept off. Thus, most neurons cannot grow and divide, but they contain a rep-

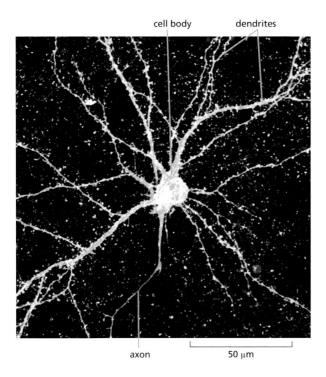

cell body   dendrites

axon   50 μm

FIGURE 4.13 **A well-differentiated neuron showing the extent of morphological specialization in this cell type.** The cell body contains the cell's nucleus and a lot of both ER and Golgi apparatuses. The dendrites receive signals *from* other neurons and the axon carries a signal *to* other neurons, activating them to send more signals. (Courtesy of Nick Galati, Colorado Medical School, Denver.)

ertoire of proteins to build the structures needed for the specific jobs of nerve cells, such as extending and maintaining the long arms that a neuron needs to transmit signals between cells that are a considerable distance apart (FIGURE 4.13).

The fact that neurons and some other well-differentiated cells have largely withdrawn from the growth and division cycle may contribute to the fact that cancers rarely arise from these cells. Most brain tumors are cancers formed from nonneuronal cell types in the brain. The cancers of neurons that do form generally arise in neuroblasts, which are cells in the process of differentiating to become neurons These immature neurons are common in fetuses and children, so this is the time of life when such cancers will more commonly arise.

One aspect of differentiation is the synthesis of proteins and polysaccharides that comprise the extracellular matrix.

Most cells make and secrete proteins that contribute to their environment in the tissue where they live. These proteins and associated polysaccharides make up the **extracellular matrix** (**ECM**) mentioned in Chapter 3. Bone is a tissue in which cells form a particularly hard extracellular matrix, thanks to the ability of protein fibers, made and secreted by the bone cells, to precipitate minerals (calcium and phosphorus). As a more general example of ECM, epithelial cells make a matrix of proteins and sugars that assembles into a layer called the **basal lamina**. This layer separates the epithelium from neighboring tissues (FIGURE 4.14). The basal lamina is sufficiently compact that most cells cannot pass through it. Epithelial cells are therefore segregated from the rest of the body and stay close to their place of origin, even if the bonds between neighboring cells should weaken. It follows that the cells of a metastatic carcinoma, which do leave the epithelium and wander throughout their host, must be able to penetrate the basal lamina, tough as it is. Metastatic cells acquire this ability by mutating in ways that let them secrete increased quantities of enzymes called **matrix metalloproteinases** (**MMPs**). These enzymes

You are a cancer biologist who has just invented a powerful inhibitor of MMPs. You propose to use this chemical as a drug to block metastasis. What arguments would you use to convince doctors that your chemical is likely to be a valuable treatment for their cancer patients?

SIDE QUESTION 4.8

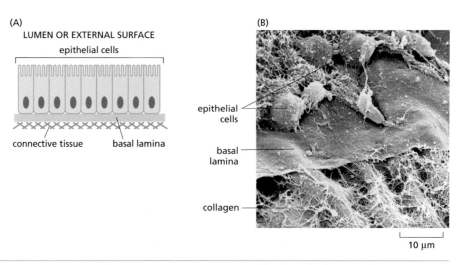

FIGURE 4.14 A specialized area of extracellular matrix called the basal lamina separates epithelial cells from underlying tissues. (A) Diagram of the relationships between a layer of epithelial cells, a basal lamina, and the connective tissue that underlies the epithelium. (B) A scanning electron micrograph of some epithelial cells, the lamina that lies beneath them, and the underlying connective tissue, which in this case is composed largely of fibers made from the protein collagen. Collagen makes strong fibers that help to define the mechanical properties of both the skin that lies beneath the outermost epithelium and the tendons that connect muscles to bones. (B, courtesy of Robert Trelstad.)

can break down the proteins of the lamina and allow the epithelial cells to pass through.

The secretion of MMPs by cancerous cells enhances several kinds of cellular misbehavior. ECM often binds signaling molecules, such as growth-enhancing factors. When MMPs are secreted by a cancerous cell, they degrade fibers of the ECM and release these growth factors, which act on the cancer cells and promote excessive cell division.

SIGNALING BETWEEN CELLS

At several places in the preceding account of cell growth and differentiation, you have read that cells receive signals that help to direct their choices among several possible behaviors. In this section, you will read more about these signals: what they are in chemical terms, how a cell perceives signals, and how it responds to them. In recent years, scientists have learned a large amount about this subject, but the material presented here is trimmed down to the facts and properties that are essential to build your understanding of cancer.

Cells sense aspects of their external environment with membrane-bound receptors.

One example of a cell's ability to receive information from its environment is the ability of skin cells to sense whether they are properly attached to their neighbors. These cues come in part from membrane proteins that form bonds between neighboring cells. If contact is broken, for example, by a wound, the cells on the edge of the cut sense that their neighbors are missing. They respond by migrating in the direction of the cut and making new proteins. These actions help the skin cells to fill the gap left by the wound (FIGURE 4.15). The ability of our bodies to heal wounds is an important subject, so it has received a lot of attention from both doctors and scientists. Cell biologists have invented systems with which to study wound healing in cell culture (FIGURE 4.16). The cultured cells

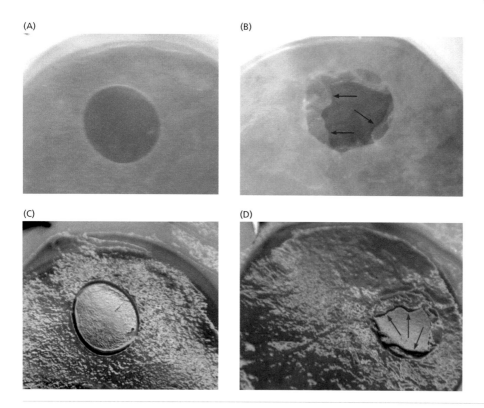

FIGURE 4.15 **Wound healing includes the migration of epithelial cells to replace missing tissue.** A wound was made in a layer of epithelial cells by using a biopsy tool to cut a hole 6 mm in diameter. Immediately after the wound was made (A, C), the margins of the hole are sharp. Two days later (B, D), the hole has begun to fill in as a result of inward migration by epithelial cells from the edge of the wound (red arrows). Images in A and B differ from those in C and D by the orientation of the lighting. (From Carrier P, Deschambeault A, Talbot M et al. [2008] *Invest Ophthalmol Visual Sci* 49:1376–1385. With permission from Association for Research in Vision and Ophthalmology.)

respond to a wound by becoming motile and migrating into the empty space. The lack of neighboring cells, brought about by the wounding process, generates signals that activate cell motility, which helps to repair the damage.

Cells in the neighborhood of a wound also initiate cycles of growth and division; they are instructed to do so both by loss of their tightly bound neighbors and by signals that come from cells at the site of damage. The resulting cell divisions generate new cells to replace the ones that were

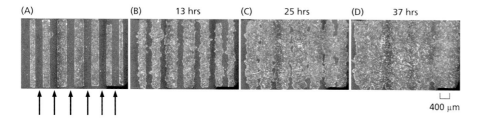

FIGURE 4.16 **The use of cultured cells to study cell behavior in response to a wound.** The bottom of a cell culture dish was covered with a mask made from several thin, parallel plastic strips. Cells were plated onto the dish and allowed to grow and fill the spaces between the strips. The plastic strips were then pulled off the bottom of the dish, leaving parallel bars of empty surface where no cells were present (arrows). The culture dish was returned to the incubator, and the cells were left to respond to the empty spaces for the times shown in parts B–D. During this time, many cells migrated into the regions of the dish where no cells were previously present, showing a behavior like that of epithelial cells migrating into a wounded region to help repair damage, as shown in Figure 4.15. Close examination of the cells in each of these images showed that there were few if any cell divisions, so this filling in of empty space was accomplished more by cell motion than by cell proliferation. (From Poujade M, Grasland-Mongrain E, Hertzog A et al. [2007] *Proc Natl Acad Sci USA* 104:15988–15993. With permission from National Academy of Sciences.)

SIDE QUESTION 4.9

If a wound induces nearby cells to migrate into the damaged area, shouldn't that heal the wound? What is the advantage for wound healing of other signals that induce cell growth and division?

harmed by the wound. Moreover, a big cut will induce changes in the behavior of cells that line nearby blood vessels; they grow and divide and then organize new blood vessels that can provide better blood circulation for the cells in the damaged region. If the wound has caused an infection, that is, if bacteria have gotten into the body, then there is also a mobilization of the body's leukocytes, which can consume bacteria and keep the body sterile. Thus, even when cells look stable and fixed in place, they are capable of moving, growing, and dividing when their environment changes in ways that require it.

All these changes are particularly interesting in the study of cancer, because they show that several aspects of cancer cell behavior (growth and division, migration, and induction of new blood vessels) are all aspects of normal cell behavior. The fact that they can occur for the benefit of the body when responding to a wound is an excellent example of why we must think of cancer as a loss of proper regulation rather than the emergence of novel properties. It is not novel behavior of cancer cells that makes them dangerous; most of the things that cancer cells do are perfectly normal. Cancers are dangerous because their component cells are doing things without proper regulation and without a sense of what is good for the organism as a whole.

Many factors help to regulate cell behavior.

Circulating blood contains more than just the food and oxygen that are important for making ATP and synthesizing macromolecules. It also carries hormones with familiar names, like adrenalin, insulin, estrogen, and testosterone, plus many additional signaling molecules. These molecules carry instructions that tell particular cells how to behave in the interests of the body as a whole. Many of these signaling molecules are not called hormones, though they have analogous functions. For example, **mitogens** are molecules that encourage cells to grow and divide. There are also some signaling molecules that suppress cell growth, so most cells are continuously assessing the concentrations of multiple factors in their environment and then making decisions on the basis of the relative strengths of the various signals, just as you saw in Figure 3.36B. Many such signals are perceived by membrane-associated receptors. Several mitogen receptors are specific examples of such membrane-bound receptors. These receptors are protein kinases whose enzyme activity is turned on by the mitogen they bind. Once they are activated, the receptors initiate a cascade of events that leads to the synthesis of specific mRNA molecules, whose protein products will now initiate important events in cell cycle progression (FIGURE 4.17).

In Figure 4.17, the mitogen is a protein called **epidermal growth factor** or **EGF**. The receptor to which it binds is **epidermal growth factor receptor** or **EGF-R**. Activated EGF-R, that is, the receptor bound to its activating signal, becomes a protein kinase that phosphorylates itself, changing its structure so it can bind other proteins, leading to the activation of a protein called **Ras**. Active Ras leads to the activation of another protein kinase called mitogen-activated protein kinase or **MAP kinase**. This enzyme phosphorylates and thereby activates a protein that enters the nucleus where it works as a sequence-specific transcription factor and turns on expression of a gene called *myc* (biologists commonly use italic type to name a gene). The protein product of this gene, called **Myc** (capital M and no italic type to indicate a protein), is itself a transcription factor, so by activating the expression of the *myc* gene, the expression of many other genes is modified. Several of the genes regulated by Myc are involved in the control of cell cycle. Indeed, mutations that lead to inappropriate synthesis of Myc, or the synthesis of a hyperactive form of Myc,

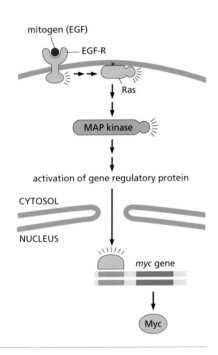

FIGURE 4.17 Diagram of intracellular signaling by a membrane-bound receptor. The signal, here a mitogen called epidermal growth factor (EGF), binds to a specific membrane-bound receptor, here epidermal growth factor receptor (EGF-R). Mitogen binding activates the receptor (red lines) to become an enzyme that can now activate other proteins, such as Ras. Ras activates the protein kinase called MAP kinase, setting off a cascade of activities that results in the expression of several genes whose products (here the protein Myc) are important for initiation of DNA synthesis. In this way, an external signal is transduced into the activation of important cellular processes that alter cell behavior.

are among the most common mutations in human cancer. So now your understanding of the inner workings of cells is reaching the point that you can learn about some of the specific molecular changes involved in cancerous transformation.

One way to think about a cell's receptors is to imagine them as sentinels who are scanning the area around a city, looking for an event that needs attention. If something noteworthy happens, like the start of a fire, the sentinel gives the alarm and sets in motion a chain of events that was organized in advance, such as alerting the fire department. A cell is similarly organized, because the necessary kinases and transcription factors are already present in the cell, but they are inactive until a signal is received. Responses like this require significant advance planning and a willingness to spend energy on information transfer and interpretation. Clearly, cells are adept at this kind of advance planning.

The signals passed among cells are commonly grouped into three categories: long-range, short-range, and contact-dependent. Hormones and many growth factors are examples of long-range signaling; the signal is made by **endocrine** cells, which usually reside in a specialized organ called a gland, which is differentiated to make and secrete hormones (FIGURE 4.18). An example is the adrenal gland that resides atop your kidneys and controls the level of adrenaline in your blood. This signaling molecule circulates in the blood to reach all the cells in your body, stimulating them to behave in a way that will help your body respond to excitement or danger. Any cell with the right receptor can respond, but cells that lack the necessary receptor cannot.

The responses of cells to a wound, on the other hand, exemplify two different kinds of signaling. When contact between neighboring cells is broken, the cells sense that a neighbor is no longer there; this response depends on direct cell–cell interaction and is called contact-dependent signaling. When cells at the wound signal to nearby blood vessels that they need better circulation of blood, that short-range action is called **paracrine** signaling (Figure 4.18). Paracrine signals do not travel in the bloodstream; they simply diffuse through the extracellular space. Paracrine signaling is effective only over short distances.

Signal transduction is the name for the many processes by which cells react to and interpret signals.

Cells devote a large number of proteins to the task of receiving and interpreting signals, an example of their ability to carry out advance planning. Some of these proteins are receptors; others are enzymes and regulatory proteins organized in chains of interdependent reactions that respond to a receptor when the right signaling molecule is bound to it. Human cells have the genes to make more than 500 different protein kinases, all of which are involved in some aspect of responding to signals, thanks to their ability to alter the activities of other proteins. The processes by which signals are received and acted upon are called **signal transduction**. At any one time, a given cell is commonly responding to many signals, perhaps five or ten. It chooses its behavior on the basis of the relative strength and character of the various messages coming in. If two signals say, Break down your energy stores so you will be able to make more ATP, while three signals give the opposite message, cells are able to balance the stimulatory signals with the repressive ones and act in a way that looks very much as if they had made a reasoned decision. Amazing!

Many cells receive signals through receptors that are not part of the plasma membrane. The signals for these receptors are carried by molecules that

SIDE QUESTION 4.10

Why does adding a phosphate group to the surface of a protein, as is accomplished by a protein kinase, change that protein's function?

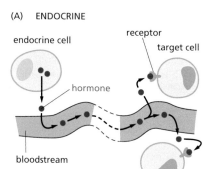

(A)   ENDOCRINE

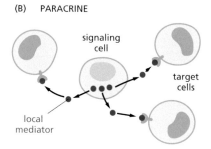

(B)   PARACRINE

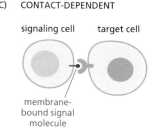

(C)   CONTACT-DEPENDENT

FIGURE 4.18 Three types of signaling between cells. (A) Endocrine signaling is the most familiar signaling type, in which a cell specialized for sending signals makes and secretes a molecule, such as a hormone, which is transmitted to other cells through the bloodstream. Cells that do this are called *endocrine cells*, and a cell that makes a relevant receptor is called a target cell. (B) Paracrine signaling is similar to endocrine signaling, except that the signaling cell and its target are close to one another. The signal does not have to travel through the blood; it can simply diffuse from one cell to its neighbors. (C) In some cases, the signaling cell and its target are touching, so a signal is passed as a result of direct contact between molecules in the plasma membranes of the two cells.

FIGURE 4.19 Some hormones can enter a cell, bind to their receptors, and turn them into activators of gene expression. Steroid hormones can pass through the plasma membrane and bind to their receptors, which reside in the cytoplasm. The receptor with signal bound then migrates into the nucleus, where it activates the transcription of specific genes. Certain mRNAs are then read and translated to make proteins that help the cell to accomplish functions in response to the signal. Commonly, the proteins whose synthesis is stimulated in this way are themselves controllers of the expression of other genes, leading to another kind of cascading response to a signal. Steroid-induced changes are commonly slow when compared with the changes induced by a membrane-bound receptor. (Adapted from Alberts B, Johnson AD, Lewis J et al [2002] Molecular Biology of the Cell, 4th ed. Garland Science.)

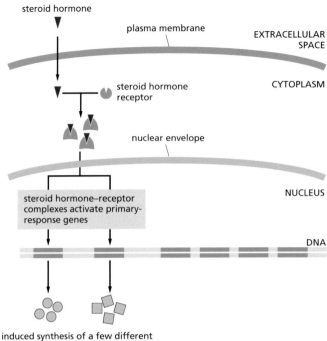

can pass right through the plasma membrane, because they are soluble in lipid as well as water. Steroid hormones behave this way; they bind to receptor proteins in cytoplasm. When a steroid binds its receptor, the activated receptor, commonly with the steroid still bound, enters the nucleus, where it changes the cell's decisions about which genes to express (FIGURE 4.19).

Estrogen and testosterone are familiar examples of steroids that convey their signals in this way. However, signaling molecules, including steroids, are effective only on the cells that make receptors that can bind the signal. Many of the cells in a woman's mammary gland make receptors for estrogen, and several cell types in parts of the male reproductive system, like the prostate gland, make receptors for testosterone. For these kinds of cells, the signal from a steroid is necessary not only to initiate cell division but also to survive.

While estrogen and testosterone function in much the same way, they convey very different messages. One steroid might up-regulate the expression of genes A, B, and C, which are involved in the development of female characteristics; the other steroid, acting on a different receptor, might up-regulate genes C, D, and E, which are involved in the development of male characteristics. Moreover, either hormone acting on a different cell type might bring about a quite different response. Put another way, the messages given by a signaling molecule can mean different things in different contexts. One way to think about this is to consider each signal as a simple statement, like GO, RUN FASTER, or STOP. If you tell a sprinter to go, you get one result; if you tell a banjo player to go, you get a different one. Many cells in an adult body have differentiated so they are analogous to a sprinter or a musician; each is very capable, but they are good at different things. Thus, understanding biological signal transduction requires not only a knowledge of the signal and its receptor but also the context in which the signal transduction is taking place. We will come back to signal transduction in several later chapters, because it is at the heart of many aspects of cancer cell behavior.

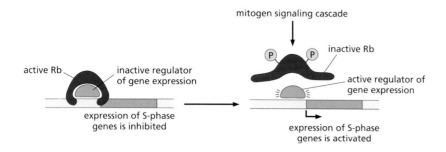

FIGURE 4.20 Mitogenic signals act on the retinoblastoma (Rb) protein. Rb normally inhibits DNA synthesis by blocking the action of a transcription factor (blue) that controls the expression of genes necessary for DNA replication. When a mitogen binds to its receptor, it sets off a cascade of events that activates a protein kinase that phosphorylates Rb, changing its shape. Now the ability of Rb to inhibit gene expression is turned off, which allows the transcription of genes necessary for DNA replication.

Regulation of the human cell cycle exemplifies a cell's ability to interpret signals.

At the start of this chapter, you read about the stages of a cell's growth and division cycle and the fact that signals received during G1 control cellular behavior at start. These signals tell a cell whether to initiate S phase and go through another round of growth and division or to leave the cycle and differentiate. Now we can describe some aspects of what these signals are and how a cell transduces them to behave in ways that are valuable for the body as a whole.

One important aspect of cellular growth control is the action of mitogens on a molecule called **retinoblastoma protein** (**Rb protein**). Normally, this protein inhibits the expression of genes whose protein products are needed for the onset of DNA synthesis. In essence, the Rb protein acts as a brake on a cell's ability to enter S phase. When a cell is stimulated by a mitogen, the inhibitory action of Rb protein is turned off. The inhibition of Rb protein is accomplished by activating a protein kinase that attaches phosphate groups to Rb, changing its shape, so the expression of the genes whose protein products are needed for DNA replication is no longer inhibited (FIGURE 4.20). The cell can now make these proteins, and DNA replication can proceed.

The inactivation of an inhibitor sounds like an unnecessarily fancy way to initiate a process, but cells often use inhibition as a way to control important events. This strategy makes sense because it means that many of the molecules necessary for an event, for example, the start of DNA synthesis, can be present in the cell and ready to go. The process does not begin because the expression or action of one or a few key genes has been inhibited. In these circumstances, a mitogen can act through signal transduction to turn off that inhibition, and now DNA synthesis can start. We will encounter other examples of this kind of control as we work our way through the biology of cancer.

One of the genes whose expression is turned on when the inhibitory activity of Rb is turned off is *myc*, which encodes a transcription factor of the kind shown in Figure 4.9. It stimulates the action of RNA polymerase at the TATA boxes of multiple specific genes, bringing on the synthesis of the mRNAs and then of the proteins that are the products of these genes. With this example, you can see an important aspect of mutations that give rise to cancer. Inactivation of the gene for Rb removes an inhibition of starting S phase, whereas a gain in the activity of the gene for Myc would have the same effect. These two changes in completely different genes have quite similar effects on cell behavior. Both of these mutations are found quite frequently in human cancers.

To help you develop a better feeling for cellular control pathways, there is a fuller description of signal transduction in SIDEBAR 4.2.

Do you see any fundamental difference between a control that starts a process by shutting off an inhibitor versus one that turns on an initiator? Explain.

SIDE QUESTION 4.11

A CLOSER LOOK: Signaling pathways and networks

Since cells receive multiple signals at any one time, they must have mechanisms that allow them to interpret the multiple inputs and use them to choose appropriate behaviors. You have just seen an example in which a cell initiated DNA replication as a result of a signal that turned off an inhibitor of that process. To help keep track of this kind of complexity, biologists have developed a notation that will help you both to diagram regulatory signals and to understand them. Here are some of the basics for how this notation works.

When some factor, called A, stimulates the activity of a process B, it is indicated like this: A → B. If A acts to inhibit B, it looks like this: A ⊣ B. To depict the example of Rb protein, just described in the text, we need three components: the process C, the factor B that inhibits C, and the factor A that inhibits B. A diagram of this arrangement looks like this: A ⊣ B ⊣ C. In this situation, if A increases, B is more strongly inhibited, so its ability to inhibit C is reduced. Now the process C will go forward more strongly than before A was turned up. This is a bit like two negatives making a positive, because the resulting increase in C is the same as would have happened if A were a stimulus to C, rather than an inhibitor of B.

In this situation you might well ask, Why be so indirect about it? That question is even more obvious in Figure 4.17, where the activated EGF-R activates RAS, which activates MAP kinase, which activates a factor controlling gene expression. Why doesn't the activated EGF-R simply activate gene expression without going through all those intermediate? (Note that this figure is a significant oversimplification of the real situation; when the whole pathway is spelled out, it involves eight steps, not four.)

There are at least two reasons for multi-step pathways; both are worth knowing about. One reason is that each step can involve some amplification of the signal, so what starts as a small stimulus can lead to a very big effect. Imagine, for example, that the binding of a single hormone molecule to a single receptor must lead the cell to commit thousands of copies of an enzyme to speeding up a particular process. This actually happens in the case of adrenaline acting on muscle or liver cells to get them to convert their stores of food energy into a form that can be used by mitochondria to make ATP. In this case, step A is the binding of one molecule of adrenaline to the adrenaline receptor. This activates a chemical reaction that generates many copies of a signaling molecule called a second messenger; each second messenger can activate a copy of a protein kinase. Each of these kinases now goes on to activate mul-

tiple copies of a different protein kinase, leading to hundreds of active kinases of the second kind. These kinases in turn activate yet another enzyme, leading to thousands of molecules that are active in breaking down stored food energy. This means that energy is made available quickly in times of need, such as running away from a threat. This kind of amplification can be very valuable when one wants a big result from a small stimulus. Diagramming this pathway with the new notation:

hormone + receptor becomes an active receptor → second messengers → protein kinase 1 → protein kinase 2 → enzyme for breakdown of stored energy

The example of EGF interacting with EGF-R also involves a chain of events but for a different reason. Here the desired final response is not a big biochemical event involving many thousands of enzymes; all the cell needs to do is turn on the expression of a few genes. However, it is very important to make sure that this event happens only when it is right for the cell; that is, the multiple signals relevant to the final act have all been taken into account. In this case there are multiple steps in the signal transduction pathway so other inputs can be weighed into the final biochemical event. Using the new notation to describe the information in Figure 4.18, we could write:

EGF → EGF-R becomes EGF-R* (active) → RAS → MAP kinase → expression of genes for DNA synthesis

However, a more realistic version of this pathway would include additional inputs of information:

different activator   another activator
        ↓   ↓
EGF → EGF-R* → RAS → MAP kinase → DNA synth.
        ⊤   ⊤   ⊤
a phosphatase   a RAS   another phosphatase
activated by a   inhibitor   activated by yet a
different signal   different signal

These additional activating and inhibiting factors could be the result of signals coming into the cell from other receptors, such as other growth factors or membrane proteins that sense cell–cell contact. If all neighboring cells are adhering tightly to the cell in question, they may provide enough inhibiting signal to block the cell from entering S phase. This could be achieved by any of the three inhibitory signals diagrammed above. The excitatory signals (indicated by arrows) could be coming from other growth factors (there are lots of them) that have bound to their own receptors and are encouraging this cell to initiate DNA synthesis and divide. In short, the presence of multiple steps along a pathway provides multiple places at which additional inputs can feed information into the final step of activating the expression of specific genes. This is an example of the mechanisms for cellular logic.

In a final example of using this new notation, we can look at one of the ways cells control the amount of a given chemical that is produced by a set of chemical reactions catalyzed by enzymes. For example, cells must make the nucleotides that are used to synthesize DNA and RNA. It makes sense to have a store of these molecules around, but one doesn't want too much of any one of them; that would be a waste of materials and energy. The pathway by which cells make the nucleotide G can be diagrammed:

$$C \rightarrow D \rightarrow E \rightarrow F \rightarrow G$$
enzyme 1   enzyme 2   enzyme 3   enzyme 4

If the concentration of G has an effect on the activity of enzyme 1, then this pathway is controlled in an interesting way. Say that G inhibits enzyme 1, written G ⊣ enzyme 1. Now the pathway would be described as

$$C \rightarrow D \rightarrow E \rightarrow F \rightarrow G$$
enzyme 1   enzyme 2   enzyme 3   enzyme 4

This kind of effect is called negative feedback, because the amount of end product, G, has a negative effect on the reaction that produces G. Such feedback will limit the amount of G that is produced, because the more G the cell makes, the more slowly it will make more G. This is one of the ways cells keep the concentrations of important molecules at the right level, so they don't waste time, energy, and materials making things they don't need. Cool!

An alternative kind of feedback is positive, meaning that the more G a cell makes, the faster it makes more G. This situation would be diagrammed as follows:

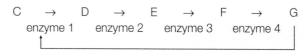

$$C \rightarrow D \rightarrow E \rightarrow F \rightarrow G$$
enzyme 1   enzyme 2   enzyme 3   enzyme 4

Positive feedback leads to a very different result. Now the cell is producing ever more G, a situation that is unlikely to be good for a cell's well-being. There are rare cases where cells use positive feedback, but most examples of cellular feedback control are negative.

HOW CELLS COPE WITH MISTAKES

For a cell to grow and divide, a large number of molecules must be synthesized. As in any construction process, there is a possibility that some of the products will not be perfect. When RNA polymerase assembles the nucleotides to make a particular mRNA, or when a ribosome assembles the amino acids to make the corresponding protein, the resulting protein may contain errors and not be able to do its job correctly. In the case of individual proteins, this is not an issue of critical importance. The inactivity of a single enzyme will generally not affect cell function, because there are usually many copies of the same protein. Cells contain proteinases to degrade proteins after they have been in cytoplasm for a while, so normal protein degradation and synthesis will eliminate the defective gene product and take care of the problem. Inheritable mistakes are, however, a much greater problem. Therefore, we now look at both DNA synthesis and chromosome segregation to see how cells deal with possible errors in these critical processes.

DNA replication is accurate but not perfect.

As DNA polymerase moves along an old strand of DNA, inserting the As, Ts, Gs, and Cs from solution that are complementary to each nucleotide on the template strand, it does the job very well. However, mistakes are occasionally made. A mistake in DNA synthesis introduces a change in the DNA sequence of the new strand, and that is a mutation. Avoiding mistakes is therefore important. To minimize the chances of mutation, the overall process of DNA replication includes additional steps to minimize mistake frequency. There is a system for proofreading, which double-checks the complementarity of bases lying opposite one another on the DNA strands, and there are enzymes that scan double-stranded DNA to make sure there are no bumps of the kind that occur when the bases are not complementary. When all the processes that support the accuracy of DNA replication are taken into account, the frequency of mistakes that persist is about 1 in 100 million. This level of precision is spectacularly accurate, much better than most human-designed processes of manufacture. However, there are about 6.4 billion nucleotides in the two sets of chromosomes of a human cell, so several mistakes are commonly made with each round of DNA replication. This means that mutations are bound to accumulate as a human grows and ages.

Mutations are generally inherited when a cell goes through DNA replication and divides. If a spontaneous error or an external insult has altered a nucleotide on one strand, DNA synthesis will copy this situation in a way that produces one DNA duplex that is now mutant on both strands and another duplex that is normal (FIGURE 4.21). Thus, a mutation that is propagated by cell growth and division will lead to two populations of cells, one that is normal and one that is mutated on both DNA strands.

Some important features of human DNA mean, however, that most mutations make no difference to the individual who hosts that DNA. Because we are diploid, our cells have two copies of every gene. If one gene is inactivated by a mutation, the other copy can often cover for the no-longer-functional gene. As a result, many mutations that cause a loss of gene function do not show up in cell structure or behavior. An additional reason why some mutations can be tolerated is that the human genome includes long stretches of DNA between genes that are replicated but are never transcribed into RNA. Most mutations in such regions will never show up in the structure of gene products, so they have no consequence. In addition, many segments of DNA that are transcribed contain sequences that are spliced out of the RNA during the preparation of mRNA. Mutations in these regions, too, are commonly of no consequence. Moreover, some changes in a DNA sequence that does encode a protein do not change the amino acid that is inserted at that site during protein synthesis. For example, the codons CUU, CUC, CUA, and CUG all encode the amino acid

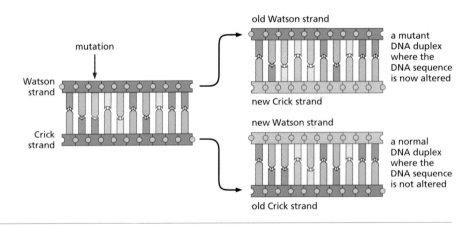

FIGURE 4.21 **The replication and inheritance of a mutation.** In this representation of a DNA duplex, pink bases normally pair with green bases, and blue bases normally pair with yellow bases. However, this stretch of DNA has been damaged, resulting in a mutation. At the third nucleotide from the left, marked with an arrow, a green nucleotide has been altered, so it is now more like a blue nucleotide. It is opposite a pink nucleotide but cannot base-pair with it (shown by the absence of short red bonds). As a result of this change, the two DNA duplexes formed by DNA synthesis are not identical. During this replication event, DNA polymerase is making no mistakes. After DNA synthesis, the mutant nucleotide is accurately paired with the complementary nucleotide on the opposite strand, so now both strands in the upper DNA duplex have been altered; a heritable mutation has been made. The other DNA duplex is, however, normal.

leucine. The codons AUU, AUC, and AUA encode the very similar amino acid isoleucine. In cases where a mutation alters the DNA sequence to yield either an identical or a very similar amino acid, the mutation will commonly be of little or no consequence. For all these reasons, a modest level of error in DNA synthesis and a low level of mutagenesis coming from the environment make no trouble.

Some mutations, on the other hand, really do matter. We can see this by looking at some of the proteins you have just met as components of the signal transduction pathways that regulate cell growth and division. Even here, however, some genes and their products are more important than others. Imagine a pair of mutations that inactivate both alleles of the gene encoding the EGF receptor. If this protein is not made, or if the protein is altered so it doesn't work, then that cell cannot respond to EGF. If those mutations occurred in an epithelial cell, that particular cell will be unable to respond to EGF, initiate DNA synthesis, and go on to divide. This kind of mutation matters very much to the cell itself: it makes no progeny. But for the organism as a whole, such mutations make no detectable difference. Other epithelial cells whose EGF receptors are still active can engage in wound healing or tissue repair, so the loss of one cell is inconsequential.

Now imagine a pair of mutations that inactivate both the genes encoding the Rb protein. If this protein is not made, or if it is made in a faulty form, the mutated cell is now missing an important protein that *prevents* the onset of DNA synthesis until a mitogenic signal is received. A cell that lacks functional Rb protein is prone to initiate DNA synthesis without receiving a signal to do so. This cell is likely to replicate frequently. Moreover, all its daughter cells will have the same mutation and thus the same behavior. This kind of uncontrolled cell reproduction is just what happens in cancer. This is the kind of mutation that can matter a great deal for the fate of not only the cell harboring the altered DNA but also the organism hosting that cell. To prevent events like this, cells have safeguards to minimize the number of mutations that persist.

Some environmental factors can damage DNA, but cells can repair it.

Agents that damage DNA are called **mutagens**. Many mutagens are also carcinogens, because they increase the likelihood of cancer. They are a central subject in understanding the initiation of cancerous growth, so they are treated more fully in Chapter 5. Here, we will consider only a single example of DNA damage from external sources and the ways that cells correct the problem. When X-rays pass through a cell, they can break a piece of double-stranded DNA in two. This kind of damage, called a **double-strand break**, can occur in the middle of a gene, meaning that a normal gene product can no longer be made from that chromosome. Clearly, the diploid condition of our cells is important here, because the inactivation of one allele by such a break may have only modest consequences for the expression of that gene. Nonetheless, cells work in several ways to repair such breaks and do it as quickly as possible.

There are two processes for double-strand break repair. One uses an unbroken copy of that chromosome to help line up the two broken pieces. The nucleotide sequences in the unbroken chromosome guide the activity of special DNA polymerases, so they can repair the broken chromosome. There are two possible sources of an undamaged chromosome to help in this kind of repair. If the cell is in the G2 phase of its cycle, there is already a sister chromosome, tethered by cohesin to the one that was damaged. It contains a sequence identical to the one that was damaged, allowing a

SIDE QUESTION 4.12

In the above discussion, a mutation is presumed to cause a loss of function in the product from that gene. Some mutations result in a GAIN of function in the gene product; for example, a change in the promoter can lead to the gene's overexpression. What would be the result of a gain-of-function mutation in the gene for Rb? Would it promote the formation of a cancer? What about a gain-of-function mutation in the gene for EGF-R? How about *myc*?

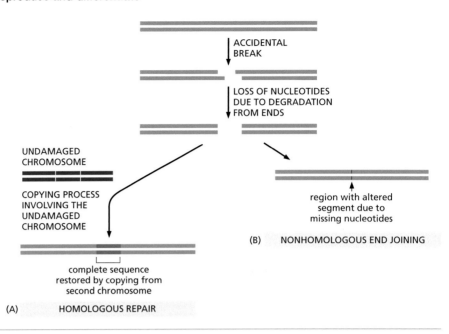

FIGURE 4.22 **Two pathways for repair of a double-strand break in DNA.** (A) In homologous repair, an identical or homologous chromosome is used to guide a repair process in which the missing DNA sequence is replaced, and the chromosome comes out of the event with a perfect repair. (B) In nonhomologous end joining, the two ends made by a break in DNA are simply put back together without regard for any changes in DNA sequence that might occur. Thus, a mutation may be introduced.

perfect repair. If the cell is in G1, there is no sister available, but there is a very similar chromosome, thanks to the cell's diploid condition. Because the undamaged chromosome contains the same genes as the broken one, albeit with different alleles, these two chromosomes are said to be homologous, which means related. The guiding provided by the homologous chromosomes commonly leads to a successful repair (FIGURE 4.22A).

The other kind of double-strand break repair simply puts the two broken ends together. This is called **nonhomologous end joining** (FIGURE 4.22B). End joining without the help of an identical or homologous chromosome is unable to get the broken chromosome back to its initial structure; a few nucleotides are commonly lost from either strand as they are put back together. Now the DNA sequence is altered, so a mutation has been introduced. Whether the mutation matters or not for cell behavior will depend on the location of the break and the details of what happened during end joining.

There are several other DNA repair processes, all of which are important for preventing or limiting cancer. They are described in more detail in Chapter 7, when we consider some specific cancers in which loss of function in a gene that encodes a relevant enzyme contributes to the likelihood of carcinogenic mutations as results of DNA damage.

Our understanding of DNA and the factors that affect it has grown tremendously in the last 50 years. The majority of progress has been a result of work on bacteria, because these cells are easy and inexpensive to grow, their DNA is very similar to human DNA, and many of the processes of DNA synthesis and repair are far easier to study in these comparatively simple cells (FIGURE 4.23). During the last few decades, this understanding has been extended to the discovery of many genes and gene products whose alterations bring on cancerous behavior. The

Sunlight promotes the formation of skin cancer through the action of its rays on DNA in the cells of our skin. If a person carried a loss-of-function mutation in a gene for DNA repair, would they be more or less susceptible to developing skin cancer from the mutagenic action of sunlight? Explain.

work has been possible because of tremendous progress in methods available for manipulating DNA in the laboratory. For a description of some of these methods, see SIDEBAR 4.3.

Mitosis, like DNA replication, is quite accurate but not perfect.

The mitotic spindle is a remarkably accurate machine, but it too makes mistakes occasionally. A daughter cell will get too many or too few chromosomes once in about every 10,000 divisions of a healthy human cell. Mistakes of this kind are important, because getting an extra copy of even one chromosome leads to an imbalance: the cell has too many copies of all the genes on that chromosome. When expression of those genes is turned on, the cell can produce too much of the resulting gene products. Sometimes this has damaging effects on cell performance. The other daughter cell, however, got one too few chromosomes, and this may be even worse. Thus, doing mitosis right is important.

CHECKPOINTS ARREST CELL CYCLE PROGRESSION TO IMPROVE THE FIDELITY OF CELL REPRODUCTION

Mitosis and DNA replication are obviously key events in cell reproduction. You now know that both processes are imperfect. Do cells simply live with this imperfection, or do they have some way to improve the chances that cell reproduction will be accurate? It turns out that the fidelity of each of these processes is enhanced by a cell's ability to stall cell cycle progression when something is going wrong, allowing additional time for corrective processes and increasing the chances that things will ultimately be done right.

Chromosome segregation is delayed until all chromosomes are properly attached to the spindle.

The job of the mitotic spindle is difficult, because it must organize and segregate all 46 of the already duplicated chromosomes. The microtubules that accomplish this task are made from many copies of only a few proteins. Each of these proteins is very small relative to a cell. For example, tubulin, the protein subunit of microtubules, is about $4 \times 4 \times 8$ nanometers (a nanometer, abbreviated nm, is one-billionth of a meter). The smallest chromosome is about $500 \times 500 \times 1000$ nm in size, and a human cell at mitosis is a sphere about 15,000 nm in diameter. If you imagine yourself as a tubulin molecule, the job of moving a chromosome is analogous to your having to move a dead whale from one place to another on a 2-mile stretch of beach. Doing such a job by yourself would be impossible. Even with lots of friends to help, doing it right would be very difficult and would take time. How does a cell do it, and how does it know when all the chromosomes are correctly attached to the spindle (Figure 4.6D), so it is time to start chromosome segregation (Figure 4.6E)?

Spindle assembly includes a process that monitors chromosome attachment and prevents chromosome segregation until all the chromosomes are properly bound to microtubules. Early in mitosis, each kinetochore emits a signal that inhibits the start of chromosome segregation. Proper attachment of each chromosome to spindle microtubules silences this inhibition. When all the chromosomes are properly attached, all the inhibitory signals are silenced, and the cell can start segregation. This control device is called the spindle assembly **checkpoint**, because it evaluates

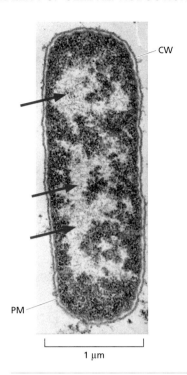

CW

PM

1 μm

FIGURE 4.23 **The structure of a bacterial cell.** An electron micrograph of the common bacterium *Escherichia coli* (*E. coli*), which thrives in the human intestine (hence its name coli, referring to the colon, which is the lower intestine). This prokaryotic cell is only about 1 μm in diameter, versus the 10 μm or more that is a common minimum dimension of a human cell. As a result, its volume is less than 1/200 the volume of a human cell. Nonetheless, it is packed with ribosomes (darkly staining material), it contains DNA (lighter regions indicated with red arrows), it is surrounded by both a plasma membrane (PM) and a cell wall (CW), and it is capable of growing on comparatively simple food. In a rich medium, these cells grow and divide every 20 min, meaning that if one *E. coli* cell were maintained in fast growth conditions for a whole day, it could become about 4,700,000,000,000,000,000,000 cells. The resulting mass of cells would weigh more than 1000 tons! These remarkable cells have been tremendously useful in the development of our understanding of DNA and other aspects of molecular biology. (Courtesy of E. Kellenberger.)

A CLOSER LOOK: Manipulation of DNA

Our understanding of DNA took a giant step forward with the discovery of its structure by James Watson and Francis Crick. The double-helical arrangement of nucleotide polymers has turned out to be of central importance in understanding both how such a molecule could be replicated and how information stored in DNA's structure could be interpreted by the cell through the synthesis of RNAs complementary to specific regions on one of the two DNA strands. Wide-ranging studies of DNA in many organisms have revealed that the overall structure of DNA is the same in all forms of life on earth, so the double helix is a fundamental aspect of biology. Only the nucleotide sequences differ between our own DNA and that of every other living thing.

Another important advance came with research on the DNA of bacteria, including the most common bacterium in our own gut, called *Escherichia coli* (or simply *E. coli*). During the course of this work, it became clear that these cells commonly contain two pieces of DNA: their main chromosome and a small, circular piece of DNA called a **plasmid**. Bacteria can transfer some plasmids from one cell to another, and with the DNA go the genes encoded on the plasmid. By playing with the conditions for bacterial growth, scientists were able to isolate a plasmid from one bacterium and then insert it experimentally into a different bacterium, transferring with it any genes that were on that piece of DNA.

The simple medium needed for bacterial growth have made these cells much easier and cheaper to use than cultured human cells for study of DNA behavior. Human cells contain nuclei, so they are called **eukaryotic** cells. (The word *karyon* is Greek for nut or ker-

nel.) Bacterial cells lack a nucleus; their DNA is simply packed into one or more regions of the cell (arrows in Figure 4.23), so they are often referred to as **prokaryotic**. In addition to having easily transferred DNA, bacteria are commonly master biochemists. Small as they are, they can make all the amino acids needed for protein synthesis, and they do not require vitamins for growth; they thrive on a simple medium that contains only water, sugar for energy, ammonia for nitrogen, and the same set of minerals that are necessary for all living things to thrive. Because of this simplicity, they are easy and cheap to grow in large numbers; bacteria have been a major workhorse of both molecular biology and biochemistry since the inception of those scientific fields.

The next leap forward in manipulating DNA came with the discovery of bacterial enzymes that can cut double-stranded DNA at sites defined by specific nucleotide sequences (Figure 1). Enzymes that cut DNA are called **nucleases**; many of these had been known for years. However, enzymes that cut DNA only at certain nucleotide sequences were new and demanded close study. We now know that such enzymes are made by many bacteria as devices to protect themselves from viruses. Viruses reproduce by inserting their DNA into a cell, and yes, there are viruses that are specialized to infect bacteria. If a bacterium has the right enzyme already expressed in its cytoplasm, the viral DNA is cut to pieces before it can do any harm, thereby restricting virus growth. The trick, then, is for the bacterium to make nucleases that are specific for DNA sequences that are not in its own DNA but are in the DNA of infecting viruses. Enzymes that can do this are

called restriction endonucleases or simply **restriction enzymes**, because they restrict virus growth. By isolating restriction enzymes from many different bacteria, scientists now have a toolbox of over 3000 enzymes that will cut DNA at many different and specific sequences.

With restriction enzymes as laboratory tools, it became possible to cut DNA isolated from the plasmid of any given bacterium and combine it with DNA from some other source, such as the DNA that encodes a protein of particular interest (Figures 2 and 3). The resulting piece of DNA is called **recombinant** DNA. When a piece of recombinant DNA is inserted into a bacterial host, the bacterium now contains both the plasmid and a foreign gene, commonly called a **transgene**. In the right culture medium, these cells can be grown in quantity to prepare large amounts of the recombinant DNA (Figure 4).

More recent advances have come from methods that insert recombinant DNA into cultured mammalian cells, even human cells. This is usually done by making recombinant DNA in bacteria, adding promoters that work in human cells, and then inserting these transgenes into cultured cells. This can be achieved by mixing the recombinant DNA with the right lipids and detergents to get the DNA through the plasma membrane. Scientists also use viruses that infect human cells as tools for transgene insertion, but one must be very careful that the viral DNA has been altered so it is no longer infectious; otherwise one might make a novel kind of germ. As this work has gone forward, scientists have evolved rules for how such work should be carried out, so no harm is done through the construction of novel DNA.

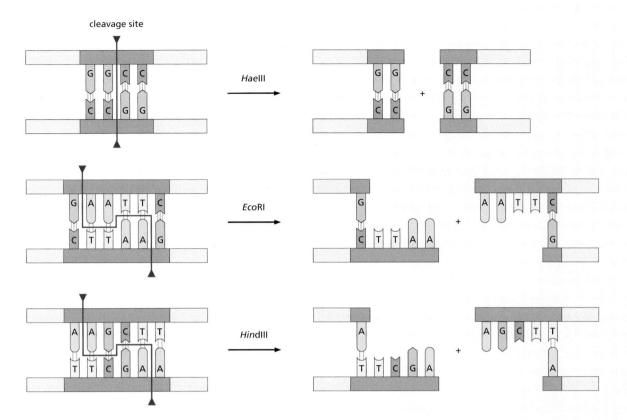

Figure 1 Restriction endonucleases cut double-stranded DNA at specific nucleotide sequences. The gray bars represent the strands of DNA duplexes that extend far beyond the short segments shown. Orange identifies the nucleotide sequence that a given restriction enzyme will cut. The bases that define the cleavage site for a particular restriction endonuclease are shown at each site. The name of each restriction enzyme that will cut DNA at that sequence is given above the black arrow. These names come from the name of the bacterium where each enzyme was first found: *Hae*III (pronounced hay three) was the third restriction enzyme found in the bacterium *Haemophilus influenzae* biogroup *aegyptius. Eco*RI (pronounced eco-R one) was the first restriction enzyme found in *Escherichia coli*, and so forth. At the right are diagrams of the ends of the DNA segments formed when these enzymes cut a DNA duplex. Note that the DNA sequences at each restriction site are the same in opposite directions on the two DNA strands. This symmetry comes from the fact that restriction enzymes are commonly dimers with the two protein subunits facing each other. When the enzyme interacts with DNA, there is one protein to cut each DNA strand, resulting in the double cut at a place where the nucleotide sequence is right for that enzyme.

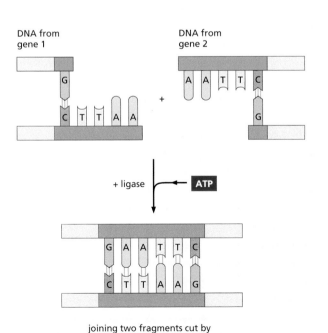

Figure 2 Two pieces of DNA that have been cut by the same restriction enzyme can be linked to form continuous, double-stranded DNA. The gray and orange DNA on the left comes from one piece of DNA, and the similar structure on the right comes from a different piece of DNA. Both pieces of DNA have been cut with the same restriction enzyme, so they have end structures that are complementary, as in Figure 1. Because the sequences are complementary, the ends of these two pieces will stick to one another through the bonds formed between complementary bases. One can then add the enzyme ligase (plus ATP for energy), and the sugar–phosphate backbones will be connected, leading to double-stranded DNA. Thus, two sequences that have never previously been contiguous are now linked, making a novel piece of recombinant DNA.

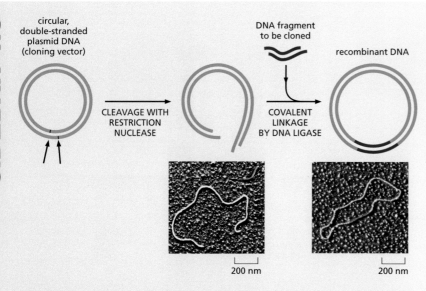

Figure 3 A circular piece of DNA, like a plasmid, can be modified to carry foreign DNA. The two concentric circles (left) represent the two strands of DNA in a plasmid. This plasmid will be used to receive a piece of foreign DNA, for example, DNA from a mammalian gene, and insert it into a bacterial cell. It is therefore called a cloning vector. The two black lines indicated by arrows are places where a restriction enzyme can cut this DNA. After cleavage, the double-stranded DNA has sticky ends, given the symmetry of the cut made by a restriction enzyme. If a different fragment of DNA has similarly sticky ends, as a result of having been cut by the same restriction enzyme, it can hybridize with the sticky ends on the plasmid, forming a larger circle (right). The lower pictures are electron micrographs of comparable molecules of DNA, made visible by shadowing with a heavy metal. The short, linear segment colored red represents foreign DNA that is being added to the plasmid.

As an example, one can make recombinant DNA that contains both the gene for a fluorescent protein and the gene for a protein of interest, like a histone that binds to DNA. This engineered piece of DNA can be inserted into a population of cultured human cells. The recombinant DNA is transcribed and translated to make the foreign, engineered protein, which you can see in a fluorescence microscope (**Figure 5**). Now you can watch the cell as

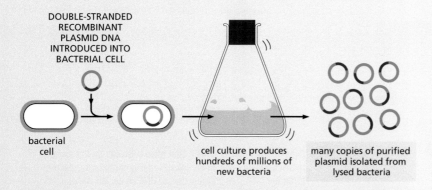

Figure 4 Once a recombinant plasmid has been constructed, it can be made in quantity for further study. To obtain large quantities of a given piece of recombinant DNA, a plasmid that contains it can be introduced into a few bacteria, which are then placed in a culture medium that supports bacterial growth. The few bacteria that were transformed by the addition of the plasmid will now grow and divide to form many millions of daughter cells. If the plasmid is designed to contain a gene that gives these bacteria resistance to a particular antibiotic drug, then that drug can be added to the culture, so only bacteria that contain the plasmid will grow and divide. The bacteria can now be broken open and the plasmid DNA can be isolated for further use.

it goes through its growth and division cycle, using the fluorescence of the engineered gene product, expressed from the inserted DNA, to observe the localization of this protein. The histone, which appears red in this figure, is located in the nucleus, but during mitosis it would be concentrated in the chromosomes, and you could watch individual chromosomes as they are moved around by the mitotic spindle. This and many other kinds of issues can be studied with a combination of molecular biology and cell culture.

Following improvements in the methods for growing and expressing recombinant DNA in cultured cells, scientists have also been able to develop methods for expressing transgenes in experimental animals, like mice. These procedures usually start with a very young mouse embryo, even a single fertilized egg, which is used as a host for the transgene developed in bacteria. If the foreign DNA integrates into the DNA of the host cell, this egg or early embryo can be placed in the uterus of a female mouse, where it will grow up to be born as an apparently normal mouse that contains the transgene in all its cells. If the regulatory part of the transgene was assembled correctly, the transgene will be expressed. It is even possible to use regulatory elements that are active only in certain tissues, where transcription factors, etc., induce strong expression of the RNA encoded by that gene in only one part of the body or one cell type. In this way, recombinant DNA can be used to study gene function and regulation in animals.

Another key advance in manipulation of DNA came with the development of methods for determining DNA sequence. The first efficient methods were invented by two groups, those of Fred Sanger

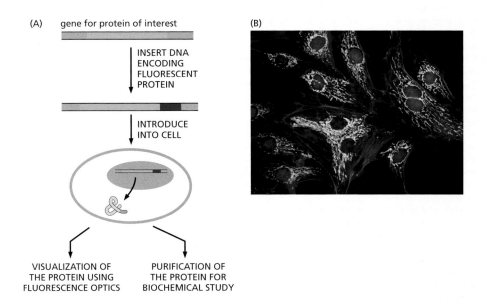

(A)   gene for protein of interest

INSERT DNA ENCODING FLUORESCENT PROTEIN

INTRODUCE INTO CELL

VISUALIZATION OF THE PROTEIN USING FLUORESCENCE OPTICS

PURIFICATION OF THE PROTEIN FOR BIOCHEMICAL STUDY

(B)

Figure 5 Using methods from molecular biology to study protein behavior in cultured cells. (A) DNA that encodes a protein of interest (orange bar), like a histone that binds DNA, can be modified by adding to it the DNA that encodes a fluorescent protein (red). This piece of recombinant DNA can be coaxed into a cell, where it will enter the cell's nucleus and be transcribed by the host cell's enzymes. The resulting mRNA goes into the cytoplasm where it is translated into a novel, two-part protein: one part is the histone, and the other part is the fluorescent protein. These two functions are linked together in the same protein molecule. (B) Assuming that the presence of the fluorescent protein doesn't harm the activity of the histone, the histone's position in the cell can be seen by observing the transformed cells with a fluorescence microscope. In this image, a histone-fluorescent protein combination is making the nuclei red. The mitochondria are stained by a protein tagged with a yellow fluorescent protein, and the cytoplasm contains a modified actin that stains the microfilaments blue. (B, courtesy of Jennifer Lippincott-Schwartz, National Institutes of Health.)

in England (who had previously discovered a way to sequence proteins; he earned a Nobel Prize for *each* discovery) and Walter Gilbert in the United States. These methods have been extended and automated to the point that it is now quite easy and cheap to get the sequence of DNA, even all the DNA in a human cell! With sequence information in hand, you can choose the right restriction enzymes to make cuts just where you want them, and thus you can construct pieces of DNA to be inserted into a plasmid (or a virus) and then passed into a bacterial host for replication, expression, and study.

An even more recent discovery has built upon yet another aspect of the ways in which bacteria protect themselves from virus infection. Many bacteria can cut

short segments of DNA from an infecting virus and then insert them into a particular spot in their own chromosome. From these DNA segments, they make RNA that serves to guide the activity of a unique nuclease called Cas9. Guided by its associated RNA, Cas9 will cut any DNA with the same sequence as that of the guide RNA. The guide RNA makes Cas9 like an adjustable restriction enzyme that will cut any DNA for which it has a guide. If a new virus infects a bacterial cell, Cas9 can use all the guide RNAs now available to identify one that matches the viral DNA and then cut it, blocking a new round of virus infection.

This whole system, called **CRISPR-Cas**, is now being re-engineered to make cuts in the DNA of human cells in culture as

well as the DNA of experimental animals; the process allows the animal's genes to be altered or deleted. Most recently, the method has been used to correct a genetic defect in a human embryo, although the resulting cells were not allowed to develop into a fetus. Visionaries are talking about developing cures for muscular dystrophy, AIDS, and other diseases from such technology, but this is still over the horizon. CRISPR technology is currently very hot, and articles about it appear almost daily in the press and on the Internet. The point is that the tools for making and using recombinant DNA are still advancing. In Chapter 10, after the complexities of cancer have been more fully described, you will see that such advances may help us to cope with several aspects of the cancer problem.

FIGURE 4.24 Cell cycle progression can be stopped by checkpoints. Diagram of a cell cycle showing the relative positions of G1, S, and G2 and the possibility of a cell leaving the cycle to enter G0 (pronounced gee zero), where it can differentiate. At four times in the cycle (marked by red dots), the cell can arrest its cycle as a result of a checkpoint, activated by a significant problem, such as DNA damage, incomplete replication of DNA, or formation of a defective spindle.

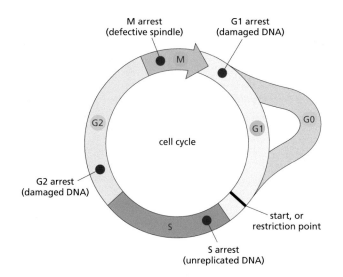

the quality of the mitotic process and won't let the chromosomes segregate until the spindle is properly assembled (FIGURE 4.24). This cell cycle checkpoint helps to assure that chromosome segregation is accurate.

Replicating and repairing DNA take time, so checkpoints make cell division wait for these processes to be completed.

When DNA replication has gone well, the cell has achieved one of the most important aspects of its preparation for division. However, things don't always go as they should. If DNA becomes damaged, cells can commonly sense this situation and generate a signal that overrides signals from mitogens, halting cell growth and division. This is another kind of cell cycle checkpoint. Either unfinished DNA replication or DNA damage can arrest cell cycle progression (Figure 4.24). Using the notation for cell signaling described in Sidebar 4.2, we can write:

$$\text{DNA damage} \dashv \text{mitosis}$$

$$\text{incomplete DNA synthesis} \dashv \text{mitosis}$$

In addition, DNA damage can also inhibit DNA synthesis. This inhibition is also important because the faulty material should be repaired before it is replicated. Thus, we can also write:

$$\text{DNA damage} \dashv \text{DNA synthesis}$$

These controls are all examples of checkpoint activity. They are important for the maintenance of an undamaged genome. In addition, the processes of DNA repair and checkpoint control on the cell growth and division cycle are important for the biology of cancer. The genes that encode checkpoint proteins are among the genes that are mutated to a loss of function in cancer cells. This big subject is central for understanding the mutations that are important in the emergence of cancerous behavior in previously normal cells. It is covered in more depth in Chapters 6 and 7.

The ultimate checkpoint is a cell's decision to kill itself.

In addition to the signals that induce cells to divide or differentiate, there are signals that induce a cell to destroy itself. Why would a cell ever do such a thing? In a multicellular organism, cells must come and go for the benefit of the organism as a whole, not just for their own well-being. In multicellular organisms, such as ourselves, each cell must do its part, and

(A)
(B)

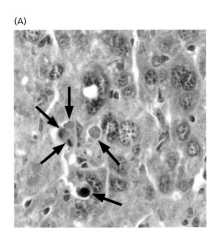
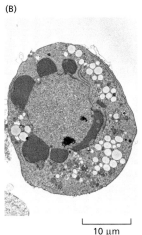

10 μm

FIGURE 4.25 **Images of cells showing steps in apoptosis.** (A) Light micrograph of mouse liver cells, five of which are undergoing apoptosis (arrows). The differences in cell shape and staining are a result of their being at different stages in the process of programmed cell death. (B) An electron micrograph of a single cell prepared for microscopy in the middle of apoptosis. The cytoplasm is filled with vesicles (light circles). The DNA within the nucleus is hypercondensed (visible as dark lobes projecting out from the nucleus) and will soon be degraded. (A, from Nat. Inst. Environ. Health Sci, NIH; B, courtesy of Julia Burne.)

sometimes a cell outlives its usefulness. For example, if a cell accumulates more DNA damage than it can repair, or if it makes serious mistakes in mitosis, it will often initiate the process called **apoptosis** (from a Greek word that means falling off).

Apoptosis, sometimes called programmed cell death, starts with the activation of a specific class of proteinases, that is, enzymes that chew up proteins. After some proteins have been degraded, there is a second wave of enzyme activity that chews up DNA (FIGURE 4.25). For a dramatic and informative view of this process, watch MOVIE 4.4. The net result of this degradation is a dead cell whose remains can be taken up by other cells, such as leukocytes, so the materials of the now-dead cell are recycled by their still-living neighbors.

Apoptosis is essential for the maintenance of healthy tissues and even for normal human development. For example, when hands and feet are forming in a human fetus, each structure starts out as a paddle-shaped object at the end of a tiny limb. Fingers and toes are made by the controlled death of some cells to form a space between each pair of digits. Apoptosis is also important in adults, for example, in assuring the death of a virus-infected cell, so the viruses cannot replicate so easily and make more cells sick. You'll learn more about this kind of process in Chapter 9, which describes the immune system.

Apoptosis is important in the biology of cancer for a key but subtle reason. If a normal cell's DNA has experienced more damage than can be repaired, signals activate the relevant proteinases; apoptosis begins, and the cell dies. If this were not to happen, cell division would go forward, producing two cells with damaged DNA. Continued growth and division of these cells could lead to the formation of multiple mutant cells, each of which might have dangerous properties. Here, the death of an individual cell is good for the well-being of the organism as a whole. If apoptosis fails, however, the rogue cell can go on to grow and divide, making many more of its own kind. Cancer cells find a way to block apoptosis, so even if their DNA is significantly altered, they do not die. This immortality of cancer cells makes them a mortal danger to their host.

SIDE QUESTION 4.14

If one population of cells was capable of apoptosis and a second population was not, which population would be more likely to die off in a viral infection? Why?

SUMMARY

Cells reproduce by cycles of growth and division. The period between divisions is called interphase. The most important event of interphase is the replication of the cell's DNA during the period called S phase. Between the end of cell division, when a new cell is born, and the onset of S phase, there is a gap in time called G1; after S phase, there is another gap called G2 before the next division starts. Throughout interphase, cells are making multiple copies of important RNAs, proteins, lipids, and polysaccharides. These assemble into enzyme complexes and organelles, so when the cell divides, each daughter cell has all the materials it needs to grow and divide again.

DNA synthesis generates two identical DNA duplexes from one. It begins at origins of replication, where the two strands of DNA separate. Then enzymes called DNA polymerase bind to the existing DNA strands. These enzymes move along the old DNA strands, taking nucleotides from solution that are complementary to the base at each position on the old strand and linking them together to form the sugar–phosphate backbone of a new DNA strand. As a result, the sequence of bases in the new strand is exactly complementary to that on the old strand, which was used to template synthesis.

Mitosis is the process by which cells segregate their already-replicated DNA into two equal sets, one for each daughter cell. Now the cell can divide by cytokinesis, and both daughter cells will have all the information necessary to grow and divide again. Cells decide whether to initiate another round of growth and division at a time in G1 called start or the restriction point. Prior to start, a cell can leave the cycle and differentiate. Once start has occurred, the cell is committed to a round of growth and division.

Differentiation is accomplished by the regulated expression of genes. The products of specific genes are needed for the formation of each differentiated cell type. Each gene has a structural part, which encodes the gene product, and a regulatory part, which can bind RNA polymerase and initiate gene expression. Gene expression happens only when proteins called transcription factors have bound to DNA sequences in the regulatory part of the gene and made a stable and productive site for binding RNA polymerase. The regulatory part of the gene controls both when and how often the gene is expressed.

Most tissues are made largely from differentiated cells that are no longer growing and dividing. The health of such tissues is maintained by stem cells that are specialized to grow and divide, producing one daughter cell that retains its stem cell character and another that will differentiate. The initial stages of differentiation may involve additional cycles of growth and division, during which the cell is called a transit amplifying cell. Stem cell growth, division, and differentiation in the intestinal epithelium and in bone marrow for the formation of blood are described as examples of these events.

One aspect of cell differentiation is the synthesis of proteins and polysaccharides that are secreted from the cell to make an extracellular matrix. This matrix helps to define the environment for the cell, and it can bind molecules that regulate cell growth.

Cells grow and divide or differentiate in response to signals they receive from elsewhere in the body. Some signals are hormones, such as adrenaline, insulin, and steroids. Others are mitogens that stimulate cell division. Signals affect cell behavior by binding to proteins that serve as receptors and initiate a chain of events to alter cell behavior. The processes of signal reception and interpretation are called signal transduction.

Epidermal growth factor is an example of a mitogen. Its receptor initiates a signal transduction cascade. First, the receptor becomes a protein kinase and phosphorylates itself, altering its structure. It can then bind and activate proteins that activate the protein Ras, which activates MAP kinase, which in turn activates a transcription factor, leading to the expression of genes important for the initiation of DNA synthesis.

Neither DNA replication nor mitosis is perfectly accurate, so mistakes are occasionally made. Moreover, external factors, such as hazardous chemicals or radiation, can damage DNA. Cells can repair DNA by several processes that help to keep the stored information accurate.

DNA repair takes time, but cells have checkpoints that identify the fact that repair is going on, and they slow or stop cell cycle progression to allow time for its completion. Likewise, getting all the chromosomes properly attached to the mitotic spindle takes time. There is another checkpoint that postpones chromosome segregation until spindle assembly is complete. These checkpoints help to ensure the integrity of a cell's genome.

If DNA damage cannot be repaired, a cell can initiate apoptosis, the process of programmed cell death. Apoptosis assures that cells with damaged DNA do not reproduce. Both cell cycle checkpoints and the ability to initiate apoptosis are commonly lost in cancer cells. These deficiencies contribute to their genetic instability.

ESSENTIAL CONCEPTS

- Cells reproduce by cycles of growth and division. Interphase is the time between cell divisions. During interphase, cells synthesize all the materials necessary for two cells. Thus, the parent cell prepares a dowry of materials for its daughter cells, enabling them to grow and divide again.

- During interphase a cell replicates its DNA by going through a round of DNA synthesis (S phase). After cell division but before S phase, there is a gap in time called G1. After S phase and before chromosome segregation, there is a second gap called G2. Human cells commit to either differentiate or replicate during G1 phase.

- The mitotic spindle is a machine made from microtubules and associated proteins that organizes and segregates both the chromosomes and the centrosomes to opposite ends of the parent cell. The parent cell then builds a ring of microfilaments around its center and pulls the ring tight to divide itself into two.

- Mitosis is quite accurate but not perfect. Mistakes in mitosis can lead the daughter cells to contain the wrong number of chromosomes. This condition reflects an instability in genome management that is characteristic of cancer cells.

- Many of the cells in an adult have withdrawn from the cell cycle. They have turned on the expression of genes that allow them to differentiate, so they can perform a particular function for the good of the body as a whole.

- Cell differentiation depends on the regulated expression of specific genes. Specificity of gene expression is achieved by transcription factors that guide the action of RNA polymerase to the right places on DNA. Gene expression results from transcription of mRNA from the appropriate stretch of DNA, followed by translation of this message into protein.

- While some transcription factors are common to all sites of gene expression, others are more specific for particular genes. These transcription factors can be modified by processes such as phosphorylation, so they act only at specific times. The protein Myc is an example of such a transcription factor; it controls the expression of genes necessary for DNA synthesis.

- Transcription factors that turn on genes for differentiation often antagonize the activity of transcription factors needed for the cell cycle. Thus, cells that commit to differentiation generally leave the cell cycle. Carcinogenic transformation disrupts this sort of commitment, allowing cells lose their differentiated state and enter the cell cycle.

- Stem cells are specialized cells in a differentiated tissue that retain the ability to grow and divide. Division of a parent stem cell produces daughter cells with different fates: one continues to be a stem cell while the other differentiates to perform tissue-specific functions. The action of stem cells is important for the continual production of differentiated cells that can replace older ones that no longer function properly. This is the body's mechanism for preventive maintenance of tissues.

- Both the intestine and blood are maintained through stem cell division and differentiation. This activity balances the continuous death of old or damaged cells.

- Cells can sense aspects of their environment through the action of receptors. Many receptors are membrane proteins with a signal-binding part outside the cell and a reactive part in the cytoplasm. The binding of a signal to its receptor activates the receptor to send a signal into the cell.

- Some receptors, like those for steroids, reside in the cytoplasm. However, when they bind their signal, they enter the nucleus, where they modulate gene expression.

- Most cells have many receptors and can sense several signals at once. They balance the strength of these signals in deciding what to do. This is almost like a reasoned judgement.

- Some signals are mitogens; they urge a cell into its growth and division cycle.

- Epidermal growth factor (EGF) is a mitogen that can be important for cancer. EGF binds its receptor, EGF-R, and activates it. This activation turns on protein kinases that up-regulate the expression of genes important for the onset of DNA synthesis. This kind of response is called signal transduction.

- Rb protein inhibits the onset of DNA synthesis. In its normal state, Rb blocks the expression of genes whose products are necessary for DNA synthesis. When Rb is phosphorylated, as a result of a mitogen-initiated signal transduction cascade, it is inactivated, so now the genes necessary for DNA synthesis can be expressed.

- DNA synthesis is a complicated process but remarkably accurate; it makes only one mistake in about 100 million nucleotides copied. However, there is a lot of DNA to replicate, so mistakes are made with every round of synthesis.

- Many mutations don't matter, in part because our cells are diploid—they have two copies of every chromosome—and in part because much of the DNA in our cells does not encode gene products.

- Cells carry out processes that recognize and correct errors in DNA, both errors that came from faulty synthesis and those that resulted from external insults, such as exposure to X-rays. A cell's ability to recognize and correct errors in DNA limits the likelihood of cancer.

- If the cell senses DNA damage, it can stop cell cycle progression through the action of checkpoints.

- There is also a checkpoint that allows a cell to recognize when not all chromosomes are properly attached to the spindle. This activity delays chromosome segregation until all the chromosomes are ready to separate and segregate.

- Some of the signals a cell can receive control the onset of apoptosis.

KEY TERMS

apoptosis

basal lamina

cell culture

centrosome

checkpoint

CRISPR-Cas

culture medium

cytokinesis

DNA polymerase

double-strand break

extracellular matrix (ECM)

endocrine

epidermal growth factor (EGF)

epidermal growth factor receptor (EGF-R)

erythrocyte

eukaryotic

extracellular matrix

G1

G2

homeostasis

interphase

leukocyte

lumen

MAP kinase

matrix metalloproteinase (MMP)

mitogen

mitosis

mutagen

Myc

nonhomologous end joining

nuclease

paracrine

plasmid

prokaryotic

promoter

proteinase

Ras

replication origin

restriction enzyme-retinoblastoma protein (Rb)

S phase

serum

signal transduction

sterile technique

TATA box

transcription factor

transit amplifying cells

tran

FURTHER READINGS

This chapter and Chapter 3 are essentially a primer in cell biology. There are descriptions of basic cell biology in any good introduction to biology, and there are also many excellent cell biology texts that describe all processes treated in this book in significantly more detail. As stated for Chapter 3, one useful text is *Essential Cell Biology*, 4th ed., by Alberts et al., published by Garland Science. *Molecular Biology of the Cell*, 6th ed., by the same authors and publisher, is a more in-depth text on the topics covered in this book. Other excellent texts include *Molecular Cell Biology*, 8th ed., by Lodish et al., published by Freeman, and *Cell Biology*, 2nd ed., by Pollard et al., published by Saunders.

Many aspects of cell biology are treated in the series of free videos available on the Internet at the iBiology Website http://www.ibiology.org/.

QUESTIONS FOR FURTHER THOUGHT

1. When a cell is receiving multiple signals, what do you suppose is going on biochemically in its cytoplasm?

2. Why is DNA synthesis imperfect?

3. If DNA synthesis were perfect, so the sequences of DNA never changed from one cell generation to the next, what would be the impact on the evolution of organisms?

4. Cells become genetically unstable through an increase in the frequency of mistakes in mitosis but also through failure of the cell cycle to correct errors in DNA synthesis. What do you suppose genetic instability does to the rate of evolution for cancer cells? Thinking about a population of cells with similar genetic instability, is there such a thing as too much genetic instability? What do you suppose is the fate of cells that become very genetically unstable?

ANSWERS TO SIDE QUESTIONS

4.1 Fetal calf serum works better because in contains more growth factors. Many cells in a fetal calf must be stimulated to grow and divide because the fetus is developing into a calf. In adult cows, the need for circulating growth factors is much less

4.2 To be ready for division, a cell must duplicate everything in a cell. An abbreviated list includes complete replication of DNA, proteins to make new mRNAs, sufficient mRNAs to get any essential proteins made, some copies of all cytoplasmic organelles, and sufficient plasma membrane to cover two cells and to provide all essential pumps, channels, and receptors, so the daughter cell can maintain its level of nourishment and an accurate awareness of its neighbors. Back in the nineteenth century, it was recognized that all cells come from cells, so an appropriate dowry for a cell is really everything in the cell.

4.3 Early activation of this proteinase would allow the identical copies of each chromosome to separate before the spindle was fully formed. As a result, the chromosomes might not separate properly. This would lead to aneuploidy, which is a form of genetic instability. It is likely to lead to serious trouble.

4.4 I would expect the expression of transcription factors to be regulated, so one is not making a large amount of any of them all the time. This implies that there are transcription factors that regulate the expression of other transcription factors. What an interesting and complicated situation! Just like a cell.

4.5 To differentiate, a cell withdraws from the cell cycle. If it does not differentiate, it can continue through the cell cycle. Thus, undifferentiated cells are more likely than a differentiated cell to continue the growth and division cycle unceasingly, just like a cancer cell.

4.6 The cellular event likely to lead to intestinal polyps is too many cycles of growth and division in the cells that provide replacements for dying epithelial cells, particularly the transit amplifying cells in the crypts (see Figure 4.11). A mutation that could lead to this behavior is a loss of function in a gene that encodes some protein involved in shutting off cell division as the transit amplifying cells reach the top of the crypt and start to differentiate fully. This loss would lead cells that carry such a mutation to keep on dividing.

4.7 Leukemias are always metastatic in the sense that the cells of the tumor are not sticking to their neighbors. Because they are floating in blood, they are not capable of staying in a single tissue (at one site) but instead are constantly wandering throughout the body.

4.8 Carcinomas (cancers of epithelial cells) are the most common kind of human cancer. All epithelia are separated from the underlying tissues by a basal lamina, which is one form of the extracellular matrix. Matrix metalloproteinases are enzymes that can digest this layer and allow cells to crawl through. Cancer cells often make and secrete MMPs to get through the basal lamina and initiate metastasis. If MMPs were effectively blocked, carcinomas should not be able to wander into the rest of the body; they would be confined to the epithelium in which they started. This should help to prevent the spread of cancer.

4.9 Cell division is likely to be required because the wound may have killed cells around the wound site. Simply signaling for cell migration will not produce enough cells to replace damaged or dead cells at the wound site.

4.10 You will remember from Chapter 3 that a phosphate group carries two negative charges. Adding these charges to the surface of a protein will alter the distribution of charges on the protein's surface and thereby change its structure. A change in structure is likely to mean a change in function.

4.11 Signaling to turn off an inhibitor in order to begin a process, versus the more direct signaling to turn on the process, will clearly achieve the same results, so in this sense the two are the same. Turning off an inhibitor is more complicated, because it adds a factor: the inhibitor. The added complexity does, however, open the possibility of an additional avenue for control, such as enhancing the inhibitor to prevent some other signal from turning the process on.

4.12 A gain in function of the Rb protein would lead to a stronger brake on the ability of the cell to initiate DNA synthesis. A cell carrying such a mutation would be less likely to divide than a normal cell, so this mutation would not promote the formation of cancer. A gain in function of EGF-R, on the other hand, would promote the onset of DNA synthesis and ensuing cell division. The same is true for a gain of function in Myc. Both of these changes sound carcinogenic.

4.13 If a person is less able than normal to repair their DNA, they are more likely to acquire mutations. Whereas many mutations are likely to be of no consequence, some may lead to a loss of normal cellular control and thus be carcinogenic. People with the loss-of-function mutation are therefore more likely to get skin cancer.

4.14 The population that cannot carry out apoptosis is more likely to die from a viral infection, because the cells in that population cannot be killed by apoptosis, so they are prevented from killing a virus-infected cell. A virus-infected cell will therefore will live long enough to produce lots of virus, which means viruses will eventually spread to all the cells in the population, killing them all. This is an example of a case where the death of a few cells (the first ones to become infected by the virus) is good for the population as a whole.

Factors that promote the cancerous transformation of cells

5

Doctors have treated and studied patients with cancer for many years. Their records show clearly that certain circumstances can greatly increase a person's risk of getting cancer. A combination of good medicine and good sleuthing has identified several hazards, such as exposure to specific substances, that undoubtedly increase the frequency of human cancer. For example, people who use tobacco or work with asbestos have a significantly greater chance of getting lung cancer than people who have lacked those exposures. Scientists have also explored the role of radiation in carcinogenesis, as mentioned in Chapter 2. Other scientists have found that certain viruses induce cancer in experimental animals at alarmingly high rates. Further work has identified a few viruses that promote cancer in humans, although it is very clear that most human-infecting viruses are not carcinogenic. Thus, there are multiple factors in our environment that can have a significant effect on the likelihood that cancer will occur.

Meanwhile, students of cancer have long known that some cancers run in families, suggesting that cancer is a heritable disease. There are also data showing that the prevalence of some cancers varies with diet, not genetics. For many years, this diversity of causes for cancer was confusing. Neither scientists nor doctors could understand how such diverse factors could increase cancer risk. In this chapter, we examine all these cancer-inducing factors, starting from a historical perspective. We begin with radiation, because the ways in which it is carcinogenic are comparatively straightforward and well understood. Then we will look at other cancer-causing factors and relate their carcinogenic activity to that of radiation. Finally, we will see how the work of generations of scientists allows us now to propose a unifying hypothesis about how these diverse phenomena contribute to cancer **risk**; that is, the likelihood that cancer will occur.

CARCINOGENIC FORMS OF RADIATION

CARCINOGENIC CHEMICALS

CARCINOGENIC VIRUSES

CANCER RISK THROUGH INHERITANCE AND LIFESTYLE

LEARNING GOALS

1. Describe why certain kinds of radiation are carcinogenic while others are not.
2. Explain how different chemicals can induce the same or similar types of cancers.
3. Compare and contrast the carcinogenic effects of radiation, chemicals, and viruses.
4. Identify three features of the human genome that help to minimize the likelihood that a mutation will lead to a change in an individual's phenotype.
5. Distinguish between carcinogenic factors that are mutagens and those that are not.

6. Explain how cancers can run in families, even though the disease itself is not inherited.

7. Describe how lifestyle choices can change the likelihood of cancer.

CARCINOGENIC RADIATION

The first form of carcinogenic radiation to be identified was X-rays. X-rays were discovered near the end of the nineteenth century by a German physicist named Wilhelm Röntgen. While working with electric currents, he generated an invisible form of radiation that could pass through air, glass, and human tissue, but it was stopped by heavy metals like lead. Shortly after this discovery, X-rays came into widespread popular use, resulting in the exposure of some people to quite high doses. Many of them became afflicted with cancer, providing evidence that X-rays are carcinogenic. The history of X-ray discovery and overuse is described in SIDEBAR 5.1.

Several kinds of radiation are now well known to be carcinogenic. This knowledge has been acquired gradually over many years and in most cases at considerable human cost. A familiar form of radiation with weak carcinogenic activity is the ultraviolet (UV) rays in sunlight. Less familiar are the three forms of rays emitted by unstable atomic nuclei, the kind of radiation termed **radioactivity**. These rays are much more carcinogenic than UV light, but to understand their carcinogenicity, you need to know a little more about each of these rays and how they interact with biological material.

An atom with an unstable nucleus is radioactive.

While Röntgen was working on X-rays in Germany, several other physicists were at work in other parts of Europe, studying the radiation emitted by some naturally occurring substances, like uranium. These scientists discovered and helped to characterize the phenomenon of radioactivity, as described in SIDEBAR 5.2. The chemists and physicists who made these discoveries built on knowledge about minerals that had been acquired over many previous decades. Thanks to this work, they knew that most natural materials, such as rocks, were composed of several different substances. They also knew that there were only a limited number of materials that were pure in the sense of being made from only one component. These pure substances were called elements, and the components from which they were made were called atoms. About 90 naturally occurring elements were recognized in the late nineteenth century, when this work was being done. Several of these elements were found to be radioactive.

Radioactivity is emitted by an element, such as uranium, that has an unstable atomic nucleus. The nucleus of every atom contains protons, which are tiny particles with a positive charge. The number of protons in an atom's nucleus is called its **atomic number**. The electric charges of these protons are balanced by an equal number of electrons, which are lightweight, negatively charged particles that circulate around the atom's nucleus (FIGURE 5.1A). Thanks to this balance of electrons and protons, atoms are normally uncharged. If an electron is added to or knocked away from the cloud surrounding the nucleus, that atom becomes charged, whereupon it is called an **ion**. Since like charges repel one another, the positive charges of protons in the atom's nucleus make them tend to fly apart. They are kept together by intranuclear forces that depend in part on the other principal particles in the nucleus, the neutrons. These particles

A CLOSER LOOK: The discovery of X-rays.

Wilhelm Röntgen was fascinated by a strange glow that occurred when he applied a high voltage to pieces of metal placed at opposite ends of a glass tube from which most of the air had been pumped out. By putting various targets in and around the tube, Röntgen discovered that not only light but also a strange and invisible form of energy was emerging from the tube (Figure 1A). This energy could travel long distances and penetrate cardboard, skin, and many other objects, but it was attenuated by bone and heavy metals. In 1895 Röntgen used this energy to capture the first image of bones inside the hand of a living human (Figure 1B). He called this form of energy X-rays because he didn't know what it was.

It took a long time before the scientific and medical communities realized the dangers of X-rays. X-rays were so interesting and useful that initially they were overused. Even as late as the 1940s, shoe stores often had an X-ray machine with which customers could see the bones of their feet as they fit into a new pair of shoes. The American inventor Thomas Edison developed a convenient device for direct visualization of people's bones. X-rays passed through the patient's body and landed on a screen that generated visible light when X-rays hit it, a so-called **fluoroscope** (Figure 2). Edison's lab assistant, Clarence Dolly, frequently demonstrated this device to friends and visitors, exposing himself to large doses of X-rays. Soon, he developed ulcers on his hands. Later, his hands developed cancer, which was treated by amputation of both arms. Dolly died of metastatic cancer in 1904. His condition may have been the first instance of death due to radiation-induced cancer.

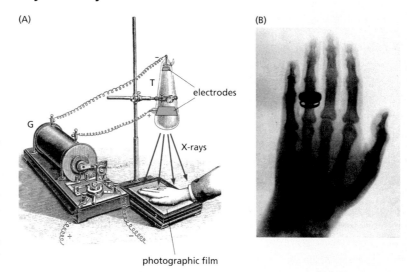

Figure 1 A device similar to the one used by Röntgen when he first discovered X-rays. (A) When a high voltage is applied to separated electrodes in an evacuated glass tube, X-rays are produced. This drawing includes a generator of high voltage (G), which is carried by wires to electrodes placed at opposite ends of a glass tube (T) from which most of the air has been pumped. The voltage causes the gas that remains in the tube to form ions (charged atoms or molecules), which are pushed by the voltage on the plates, so they carry current between electrodes. When the moving ions bang into the electrodes, they lose or acquire electrons, and this produces light of many wavelengths, some of which is the invisible light that Röntgen called X-rays. These rays are sufficiently energetic that they can penetrate the walls of the glass tube and pass through the surrounding air. If a photographic film is exposed to X-rays, it turns dark, allowing an image to be seen. Röntgen used this process to take a picture of his wife's hand, the first X-ray picture ever taken. (B) A negative of an X-ray picture, much like the one made by Röntgen and his wife. It shows finger bones and a metal ring, like the one she was wearing.

Figure 2 Thomas Edison invented the fluoroscope, a device that permits visualization of X-ray images. (A) X-rays are invisible to the eye, but they can be detected either with photographic film, as in Figure 1, or by allowing them to encounter a material that is fluorescent. The fluorescent material absorbs energy of one wavelength (for example, X-rays) and emits energy of a longer wavelength (for example, visible light). The fluoroscope that Edison designed was strapped to a viewer's head, shielding his eyes from room light. The end of a pyramid-shaped box far from the head held a glass screen, placed just far enough from the eyes to be easy to view. On the glass was a fluorescent material that converted X-rays into visible light. In this picture, Edison is using his fluoroscope to see the X-rays coming up from below and passing through the hand of his assistant, Clarence Dolly. Edison would have seen the bones in Dolly's hand in an image much like Figure 1B.

SIDEBAR 5.2

A CLOSER LOOK: The discovery and characterization of radioactivity.

The French physicist Henri Becquerel discovered that salts of uranium produced a form of radiation that displayed some of the properties of X-rays. Maria Sklowdowska, a Polish scientist then working in Paris, carried out experiments to understand the radiation-producing properties of several kinds of uranium ore. She was a good chemist and was able to purify uranium from these ores, only to find that pure uranium produced less radiation than the crude ores. She inferred that other components of each ore were responsible for much of its radiation. Together with her colleague and future husband, Pierre Curie, she purified these radiation-producing materials from the same ores. The first such substance she called polonium (in honor of her native Poland), and the second she called radium. Each of these substances is an element that emits strong radiation analogous to the X-rays discovered by Röntgen. Becquerel, the Curies, and Röntgen shared several Nobel Prizes in honor of their discoveries.

It took years of experience for people to realize how dangerous these various kinds of radiation could be. You cannot perceive any of these types of radiation with any of your five senses, unless the dose is so high it causes a burn. Because radiation generally caused no pain, most people presumed that it did no harm, and some enthusiasts even thought it did you good. Radithor was a trade name for radium suspended in water. William Bailey sold about 400,000 bottles of Radithor to both doctors and the general public, advertising that it would make you healthy and strong. One user drank many bottles of this tonic, but some years later he began to lose his teeth and then become anemic. Subsequently, he died from a loss of the ability to make new blood cells.

In an analogous example, many customers were delighted with wristwatches in which radium had been added in small amounts to a fluorescent paint that marked the numbers on their watch dials. The radiation from radium stimulated emission of light from the fluorescent paint; these glow-in-the-dark watches allowed to you see the time, even in the pitch black. The workers who applied the radium got a significant exposure to its radioactivity, especially because they needed a fine-tipped paint brush to do the application neatly. This involved licking the tip of the brush to get its hairs nicely bunched. Many of these painters came down with jaw problems and loose teeth. Upon their death, autopsies revealed that many of them had various kinds of leukemia. Since humans cannot feel X-rays or radioactivity, unless the radiation is intense, these workers and their bosses were unaware of exposures that would cause them so much trouble in the long run.

have no charge, but each neutron has about the same mass as a proton. The number of neutrons in an atomic nucleus can differ from one form of an atom to another; these different forms of atoms with a given number of protons (the same atomic number) are called **isotopes** of that atom type. For example, the nuclei of all oxygen atoms contain eight protons, but the nuclei of different isotopes of oxygen contain different numbers of neutrons. The most abundant isotope of oxygen is oxygen-16, with eight protons and eight neutrons. Having too many or two few neutrons can lead an atomic nucleus to be unstable; such an isotope is radioactive.

Most kinds of atoms have multiple isotopes. The nuclei of some isotopes are stable, but some are not. For example, most carbon on earth is carbon-12 (six protons and six neutrons), which is stable. A small fraction of all natural carbon, however, is carbon-14 (chemists and physicists write the names of different isotopes with a superscript, such as ^{14}C), which has two extra neutrons, making that nucleus somewhat unstable. About half of all the ^{14}C nuclei in a given sample of carbon will change every 5730 years, emitting radioactivity. This nuclear change occurs randomly, so one cannot know when any particular ^{14}C nucleus will change, but on average, half of all ^{14}C nuclei will change in the stated number of years, so this time is called the **half-life** of this isotope. Because the emission of radiation from ^{14}C occurs only occasionally, this element is not very radioactive. On the other hand, the nucleus of one isotope of the naturally occurring gas radon, known as radon-222 (^{222}Rn), is much less stable than the nucleus of ^{14}C. Half of these nuclei fall apart every 3.8

SIDE QUESTION 5.1

To get a feeling for the impact of differences in half-lives on the amount of radiation emitted by radioactive elements, solve the following arithmetic problem. You have one sample that contains one billion atoms of ^{14}C and another sample with one billion atoms of ^{222}Rn. In one hour, how many radioactive events will have occurred in each sample? (Hint: Ask yourself what fraction of a half-life is one hour for each of these two elements?)

days. Thus, even a small amount of this form of radon will emit a lot of radiation.

Radioactive elements, like uranium and radon, were found to produce three kinds of radiation called alpha, beta, and gamma rays (FIGURE 5.1B). An **alpha ray** is like the nucleus of a helium atom; it contains two protons and two neutrons. When alpha rays are emitted from an unstable atomic nucleus, they travel very fast (almost 10,000 miles per second) until they bump into something, like a nearby atom. Given their speed and mass, they can alter whatever they bump into, which is why such radiation is damaging. **Beta rays** are fast-moving electrons. They travel about 165,000 miles per second, so even though they are less massive than alpha rays, they have enough energy to change the molecules they bump into. **Gamma rays** are physically similar to X-rays and visible light, but their wavelength is even shorter and they are much more energetic than anything our eyes can see.

The relationship between gamma rays, X-rays, UV light, visible light, microwaves, and radio waves is important to understand, so a spectrum of these rays, called electromagnetic waves is diagrammed in FIGURE 5.2. The term electromagnetic comes from the fact that all these forms of energy are fast oscillations of electricity and magnetism that propagate at the speed of light (186,000 miles per second). On this figure you can see some forms of electromagnetic energy that are proven carcinogens (gamma rays, X-rays, and UV light) and others that are not (visible light, infrared, etc.). To understand these differences we must answer the question, How do these different forms of radiation interact with biological material to promote cancerous transformation?

All ionizing radiation is carcinogenic.

Four kinds of radiation are so energetic that they can knock an electron away from an atom they hit, turning the atom into an ion. Alpha, beta, and gamma rays and X-rays are therefore called ionizing radiation. When a gamma ray or X-ray passes through water, it will occasionally split a water molecule into two parts: a hydrogen atom (H), in which one electron has been displaced to make it reactive, and a **hydroxyl group** (OH), which also has a displaced electron (FIGURE 5.3). These displaced electrons are in a condition that makes them chemically very reactive. The products of water splitting will sometimes recombine to re-form ordinary water, but more commonly the pieces of water behave as **free radicals**, which means they can react with almost any molecule they encounter. The hydrogen radical (H•, where the dot to the right of the letter indicates the displaced electron that is so reactive) can interact with oxygen (O_2), which is dissolved in water, to make HO_2•, called the hydroperoxy radical. The hydroxyl radical (OH•) can react with water or another OH• to make hydrogen peroxide (H_2O_2), a chemical commonly used for bleaching clothes, wounds, or hair. All of these molecules contain oxygen in a form that is highly reactive, so collectively they are called **reactive oxygen species** (**ROS**). If ROS bump into another molecule, its components will almost always react with that molecule, altering its structure.

When X-rays or gamma rays pass through a glass of water, free radicals are generated in large numbers. According to one study, a single gamma ray can generate as many as 36,000 OH• molecules! When generated in water, these radicals react with water or with each other and ultimately die out, meaning that they form unreactive molecules. However, when these same rays pass through the water in one of our cells, the free radicals that form can react with proteins and other components of the cell. If the components of ROS interact with a protein, the latter is commonly

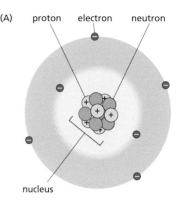

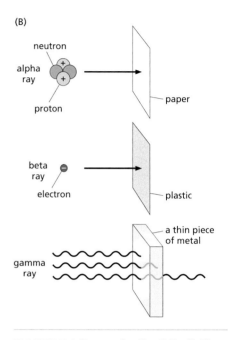

FIGURE 5.1 **Atoms and radioactivity.** (A) All atoms contain a nucleus made of protons, which are positively charged. The nuclei of most atoms also contain neutrons, which have no charge. Negatively charged electrons, equal in number to the protons in the nucleus, orbit the nucleus. (B) Many atomic nuclei are stable, but if there are either too many or too few neutrons, the nucleus may spontaneously change, emitting any of three forms of radiation: alpha rays, which contain two protons and two neutrons; beta rays, which are electrons, and gamma rays, which are high-energy light, too high in energy for our eyes to see. X-rays too are high-energy light, but gamma rays are even more energetic, meaning that they can penetrate thicker targets and do more damage. Alpha rays, also called alpha particles, are sufficiently massive and charged that they interact strongly with matter. A piece of paper or the outer layer of your skin will stop them. However, the stopping process can be damaging; if cells are exposed directly to alpha particles, their DNA can be damaged. Beta rays interact with matter less strongly, so they can penetrate more deeply into tissues. An energetic beta ray can penetrate paper and skin, but it is stopped by a half-inch-thick piece of glass or plastic. Gamma rays are the most penetrating of all these rays; they can penetrate paper, glass, plastic, and skin quite easily, but they are stopped by dense materials, like lead.

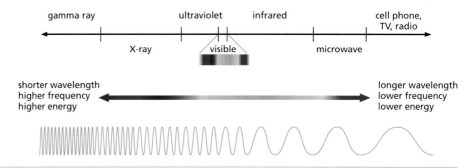

FIGURE 5.2 The full spectrum of electromagnetic energy. Visible light is one part of a wide range of similar kinds of energy. Sunlight includes visible light but also infrared light, which has longer wavelengths than visible light, and ultraviolet (UV) light, whose wavelengths are shorter. UV light is sufficiently energetic to be carcinogenic. X-rays are even more energetic; they can break molecules of water into pieces, as diagrammed in Figure 5.3. Gamma rays are even more energetic, so they too can break water molecules. Toward the other end of the electromagnetic spectrum, there are waves of longer wavelength and therefore lower energy. Here you find infrared light, microwaves, and radio waves, all forms of electromagnetic energy that are less energetic than visible light, and therefore unlikely to be carcinogenic. This region of the spectrum includes microwaves, like those used for cooking, TV, cell phones, and radio transmission. The electromagnetic waves generated by electricity transmission lines are of even longer wavelength.

SIDE QUESTION 5.2

Figure 5.4 shows an overview of what happens to a sequence of nucleotides when mutated DNA is replicated. Refer to Chapter 4, Figure 4.21, and use that figure plus Figure 5.4 to draw your own diagram of what happens when damaged DNA is replicated.

inactivated, but this is usually not important. Cells contain many copies of most proteins, so normal tasks are carried on without the help of the damaged one. If ROS bump into DNA, on the other hand, there is a good chance that they will cause a mutation. Examples of the chemical changes that ROS can make in DNA are shown in FIGURE 5.4. Both X-rays and gamma rays are therefore mutagens, that is, factors that increase the likelihood that a cell will give rise to mutant daughter cells.

Beta rays, or fast-moving electrons, are emitted by radioactive isotopes of several familiar elements, such as phosphorus. Although beta rays are generally not as energetic or as penetrating as gamma rays (see Figure 5.2), they can still damage DNA, either by direct interaction with the electrons of DNA or by generating ROS. Alpha rays, too, are mutagenic. Indeed, alpha particles are so big and reactive they can't travel far in any kind of material; they keep bumping into the atoms that make up the molecules of a cell, whereupon they lose their energy as heat.

As discussed in Chapter 2, a low level of X-ray exposure is of little or no consequence, because the chances of a mutation that severely damages the cell are small. Moreover, the irradiated cell will often detect the DNA damage and repair it. At higher levels of radiation, however, the increased risk of carcinogenesis becomes something to worry about (SIDEBAR 5.3). Thus, scientists and medical professionals who work with X-rays or gamma rays wear shielding to reduce the amount of this radiation that will enter their bodies. They also wear monitors that measure the total dose of radiation they have been exposed to. Health-safety officials check those monitors to make sure that these people are not accumulating a significant level of irradiation. With proper safeguards, the risk of cancer from this kind of exposure is very small.

FIGURE 5.3 The action of ionizing radiation on water. (A) High-energy rays, like gamma rays or X-rays, are so strong that they can alter the behavior of the electrons that form chemical bonds between the atoms in any molecule they encounter. This can be a macromolecule, but by far the most common molecule in cells is water, so water is the molecule most likely to be affected by ionizing radiation. In some cases, such interactions knock an electron off of water, making it reactive. In other cases, radiation causes water to break apart, forming a hydrogen radical (H$^\bullet$) and a hydroxyl radical (OH$^\bullet$), which are both very reactive. Two hydroxyl radicals can interact to make H_2O_2, the molecule called peroxide that is used for bleaching hair and disinfecting wounds. The hydrogen radical can interact with oxygen dissolved in water to form super oxide ($O_2^\bullet$), which in turn reacts with water to make a molecule called the hydroperoxyl radical (HO$_2^\bullet$), another reactive molecule. Collectively, these reactive molecules are called reactive oxygen species (ROS). ROS can go on to interact with and damage proteins, carbohydrates, lipids, or nucleic acids. The most important of these reactions is with DNA, because it can cause mutations.

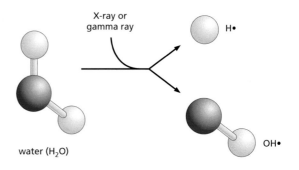

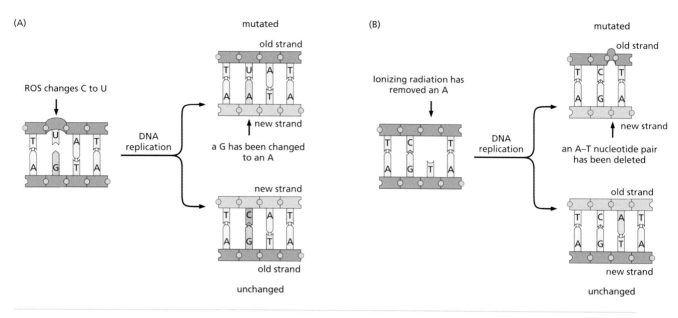

FIGURE 5.4 Radiation-induced damage to bases in DNA can lead to mutations. (A) An example of the kind of damage ROS can do to DNA. ROS can break chemical bonds. The nucleotide G was originally paired with C, but part of C got knocked off by ROS, making it U, which is normally found in RNA. U is not complementary to G, so it cannot form good bonds with it, resulting in a bulge in DNA structure. Worse, after DNA replication, one of the DNA duplexes formed is now mutant, because U pairs better with A than with G, so an A has been inserted into one of the daughter strands. (B) In this example of DNA damage by ionizing radiation, the A that used to pair with T has been removed. After DNA replication, one duplex is again correct but the other contains a mutation, this time a nucleotide deletion. Such mutations can make trouble when the RNA transcribed from this DNA is read on a ribosome. A one-base deletion results in a shift of the frame for reading the three consecutive bases that make up a codon, so the amino acids added to the protein are generally wrong from there on. Such a sequence will soon include an instruction to stop translation. The resulting gene product is almost always functionless, which can lead to changes in cell behavior. (Adapted from Alberts B, Johnson AD, Lewis J et al [2015] Molecular Biology of the Cell, 6th ed. Garland Science.)

The naturally occurring gas radon is a significant risk for lung cancer.

Radon, composed largely of ^{222}Rn, is a naturally occurring, radioactive gas that forms during the radioactive decay of the element radium. In some parts of the world, like the Czech Republic, western Scotland, the northern Midwest of the United States, and the Rocky Mountains, radon

A CLOSER LOOK: The carcinogenic effects of high-level radiation.

The role of radioactivity in the onset of human cancer was demonstrated unequivocally, though at tremendous human cost, near the end of the Second World War. Many Japanese people who lived near the sites of atom bomb explosions, detonated by the United States at Hiroshima and Nagasaki, were killed immediately. However, about 100,000 people in the vicinity of the explosions lived far enough away to survive. Their post-blast medical histories have provided inescapable evidence that the strong radioactivity generated by these blasts was carcinogenic.

The first kinds of cancer to appear among the survivors were blood cancers, like leukemia. Within 5 years, the incidence of leukemia in these populations was far above that in control groups, that is,

people who lived farther away from the blasts. At that time, their incidence of solid tumors, like carcinomas, was no higher than that of the control groups. As time passed, however, the incidence of carcinomas in people who had received significant irradiation increased to well above control levels. Moreover, the closer people lived to the blast sites, the higher their incidence of cancer, suggesting that the amount of radiation received made a difference but that it took time for the effects of radiation to show up in the formation of solid tumors. The conclusion that radiation does hidden damage and increases cancer risk was now inescapable. As soon as this was realized, strong cautions about exposure to radiation were enacted, both in law and in common medical and scientific practice.

SIDEBAR 5.3

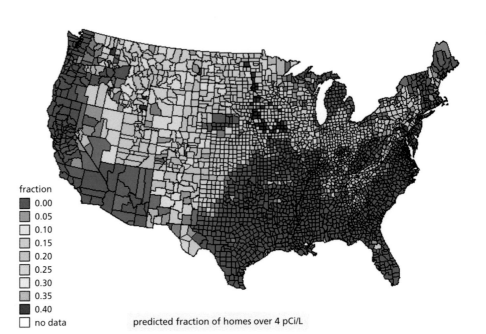

fraction
- 0.00
- 0.05
- 0.10
- 0.15
- 0.20
- 0.25
- 0.30
- 0.35
- 0.40
- no data

predicted fraction of homes over 4 pCi/L

FIGURE 5.5 Amounts of radon in different parts of the United States. A heat map showing the amount of radon found in different parts of the United States. Colors indicate the concentration of radon near the earth's surface, with blue meaning low and red meaning high. In the places where the color is not blue, radon is being emitted from the earth at a high enough rate that it is likely to contaminate houses built there. Data are in picoCuries per liter (pCi/L) of air. A level of 4 pCi/L corresponds to about 9 radioactive events per minute in each liter of air. (Courtesy of the Environmental Protection Agency.)

is found in rocks near the earth's surface, so it leaks slowly into the atmosphere (FIGURE 5.5). When these rocks are disturbed, as when people mine for ore or dig the foundation for a house, radon is released at a higher rate. If this release is outdoors, the radon gas blows away, so its concentration in any one place becomes low, and it poses no health hazard. If it is released into an enclosed space, however, it can accumulate to levels where it constitutes a danger, particularly as a carcinogen of the lungs.

^{222}Rn has an unstable atomic nucleus, so it is radioactive. A radon nucleus rearranges with a half-life of 3.8 days, emitting an alpha ray and forming another radioactive atom. There follows a cascade of alpha- and beta-ray emissions that results ultimately in the formation of lead-206, which is stable and doesn't change further (FIGURE 5.6). The four alpha rays released by one atom of radon are a health hazard because they are potentially damaging to DNA. Alpha particles are different from other kinds of ionizing radiation: because they are so reactive, they don't travel very far. An atom that emits alpha rays on the surface of your skin will blast those already dead cells, but the rays don't penetrate far enough to reach the living and dividing cells that are under the many layers of dead and dying cells near the skin's surface. In the lungs, however, the epithelium has no overlay of dead cells and is very thin. The epithelium must be thin to allow the efficient exchange of oxygen from the air we breathe with carbon dioxide in the blood. (Recall that carbon dioxide in the blood is a waste product of the metabolic activity of cells all over the body.) If radon gas gets into the lungs, the four alpha rays it emits can reach the nuclei of lung cells and damage their DNA.

Radon is now regarded as a significant carcinogen in the development of lung cancer, second in importance only to cigarette smoke. Studies of

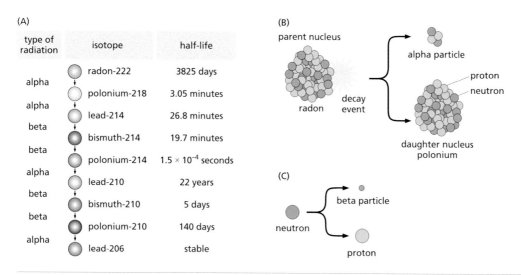

type of radiation	isotope	half-life
alpha	radon-222	3825 days
alpha	polonium-218	3.05 minutes
beta	lead-214	26.8 minutes
beta	bismuth-214	19.7 minutes
alpha	polonium-214	1.5×10^{-4} seconds
beta	lead-210	22 years
beta	bismuth-210	5 days
alpha	polonium-210	140 days
	lead-206	stable

FIGURE 5.6 **Radioactive events by which the nucleus of radon decays.** (A) Radon, which is a gas, can be inhaled with air. Its short half-life means that many of the inhaled atoms will soon decay and produce polonium, which is not a gas and can adhere to the surfaces of the cells that line the lung. The half-lives of the subsequent decays vary, but three alpha particles and two beta particles are emitted soon, meaning that after radon gas has been inhaled, the cells of the lung are exposed to rather severe ionizing radiation. (B) When an atom of radon emits an alpha particle, the numbers of both protons and neutrons in its nucleus go down by two, changing both the atomic number (which changes the name of the element) and the atomic weight. (C) Since a beta particle is a fast-moving electron, the neutron loses a negative charge and becomes a proton, again changing the atomic number of the atom.

miners who dug uranium, in which radium is a common contaminant, and of people who lived for years in a radon-contaminated house have provided clear evidence that exposure to radon gas is a significant risk in lung cancer. For these reasons, getting a new home tested for radon gas, particularly in basement rooms where children may sleep and play, is a worthwhile thing to do (**SIDEBAR** 5.4). Radon exposure is bad for anyone, but its carcinogenic effects are even stronger for people who smoke tobacco.

CASE STUDY: Rob

My own experience with radon testing is worth reporting. When my wife and I moved to Boulder, Colorado, in 1970, I had never heard the word radon outside of a physics course, and the testing of houses for radon was not done. I spent the first summer after our move fixing up the downstairs of our house, so there would be a utility room for laundry, a playroom for our kids, and a room for Rob, then 4 years old. He chose the color scheme (racing car red on one wall, ivory on the trim) and together we put in two closets, one with shelves. He stayed happily in that room for about 10 years, enjoying the independence of being one floor down from everyone else and right next to both the playroom and an outside door. It was only a couple of years after his cancer diagnosis that we thought about radon and the fact that our house had never been tested. We found a local company that would do the job by the simple

method of putting some gas-absorbing charcoal into Rob's old room for a defined period of time and then taking it off for the detection of any radioactivity. The test came back with a higher-than-acceptable radon count, indicating that throughout his childhood, Rob had been exposed to this damaging gas at a level that exceeded recommended levels. Marjorie and I were really upset to learn that our ignorance might have contributed to his cancer, but when we called the Colorado Health Authorities (it was then 2002), they were clear that there was no evidence to show that radon was a human carcinogen. Now we know that it is. This story shows that our knowledge of the breadth and complexity of the cancer problem is increasing all the time, but it is also a poignant fable about the fallibility of our wishes to do right by our families.

SIDEBAR 5.4

The ultraviolet rays of sunlight are a significant risk for skin cancer.

Sunshine contains light whose wavelengths are too short for our eyes to see (ultraviolet light) as well as light with wavelengths too long for visualization (infrared light) (see Figure 5.2). Infrared light provides our bodies with heat but has no known damaging effects upon our cells. Ultraviolet (UV) light is different, because a person's risk of skin cancer increases with exposure to three different ranges of ultraviolet light: the shortest wavelength, UV-A; the intermediate wavelength, UV-B; and the longer wavelength, UV-C. All this radiation is less energetic than either gamma rays or X-rays, but UV-A is still sufficiently energetic to be ionizing. Its mechanism of carcinogenesis is the same as that described above for X-rays and gamma rays. UV-B is not ionizing, because its energy is not sufficient to knock an electron out of a molecule. However, it is still carcinogenic because its energy is absorbed directly by the bases that pair between complementary strands of DNA. When this energy is absorbed by thymine (T), the base becomes activated to carry out a chemical reaction that would not normally occur; it forms bonds with a neighboring T along one strand of a DNA double helix. These bonds distort the structure of DNA (FIGURE 5.7), and if such DNA is replicated before the distortion is repaired, a mutation is introduced.

UV light does not increase the risk of cancer in the liver, colon, or other internal organs, because it is not sufficiently energetic to penetrate into the body; it interacts only with skin cells near the body's surface. This is the reason that people whose skin lacks the protection of melanin should use sunblock when they spend a lot of time in bright sunshine. Recognize, though, that the sun protection factor (SPF) of sunblock is not a reliable indicator of protection against UV mutagenesis, and there is controversy about the relative roles of different wavelengths of UV light in skin cancer. Regardless of the controversy, taking care to limit sunlight exposure is particularly important for children, not because their DNA is more sensitive than that of adults but because they probably have more years of life ahead of them than an adult. There will therefore be more time for them to accumulate additional unlucky mutations, which could be steps along the road to cancer. Note, however, that there is a downside to limiting one's exposure to UV-B. This part of sunlight has a beneficial effect on human health; it converts a steroid that our bodies can make into the essential nutrient vitamin D, which we cannot make by any other route. Thus, depriving people of *all* bright sunlight is bad for their health.

The need for this vitamin-producing, UV-B-dependent reaction has led human populations that have lived for many generations in different parts of the world to have different levels of the light-absorbing pigment melanin in their skin. Near the equator, sunlight is so strong that humans

FIGURE 5.7 UV-B can induce adjacent Ts to form abnormal bonds, distorting the DNA duplex. A model of DNA, showing the backbones of two DNA strands in a double helix. The structure of some nucleotides is normal, and complementary bases form bonds from one helix to the other (dark gray and blue). As a result of energy absorbed from UV-B, parts of two adjacent bases (Ts shown in purple) have bonded along one DNA strand, distorting the double helix and preventing proper bonding between strands. This distortion will interfere with proper base pairing during subsequent DNA replication. If this improper linkage of Ts is not repaired before S phase, mutations will result. (From Goodsell DS [2001] *The Oncologist* 6:298–299. With permission from the Society for Translational Oncology.)

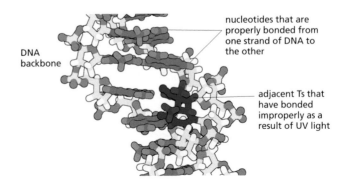

nucleotides that are properly bonded from one strand of DNA to the other

DNA backbone

adjacent Ts that have bonded improperly as a result of UV light

make lots of melanin to protect their skin cells from too much UV; being dark-skinned is a significant advantage in avoiding the carcinogenic effects of UV-B. The equatorial sun is so bright that enough UV-B gets past the melanin to make sufficient amounts of vitamin D. Far from the equator, it is advantageous to let more UV through the skin to keep vitamin D levels up. Therefore, peoples who have lived for many generations in the far north have lost much of the pigmentation needed when living nearer the equator. Only those people who had reduced the darkness of their skin could produce enough vitamin D to thrive. Again, we see the importance of balance in human health: some sun exposure is good and healthy, and too much can be bad for you.

Not all forms of radiation are carcinogenic.

What about other kinds of radiation to which we are commonly exposed? Microwaves are used in ovens to heat and cook food, while radio waves are used to transmit TV, radio, and cell phone signals. High-tension power lines radiate very low frequency radiation of high intensity. All these forms of radiation are electromagnetic waves, as indicated on the spectrum shown in Figure 5.2. Then there are sound waves, which are not electromagnetic; they are waves of molecular vibration. Are any of these forms of radiation carcinogenic? On current evidence, sound waves are clearly not carcinogenic, but quite a few scientists have claimed that long-wavelength electromagnetic radiation can be. However, a careful consideration of the evidence so far published finds it unconvincing. None of these forms of electromagnetic radiation is sufficiently energetic to be ionizing, so they are very unlikely to generate ROS. Electromagnetic radiation with wavelengths longer than infrared light are absorbed poorly by the molecules in cells. They are not like UV-B, which interacts strongly with DNA. Thus, it is hard to see how they could cause the mutations brought about by gamma rays and X-rays, which are proven carcinogens. The balance of evidence therefore says it is unlikely that any of these forms of radiation is carcinogenic. However, one must always be prepared for a new discovery in which a clever scientist identifies a novel way to improve our understanding of nature. One cannot say that these forms of radiation will never be identified as carcinogenic, but on current evidence, they are nothing to worry about.

CARCINOGENIC CHEMICALS

The carcinogenic action of some materials has been known for a long time. Near the end of the eighteenth century, two English doctors noticed connections between exposure to certain substances and specific kinds of cancer. John Hill found that people who used snuff, a powered form of tobacco that one takes by sniffing, had a higher than normal rate of nasal cancer. Percivall Pott realized that adult men who had worked as chimney sweeps during their childhood showed an abnormally high rate of cancer in the skin of the scrotum. Pott surmised that the soot these men encountered when they were children, combined with their poor sanitation habits, led to their being exposed to a noxious substance for far longer than necessary. He proposed that improved washing would reduce this workplace hazard. Although English customs were slow to change, prompter action urging daily baths for chimney sweeps in other countries provided evidence that Pott was right. The observations that snuff and soot exposure correlated with higher-than-normal rates of cancer nurtured the idea that noxious chemicals in our environment were the cause of cancer.

SIDE QUESTION 5.3

Vitamin D is also available from pills and from some natural substances, like oil from the livers of cod fish (which tastes pretty awful). If you had an active 4-year-old child who liked to be outside, would you demand the use of sunblock at all times? If so, would you do anything to supplement his diet? Explain your answer.

SIDE QUESTION 5.4

From what you now know about electromagnetic waves, are you inclined to use your cell phone less? Why?

Even purified chemicals can be carcinogenic.

Since this pioneering work, many substances have been identified as carcinogenic. Some evidence for this statement comes from work with experimental animals, like laboratory mice or rabbits. For example, in the early twentieth century a Japanese scientist, Katsusaburo Yamagiwa, applied coal tar, a crude extract analogous to the soot that plagued chimney sweeps, to the ears of rabbits. Repeated exposures led to the development of carcinomas in the treated skin. As methods of chemistry improved during the subsequent decades, specific molecules were purified from coal tar and other crude substances. When these chemicals were tested by repeated application to the skin of laboratory animals, several of the molecules called PAHs (polycyclic aromatic hydrocarbons) were found to be carcinogenic (FIGURE 5.8). The case for chemical carcinogenesis became strong, and the new knowledge raised the concern that many chemicals might be carcinogenic.

Companies that invented or purified chemicals for human use (cosmetics, pharmaceuticals, and household products like cleaning supplies, paints, glues, lubricants, and solvents) came to recognize the importance of chemical carcinogenesis, so they developed tests to identify carcinogenic properties in their products. For example, two equivalent groups of small mammals (laboratory mice, rats, or rabbits) would be maintained under healthy conditions. One group would be treated frequently with the chemical to be tested, while another would be left untreated (the control population). If no cancers emerged over months or even years of treatment, the chemical could be described as noncarcinogenic. Experience showed, however, that to say a given chemical is NOT carcinogenic, one must apply it to an experimental animal over a prolonged time. Moreover, the appropriate dose and duration of treatment were not well defined. These issues led to uncertainty about results of such testing, and since the costs of this work are high, companies have sought alternatives.

It must be added that some well-informed scientists have argued that results from animal tests cannot be taken as proof that a given chemical is or is not carcinogenic in humans. There are clear differences in the ways different animals respond to toxic chemicals. Because the United States is such a litigious society, companies that invent and/or produce chemicals for public consumption have a financial motivation to show that they have done all that can reasonably be expected to protect the

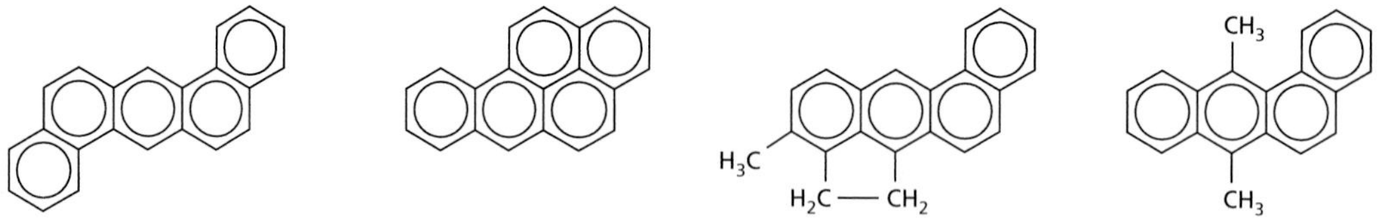

dibenz[a,h]anthracene benzo[a]pyrene 3-methylcholanthrene 7,12-dimethylbenz[a]-anthracene

FIGURE 5.8 Structures of some chemicals formed during incomplete burning of materials like wood or coal. Each diagram shows the structure of a molecule that has been purified from a material, like soot or coal tar, and then shown to be carcinogenic in experiments like those conducted by Katsusaburo Yamagiwa. These molecules are all of a type called polycyclic aromatic hydrocarbons (PAHs). A substance like soot, which contains these and many other chemicals, is obviously complex. Several of the chemicals in soot or coal tar are carcinogenic, particularly the PAHs. Such common and naturally occurring materials are therefore carcinogenic. These diagrams are characteristic of the ones used by chemists to display the arrangement of atoms in a molecule's structure. As described in previous chapters, there is a carbon atom at the vertex of each hexagon, usually represented by C but here omitted for brevity. Each edge of a hexagon indicates a chemical bond between carbons. The circle within a hexagon indicates additional chemical bonds, because many of these carbon atoms are joined by what chemists call a double bond. The chemical name for each molecule appears complex, but there is a logic to how each name is given. The name tells a chemist what the structure looks like, so the diagram is not necessary to convey structural information. Benzo[a]pyrene was one of the first chemical carcinogens to be purified; it is a significant molecule in the carcinogenic activity of both soot and cigarette smoke.

(A)

(B)

(C)

FIGURE 5.9 **Asbestos is prepared from a mineral, yielding strong, chemically inert fibers with multiple uses.** (A) A piece of tremolite asbestos from California, one of the ores from which asbestos can be purified. (B) The white material is asbestos. While this worker is protected by an air-circulating system from breathing the asbestos fibers that might fluff up while he works, his clothes and hands may become contaminated in ways that might lead to his breathing this material later. We now know that gloves and protective clothing should have been worn. (C) Asbestos in an attic, where it is helping to insulate the lower rooms against heat loss. (A, courtesy of the California Department of Conservation; B, courtesy of John Austin; C, courtesy of the Environmental Protection Agency.)

public from health hazards. However, the resulting research on carcinogenesis increases costs, which are passed on to the consumer. The identification of danger in new chemicals is an important issue that will soon be described in more detail.

In a few cases, however, people have been exposed to carcinogenic chemicals without anyone planning it. Asbestos is a mineral composed of several chemical arrangements of the elements silicon and oxygen (FIGURE 5.9). Lumps of asbestos can be broken up into inert fibers that can be woven into cloth or packed into a batting that was widely used as a fireproof insulation for home and office construction until the mid-1970s. About that time, however, asbestos was recognized as a human carcinogen. Fibers of asbestos are small enough that they can become airborne, so one can internalize them by breathing. The fibers stick to cells in a particular lung tissue called the mesothelium (FIGURE 5.10). Small asbestos fibers are then brought into the cells by endocytosis, whereupon they can interfere with intracellular motions, like the segregation of chromosomes during mitosis. It is probably this phenomenon

(A)   (B)

(C)   (D)

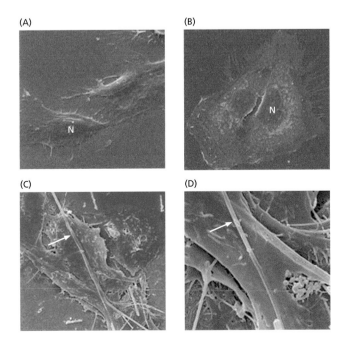

FIGURE 5.10 **Human mesothelial cells in culture treated with asbestos.** (A, B) Normal, untreated mesothelial cells growing on a glass surface and imaged with a scanning electron microscope. The cells have spread out on this surface, becoming thin except near their centers, where the nuclei of the cells (marked N) maintain some thickness. (C, D) An identical culture of mesothelial cells treated with asbestos fibers (indicated by arrows). The fibers stick to the cells and modify their behavior. Tiny fibers of asbestos can even be brought into these cells by endocytosis (not shown). (From Andolfi L, Trevisan E, Zweyer M et al. [2013] *J Microsc* 249:173–183 [doi:10.1111/jmi.12006]. With permission from Wiley.)

that leads asbestos to be a carcinogen. Many scientists have tried to learn exactly how asbestos carcinogenesis works. One publication obtained evidence that treatment with asbestos fibers caused a loss-of-function mutation in the Rb gene, which you know from Chapter 4 helps to prevent the onset of DNA replication. Perhaps the presence of asbestos fibers in dividing cells causes errors in mitosis, and some of these lead to loss of proper cell cycle control. Further research will be required to learn the real mechanism of asbestos carcinogenesis.

The data showing that asbestos fibers are carcinogenic have been painfully acquired by the study of unfortunate people who worked in the mining, milling, and processing of asbestos or who laundered the clothes of those who did. All this happened before medical science had discovered the strong correlation between asbestos exposure and a cancer called **mesothelioma**. Initially, neither the workers nor their employers knew the precautions that should have been taken to prevent dangerous exposure, but now great care is taken by people who work with asbestos to prevent them from inhaling the fibers. There is much less evidence about danger from asbestos that is already in the walls of buildings. Nonetheless, there is now a significant industry that removes asbestos from older buildings. This is probably a good thing, but asbestos insulation that is undisturbed, so its fibers are not airborne, does not present the same risk as that encountered by people who mined or milled it all day long.

Identifying a carcinogen can take a long time.

Why did the asbestos industry, and other companies that worked with materials now known be carcinogens, allow their workers to be exposed to these dangerous substances? The answer in almost every case is that employers did not know of the danger because the action of a carcinogen is slow. It is generally 20 or more years between the initial exposure to a carcinogen and the first appearance of an increased frequency of cancer, particularly of carcinomas. This phenomenology has made it hard for even the most responsible employers to identify carcinogens they may have used in the past and to be certain there is no carcinogenic substance or process in the current workday experiences of their employees.

For a long time, no one understood this delay between exposure to a carcinogen and the appearance of cancer. People are certainly more accustomed to worry about rapid impact dangers, like overexposure to the heat from a furnace that results in a burn, or the ingestion of an acute poison that results in internal damage. In these rapid-impact cases, cause and effect are connected within seconds to hours. The interval of years between exposure to a carcinogen and the onset of cancer was puzzling, not only for the doctors who sought to improve health in the workplace but also for scientists who were trying to understand carcinogenesis.

Substances that are proven human carcinogens are called class 1 carcinogens. Class 2 carcinogens are proven carcinogens in nonhuman animals but not in humans. The latter substances are reasonably expected to be human carcinogens, but good evidence on the issue is not yet available. Names of some carcinogens in each class are given in TABLE 5.1. They represent a wide range of compounds, wide enough that it is not obvious what they have in common. However, several compounds in both classes have been studied carefully, and from this work, a pattern emerges. These substances, or derivatives of them, can react with the nucleotides in DNA. When they do, they make accurate DNA replication impossible. In short, the carcinogens are mutagens.

SIDE QUESTION 5.5

On the basis of what you now know about carcinogenic transformation, propose a hypothesis for why there is a long interval between exposure to a carcinogen and the appearance of cancer.

Table 5.1 Known Human Carcinogens	
Aflatoxins	Hepatitis B virus
Alcoholic beverage consumption	Hepatitis C virus
4-Aminobiphenyl	Human papilloma viruses: some genital-mucosal types
Analgesic mixtures containing phenacetin	Melphalan
Aristolochic acids	Methoxsalen with ultraviolet A therapy
Arsenic and inorganic arsenic compounds	Mineral oils: untreated and mildly treated
Asbestos	Mustard gas
Azathioprine	2-Naphthylamine
Benzene	Neutrons (or ionizing radiation)
Benzidine	Nickel compounds
Beryllium and beryllium compounds	Radon
Bis(chloromethyl) ether	Silica, crystalline (respirable size)
1,3-Butadiene	Solar radiation
1,4-Butanediol dimethanesulfonate	Soots
Cadmium and cadmium compounds	Strong inorganic acid mists containing sulfuric acid
Chlorambucil	Sunlamp or sunbed exposure
1-(2-Chloroethyl)-3-(4-methylcyclohexyl)-1-nitrosourea	Tamoxifen
Chromium hexavalent compounds	2,3,7,8-Tetrachlorodibenzo-p-dioxin
Coal tars and coal-tar pitches	Thiotepa
Coke-oven emissions	Thorium dioxide
Cyclophosphamide	Tobacco smoke, environmental
Cyclosporin A	Tobacco smoking
Diethylstilbestrol	Tobacco, smokeless
Dyes metabolized to benzidine	o-Toluidine
Erionite	Ultraviolet radiation, broad-spectrum
Estrogens, steroidal	Vinyl chloride
Ethylene oxide	Wood dust
Formaldehyde	X-radiation and gamma radiation

Substances listed in the *13th Report on Carcinogens*, **U.S. Department of Health and Human Services,** released on October 2, 2014.

The actions of carcinogens can be either direct or indirect.

Some carcinogens will enter a cell and act directly on DNA, changing its structure and causing mutations; these are called **direct carcinogens**. For example, the class 2 carcinogen *N*-methyl-*N*-nitrosourea reacts with guanine (G) or thymine (T) by adding the group of atoms CH_3 to these bases (FIGURE 5.11). This chemical process is called alkylation, and the atoms added are called an alkyl group. There are many alkyl groups; they all contain carbon (C) and hydrogen (H) in the ratio C_nH_{2n+1}, where n can be any integer. You will encounter alkylation again when we talk in Chapter 8 about chemicals that are used to treat cancer.

Simply adding one C and three H atoms to a base doesn't sound very important, but if this addition is in the right place, it is big enough to interfere with the pairing of complementary nucleotides during DNA synthesis.

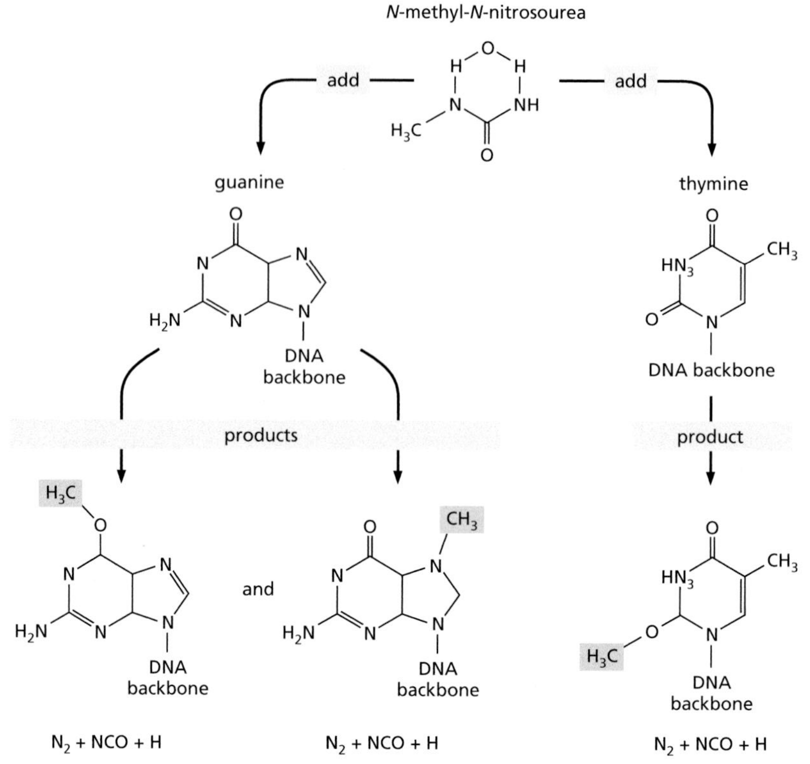

FIGURE 5.11 The chemical reaction of a known carcinogen with two nucleotides from DNA.

FIGURE 5.11 The chemical reaction of a known carcinogen with two nucleotides from DNA. *N*-Methyl-*N*-nitrosourea can react with either guanine (G) or thymine (T). The reactions add a methyl group (CH₃, shaded red box) to either an oxygen or a nitrogen atom in one of these bases. During the reaction, the carcinogen comes apart into pieces (depicted as N₂, NCO, and H). The modified nucleotides are called DNA adducts. They are unable to pair properly with their complementary bases during DNA replication, resulting in a mutation.

Remember that when DNA polymerase works its way along an old strand, it makes a new DNA strand that is perfectly complementary to its template. The choice of which nucleotide to add is defined by the several weak bonds that form between complementary nucleotides (see Figures 4.2 and 4.3). Even a few atoms attached to a base in the wrong place will disrupt this bonding and thus the accuracy of the pairing, making it likely that the wrong nucleotide will be added, which introduces a mutation. The general term for the product of a reaction between a nucleotide in DNA and a chemical that becomes attached to it is a **DNA adduct**. Any chemical, like an alkylating agent, that forms a DNA adduct is a mutagen, and many such molecules are now known to be class 1 carcinogens.

A striking feature of the list of carcinogens in Table 5.1 is that many of these substances are natural. This means that animal life has been subject to the effects of these materials for a long time but has survived perfectly well. This situation is probably a result of four factors: (1) that amount of exposure to naturally occurring carcinogens is generally small; (2) cells contain several mechanisms for repairing DNA (as we will discuss in Chapters 6 and 7), which reduces the effects of mutagenic chemicals; (3) animals with short life spans are less likely to get cancer than others because cancer takes a long time to develop; and (4) since cancer is slow to develop, many organisms that are exposed to carcinogens and may ultimately get cancer are able to reproduce before their cancer-induced death. Thus, their susceptibility to this carcinogen is not selected against. Humans nowadays have an unusually long life span, making carcinogens a bigger problem for us than for many other animal species. These facts lead, however, to an important point to consider: one cannot maintain the all-too-common modern faith that if something is natural it is safe and good. The list of natural poisons is even longer than the list of natural carcinogens, so be careful about people suggesting that natural equals good for you.

Aflatoxin is an excellent example of a naturally occurring carcinogen and an example of an additional principle of carcinogenesis. Aflatoxin is a

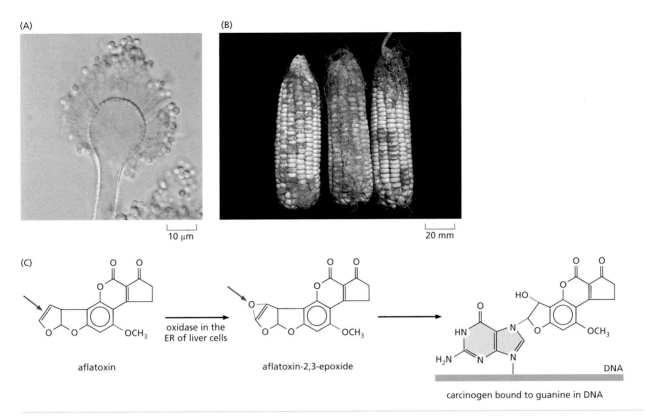

FIGURE 5.12 **The mold *Aspergillus*, an infected host, and a medically important chemical produced by *Aspergillus*.** (A) *Aspergillus flavus* is a mold that grows well on several forms of food, such as corn, rice, and wheat, if they are not stored in a properly dry environment. (B) Three ears of corn that have been badly infected with *Aspergillus*. (C) *Aspergillus* makes and secretes aflatoxin, a polycyclic aromatic hydrocarbon (PAH). If aflatoxin is eaten, it passes from the intestine into the blood and then to the liver. Human liver cells modify it in their effort to detoxify it. In doing so, they actually make it more reactive. Enzymes in the smooth endoplasmic reticulum (ER) of liver cells convert a double bond in aflatoxin (red arrow at left) into an epoxide (second red arrow), which makes aflatoxin sufficiently reactive to bind strongly with guanine (G) (red line at right), making a DNA adduct. At the next round of DNA replication, a mutation will be introduced. (A, Courtesy of Medmyco CC BY-SA 4.0. B, courtesy of National Institutes of Health.)

chemical produced and secreted by some molds of the genus *Aspergillus*. This mold grows on some rotting foods, such as corn, rice, and peanuts (FIGURE 5.12). If people eat the *Aspergillus*-infected food, they ingest aflatoxin, which gets into their bloodstream and then into the liver. The smooth endoplasmic reticulum (ER) of liver cells contains enzymes that react with and alter many foreign chemicals, including aflatoxin. The purpose of these reactions is to detoxify the foreign chemicals by changing their structures through enzyme-catalyzed reactions that alter chemical bonds or add atoms. Unfortunately, this system sometimes makes mistakes, so the changes made alter the foreign molecule into something that reacts with DNA, leading to the formation of a DNA adduct. Liver enzymes alter aflatoxin in just this way, leading it to be able to react with guanine (G) (Figure 5.12C). The resulting DNA adduct will lead to a mutation in the next round of DNA synthesis. Chemicals that are not in themselves carcinogenic but are made into carcinogens by our body's enzymes are called **indirect carcinogens** to distinguish them from direct carcinogens, like *N*-methyl-*N*-nitrosourea, which are carcinogenic in their own right.

The formation of carcinogenic substances from indirect carcinogens is a big subject that involves lots of chemistry, so we won't pursue it here. It is important to note, however, that in some cases the enzymes that do these self-destructive modifications reside in different parts of our body. For example, some indirect carcinogens are activated by one or more enzymes in the liver and then by additional enzymes in the bladder.

Think about how carcinogenic chemicals get into the human body. What would you expect to be the two most common routes? Does this situation play a role in which human tissues are most likely to develop cancers? Explain your answer.

SIDE QUESTION 5.6

Many chemical carcinogens are avoidable.

Most carcinogenic chemicals in Table 5.1 are readily avoidable, since they are not part of our daily life. A few, however, can be encountered at home or in one's workplace. Epoxides, which are chemicals used in varnish, plastics, and glue, are carcinogenic when in liquid form. After they have polymerized (a chemical reaction in which many molecules combine to form a larger molecule), they become solid and are much less dangerous. However, the dust created when an epoxy-varnished floor is sanded should be avoided. Also, there are toxic chemicals, such as bisphenol A, which can leach out of some plastics, even though the plastics themselves are solid. Thus, it makes sense to learn about the materials you use in daily life to assure yourself that they do not pose hidden problems.

Chemical companies and some manufacturers are constantly inventing new molecules for human use; this is a major industry. How can such companies learn at reasonable cost whether a new class of compound will be carcinogenic and therefore impractical to develop? In the early 1970s, Bruce Ames, a biochemist at the University of California, Berkeley, published a paper that described a cheap, rapid, and simple test for the mutagenic activity of any chemical. He applied his test to a significant number of proven carcinogens and found a very high correlation between carcinogenesis and mutagenesis. The properties of his test have led to its being adopted by many companies as a way to screen products for possible carcinogenic activity in an efficient and economical way. SIDEBAR 5.5 describes the Ames test and how it works.

The action of a carcinogen does not immediately lead to cancer.

You have read about the lag between the time of exposure to a carcinogen and the emergence of cancer. One reason for this lag is that cells have multiple safeguards against misbehavior, so multiple genes must be altered by mutations before a cancer cell will emerge. This takes time. Thus, exposure to a carcinogen and the ensuing formation of DNA damage can precede the appearance of disease by years.

An additional reason for this long delay is the **diploid** organization of the genome in almost every cell of your body, meaning it contains two copies of almost every gene. If one copy of a gene is inactivated by mutation, you still have a good copy; this commonly means that the mutation won't show up in that individual's **phenotype**. It is worth noting, though, that some mutations produce alleles that do appear in phenotype, even in the presence of a wild-type allele. Such alleles are said to be dominant, For example, a single mutation that produces a gain in the function of a gene that promotes cell growth and division will promote a cancerous transformation. One way in which a gain of function can occur is through a mutation that causes overexpression of the gene product; for example, a mutation that increases the activity of that gene's promoter. Such mutations turn out to be important in understanding cancerous transformation, so they are dealt with more thoroughly in the next chapter.

Some chemicals and physical processes act as tumor promoters.

Several factors are cancer inducing, even though they are not mutagenic. Exposure to these factors on their own does not lead to cancer, but if other events have caused mutations and thereby initiated cancerous transformation, they accelerate the completion of the transformation. These

A CLOSER LOOK: The Ames test is a way to identify possible carcinogens.

The Ames test is based on the assumption that any chemical that increases the frequency of mutations in DNA is likely to be a carcinogen. It uses the fact that the DNA from a bacterium is biochemically the same as our own DNA. (Remember that bacteria too are cells, and they use DNA as their memory for how to make the proteins they need to live. In fact, some pieces of human DNA work fine in a bacterial cell, and vice versa!) Ames reasoned that if a chemical was mutagenic for bacteria, it was probably a human mutagen and therefore a likely carcinogen. He designed a strain of bacteria containing several mutations that blocked the production of a particular amino acid necessary for protein synthesis. These bacteria cannot grow and divide unless this amino acid is part of their food, though normal bacteria can live on very simple food, for example, water, minerals, one sugar, and a simple, nitrogen-containing compound, like ammonia.

One can easily spread thousands of these mutant bacteria on a small plate, ~4 inches in diameter. The trick is to make a special plate that contains nutrients that will allow normal bacteria to grow well but includes only a small amount of the amino acid needed for these mutants to grow. As a result, they will grow and divide only a couple of times before they are starved for that amino acid and become dormant. If, however, one adds a mutagenic chemical that alters the DNA of these bacteria, a change may occur that allows them to grow like wild–type (unmutated) bacteria. Such a mutation is called a reversion, because it reverts the cell to the wild phenotype for growth, and such a cell is called a revertant.

As you can imagine, the chances of a reversion are low, but Ames set things up so any one of several mutations would allow the bacteria to revert and grow on the simple medium provided. Moreover, there are millions of bacteria on the plate, so if mutations are occurring, as a result of the added chemical, the chances are good that one or more of these changes will enable the bacteria to grow. Each reverting mutation will lead to a clone of bacteria that can grow up to form a cluster of several thousand cells, called a colony, which is easy to see. Thus, a mutagen will lead to the formation of colonies and be called positive in the Ames test. The stronger the mutagenic activity of the chemical, the more colonies will form, so the test is quantitative as well as qualitative.

A limitation of this test as described is that it won't identify indirect mutagens. To deal with this issue, Ames ground up rat liver, making a paste that contained essentially all liver enzymes, including the ones on the endoplasmic reticulum (ER) that might convert an indirect carcinogen into its active form. With this paste he made a second kind of plate. Now he could test chemicals not only for their ability to mutate DNA but also for their ability to *become* mutagens when in the presence of liver enzymes (Figure 1). This trick made the test much more valuable,

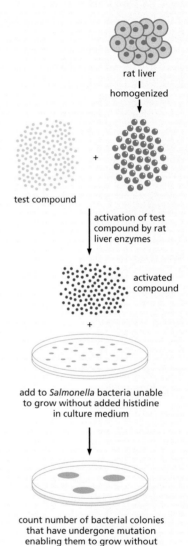

rat liver
|
homogenized

+

test compound

activation of test compound by rat liver enzymes

activated compound

+

add to *Salmonella* bacteria unable to grow without added histidine in culture medium

count number of bacterial colonies that have undergone mutation enabling them to grow without added histidine

Figure 1 The Ames test. A chemical to be tested for its ability to cause mutations is first mixed with homogenized liver, a cell extract that provides enzymes, which might convert an indirect carcinogen to its active form. This mixture is added to a plate where thousands of bacteria are already present. Here, the bacteria used are *Salmonella*. These bacteria have previously been mutated and are unable to make the amino acid histidine, so this nutrient must be present for them to grow and divide. Previously, histidine was supplied, so the bacteria could increase in number, but the medium on the plate lacks histidine, so the bacteria can no longer grow. (Bacteria are represented by small green circles on the plate.) If the activated mixture is mutagenic, many of the thousands of bacteria will become mutated. A small fraction of these mutations will change the bacterial DNA in such a way that the cell reverts to the wild phenotype and can now grow without histidine. These so-called revertants go through multiple cycles of growth and division, making enough progeny to form a visible clone, called a colony, represented by the large green circles on the bottom plate. The greater the mutagenic activity of the tested chemical, the greater the number of colonies that will form. (Adapted from Weinberg R [2013] The Biology of Cancer, 2nd ed. Garland Science.)

because it now recognized chemicals that might be indirect carcinogens as well as direct ones.

The great advantages of the Ames test are its speed (a few days to assess the mutagenic properties of a compound versus years with lab animals) and economy (a few dollars rather than hundreds of thousands of dollars for animal care). The question, then, is its reliability in identifying human carcinogens. Several scientists have explored this issue, comparing activity in the Ames test (the number of bacterial colonies formed at each dose of a given chemical) with carcinogenicity (the dose of each chemical that induced cancer in 50% of the animals studied). The Ames test is not perfect, but it is very good, as indicated by the number of points, each representing a specific chemical, that lie on the diagonal line in Figure 2. The Ames test is a powerful example of the importance of basic research in biology and biochemistry to progress with the cancer problem.

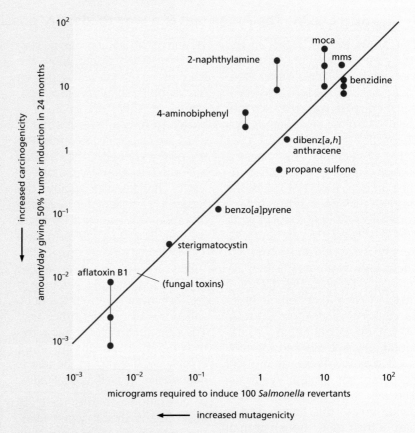

Figure 2 Studies on 11 chemicals, comparing their strengths as carcinogens in mice and as mutagens in bacteria. The vertical axis displays the strength of carcinogenesis; the position on this axis tells how much of each chemical had to be given to experimental animals over two years for half of them to form a tumor. The less material needed (the lower the point on the vertical axis), the stronger the carcinogen. The horizontal axis gives similar information about the amount of the material needed in the Ames test to produce 100 colonies of revertant bacteria on each plate. The fact that most of these points lie close to a straight line shows that the two assays give strikingly similar results. Subsequent work has shown, however, that the Ames test (and others like it) are far from perfect in identifying carcinogens. The percentage of established mouse carcinogens identified as mutagens by the Ames test is about 50%, and about two-thirds of chemicals that appear not to be carcinogens are found not to be mutagens. The differences may in part be a result of the action of some chemicals as tumor promoters, as discussed later in the text. (Adapted from Meselson M et al. [1977] In Origins of Human Cancer, Book C: Human Risk Assessment [HH Hiatt, JD Watson, JA Winsten eds]. Cold Spring Harbor Laboratory Press.)

The Ames test provides an efficient way of evaluating the extent to which a chemical causes mutations in bacteria. A skeptic could argue that since bacteria are not humans, a strong positive response in the Ames test says nothing about the extent to which that chemical will be mutagenic in people. What points would you bring out for or against this skeptic's opinion?

chemicals and processes are called **tumor promoters**. Some of them are now rather well understood, and two examples will make the point. A chemical is tumor promoting if it can pass through the cell's plasma membrane and mimic the stimulating effect of a growth factor. One example is a chemical that stimulates the activity of MAP kinase, a key part of the pathway leading from EGF-R to the initiation of DNA synthesis (see Figure 4.17). Such a chemical will promote cell division so long as it is present. This is not an inheritable change in a cell's **genotype** the way a mutation would be. The promoter must be continuously present to induce the growth and division cycle. Note, however, that if these cells had previously been mutated and are now induced by the tumor promoter to grow and divide, they form a clone of already-mutated cells. This makes multiple copies of cells that already have at least one mutation, increasing the likelihood that one of those cells will experience a second (or third) mutation and thereby move farther along the path of cancerous transformation (FIGURE 5.13).

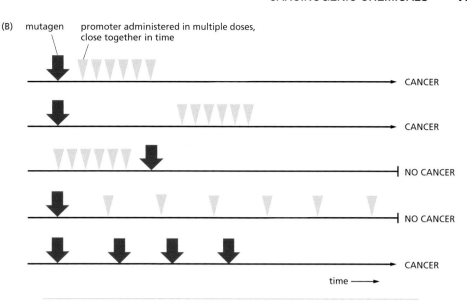

(A) normal cells

↓ MUTAGEN

isolated cell has mutation but its growth is restrained

↓ PROMOTER RELEASES RESTRAINTS

mutant cells grow into a large clone of cells, in which a further mutation may occur

FIGURE 5.13 **A tumor promoter is a chemical or a process that increases the frequency of cell growth and division.** (A) Tumor promoters can release growth constraints that keep cells from proliferating. If a particular cell had already received a carcinogenic mutation, the result of a tumor promoter's action is a clone of cells, all carrying that initial mutation. Now, multiple cells are one step along the path of carcinogenic transformation. The presence of multiple cells, all with one carcinogenic mutation, increases the likelihood that one of them will receive the additional mutations necessary for carcinogenesis. (B) Tumor promoters and mutagenic carcinogens can interact to bring on cancer. Promotion after mutagenesis is carcinogenic. Promotion before mutagenesis is not. A lower level of promoter (indicated by greater spacing between green arrowheads) is not sufficient to develop a clone of mutated cells and is therefore not carcinogenic. Repeated applications of a mutagen can be carcinogenic without the help of a tumor promoter. (Adapted from Alberts B, Johnson AD, Lewis J et al [2015] Molecular Biology of the Cell, 6th ed. Garland Science.)

Chronic inflammation is an example of a process that can be tumor promoting. Inflammation is the body's response to damage. Chronic means something that persists for a long time. Thus, a sore that stays infected for a long time and an internal organ that is badly affected by a drug or infection are two examples of chronic inflammation. The body's response is to try to heal the wound, which means cell growth and division. If one of the cells that is stimulated to divide had previously experienced a carcinogenic mutation, that cell will increase in number, forming a population of mutated cells. The result is just like that of a chemical tumor promoter: a clone of already mutated cells that increases the likelihood one cell will become multiply mutated and therefore farther along the path of cancerous transformation. An example of this kind of action is the high incidence of throat cancer among people who chew tobacco. Here the foreign substance (saliva containing many chemicals from the tobacco leaf) is both a mutagen and an irritant; the former causes mutations and the latter works as a tumor promoter. Another example is the effect of liver toxins on the development of liver cancer. Chemicals that damage liver cells induce the growth and division of other liver cells, thereby serving as tumor promoters.

Tumor promoters increase the likelihood of cancer in two ways.

Tumor promoters increase the likelihood of cancer both by increasing the number of already mutated cells, if there were any at the time of tumor promotion, and by increasing the number of dividing cells. Thus, they increase the frequency of DNA synthesis, which is the time in the cell cycle

It is very uncommon for anyone to be born with carcinogenic mutations in both copies of the same gene. Why?

SIDE QUESTION 5.8

A chemical called PMA (phorbol 12-myristate 13-acetate) is an effective tumor promoter. You saw earlier that several complex chemicals called PAHs were genotoxic carcinogens, capable of inducing skin cancers in rabbits. If you wanted to study skin carcinogenesis, would you administer PMA and then a PAH, or would you apply them in the opposite order? Why?

when DNA is most susceptible to mutation. Since a tumor promoter increases the likelihood of cancer in two ways, it may legitimately be called a carcinogen. Some scientists refer to both mutagens and tumor promoters carcinogens but distinguish them by calling the former a *genotoxic* carcinogen.

Doctors have long thought about carcinogenesis as a three-step process: initiation, which is the result of mutation; promotion, which is the multiplication of precancerous cells; and progression, during which the precancerous cells change even farther from normal as they become metastatic. This view has merit, but studies of cancer progression under a variety of circumstances have shown that genotoxic carcinogens can also serve as tumor promoters. This makes the three-step model seem too narrow to be satisfactory. The key point is that at least one of the body's cells must accumulate multiple mutations to overcome all the natural constraints on their misbehavior, allowing them to acquire all the properties we associate with metastatic cancer. A tumor promoter can help in cancerous progression by expanding the numbers of already mutated cells, but this function can also be achieved by a mutation that achieves the same result.

CARCINOGENIC VIRUSES

The evidence is now unambiguous that some viruses can cause cancer. Viruses that are carcinogenic to animals have been known for many years, so it was not surprising to find that some human-infecting viruses are also carcinogens. Indeed, a lot of what we now know about cancerous progression in humans has come from detailed studies of viruses that cause cancer in mice and chickens, as described in Chapters 6 and 7. Thus, it is remarkable that so few viruses are carcinogenic in people. You know many human-infecting viruses, such as those that cause colds, measles, influenza, or mumps. These viruses are not carcinogenic, so what allows certain viruses to cause cancer?

A virus is a small but highly effective parasite.

A virus is a tiny infectious agent, much smaller than a cell. It is not alive, like a bacterial cell, because it relies absolutely on the cells it infects for the materials and energy that are necessary for it to reproduce. Most viruses have a reproductive cycle that includes (1) infection of a host cell, (2) reproduction within that cell, and (3) release from that host cell so the virus can spread and infect other host cells (FIGURE 5.14A). During the stage when the virus is outside its host, it is a well-defined particle, called a **virion.** A virion is a package of either DNA or RNA, depending on the virus. This nucleic acid serves as the genome for the virus. It is encased in a protein coat, called the virus capsid (FIGURE 5.14B). Some virions also have a membrane that envelops their capsid (FIGURE 5.14C). Viruses infect cells by inserting their DNA or RNA into the cell, which is now the host for virus replication. This nucleic acid takes advantage of the many molecules, structures, and processes already built by the host cell. The viral genome is replicated and initiates the synthesis of the proteins needed to make more virions that will escape the host cell and go on to infect other cells.

A few human viruses are known to cause specific cancers.

Human viruses with well-established carcinogenic action include hepatitis B and C viruses (HBV and HCV). These viruses can promote liver cancer long after the symptoms of hepatitis infection are gone. Human T-cell

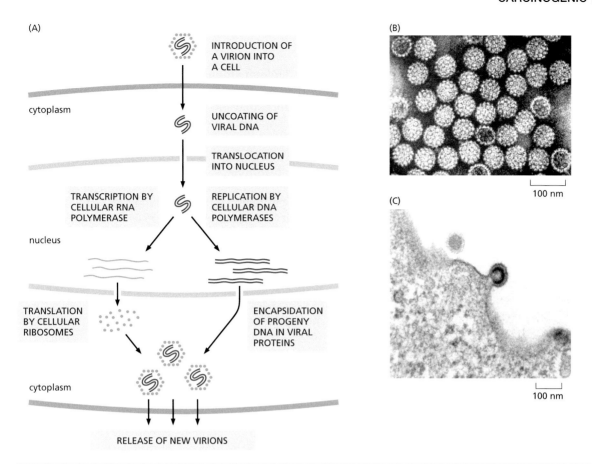

FIGURE 5.14 The reproductive cycle and structure of a virus. (A) A virus can exist as a particle, called a virion, which is capable of infecting cells. A virion enters the cell that will be its host for replication. The viral DNA is released into the cytoplasm and then migrates into the nucleus, where it can be transcribed into RNA that encodes virus-specific proteins. The viral DNA is also replicated, so the cell comes to contain multiple copies of the viral DNA. Now the viral DNA interacts with some of the viral proteins to form more virions that can go on to infect other cells. Some viruses are released because their host cell bursts. Other viruses are released by budding from the cell's surface. (B) Electron micrograph of virions from a virus that can infect people and promote the onset of cervical cancer. (C) A virus budding from the surface of an infected cell so it can leave its host and wander away to find another host to infect. (A, adapted from Alberts B, Johnson AD, Lewis J et al [2015] Molecular Biology of the Cell, 6th ed. Garland Science. B, courtesy of D. DiMaio. C, courtesy of Laboratoire de Biologie Moleculaire.)

leukemia virus (HTLV) promotes a kind of blood cancer. A herpes viruses, called Epstein–Barr virus (EBV), causes infectious mononucleosis (commonly called mono or kissing disease), but persistent infection with this virus can lead to nasopharyngeal carcinoma, a nose and throat cancer. Occasionally, EBV infection leads to a rare kind of lymphoma. There are some indications that EBV is also linked to about one-third of the cases of the more common Hodgkin's lymphoma, but doctors are not sure about this, because infection with EBV is so common that it is hard to know whether the signs of EBV infection in patients with this blood cancer are significant for the formation of the cancer. However, EBV certainly contributes to Burkitt's lymphoma in African children. This is a cancer of white blood cells that causes major swelling in the jaw. Burkitt's lymphoma will occur only if the patient is infected simultaneously by EBV and the parasite that causes the mosquito-borne disease malaria, so Burkitt's lymphoma is found only in tropical parts of the world.

How can a virus be a factor in cancer progression? Are these viruses mutagens? Are they tumor promoters? Do they work by some novel mechanism that remains to be discovered? We can get insight into these questions by considering a common, sexually transmitted infection that is caused by a carcinogenic virus known as **human papilloma virus** (**HPV**).

Human papilloma virus is both mutagenic and a tumor promoter.

There are more than 50 different strains of HPV, each slightly different in the proteins that surround the viral DNA. About one-third of these viruses are now recognized as important factors in human cancer worldwide. These viruses are efficiently transmitted by sexual intercourse, so promiscuous sexual activity, particularly without the protection offered by a condom, is an all too effective way to pass this virus around. The virus promotes cervical cancer in women, penile cancer in men, and anal cancer in both men and women. It has also been linked to cancer at the back of the throat. Not all people with an HPV infection will get cancer, but there is clear evidence that HPV increases cancer risk. Indeed, a large fraction of the women who get cervical cancer do have an HPV infection.

HPV and other carcinogenic viruses are mutagenic because their DNA can insert right into a human chromosome (FIGURE 5.15A). They can do this as a result of special DNA sequences in their genomes that promote this integration event (FIGURE 5.15B). The DNA of most viruses lacks these special sequences and therefore cannot integrate into its host cell's chromosomes. Such viruses are neither mutagenic nor carcinogenic. The insertion of viral DNA into one of our chromosomes is mutagenic for two reasons. First, when viral DNA inserts into host DNA, it separates nucleotides that used to be adjacent, and this alters the structure of that stretch of DNA (Figure 5.15B). If that DNA includes a gene, the gene is mutated. Moreover, the DNA of cancer viruses can also leave the host DNA, and the process by which it does so is imprecise. It can either leave a bit of viral DNA behind or take a bit of the host cell chromosome along with it; in either case, the sequence of the host cell chromosome is altered. Such alterations are, by definition, mutations, and with bad luck, they can be carcinogenic.

The second way in which a virus can cause cancer is not through mutagenesis but through information encoded in some of the genes it carries. For example, the DNA of many HPVs encodes a protein that inactivates the Rb (retinoblastoma) protein, which you know from Chapter 4 acts as a brake on a cell's progression into DNA synthesis. This inactivation of Rb produces results similar to what happens when the gene for Rb suffers a

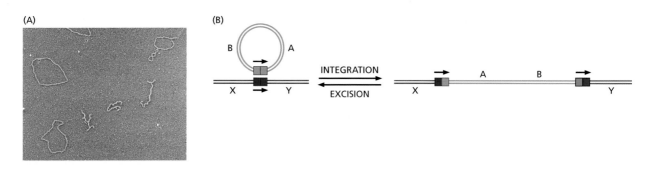

FIGURE 5.15 The DNA of many tumor viruses forms a closed loop that can insert into the DNA of a host cell. (A) Electron micrograph of DNA from a tumor virus. These pieces are all loops, often called circles, though many of the loops are twisted upon themselves to make more complicated-looking structures. (B) Circular viral DNA (green) can integrate into host cell DNA (red) by a process called crossing over. This is the same mechanism that our bodies use during the formation of sperm or egg cells to cause some benign rearrangements of genes, which is one cause of the genetic differences between parents and their offspring. The green and red boxes represent DNA sequences that are similar and therefore particularly likely to cross over. Note that host genes, labeled X and Y, become more distant from one another when viral genes A and B are inserted. If crossing over occurred within a host cell gene, the continuity of that gene would be disrupted, causing a mutation that might or might not have a noticeable phenotype. Note also that when the viral DNA comes back out of the host chromosome, as viral DNA will do, the excision is not necessarily the exact opposite of the integration process. Excision can result in the virus either taking some host cell DNA with it or leaving some viral DNA behind in the host chromosome. Either event leaves a permanent change in the host cell DNA. (A, courtesy of Jack D. Griffith, University of North Carolina.)

loss-of-function mutation; the cell cycle progresses when it should not. If HPV is infecting cells of the uterine cervix, it can induce the synthesis of this protein on host cell ribosomes, and this causes inappropriate cell division. In this way, HPV works as a tumor promoter. Given this pair of carcinogenic effects, it is fortunate that medical science has already been able to make a vaccine that offers significant protection against HPV infection. The vaccines currently available do not prevent infections by all carcinogenic types of HPV, but protection is good enough to reduce dramatically the danger of HPV infection. The National Cancer Institute estimates that current HPV vaccines can prevent 70% of cervical cancers, and medical scientists are working on better vaccines whose protection may reach 90%. Immunization with this vaccine is now available to teenagers and young adult men and women over much of the world. The global medical community agrees that HPV vaccination is a cancer prevention step well worth taking.

Other cancer-causing viruses, commonly known as **tumor viruses**, carry different genes that promote cancer in different ways. This subject is complicated and hard to explain without more background information, so it will come up again after you have learned more about the particular genes that are important for cancer development, which is the subject of the next two chapters. Meanwhile, we can touch briefly on additional factors important in carcinogenesis.

CANCER RISK THROUGH INHERITANCE AND LIFESTYLE

Doctors have long known that cancer can run in families. For example, about 12% of all women in the United States develop breast cancer at some time during their lives. Of these women, 5–10% have inherited a gene from their father or mother that increased their risk of the disease. Meanwhile, the incidence of breast cancer in men is only about 0.1%. These differences in the likelihood of cancer, due to either gender or inheritance, seem to be at odds with the idea that environmental factors (carcinogens) are the source of cancer risk. Research is also showing that certain human behaviors, like diet and exercise, have an effect on cancer incidence. This too seems to suggest novel modes of carcinogenesis. It is now possible, however, to put all these facts together into a single picture of cancerous transformation.

Mutations that increase cancer risk can be inherited.

In 1971, a medical geneticist named Alfred Knudson proposed a two-hit model for the emergence of human cancers. His logic was that the diploid nature of the human genome meant that mutations had to inactivate both the maternal and the paternal copies of a gene before cancerous behavior would emerge. In 1973, another geneticist named David Comings published the idea that soon became the framework by which scientists understand the inheritance of cancer risk. He proposed that a child could inherit from one parent a single loss-of-function mutation in a gene that inhibited cancerous cellular behavior; s/he would then be **heterozygous** for that gene. There was a functional copy of the gene from either mom or dad and a nonfunctional copy from the other parent. Comings argued that this was one step along the pathway to cancerous transformation, because now it would take only a single loss-of-function mutation in the wild-type copy of that same gene in any of the body's cells to achieve Knudson's two-hit condition. A single mutation in a single cell would lead that cell to lack the function completely. This important idea was

SIDE QUESTION 5.10

It is unusual for a virus to insert its DNA into the DNA of its host's chromosomes, but cancer-causing viruses can do it. Once in the host's DNA, the viral DNA can express genes that encourage the host cells to grow and divide. Do you see this behavior as advantageous or disadvantageous for the virus? Why?

FIGURE 5.16 Inheritance can increase cancer risk, as exemplified by the *Rb* gene. The diploid condition of our cells means that two mutations are often required for a phenotype to appear. This is because mutations that inactivate a gene are likely to be recessive, so the unmutated gene, which is still wild type, defines the phenotype. Here you see a comparison of responses to mutation in an individual who is genetically normal at the *Rb* gene (left) and in someone who has inherited a loss-of-function mutation in one copy of *Rb* (right). Inheritance of the mutation means that all the cells in that person's body carry the mutation in one chromosome, so the formation of a cell with *no* normal allele of the gene requires only a single mutation. (Adapted from Alberts B, Johnson AD, Lewis J et al [2015] Molecular Biology of the Cell, 6th ed. Garland Science.)

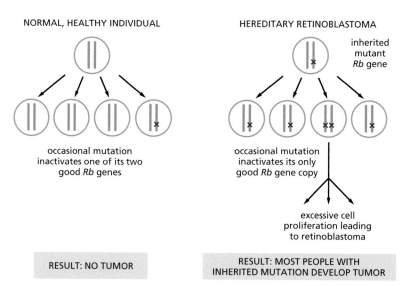

strongly supported by research soon carried out by Knudson and others on children who came down with the serious eye cancer **retinoblastoma**.

From your reading in Chapter 4 about cell cycle control and from the description of carcinogenesis by HPV, you know that the product of the retinoblastoma gene, the Rb protein, helps to keep cells from starting inappropriate DNA synthesis. The *Rb* gene was discovered and named through studies of a form of retinoblastoma that appeared in very young children and often affected both their eyes. This form of retinoblastoma was distinct from a form that appeared later in life. Through extensive studies of children who showed the early-onset form of retinoblastoma, Comings' ideas about the inheritance of a loss-of-function mutation were validated with experimental facts. FIGURE 5.16 compares the genotypes of a healthy individual with someone who inherited a loss-of-function mutation in the *Rb* gene. A cell in the individual on the left can acquire a loss-of-function mutation in one of his/her *Rb* genes, but s/he still has a wild-type copy of the gene, so there is no change in cell behavior.

In Figure 5.16, the individual on the right was unlucky enough to *inherit* a loss-of-function mutation in the Rb gene. Every cell in his/her body is heterozygous for *Rb*, containing one good copy and one mutant copy of the gene. If any of the cells in this individual now acquires a loss-of-function mutation in the other copy of the *Rb* gene, the resulting cell has no good copy of the gene, so it will behave as if the *Rb* gene was not there; it will start DNA replication when no signal has been received to do so. This is cancerous behavior.

Patient and skillful study of families with high inherited cancer risk has identified several genes analogous to Rb that are important for cancer. *BRCA1* is a gene whose mutation to loss of function is important for the likelihood that a woman will get breast cancer. To understand the idea of inherited cancer risk, imagine a woman who lived many years ago and by chance acquired a mutation in the gene *BRCA1*, not in the DNA of a cell in her body but in one of her **germ-line cells**, that is, a cell that would become an egg. That egg happened to get fertilized, and the offspring was a girl who inherited her mother's mutant allele of the *BRCA1* gene. Fortunately, her father's *BRCA1* gene was not mutated (wild type), so she had one good copy of this gene. Since she lived long ago, her life span was rather short, and throughout her life, she was cancer free. Imagine further that she had a son who inherited her mutated *BRCA1* gene. He never had any problem with breast cancer, because this disease

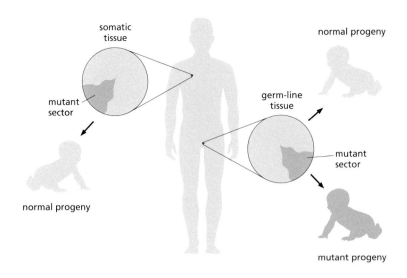

FIGURE 5.17 Differences in patterns of inheritance for mutations in germ-line cells versus somatic cells. Cells that contribute directly to inheritance, that is, those that become eggs in females or sperm in males, are called germ-line cells. Most of our body's cells have nothing direct to do with inheritance; they are called somatic cells (*soma* is Greek for body). A mutation in a somatic cell can affect the behavior of that cell and of its daughter cells, but not the genes in descendants of that person. Mutations in germ-line cells, on the other hand, can be passed on to the next generation. Note that women mature a new egg every month, but only a few of those eggs get fertilized, so a mutation is not necessarily passed on. Likewise, men make billions of sperm during the course of their lives, but most of those sperm never fertilize an egg, so again, a mutation in the germ line does not assure that a child will inherit it. (From Griffiths AJF, Miller JH, Suzuki DT et al. [2000] An Introduction to Genetic Analysis, 7th ed. W. H. Freeman.)

is uncommon in men. He did, however, have daughters, and one of them inherited the mutated *BRCA1* gene. This girl lived long and she did get breast cancer; some of her daughters did too. The longer life span of these individuals gave them the time to accumulate additional mutations in one or more of their body's cells, taking one of those cells through a complete cancerous transformation.

As shown in Figure 5.16, if an individual inherits a loss-of-function mutation of a gene that normally helps to prevent cancer, every cell in their body has only one good copy of that gene. If any single cell in that person's body happens to receive a mutagen that alters its one good copy of the gene, that cell is now on the road toward becoming a cancer cell. One mutation is more likely to occur than the two that we have previously said were necessary to take out both copies of a normal gene, given that people are diploid. Thus, the fact of inheriting a loss-of-function mutation in a cancer-relevant gene *does* raise the likelihood that an individual will get cancer. Inheritance of a mutant form of one of these cancer-preventing genes explains the phenomenon of cancers running in families. The difference between mutations in cells of the body, called **somatic** cells, and cells that contribute to the **germ line** is diagrammed in FIGURE 5.17.

Lifestyle choices, like the decision to use tobacco, greatly increase cancer risk.

Some aspects of cancer risk are under our personal control. Perhaps the most notable of these is the use of tobacco. The case that tobacco smoke increases cancer risk has now been demonstrated by laboratory tests on several kinds of animals and by clinical work with people who do and don't smoke. A list of currently known carcinogens in tobacco smoke is included in TABLE 5.2. Inhaling smoke from 30 cigarettes per day increases one's risk of getting lung cancer by about 27-fold relative to someone who has never smoked. That is a 2700% risk increase! Bear that in mind when the newspapers are talking about a 10% risk increase associated with a particular food or lifestyle choice that is under current study. Indeed, lung cancer currently accounts for 26% of all cancer deaths in American males and 25% in females. If you smoke, stopping will definitely improve your chances of avoiding cancer, although it will not bring you back to the cancer risk of someone who has never smoked.

A striking feature of tobacco smoke is that current medical evidence shows its impact not only on the incidence of cancer in the lungs but in many other parts of our bodies as well. Cancers of the tubes by which we

Cancer is commonly a lethal disease. If there are inheritable forms of genes that promote cancer in the people who carry them, why haven't those alleles been selected against through the course of human evolution?

SIDE QUESTION 5.11

Table 5.2 Carcinogens in Cigarette Smoke	
Chemical	**Amount (per cigarette)**
Acetaldehyde	980 µg–1.37 mg
4-Aminobiphenyl	0.2–23 ng
o-Anisidine hydrochloride	Unknown
Arsenic	Unknown
Benzene	5.9–75 µg
Beryllium	0.5 ng
1,3-Butadiene	152–400 µg
Cadmium	1.7 µg
1,1-Dimethylhydrazine	Unknown
Ethylene oxide	Unknown
Formaldehyde	Unknown
Furan	Unknown
Heterocyclic amines	Unknown
Hydrazine	32 µg
Isoprene	3.1 mg
Lead	Unknown
2-Naphthylamine	1.5–35 ng
Nitromethane	Unknown
N-Nitrosodi-n-butylamine	3 ng
N-Nitrosodiethanolamine	24–36 ng
N-Nitrosodiethylamine	up to 8.3 ng
N-Nitrosodimethylamine	5.7–43 ng
N-Nitrosodi-n-propylamine	1 ng
4-(N-Nitrosomethylamino)-1-(3-pyridyl)-1-butanone	up to 4.2 µg
N-Nitrosonornicotine	14 µg
N-Nitrosopiperidine	unknown
N-Nitrosopyrrolidine	113 ng
N-Nitrososarcosine	22–460 ng
Polonium-210	variable
Polycyclic aromatic hydrocarbons	28–100 mg
o-Toluidine	32 ng
Vinyl chloride	5.6–27 ng

From the National Toxicology Program of the U.S. Department of Health and Human Services, Report on Carcinogens, 2014. A milligram (mg) = 10^{-3} g = one-thousandth of a gram; a microgram (µg) = 10^{-6} g = one-millionth of a gram; and a nanogram (ng) = 10^{-9} g = one-billionth of a gram.

breathe (the bronchi) or swallow (the esophagus) and cancer of the kidneys, bladder, pancreas, stomach, and ovaries have all been linked to tobacco use. It appears that chemicals from inhaled tobacco smoke enter one's bloodstream and from there can affect the health of multiple organs. This wide range of carcinogenesis is probably due to the large number of carcinogens and tumor promoters that are present in tobacco smoke; it is likely that these multiple factors work together to achieve their significant

A CLOSER LOOK: The Cigarette Papers

The efforts of tobacco companies to prevent the public from learning the connections between smoking and cancer present a sad chapter in the history of American industry. We can be thankful to organizations like the American Cancer Society, which persisted in studies of tobacco smoke at a time when government labs that might have studied the issue objectively were silenced through effective spending by the tobacco lobbies. A clear case for the deceitful activities of large tobacco companies can be made, thanks to a set of tobacco company documents that was mysteriously sent to a scholar interested in this problem. Analysis of these papers has shown that tobacco companies were fully aware of both the harmful effects of tobacco smoke on human health and the addictive properties of the nicotine in this smoke. They suppressed this knowledge in all their negotiations with the American government

and spent significantly to maintain their right to continue production, as documented in the book *The Cigarette Papers* by Glantz et al. In his foreword to that book, the then-current surgeon general of the United States wrote the following:

"Analysis of the previously secret papers from a major tobacco company [which is] presented in this book demonstrates that the tobacco industry was even more cynical than even I had previously dared believe ... The 1988 report to the Surgeon General, 'Nicotine Addiction', concluded that nicotine is an addictive drug similar to heroin and cocaine. At the time—as it does today—the tobacco industry vigorously attacked the report (and me) for going beyond the scientific evidence. But now, this book confirms that scientists and executives from Brown and Williamson and British American Tobacco routinely appreciated the addictive nature of nicotine a quarter century earlier, in the early 1960s. . . The tobacco industry's own scientists . . . considered nicotine an addictive drug. This

information—as well as a wealth of other important information the tobacco industry possessed—was simply not made available to the Surgeon General's Advisory Committee and the public. One can speculate, with enormous regret, how different that 1964 Surgeon General's report would have been had the tobacco companies shared their research with the Surgeon General's Advisory Committee. What would have been the history of the tobacco issue in the United States—and the world—if that report had had the benefit of all of the information available on tobacco and held privy to the inner circles of the cigarette manufacturing companies? The contrast of public and private statements from the tobacco industry reveals their deceit. . .

This book is a vital weapon in the battle against tobacco. I do not believe that anyone who reads it can remain passive in the struggle against tobacco. We all need to raise our voices to clear the air for a healthier America."

C. EVERETT KOOP, M.D., Sc.D.

SURGEON GENERAL USPHS 1981–1989

effect, a poignant example of how complex the chemistry of natural products can be.

Living with or working among smokers also poses risks, because one cannot avoid breathing secondhand smoke. Evidence for the connection between tobacco and cancer is now strong enough that one can say with confidence that anyone who cares about not getting cancer and who continues to smoke tobacco is self-delusional. The sad fact is, of course, that many young people continue to take up smoking, not simply because of advertising by the tobacco companies, though this is certainly a factor, but also because the nicotine in tobacco smoke can make you feel good, and it is fiercely addictive. Moreover, smoking is thought by many to be cool. In addition, young people often believe at some level that they are immortal; they cannot imagine that they will die from smoking. In the context of all the evidence about tobacco's dangers, it is important to know that millions of dollars were spent by tobacco companies to combat the evidence for a link between smoking and cancer (**SIDEBAR 5.6**).

Diet can influence the frequencies of several kinds of cancer.

Studies of cancer rates in different countries from around the world have shown that some kinds of cancer are more common in certain countries than in others. For example, stomach cancer is prevalent in Japan but colon cancer is rare; in the United States, just the opposite is true. Studies on people who have moved from one part of the world to another show that these specific cancer risks depend more strongly on where and how you live than on your ethnicity. Thus, lifestyle *does* matter to one's risk of getting certain kinds of cancer. You might therefore like to know what you should do, or avoid doing, to reduce your lifetime chances of getting any kind of cancer.

Some carcinogens are produced from natural products by the things people do. For example, animal fat is turned into some quite powerful carcinogens by being cooked over high heat, as when one grills a steak. The result is not just a single carcinogen but a chemical zoo that contains multiple carcinogens of related chemical structure. Thus, one can say with confidence that limiting your consumption of roasted meat is a way to reduce your exposure to chemical carcinogens. However, this idea leads to some complex thoughts. Meat is an excellent nutrient, arguing that it is good to eat. Roasted meat tastes good to many people, which makes eating it a pleasure. Moreover, barbecuing is fun! Is it something you should avoid? Decisions on things like this are matters of personal judgement. There are numerous activities with associated cancer risk, so one needs to think seriously about how best to respond to this knowledge.

However, the problem is made complicated by our incomplete knowledge of the chemicals that are truly carcinogenic in humans. The Ames test is a valuable way to evaluate the carcinogenic activity of specific substances, but it is not perfect. Additional laboratory tests have been developed, but they too fail to identify all the chemicals now known to be human carcinogens, like asbestos. It is also noteworthy that some lifestyle choices appear to reduce cancer risk. Fresh fruits and vegetables seem to be beneficial, and exercise and maintaining a modest weight reduce cancer risk for reasons discussed in Chapter 11. The complex issue of adjusting your behavior to minimize your cancer risk, while still maintaining a life you want, is a theme of several later chapters in this book.

SUMMARY

Any chemical or physical process that induces mutations can contribute to the transformation of a normal cell into a cancerous one. Ionizing radiation, UV light, chemicals that alter DNA, and certain viruses are all carcinogenic through their action as mutagens. However, the inheritance of a mutant gene that increases the likelihood of cell division or reduces genetic stability will also increase the likelihood of cancer. Thus, the causes of cancer are numerous (FIGURE 5.18). They all act either to intro-

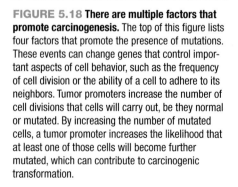
SIDE QUESTION 5.12 From what you have read in this chapter, do you think behaviors that expose an individual to cancer-promoting factors are more dangerous for young people or for the elderly? Why?

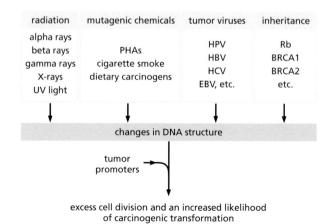

FIGURE 5.18 **There are multiple factors that promote carcinogenesis.** The top of this figure lists four factors that promote the presence of mutations. These events can change genes that control important aspects of cell behavior, such as the frequency of cell division or the ability of a cell to adhere to its neighbors. Tumor promoters increase the number of cell divisions that cells will carry out, be they normal or mutated. By increasing the number of mutated cells, a tumor promoter increases the likelihood that at least one of those cells will become further mutated, which can contribute to carcinogenic transformation.

duce mutations that alter cellular control or to increase cell division, promoting the proliferation of already mutated cells. The effectiveness of any one carcinogenic agent depends both on an individual's ability to resist mutations (for example, by accurately repairing damaged DNA) and on the presence of other factors, like tumor promoters, that stimulate cell division. Such cell divisions increase the number of cells in which mutations are already present, thus increasing the number of cells that can be further mutated. This chain of events increases the likelihood of an improbable event, like the inactivation of multiple controls on cell behavior and the development of metastatic cancer.

ESSENTIAL CONCEPTS

- Ionizing radiations, such as gamma rays, X-rays, UV-A, and radioactivity, are mutagenic because it either damages DNA directly or generates ROS.

- ROS can interact with DNA, leading to changes in the structure of a nucleotide, which in turn leads to a change in DNA sequence when that DNA is replicated. Therefore, it is mutagenic.

- ROS can interact with proteins, but while this may kill the activity of a particular protein, the result is rarely lethal for a cell, and no permanent harm is done.

- Ultraviolet light (UV-B) is mutagenic because it is absorbed by the components of DNA. Other forms of radiation, such as microwaves, radio waves, and sound waves, are not mutagenic and therefore are unlikely to be carcinogenic.

- Some chemicals are mutagenic. This is usually because they can react with the nucleotides of DNA and make adducts, which interfere with faithful DNA replication.

- Chemicals that are mutagenic are commonly, though not invariably, carcinogenic. These are called genotoxic carcinogens.

- Some chemicals, called tumor promoters, promote cancerous transformation because they induce cell divisions. If already mutated cells are induced to divide, more mutated cells are now present, and this increases the chances that additional mutations can occur in one or more of those cells, therefore promoting the formation of cancer.

- Chemical carcinogens are divided into two categories: those known to be carcinogenic in people (class 1) and those known to be carcinogenic in nonhuman animals (class 2). The latter are likely to be carcinogenic in people, but there is not yet convincing evidence on the matter.

- The Ames test uses bacteria to assess the mutagenicity of chemicals. This approach greatly reduces the time and expense of testing novel chemicals for the possibility of their having genotoxic effects on people.

- Some viruses are carcinogenic for at least two reasons: (1) They can insert their DNA into the chromosomes of their host cell, and in the process they can alter nucleotide sequences along a host DNA molecule. Thus, they are mutagens. (2) The genomes of some viruses include genes that encode proteins whose action interferes with normal control of cell cycle progression. These protein products make cells misbehave as they do after tumor promotion or even carcinogenic transformation.

- The likelihood of getting cancer can run in families, because one can inherit a mutant allele of a gene whose normal function is to help control cell behavior. Now a single mutation in the other copy of that gene will lead to a phenotype that could, for example, start a cell in your body down the path toward cancerous behavior.

- Some elements of lifestyle can increase the likelihood of cancer. Smoking tobacco is the most important of these, but diet and exercise too can be important.

KEY TERMS

alpha ray	germ line	phenotype
atomic number	germ-line cell	radioactivity
beta ray	half-life	reactive oxygen species (ROS)
diploid	heterozygous	
direct carcinogen	human papilloma virus (HPV)	retinoblastoma
DNA adduct	hydroxyl group	risk
fluoroscope	indirect carcinogen	somatic
free radical	ion	tumor promoter
gamma ray	isotope	tumor virus
genotype	mesothelioma	virion

FURTHER READINGS

Knudson AG Jr (1971) *Proc Natl Acad Sci USA* 68:820–823.

Weinberg R (2007) The Biology of Cancer. Garland Science.

   Chapter 2, The Nature of Cancer, and Chapter 3, Tumor Viruses, contain extensive discussions of topics introduced in the current chapter.

Ames BN, Durston WE, Yamasaki E & Lee FD (1973) Carcinogens Are Mutagens: A Simple Test System Combining Liver Homogenates for Activation and Bacteria for Detection. *Proc Natl Acad Sci USA* 70:2281–2285; PMC 433718.

   Note: PMC stands for PubMed Central, a service of the National Institutes of Health that makes a huge number of published scientific articles available to the public, free of charge. If you copy PMC, followed by the number, into a Web search tool, it will take you to PubMed Central and then to that specific article, so you can download it for reading. This is a research paper written for scientists, so it is not easy going, but it is well worth looking at.

Glantz SA, Slade J, Bero LA et al. (eds) (1998) The Cigarette Papers. University of California Press.

Lodish H, Berk A, Kaiser CA et al. (2012) Molecular Cell Biology, 7th ed. W. H. Freeman Co.

   Chapter 24, pp 1114–1153, includes an excellent discussion of carcinogenesis and the genes that must be altered for a cell to achieve complete carcinogenic transformation.

QUESTIONS FOR FURTHER THOUGHT

1. Why does an adduct on DNA lead to an error in the subsequent round of DNA replication? Draw out a diagram of what happens during DNA replication when a DNA adduct is present.

2. A number of carcinogenic chemicals are quite hydrophobic, meaning that they are poorly soluble in water. Where are they likely to concentrate in your body?

3. The existence of indirect carcinogens is surprising. Why would the cells in your body make enzymes that would act on foreign chemicals to make them MORE dangerous to your system? (Note that an answer to a question like this is bound to be speculative, but think of reasons to back up your answer.)

4. If it is so clearly demonstrated that smoking tobacco causes a large increase in the likelihood of cancer, why do people continue to take up smoking?

ANSWERS TO SIDE QUESTIONS

5.1. The half-life for ^{14}C is 5730 years. There are 365 days in a year, so this period of years is 2,091,450 days or 50,194,800 hours. One hour is 1/50,194,800 of a carbon-14 half-life. In one half-life, half of all the atoms will have emitted radioactivity, which means 500,000,000 events. So in one hour the number of radioactive events will be 500,000,000 / 50,194,800, which is approximately 10. For radon, on the other hand, the half-life is 3.8 days or 91.2 hours. One hour is 1/91.2 of a radon-222 half-life. In one half-life, half of all the atomic nuclei will have emitted radiation. Again, we take half of all the atoms and find out how many would have emitted radiation in a single hour: 500,000,000/91.2, which is about 5,482,456. So radon, which has a MUCH shorter half-life, would have produced about 548,246 times more radioactive events than the same amount of radioactive carbon. Impressive! (Note that this ratio is approximately the same as the ratio of the two half-lives.)

5.2. This question does not have a verbal answer, but the recommended figures should make it easy to do what is requested, allowing you to see how DNA that is damaged cannot form the bonds necessary to bring in the correct nucleotide and maintain DNA sequence. The point of this exercise is to do it yourself, so you are not simply looking at someone else's conception.

5.3. This is a question with no right answer, but it is interesting to think about. Demanding that a child always wear sunblock can be difficult to enforce and lead to controversy over an issue that may not be that important. Never using sunblock seems unwise. Taking vitamin D supplements is okay, but it too may encounter resistance. Here is a human situation where balance seems the best way to go: use sunblock as much as is convenient, use vitamin D supplements occasionally, particularly if there is a long stretch of weather with no sun, and get on with the more interesting things in life.

5.4. The information provided suggests that the radiation from a cell phone is not carcinogenic. This says that issues related to cancer have no relevance to cell phone use. There may be other reasons to use a cell phone less, but the likelihood of cancer is not one of them.

5.5. At several places in the book, it has been mentioned that cells include multiple safeguards against the loss of their normal behavior. This implies that multiple mutations must accumulate before a cell begins to show cancerous behavior. These mutations must be in the very same cell, because only then will that cell approach a full carcinogenic transformation. The accumulation of all those mutations takes time, even in the presence of a genotoxic carcinogen.

5.6. The two most common routes for entry of carcinogens are by mouth with food and by inhalation with air. Food goes from mouth to stomach to intestines, and it is near the bottom of the intestine (the colon) that one finds the residue of materials in food that did not get taken into the blood through the walls of the intestine. This concentration helps to explain why the colon is a common site for human cancer. Inhalation, on the other hand, leads to the lungs. The entry into our bodies of airborne chemicals probably contributes to the relatively high frequency of lung cancers.

5.7. An argument against the skeptic's position is that the chemistry of DNA in bacteria is essentially identical to that of DNA in humans; both are made from the same nucleotides, and they are replicated by closely similar processes. Thus, anything that harms bacterial DNA is very likely to harm human DNA too. An argument for the skeptic's position is that DNA in the cells of humans and other eukaryotic organisms is wrapped on histones, whereas bacterial DNA is not. The chemistry that goes on in bacterial cells is different from that in our own cells. Thus, the effects of genotoxic chemicals could be different in bacteria from our cells. A rigid acceptance of results from the Ames test could be misleading.

5.8. A carcinogenic mutation means a loss of normal cellular control. If a fertilized egg has by chance received carcinogenic alleles of a given gene from both parents, the resulting egg and the cells that grow from it will have no wild-type copy of that gene, so all the cells in the embryo will misbehave in a way that results from their homozygous condition at that gene. This misbehavior is very likely to cause the embryo to die before it can grow into a viable fetus.

5.9. Apply them in the order PAH then PMA, not the reverse. If a tumor promoter is administered before a genotoxic carcinogen, it does nothing to promote a tumor, because the mutations that come from the carcinogen have not yet occurred. If the order is first carcinogen and then promoter, some cells may have acquired a carcinogenic mutation, and then their number is increased by the promoter, making the chances for further mutation greater.

5.10. The behavior is advantageous for the virus, because by encouraging the host cell to grow and divide, the viral DNA is replicated along with the host DNA; the increase in cell number means an increase in the number of copies of viral DNA as well. Thus, the virus replicates with the host.

5.11. There are several reasons these cancer-inducing genes have not been selected out of the human population. One is that cancer is far more common in older people than in the young, so a person carrying such a mutant allele may well have produced progeny before cancer occurs. A second is that the need for multiple mutations for cancer to arise means that carrying only a single carcinogenic allele (and in a heterozygous state at that) gives only a mild negative selection. Its effect may be too small to detect.

5.12. They are more dangerous for the young because a carcinogenic mutation introduced in a young person will be present for a longer time than in an older person. This gives more time for the accumulation of additional mutations in that same cell, which could lead to a complete cancerous transformation.

Oncogenes and their roles in cancerous transformation

CHAPTER 6

6

The previous chapters have emphasized mutation as an important part of cancerous transformation. Now we can look at specific genes that are commonly mutated in the development of cancer and learn why these changes matter so much for cell behavior. Genes important for cancer are described in two categories: **oncogenes** and tumor suppressors. You have previously met the term oncologist, referring to a doctor who studies and treats tumors. An oncogene is a gene whose product promotes cancerous behavior in cells. A tumor suppressor is a gene whose product inhibits cancerous behavior, so its loss is carcinogenic. Tumor suppressors are the subject of Chapter 7.

Surprising as it may be, the first real progress in a deep understanding of human **oncogenesis** began with studies on the transmission of tumors between chickens. This work and analogous studies on other experimental animals led to the discovery of **oncogenic viruses**, also known as **tumor viruses**. Experiments on animals demonstrated beyond doubt that these infectious agents could cause cancers. Prolonged study of such viruses led to the identification of numerous genes important for oncogenesis. These studies occurred just as the fields of molecular and cellular biology were developing, so some of the experiments with the biggest impact on cancer research were also important in the development of scientific knowledge more generally. Also important were the inventions of technologies that made all this scientific work possible. The work described here represents a remarkable series of achievements in understanding how cells work and how their normal workings can be led astray.

LEARNING GOALS

1. Describe how oncogenic viruses can bring about inheritable changes in cell behavior.

2. Describe the essential features of an oncogene and how a tumor virus can use such a gene to upset the normal pathways for cell cycle control.

3. Compare the cancerous transformations achieved by DNA and RNA tumor viruses.

4. Specify the differences between proto-oncogenes and oncogenes.

5. Explain why there are so many oncogenes.

6. Describe the similarities and differences between chemical or radiation-induced carcinogenesis and the cancer-causing ability of an acutely transforming tumor virus. Make the same comparison with a tumor virus that is not acutely transforming.

7. Diagram ways by which gain-of-function mutations in genes that control the extracellular matrix can promote cancer.

8. Diagram ways by which gain-of-function mutations in genes that control the cell's ability to undergo apoptosis can promote cancer.

9. Specify the cellular behaviors that are necessary for a human somatic cell to become immortal, in the sense that it can grow and divide without limit.

CANCER TRANSMISSION BY CERTAIN ANIMAL VIRUSES

At about the time doctors were discovering the ability of specific substances and forms of radiation to increase the likelihood of cancer, a few scientists were exploring the possibility that cancer could be transmitted from one animal to another by infection. In 1908, Vilhelm Ellerman and Olaf Bang, working at the University of Copenhagen in Denmark, demonstrated that leukemia could be transmitted between chickens by taking blood from a diseased animal, removing all the blood cells by forcing the liquid through a very fine filter, and then injecting the resulting filtrate (the liquid that came through the filter) into a healthy chicken. In 1911, Peyton Rous, working at the Rockefeller Institute for Medical Research in New York City, was able to transmit the formation of sarcomas (solid tumors in chicken breast muscle) from one animal to another by injecting a healthy chicken with a cell-free fluid obtained by filtering ground-up muscle from an animal with a breast muscle sarcoma. Experimental evidence for the transmission of cancer from one animal to another by one or more filterable, infectious agents was now unquestionable. However, the mechanism by which this process worked was mysterious. In the early twentieth century, medical science was just beginning to understand the diversity of microorganisms that cause disease. Through the study of viruses that infected plants and bacteria, knowledge about these pathogens grew to the point that their cycles of infection could be understood (SIDEBAR 6.1).

Identification of tumor viruses initiated new ways to study carcinogenesis in cell culture.

In the 1950s and 1960s, several closely related **papilloma viruses** were identified as causative agents of warts and similar benign lumps on the skin of humans and other mammals. These viruses enter a host cell's nucleus and replicate extensively, forming many virions that can subsequently leave this host and infect other cells (FIGURE 6.1A). (For a reminder about the life cycle of many viruses, see Figure 5.14.) Papilloma viruses are simple in structure and composition. Their genome is a single piece of double-stranded DNA, and they include proteins that assemble into a capsid that protects the viral genome (FIGURE 6.1B). More recent work has shown that papilloma virus is a major factor in the initiation of cervical and penile carcinomas and some other human cancers, as mentioned in Chapter 5. In the 1970s, a very similar pathogen, **polyoma virus**, was identified as a cause for cancer in several different mouse tissues, hence its name (Latin for many cancers). Also early to be discovered was a virus

A CLOSER LOOK: Early discoveries about disease-causing microorganisms.

Roles of bacteria in human infections had been established by the mid-nineteenth century. Louis Pasteur (after whom the pasteurization of milk is named) had demonstrated the roles of bacteria in the transmission of several infections, and the German bacteriologist Robert Koch had established principles for the proof that a particular microorganism was the cause of a specific malady. By 1900, the germ theory for transmission of infection was well established. There were, however, some diseases that did not conform to the patterns of bacterial infections. In 1884, Charles Chamberland, a French microbiologist studying the cause of rabies, invented a method of filtration that could remove all bacteria from liquid taken from a diseased animal, providing a bacterium-free liquid to be used for the study of disease transmission. His filter was a thin piece of hard clay, called porcelain, whose pores were so fine that no cell, not even one as small as a bacterium, could pass through. This was the filter the Danes and subsequently Dr. Rous used in their studies of cancer transmission between chickens. Any disease-causing agent that passed through Chamberland's filter was said to be filterable and was called a **virus**, a term that comes from the Latin word for a slimy, poisonous liquid. The pathogens for both rabies and yellow fever were shown to be viruses by this criterion. Thus, the discovery that cancer could be transmitted by a filterable agent suggested that this disease too was caused by a virus, although this kind of pathogen was still poorly understood.

Progress in understating the nature of viruses was slow because the tools then available were too crude. For example, light microscopic examination of a solution containing infective bacteria revealed tiny cells, many of which could swim. Similar scrutiny of the filterable agents revealed absolutely nothing; the toxic fluids looked like water. It wasn't until the invention of the electron microscope in the 1930s that there was a tool with sufficient resolving power to see a virus for what it really is. A virus that infects the leaves of tobacco and several other plants was the first to be isolated and clearly visualized (**Figure 1A**). Chemical studies showed that these particles were composed of two molecular parts: a single molecule of RNA and many copies of a single protein (**Figure 1B**). These parts could be separated, and many copies of the protein could assemble all by themselves into structures that strongly resembled the virus particle. For a particle to be infectious, however, both RNA and protein were required. The sarcoma-causing virus discovered by Rous was also found to be made from multiple copies of a protein and one piece of RNA. Other viruses, however, contained DNA and protein, not RNA and protein. This discovery allowed a distinction between DNA- and RNA-containing viruses.

An important step in understanding viral infection was the discovery that there were filterable agents that could infect bacteria. Again, some of these contained DNA and some contained RNA. The electron microscope revealed what a bacterium looked like when infected by multiple copies of one such agent (**Figure 2A**). When these pathogens were added to a plate of bacteria, like the one used in the Ames test (Chapter 5), all the bacterial cells in the vicinity of even a single added virus were killed as the viruses multiplied and infected nearby cells, resulting in a plaque, which is a circle of lysed and dead bacterial cells (compare **Figure 2B,C**). These viruses came to be called **bacteriophage** (eaters of bacteria) or often simply **phage**.

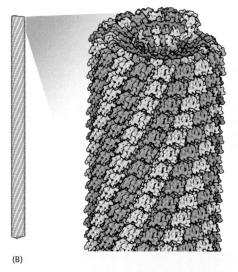

(A)   (B)

Figure 1 Electron micrograph and model of tobacco mosaic virus. (A) A modern electron micrograph of tobacco mosaic virus. A thin film of water containing many copies of this rod-shaped virus was simply frozen and then put into the microscope, using a sample holder that maintained a very low temperature. Although contrast is low, these particles contain detail at high resolution. (B) By averaging the structures in many images like the one in part A, one can visualize the shape of the individual proteins (blue and purple) that make up the viral capsid, or coat, and even the RNA molecule that is necessary for viral infectivity (red). (A, courtesy of Ruben Diaz-Avalos, HHMI, Janelia Research Campus. B, courtesy of David S. Goodsell, RCSB, PDB-101.)

(A)

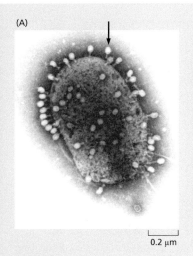

0.2 μm

(B)

(C)

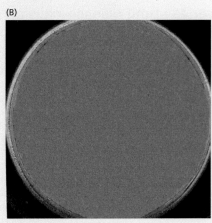

Figure 2 Bacteriophage are viruses that can infect bacteria. (A) A single bacterial cell is being infected by large numbers of the bacteriophage called T4. The light-colored circular objects (arrow) are the phage heads, which contain phage DNA; the heads are connected to the surface of the bacterium by tails through which phage DNA can pass during an infection. (B) A bacterial culture plate, often called a petri dish after the man who invented it. The dish, ~4 inches in diameter, contains a gel made from agar, which is an extract from the cell walls of the ocean alga kelp. This gel contains added nutrients, so the bacteria will grow. In this image, bacteria have formed a thin, uniform layer, called a lawn, which covers the surface of the agar so evenly that it gives no contrast. (C) A petri dish identical to that shown in part B but photographed a little while after about 500 phage had been added to the plate. Many of the phage infected bacteria in the lawn and then reproduced in quantity, killing the first bacterium infected and then spreading to all its neighbors, killing them too. Now each site of initial infection is surrounded by a circle of dead bacteria, called a plaque, which appears darker than the unperturbed lawn (arrows). One can count the number of infectious phage initially added to the plate by counting the number of plaques that have formed. (Courtesy of Stanley Maloy.)

Work on bacteriophage helped scientists to understand some surprising properties of viruses. One such discovery came from detailed study of the DNA-containing phage called lambda, the Greek letter written λ. These viruses can infect *Escherichia coli*, a commonly benign bacterium you met in Chapter 4. When the growth medium for the host bacteria contained plenty of nourishment,

λ-phage infection worked just like an infection by many other viruses (Figure 3). When nutrients were scarce, however, λ phage behaved differently. Although a phage particle still attached to a bacterial

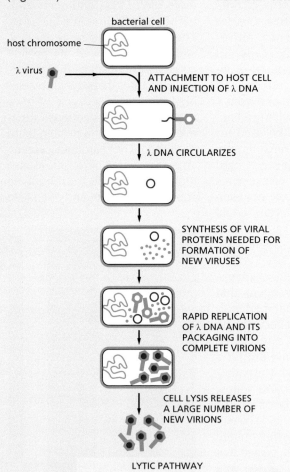

bacterial cell

host chromosome

λ virus

ATTACHMENT TO HOST CELL AND INJECTION OF λ DNA

λ DNA CIRCULARIZES

SYNTHESIS OF VIRAL PROTEINS NEEDED FOR FORMATION OF NEW VIRUSES

RAPID REPLICATION OF λ DNA AND ITS PACKAGING INTO COMPLETE VIRIONS

CELL LYSIS RELEASES A LARGE NUMBER OF NEW VIRIONS

LYTIC PATHWAY

Figure 3 A life cycle of λ phage and of many other viruses. Viruses are not alive because they are not self-replicating organisms. They require a host cell to make more copies of themselves. Viral infection involves the inoculation of viral genes (DNA or RNA, depending on the virus) into a host cell (red material entering the cell from the green phage). Some of the viral genes are then expressed, making virus-specific proteins (green dots), which help the virus take over the activities of the cell, such as DNA replication. In the presence of the virus, host DNA synthesis is usually shut down. Now viral DNA (or RNA) can replicate and interact with some of the virus-specific proteins to make new virus particles, called virions. These can escape the host, sometimes by breaking open the host cell, a process called lysis, to go on and infect other host cells.

host and injected its DNA, the infection didn't lead to the making of more phage. Only a few virus-specific proteins were made, and these helped the phage DNA to integrate into the host cell's chromosome (Figure 4). Thereafter, the integrated phage DNA is replicated every time the host cell goes through S phase and replicates its DNA. Meanwhile, no new phage particles form. This strategy for replication of phage DNA is called the prophage pathway.

At first sight this strategy looks like a failed infection, but in reality it means that phage DNA is protected when conditions for replication are poor. If the host bacterium continues to be starved, it will form a stable and long-lasting state, for example, a dormant cell or a spore that can survive harsh conditions. Thanks to this ability of the bacterium, the phage DNA is protected. When a new supply of nutrients comes along, allowing the infected bacterium to start growing and dividing again, the phage DNA is again replicated along with the host cell DNA. If the conditions became really good, so the bacterium can grow and divide quickly, the phage then undergoes induction. The phage DNA comes out of the host cell DNA and replicates extensively in the host cell, switching back to the normal viral infection pathway (Figure 4, right side). In this lytic pathway, the infected bacterium makes lots of virus-specific proteins and more copies

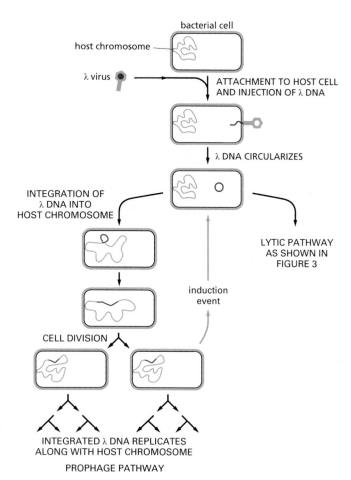

Figure 4 A different kind of infection cycle for λ phage. This diagram shows an alternative behavior of λ phage. λ phage can either carry out a typical host cell infection (the lytic cycle shown in Figure 3) or it can integrate its DNA into the host cell's chromosome, as shown here (sometimes called the prophage pathway). Now, every time the host grows and divides, the viral DNA is replicated. Note, however, that even after the viral DNA is integrated in the host cell chromosome, it can respond to environmental changes by leaving the host cell chromosome and embarking on a more typical viral infection.

of the viral DNA, which lead ultimately to lysis of the bacterium and release of many virions that will go out to infect other hosts. The fact that the DNA of λ phage can integrate into host cell DNA and later re-emerge became critically important some decades later for understanding the behavior of tumor viruses.

that caused leukemia in mice and led to the accumulation of virus-filled white blood cells in rats.

The discovery of these viruses was one thing; understanding how they caused cancer was something else. A whole animal is so complex that it was difficult to figure out how these viruses were inducing a cancerous transformation. This situation led several groups of scientists to improve methods for growing animal cells in culture, so they could study tumor virus infections in a situation where both the virus and the infected cells could be analyzed in more detail (for a reminder about cell culture, see Sidebar 4.1). One early success in this task was in 1958 at the California Institute of Technology, where Harry Rubin and Howard Temin were

100 nm

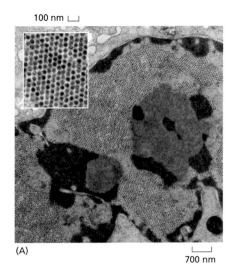

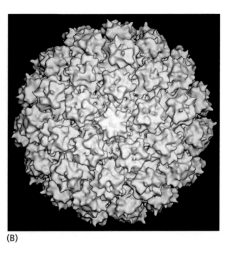

(A)

700 nm

(B)

FIGURE 6.1 Tumor viruses infect cells and duplicate in their nuclei. (A) An electron micrograph of a slice cut from the nucleus of a mouse cell infected by polyoma virus. The nucleus is almost filled with virions. When this cell dies and bursts open, it will release thousands of virions that can move by diffusion and blood flow to encounter other cells and infect them. (B) A virion of a human papilloma virus, seen at high resolution. This view, which can be obtained either by electron microscopy of frozen, hydrated virus particles or by X-ray diffraction of crystals formed from many virions, shows the proteins that make up the viral coat, called its capsid. The capsid protects the viral genome, which lies inside it. During infection of a host cell, the capsid falls off the underlying DNA, allowing the viral genome to enter the nucleus, express its genes, and be replicated. (A, courtesy of Kim Erickson and Robert Garcea, University of Colorado Boulder. B, from B.L. Trus et al. [1997] Nat. Struct. Biol. 4:413–420.)

working with the chicken sarcoma virus identified by Rous many years before. These young investigators were able to disperse cells from chicken embryos and get their fibroblasts—that is, connective tissue cells—to grow well in culture. The cultured chicken fibroblasts worked as hosts for the Rous sarcoma virus (RSV), so the virus could now be grown efficiently in a laboratory. With this system, Temin and Rubin found that RSV infection altered cell shape and induced fibroblasts to grow like cancer cells (FIGURE 6.2).

The practical implications of these findings for cancer research were profound. Cultured cells could be used to grow oncogenic viruses, but even more important, the virus induced changes in cultured cell behavior that resembled cancerous transformation. For example, when uninfected chicken fibroblasts were well fed, they grew until they filled the dish and then stopped growing and dividing. Their contact with neighboring cells was a negative signal to their growth and division, a behavior called **contact inhibition of growth**, which was described in Chapter 4. When the same cells were infected with RSV, not only did their shapes change but they now grew so well they would pile up on their neighbors, as if they had lost normal growth control (see Figure 6.2B). Moreover, they no longer required so much serum in their growth medium, as if they no longer needed all the growth factors that were essential for normal cells to continue their cycles of growth and division. Finally, the virus-infected cells would grow when they were no longer stuck to the bottom of the culture dish. This property showed up both in the way they would pile up on one another and in the fact that they would grow into a colony of cells even in a loose gel where no dish attachment was possible. These features all resembled the behaviors of cancer cells. The transition from normal growth to unrestrained growth was called **cell transformation**, and it was interpreted as a model for cancerous transformation. Cultured cells could now be used as plausible models for cancerous transformation under conditions where the process could be studied in detail. With several oncogenic viruses to study and the methods of cell culture in hand, scientists with

Which aspects of cancerous transformation can be studied in cell culture and which cannot?

normal cells   infected cells

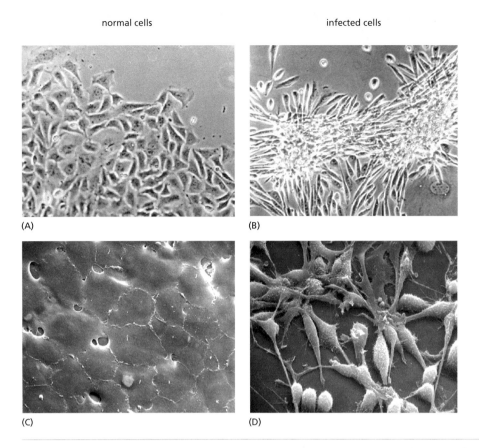

(A)   (B)

(C)   (D)

FIGURE 6.2 Micrographs of chicken embryo fibroblasts in culture, with and without infection by RSV. (A, C) Normal fibroblasts in culture, as seen (A) by light microscopy or (C) with a scanning electron microscope. (B, D) Similar cultures of the same cell type imaged after infection by Rous sarcoma virus (RSV). (B) A light micrograph shows the tendency of these cells to form clusters and grow on top of one another. (D) The scanning electron microscopic image clearly shows the changes in cell shape that are induced by infection with this virus. These behaviors of virus-infected cells resemble the behaviors of true cancer cells in culture (and in a host animal). Such similarities encouraged cancer scientists to use the transformation of cells in culture as a method to study the ways in which tumor viruses might cause cancer. (Courtesy of L.B. Chen.)

training in either the biology of cells or the molecular biology of DNA and RNA were now in a position to probe the details of this cancer-like transformation.

Cancerous transformation is associated with the continued presence of tumor virus genes.

Beginning in the mid-1960s, the pace of progress in understanding viral carcinogenesis picked up sharply. Several groups of scientists treated RSV with mutagens to alter its genes. Among the mutant viruses isolated, one was able to transform cells only at a slightly reduced temperature. At 37°C (human body temperature), the mutant virus could infect cultured chicken fibroblasts and transform them. At 41°C, the normal body temperature of a chicken, the virus-infected cells stopped showing the transformed phenotype and grew like normal fibroblasts, suggesting that the product of some viral gene was now defective and could no longer support cell transformation. This appeared to the investigators like a temperature-sensitive mutation in one of the viral genes. Cells infected with this mutant virus would grow and divide many times at the higher temperature with no sign of having been transformed. However, when their growth temperature was lowered to 37°C, these cells again displayed the transformed phenotype: rounded shape, loss of contact inhibition, and growth on top of one another. This result demonstrated that the virus's transforming principle was still present in the cell, even when its presence was not revealed in the cell's phenotype for many generations of growth and division. The cells really were permanently transformed, just as cancer cells are permanently changed by cancerous transformation, but the transformed phenotype was expressed only under conditions in which some viral gene product could function.

Analogous work with other tumor viruses showed different kinds of infection in different host cells.

Scientists studying the papilloma and polyoma viruses were also able to infect appropriate host cells in culture and bring about a cancer-like transformation of cell behavior. This result improved confidence in cultured cells as models for cancer. It also opened the way for new kinds of experiments as well as new methods to find novel tumor viruses. For example, a DNA-containing virus similar to polyoma was discovered during the big effort to raise large amounts of polio virus in monkey cells, as doctors worked to develop polio vaccines for people. It turned out that a monkey virus, called simian virus 40 (SV40), was infecting some of the cells used for polio virus culture, although the virus-infected cells showed no apparent phenotype. However, when SV40 was isolated and used to infect monkey kidney cells, it produced an obvious viral infection, forming many virions and killing the host cells (FIGURE 6.3). This behavior made it possible to grow lots of SV40, study its infective cycle, and isolate its DNA. SV40 turned out to be very similar in size, DNA content, and structure to both papilloma and polyoma tumor viruses. Each of these viruses contained a small circle of double-stranded DNA (FIGURE 6.4) made from just over 5000 pairs of complementary nucleotides. That small chromosome includes all the genes necessary for the virus to infect and multiply in a suitable host cell. Remarkably, when SV40 was used to infect cells cultured from hamster tissues (not monkey cells), the hamster cells became transformed and grew like cancer cells, even though no virus was produced. The small piece of DNA that served as a viral chromosome could not only support the development of virions when infecting one cell type, it could also induce a cancer-like transformation when infecting a different cell type, one whose genome included billions of nucleotides: astounding! The similarities among these three viruses led to their being grouped into one category. Information about the behavior of one virus

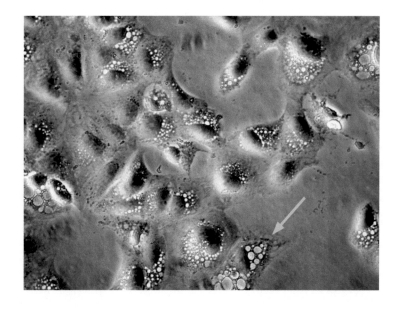

FIGURE 6.3 Monkey cells infected by the virus SV40, which can replicate in this host and kill it. The monkey kidney cells shown here were first grown in cell culture and then infected with SV40. The virus is making more virions, which leads the host cells to become sick, as seen by the white circles in the cytoplasm of most cells. These empty places are vacuoles (arrow), which will get bigger. Excessive accumulation of these vacuoles accompanies cell death. This is an example of viral infection in a permissive host cell. (From A. Gordon-Shaag et al., J. Virol. 77:4273–4282, 2003. Reprinted with permission from the American Society for Microbiology.)

was used by students of the other viruses, and progress in understanding cellular transformation by DNA-containing tumor viruses sped up. Several cellular hosts were found that would allow viral infection and reproduction; other cellular hosts did not permit viral growth but suffered cellular transformation into a cancer-like state. The first kind of virus/host combination was called permissive, because the virus could grow; the second kind was called nonpermissive, because viral growth was blocked, even though the host cells were transformed, suggesting that the virus was still present.

Viral infection of nonpermissive host cells provided essential clues for understanding virus-induced cellular transformation.

Study of nonpermissive cells transformed by DNA-containing tumor viruses showed that viral genes were present in these cells, even though virions were not being produced. This behavior was just like the cells transformed by the RNA-containing virus, RSV. It also resembled the behavior of λ phage, as described in Sidebar 6.1. In a quest for virus-specific functions that might be oncogenic, scientists infected animals with these tumor viruses and raised antibodies against the tumors that formed. Some of these antibodies reacted with a small set of virus-specific proteins that were produced in the tumor cells. Because the functions of these proteins were unknown, scientists called them simply the **T antigens**, where T indicates a tumor and antigen means a substance that will induce the formation of antibodies. When the antibodies to T antigens were used to stain either tumors growing in mice or transformed cells in culture, the nuclei of the cells stained brightly, supporting the ideas that viral infection was leading to the expression of virus-specific proteins and that transformed cells in culture were indeed a good model for cancers in animals. The same antibodies would also react with T antigens in homogenates formed by grinding up transformed cells in an aqueous solution. This property allowed scientists to begin a study of the molecules that were being produced in virus-transformed cells.

These scientists wanted to identify both the T antigens themselves and any molecules that bound to the T antigens. It seemed likely that the identity of cellular proteins that bound to the T antigens (products of tumor virus genes) might be informative about how the virus was altering cell behavior. To pursue this idea, they used antibodies to the T antigens to precipitate these proteins, hoping that the precipitate might also contain cellular proteins that bound to the T antigens. These scientists coated tiny plastic beads with the antibodies to T antigens and mixed them with homogenates of transformed cells. They then used gentle centrifugation (that is, they spun the mixture in a centrifuge at low speed) to sediment the beads and any proteins that stuck to the beads. If the idea worked, the beads should be coated not only with the antibodies and the T antigens themselves but also with any cellular molecules that bound to the T antigens (FIGURE 6.5).

Several cellular proteins were bound to the T antigens and were pulled down with the beads. Remarkably, one of them was the retinoblastoma protein, Rb, which you met in Chapters 4 and 5 as a factor that helps to keep cells from initiating DNA replication. This discovery suggested a mechanism by which these tumor viruses might induce cellular transformation. Proteins encoded by the viral DNA (the T antigens) could bind to and perhaps inactivate a cellular protein that normally inhibited cell division. This binding might inhibit the Rb-dependent inhibition of DNA replication and thereby let cells enter S phase, even when there was no

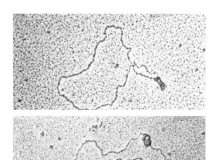

FIGURE 6.4 DNA isolated from SV40-infected cells. These electron micrographs confirm that isolated tumor viral DNA is circular, meaning that each piece of DNA has no free end (see also Figure 5.15A). Each piece of DNA shown here has a lump of material associated with it, which is interpreted as protein. One protein that binds to viral DNA is DNA polymerase, so these DNA loops could be about to duplicate. Protein is also necessary for viral DNA to interact with host cell DNA and integrate into a host cell chromosome. To get images like these, a drop of purified viral DNA in an aqueous solution is placed on a very flat surface to which the DNA will stick. The liquid is then washed off with pure water, and the surface is allowed to dry. The surface with adhering DNA is placed in a chamber from which the air can be pumped out; then a heavy metal, like platinum, is evaporated from a hot electrode, placed off to one side. This treatment spatters platinum atoms onto the surface from an angle, making shadows where the DNA and protein stick up from the surface. (From Griffith J, Dieckmann M & Berg P [1975] *J Virol* 15:167–172. With permission from American Society for Microbiology.)

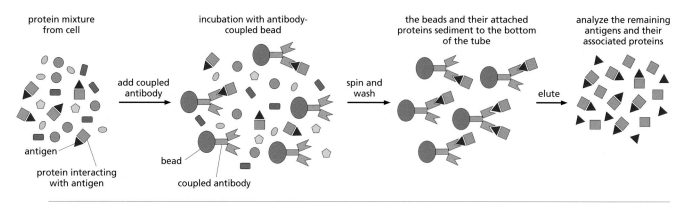

FIGURE 6.5 A method for immunoprecipitating proteins. If you break open cells, thousands of proteins are released. Isolating one or a few proteins of interest from this complex mixture is difficult. One way to do it is to use antibodies raised against the protein(s) of interest. In this example, antibodies raised against the T antigens are coupled to a tiny bead (shown here as purple ovals), making a complex that is easy to sediment in a centrifuge. (In reality, the beads are much larger than the antibodies, and each bead would be coated with many antibodies.) When the bead-coupled antibodies are mixed with a complex mixture of proteins, like that produced by breaking open cells, the antibody will bind its antigen (shown here as red triangles), tethering that protein to the bead. If the antigen is bound any other protein in the solution (shown here as blue squares), that protein too becomes stuck to the bead. When this mixture is put into a centrifuge and spun, the beads sediment quickly to the bottom of the tube, separating them from all other components of the complex mixture. After the beads have been washed, the only molecules still sticking to them are the antibodies, the antigens, and anything that bound tightly to the antigen. The antigens and their associated proteins can then be released from the antibody by making the solution more acidic. Now the beads with antibodies still attached can be spun down, and the overlying liquid will contain only the antigen and its associated proteins.

signal telling them to do so. Viral transformation of cell behavior was beginning to make sense.

A second key experiment showed how viral DNA that was not producing virions could persist in an infected cell, continually altering host cell behavior. In the first part of this experiment, DNA was isolated from SV40-infected cells permissive for viral infection (monkey cells). This preparation included DNA from both the host cells and the rapidly growing viruses. These two kinds of DNA were of very different sizes, so they could easily be separated by spinning the solution of DNA in a centrifuge; the small DNA from the virus sedimented slowly, whereas the huge pieces of DNA from the mammalian host cells sedimented more quickly (FIGURE 6.6A). Now the viral DNA could be isolated and labeled with radioactivity, so it could be used as a probe to ask whether other DNA samples contained the same or similar DNA sequences. The idea here is that when two different pieces of double-stranded DNA contain a stretch of identical nucleotide sequence, the bases in those sequences are complementary and will bind to one another, just the way the two strands of a DNA duplex bind. One can separate all the pieces of double-stranded DNA into single strands by heating them, then mix the DNA samples and cool them slowly, so complementary sequences bump into one another and bind through a process called **DNA hybridization** (SIDEBAR 6.2).

To carry out informative experiments about the presence or absence of viral genes in a virus-transformed cell, DNA was isolated from nonpermissive cells (from a hamster) that either had or had not been transformed by infection with SV40. The labeled viral DNA, isolated from a permissive host, was then allowed to hybridize with DNA from these two nonpermissive hosts, as described in Sidebar 6.2. After hybridization, the labeled viral sequences sedimented with the DNA of infected host cells (FIGURE 6.6B, red trace) but not with DNA from uninfected host cells. Further tests showed that the viral DNA was actually incorporated into one or more pieces of host DNA; the viral DNA had integrated right into a host chromosome. This finding explained how a nonpermissive host could retain

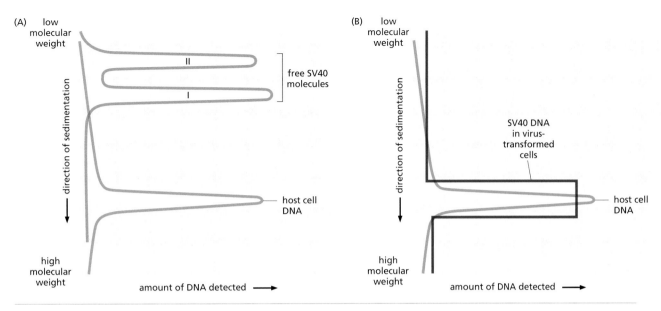

FIGURE 6.6 Experiment to study the association between host cell DNA and the DNA of an infecting tumor virus. During centrifugation, molecules with high molecular weight (greater size) move faster down the centrifuge tube than molecules with lower molecular weight. (A) The positions after centrifugation of viral DNA (green) and host cell DNA (blue) are shown. Host cell DNA (blue) is bigger, so it moves faster down the tube in the centrifuge. The two shapes of viral DNA seen in Figure 5.15A (circular versus supercoiled) account for the two peaks of viral DNA after sedimentation. (B) In an infected, nonpermissive cell, SV40 DNA (red) sediments with the host cell DNA (blue), showing that it has integrated into a host cell chromosome and is no longer free as circles or supercoils.

the transformed phenotype over many cell generations, even when virions were not being made. The DNA necessary for virus-induced cell transformation was carried along in the host chromosomes as they replicated and segregated with every cell cycle.

Initially, this discovery was surprising, because it was not obvious why a virus would develop the ability to integrate its DNA into a host cell's chromosomes. However, two factors made the finding both reasonable and informative: (1) A virus can be biologically successful by infecting a host cell and then simply inducing it to grow and divide. With every cycle of host cell DNA replication and division, the viral DNA too is replicated and carried along, wherever the host cells might go. A precedent for just such behavior had already been established, thanks to work on λ phage, as described in Sidebar 6.1, Figure 4. (2) Further work on these DNA tumor viruses showed that their ability to inhibit Rb protein, and thereby to promote the host cell's entry into S phase, was an essential part of successful infection, even in a permissive host cell. The viral DNA cannot replicate unless the host cell has turned on its own DNA replication machinery. In this light, transformation of a nonpermissive host cell appeared similar to a viral infection in which only part of the normal viral replication program was completed.

RNA tumor viruses can also integrate their genes into host cell DNA.

Integration of viral DNA into a host chromosome explained the persistence of transformation by DNA viruses, but what about tumor viruses that used RNA instead of DNA for storing their information? While the previously described discoveries about DNA tumor viruses were being made, scientists studying RNA tumor viruses were also hard at work. They too obtained valuable information about both the mechanism for cell transformation and how RNA viruses could persist as the transformed

A CLOSER LOOK: DNA Hybridization.

The essence of DNA structure is its double-helical arrangement of two strands of nucleotides that are complementary at every position along the two strands. This structural principle is the basis for faithful DNA replication and for the synthesis of RNA molecules whose sequences carry messages encoded by the DNA from which they were transcribed. It is also the idea behind a method of great simplicity and power for using a piece of DNA to seek other segments of DNA that have the same, or at least a highly similar, sequence of nucleotides, regardless of where they may be. Underlying all these phenomena is the fact that each nucleotide in a DNA sequence fits better with its complementary nucleotide than with any of the other possibilities: A fits with T but not with G or C; G fits with C but not with A or T; etc. Thus, a given DNA sequence, such as AGCTTAGC, will pair better with a piece of DNA whose sequence is TCGAATCG than with any other DNA; these two pieces of DNA are perfectly complementary.

If the double-stranded DNA from our cells is cut or broken into small pieces, these pieces are still double-stranded, and the two parts of each piece are still complementary, binding quite tightly to one another. One can, however, cause the complementary pieces to come apart, for example, by heating them to a temperature around 70°C or by treating them with a solution that is quite basic (as opposed to acidic). When the two parts of double-stranded DNA come apart, each single-stranded piece can float separately, moving under the impact of thermally driven collisions from water molecules that are diffusing around randomly. If solution conditions are brought back to normal, so the strength of thermal agitation in the solution is no longer greater

than the strength of the bonds between complementary bases, the two parts of the original double-stranded pieces can rebind one another. As the many pieces of single-stranded DNA float around in solution, complementary strands will occasionally bump into one another, and when they do, they bind more tightly to each other than to any other pieces of DNA they might encounter. Once complementary pieces find each other (by chance), they re-form a normal double-helical DNA structure (**Figure 1**). In fact, the pairing of complementary nucleotide sequences is strong enough that the two sequences don't have to be perfectly matched to bind quite well, or to **hybridize**. However, the closer their sequences are to perfect complements, the more strongly they will bind.

These ideas and procedures have allowed scientists to take an interesting piece of DNA, like the genome isolated from a virus, and use it as a probe to ask if that sequence, or a sequence like it, is present in some other DNA sample, such as DNA isolated from a virus-infected cell. To make this

kind of hunting informative, the DNA of interest has to be labeled in such a way that it can be identified, either by looking (for example, if one has made it fluorescent) or by measuring emitted radiation (if one has found a way to make the DNA of interest radioactive). The latter method was of great importance in the early days of cancer virus study, because even at that time one could buy nucleotides that contained radioactive atoms, like ^{14}C or a radioactive isotope of phosphorus, ^{32}P. Researchers could then make radioactive viral DNA or RNA by feeding cells with radioactive nucleotides, or the nutrients that cells use to make nucleotides, and then immediately infecting them with virus. As new virions are made, their DNA or RNA is made with radioactive nucleotides, so when viral DNA is isolated, it is radioactive. Once this nucleic acid has been isolated, it can be used as a probe for hybridization to similar nucleotide sequences, wherever they might be.

In such an experiment, researchers would isolate DNA or RNA from an organism to be tested. If

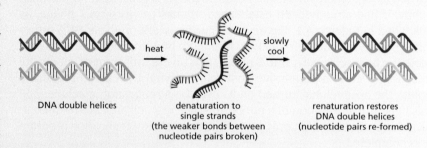

DNA double helices   heat →   denaturation to single strands (the weaker bonds between nucleotide pairs broken)   slowly cool →   renaturation restores DNA double helices (nucleotide pairs re-formed)

Figure 1 Pieces of DNA with complementary sequences will recognize each other in solution and bind. Two pieces of double-stranded DNA are shown. The sequence of bases on the yellow strand is complementary to that on the red strand, and the blue strand is complementary to the green. If these two different pieces of double-stranded DNA are denatured by being heated, so thermal energy becomes greater than the strength of the bonds between the DNA strands, the strands will separate. The now single-stranded pieces of DNA can diffuse randomly in a solution at elevated temperature, where the bonds between bases are not strong enough to hold the complementary pieces together. However, if the sample is cooled slowly, thermal energy decreases, and the bonds between complementary bases become more important. Now, when complementary sequences encounter one another by chance, they will stick together. If cooling is done at just the right rate, the complementary strands will all have a chance to find each other and zip together, re-forming their original double-helical structures.

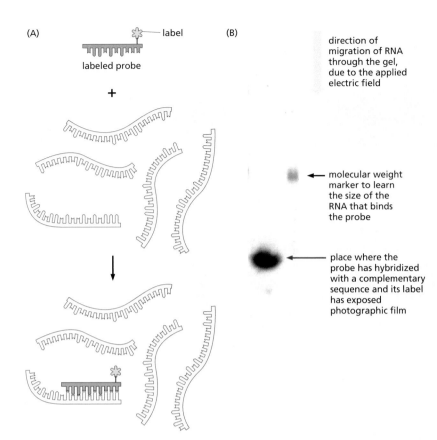

(A) label
labeled probe

+

(B) direction of migration of RNA through the gel, due to the applied electric field

molecular weight marker to learn the size of the RNA that binds the probe

place where the probe has hybridized with a complementary sequence and its label has exposed photographic film

Figure 2 Labeled DNA can find and tag complementary sequences by diffusion and base pairing. If a DNA sequence of interest has been labeled (yellow star), it can be used to find complementary sequences in any sample of DNA, for example, one made by breaking cellular DNA into small pieces. (A) The two kinds of DNA are mixed, denatured so as to become single-stranded, and allowed to mingle thoroughly by diffusion. If conditions are now returned slowly to normal, so the bonds between complementary bases become strong relative to thermal energy, any two regions that have complementary sequences will bind to one another. Through this adhesion, one can find complementary sequences in a large pool of some other DNA. (B) An actual hybridization experiment in which a DNA probe, labeled as in part A, was used to study a complex mixture of nucleic acids, in this case all the mRNAs made by a cell. The RNA was loaded onto the top of a gel slab and an electric field was applied, forcing the negatively charged RNA molecules down into the gel. The smaller the pieces of RNA, the faster they could move through the gel. After the electric field was turned off, the sample was soaked in a solution that contained a radioactive DNA probe. Hybridization was allowed to occur, and the position of the RNA that bound the probe was identified by letting the radiation from the probe expose a piece of photographic film (red arrow).

it was DNA, the molecules would come from the cell in long pieces, but they could be broken into shorter, more manageable pieces by forcing them quickly through a narrow needle at the tip of a syringe. These smaller pieces of DNA are then mixed with the radioactive probe, and the temperature or the pH is raised, causing double-stranded DNA to fall apart into single strands. After the sample has had time to let all the pieces of DNA diffuse around at random, conditions are returned slowly to normal, allowing pieces of single-stranded DNA to bump into each other. If their nucleotide sequences happen to be comple-

mentary, they will bind and stick (**Figure 2**). With a slow return to normal conditions, there is time for many pieces of nucleic acid to encounter one another by diffusion and see if they find a sequence that is truly complementary, whereupon they will be able to bind more and more strongly as conditions get closer and closer to normal. Hybridization will occur between the segments that have the most nearly complementary sequences. Of course the two strands of viral DNA will hybridize with each other, but if there are complementary sequences in the test DNA, these too will hybridize with the radioactive viral DNA.

The smaller fragments of cellular DNA made by forcing cellular polymers through a needle are still very much bigger than the viral DNA, so they are easy to separate from viral DNA by the rate at which they sediment, when spinning in a centrifuge, or migrate through a gel in response to an applied electric field. If the radioactive DNA (the original viral sequences) moves with the bigger pieces of DNA, then one knows that there are sequences in the cellular DNA that hybridized with the viral DNA, so there must be regions of complementary sequence (see Figure 6.6B).

cells grew and divided. Scientists isolated mutant forms of RSV that were replication-competent but unable to transform chicken cells in culture. They also found replication-defective strains of RSV that were still able to accomplish host cell transformation, even though they couldn't make virions. These findings suggested that viral replication and cancerous transformation were separate processes. Moreover, the strain of virus that could transform but not replicate would persist in cultured chicken cells through many generations of cell growth and division, just like DNA tumor viruses. This finding posed a puzzle: How could transformation by an RNA virus be so persistent? Were the RNA molecules of the RSV genome being replicated and faithfully segregated through multiple cellular generations, or did the virus have some other trick up its sleeve?

Around 1970, Howard Temin was working on RSV at the University of Wisconsin. He proposed the hypothesis that this RNA-containing virus somehow generated a DNA provirus, which behaved a bit like λ phage (see Sidebar 6.1). Temin suggested that virus-derived DNA was important for persistent viral transformation by an RNA tumor virus. However, this suggestion faced the problems that the viral genome was RNA, and RNA is normally made from DNA, not vice versa; there was no precedent for making DNA from RNA. Temin was proposing that RNA tumor viruses ran normal biology backwards! The scientific community was skeptical about Temin's suggestion: many scientists were scornful, and the idea did not gain traction. However, Temin undertook a search for an enzyme that could do this unprecedented job. A few years later, both he and David Baltimore, who was working at the California Institute of Technology on a different RNA tumor virus, were able to find enzymes with exactly the right properties. These virus-encoded enzymes were RNA-dependent DNA polymerases. In each case the enzyme was associated with the virions of mature RNA tumor virus, so it was carried into the host cell at the time of initial infection. These discoveries suggested a way by which RNA tumor viruses, like DNA tumor viruses, could integrate their genes into host cell chromosomes. Whereas DNA viruses could do it directly, RNA viruses could accomplish the same job by using **reverse transcriptase**, their enzyme that used an RNA sequence as a template to make a complementary strand of DNA (FIGURE 6.7). This ability of RNA tumor viruses has led to their being called **retroviruses**.

SIDE QUESTION 6.2

Do you think it is necessary for an RNA tumor virus to carry reverse transcriptase in its virion? Can you think of an alternative way the virus could still make DNA from its RNA genome and integrate this information into the host cell's chromosomes?

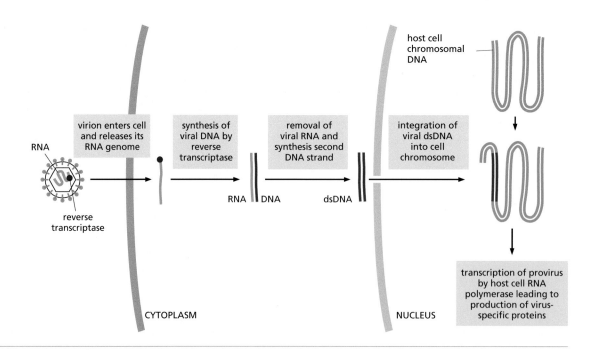

FIGURE 6.7 An RNA tumor virus can integrate its genome into host cell DNA. During infection of a host cell, the virion of an RNA tumor virus passes through the host cell's plasma membrane and releases its RNA genome into the cytoplasm. The viral protein, reverse transcriptase, now uses nucleotides from the host cell's cytoplasm and the viral RNA as a template to synthesize a single strand of complementary DNA. This DNA is used by host cell enzymes as a template to make a complementary strand of DNA, and now there is a piece of double-stranded DNA that contains the sequence of the viral genome. This is the DNA that integrates into the host cell chromosome.

The sarcoma-inducing gene of Rous sarcoma virus was the first oncogene to be discovered.

The gene in normal RSV that transformed host cells, but was missing or ineffective in the nontransforming mutants of this virus, was named *src* (pronounced sark), short for sarcoma. Note that, in talking about oncogenes, we will continue to write the names of genes in *italic* font and the product of the gene (such as the Src protein) in regular Roman font, beginning with a capital letter. The action of Src was of great interest, because it was not clear how a single protein could induce all the changes in cell behavior that one saw with cancerous transformation. Thus, *src* and Src became subjects of intense study.

In the 1970s, Ray Erikson and his colleagues at the University of Colorado School of Medicine learned that **Src** was a protein kinase. Protein kinases were already known to control a variety of cellular functions, such as the breakdown of stored food energy. (For a refresher on the action of protein kinases, see Figure 3.35.) A virus-encoded kinase, like Src, could be imagined to phosphorylate numerous cellular proteins, allowing this single gene product to modify the action of multiple host cell functions, just as one would expect for an agent that achieved cancerous transformation. Shortly thereafter, Tony Hunter at the Salk Institute for Biological Studies in California showed that Src was an unusual protein kinase: it transferred phosphate from ATP to the amino acid tyrosine, not the amino acids serine or threonine, which were the ones commonly phosphorylated by other protein kinases. The importance of Src for cellular transformation suggested that protein tyrosine kinases might be key players, in both normal cell cycle regulation and its modification during cancerous transformation.

The quest began for substrates of the Src kinase, that is, the cellular proteins that Src phosphorylated when it changed cell behavior in such remarkable and damaging ways. Many scientists pursued this goal and many papers were written on the subject, but the findings did not immediately explain how Src's protein tyrosine kinase activity induced cancerous transformation. Meanwhile, an alternative approach provided the intellectual breakthrough that helped all cancer scholars understand what was going on during oncogenic transformation.

Harold Varmus and Michael Bishop, working at the University of California, San Francisco, realized that the existence of reverse transcriptase in RSV carried a very strong implication. The genomes of chicken cells transformed by RSV should contain DNA sequences encoding the Src protein, but normal chicken cells should not. Note that this is exactly the same idea that was investigated and found to be true for DNA tumor viruses, as described above. In the case of an RNA tumor virus, however, the viral DNA would be the provirus that Temin's and Baltimore's discoveries had suggested. It would be DNA that could be transcribed and translated to make Src protein, which could maintain the transformed phenotype through multiple cell generations. These scientists reasoned that they could make a radioactive DNA probe from the RNA of transforming RSV and then use that probe to find complementary sequences in DNA isolated from RSV-transformed chicken cells.

Hybridization experiments that tracked *src* DNA found a very similar gene in untransformed chicken cells.

Varmus and Bishop found an efficient way to make the radioactive RSV DNA they needed to serve as a probe for cellular DNA with similar sequence. They grew RSV in cultured cells, isolated its RNA, and then used the newly discovered reverse transcriptase to make complementary DNA in the presence of radioactive nucleotides. They then isolated DNA from normal chicken cells (their control) and from RSV transformed chicken cells and hybridized their probe to both samples, expecting to find that the probe bound to DNA from infected cells but not to DNA from normal cells. However, the probe bound well to both samples of DNA, suggesting that the probe was not sufficiently specific. Apparently, they needed a probe that was absolutely specific for the *src* gene itself, not for DNA from the whole RSV genome.

Through a clever innovation, these investigators got rid of much background labeling, giving them a radioactive DNA probe that was highly specific for the *src* gene. With this tool, Varmus and Bishop redid their previous experiment, and the result was now much cleaner. Remarkably, though, they got the same result; there was *src*-hybridizing DNA in normal as well as transformed cells. They and their colleagues then tried the specific *src* probe on DNA from other birds and found that it reacted there, too. They then tried the *src* probe on DNA from a wide range of animals, including mammals, and found sequences hybridizing to *src* DNA in all of them. It appeared that a gene whose sequence was sufficiently similar to *src* to hybridize with the *src*-specific probe was practically universal and not limited to animals infected with RSV. These observations suggested that the transforming activity of RSV was in some way derived from normal cells and was probably a mutant form of a normal, highly conserved gene. Varmus and Bishop surmised that Src was a hyperactive form of a kinase found in normal cells.

The evidence became inescapable that oncogenes are mutant alleles of normal human genes.

Numerous researchers followed the lead suggested by the *src* hybridization experiments: they used labeled DNA from different tumor viruses to see if it would hybridize with DNA from untransformed cells. Many such examples were found. As a result, the idea took hold that there were normal genes that could be altered to become the oncogenes of tumor viruses. Varmus and Bishop coined the term **proto-oncogene** to mean a normal gene that was the forerunner of a tumor-causing gene. From this idea came a useful terminology: when a gene was discovered through the action of an oncogenic allele, like *src*, then the proto-oncogene was named to indicate that it is the normal, cellular equivalent of that gene, for example, c-*src*. This strategy helped to keep the number of overlapping names to a minimum, and it showed which form of the gene was oncogenic and which was not: c-*src* is the normal allele and is not oncogenic, while *src*, which is sometimes called v-*src* to indicate its viral origin, is oncogenic.

Scientists became fascinated with the kinds of mutation that could produce gene alterations with such drastic consequences. Was it overexpression, overactivity, a loss of regulation, or something completely unusual? Careful molecular biology and biochemistry showed that the protein c-Src is normally down-regulated, which means its activity is reduced, by being phosphorylated on an amino acid near its C-terminus. The oncogene v-*src* contains a stop codon a little too early in the gene sequence, leading to a protein product that is shortened by a few amino acids, just enough to remove the phosphorylatable amino acid from the protein's sequence. Therefore, the protein tyrosine kinase v-Src cannot be down-regulated by phosphorylation. It is always active, somehow urging cells to progress in the cell cycle, which accounts for its oncogenic activity.

THE DISCOVERY OF ONCOGENES IN CANCER CELLS

The work with animal tumor viruses provided essential advances in how to think about cancerous transformation in the laboratory, but what caused cancer in people? How could one link the findings about oncogenesis from tumor viruses with the clear evidence for carcinogenic activity of both chemicals and radiation? Some insightful experiments with cultured human cancer cells provided the information necessary to link these findings together.

The ability to transform cells in culture provided a powerful assay for DNA that could induce cancerous transformation.

In 1972, cancer researchers identified a way to force cultured cells to take up foreign DNA and express it. They called the process **transfection**, a union of the words transformation and infection (SIDEBAR 6.3). With this method in hand, several labs began asking whether one could transform normal cultured cells into tumorlike cells by transfecting them with DNA from cells that had previously been transformed into cancer cells by the action of chemical carcinogens. The answer was yes, suggesting that chemical carcinogenesis was somehow creating oncogenes in normal cells (FIGURE 6.8). This result fit beautifully with the idea that oncogenes are formed by mutations of normal DNA.

Both Robert Weinberg at Massachusetts Institute of Technology and Michael Wigler at Cold Spring Harbor Laboratory identified such pieces of cancer-inducing human DNA. The transfected cells changed their shape, lost contact inhibition of growth, and would pile up on top of one another (FIGURE 6.9). Through some expert molecular biological sleuthing, these labs were able to identify the sequences of human DNA that were responsible for this transformation. They turned out to be very similar to DNA that had previously been identified as an oncogene in a mouse tumor virus. With discoveries like this, the multiple agents that could cause cancer were beginning to form a picture that made sense.

Improved methods for both cell and molecular biology led to the discovery of new mechanisms for oncogenesis.

As more oncogenes and proto-oncogenes were discovered, most commonly by the study of animal tumor viruses, some were found to be

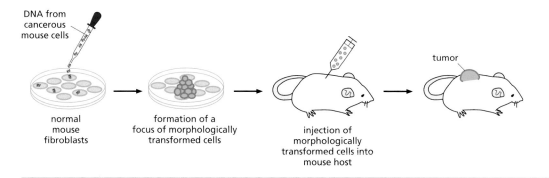

FIGURE 6.8 DNA from chemically transformed cells can be transfected into normal cells to achieve carcinogenic transformation. DNA from cancerous mouse cells can be added by transfection to normal cells. A small number of the resulting transformants acquire, by chance, a piece of DNA that is carcinogenic, inducing the cells to lose normal growth control in culture, so they pile up on their neighbors. The idea that these misbehaving cells really are cancer cells is supported by the subsequent experiment of injecting them into a mouse and finding that they will form a tumor in an animal host. (Adapted from Weinberg R [2014] The Biology of Cancer, 2nd Edition. Garland Science: New York.)

SIDEBAR 6.3

A CLOSER LOOK: Transfection of DNA into cultured mammalian cells.

The discovery of transfection, that is, methods for adding specific pieces of DNA into a cell type of choice, was an important technical breakthrough. It allowed scientists interested in a particular gene to put the relevant DNA into a cell and then study the consequences of the cell's expressing that gene product. One of the early steps in working out methods for transfection was the discovery that if DNA was dissolved in an aqueous solution containing phosphate and then mixed with a different aqueous solution containing calcium chloride, the mixing of calcium and phosphate brought on precipitation. Tiny particles of calcium phosphate formed and bound tightly to pieces of DNA. When this mixture was added to culture medium, cells would take up the particles, probably by normal endocytosis (Chapter 3), bringing into the cytoplasm a small membrane-bound vesicle that contain a drop of extracellular fluid and one or more calcium-phosphate particles containing DNA.

Once the DNA-containing particles are inside the cell, the endocytic vesicles release at least some of the DNA, which then gets into the nucleus, where it can persist for several days. The foreign DNA gets wound around histones (proteins important for the structure of chromatin; see Chapter 3) to make nucleosomes. Then it can interact with both transcription factors and RNA polymerase, but these small pieces of foreign DNA lack a kinetochore, so they are not segregated with any precision by the mitotic spindle. Thus, genes on this DNA are initially expressed, but over time they are lost from the cell. This procedure is therefore called a transient transfection. In spite of its transience, the method allows study of the impact of chosen DNA sequences on cell behavior.

Because the introduction of a particular DNA sequence into a healthy cell is an excellent way to study gene function, many scientists have worked to find ever more effective and convenient ways of achieving transfection. All these methods have to solve the problem that DNA is a big, highly charged molecule, so it does not easily cross the plasma membrane. One successful approach uses alternating electric current at quite high voltage to push and pull the charged molecules in the plasma membrane, introducing small holes into its normally continuous surface. Other methods employ lipids (fatty molecules) that carry a net positive charge. These bind negatively charged DNA very well. The resulting DNA-containing lipid droplets are taken up by the cell quite efficiently. Still other methods use positively charged, water-soluble polymers to bind DNA and promote its uptake by the cell. People have even used tiny projectiles to shoot DNA into cells that are particularly hard to transfect because of a barrier, like the cell wall that surrounds all plant cells.

With the improved transfection efficiencies, scientists have sought ways to make the effects of transfection last longer. Usually, it is just a matter of selecting again and again for cells that still contain the foreign DNA. These pieces of DNA occasionally find a way to integrate into a host chromosome. When this happens, the foreign DNA is not only replicated during S phase it is also segregated during M phase, and the transfection is permanent. If this result is strongly preferred, scientists now use parts of a mutated cancer virus to introduce the foreign DNA into the cell to be transfected. The mutated cancer virus has lost its ability to transform cells, but it retains its ability to integrate into a host cell chromosome. With this tool, the transgene is efficiently integrated into the host chromosome, so permanent transfection becomes the rule, not the exception.

carcinogenic by more than one route. The gene c-*myc*, which you met in Chapter 4, encodes a transcription factor that regulates the synthesis of RNAs that are translated into proteins important for the onset of DNA synthesis. An oncogenic form of c-*myc* occurs as v-*myc* in avian myelocytomatosis virus. Infection of birds with this virus can induce cancer by introducing an allele of *myc* that is always active, so no growth factor is required to turn it on. This is very much like the story for v-*src*. In both cases, the virus brings in an allele of a gene that encodes a protein important for the regulation of the cell cycle, but the protein product of the virus-encoded allele requires no stimulus to turn on DNA synthesis. The v-*myc* gene is behaving like a thief who hot-wires a car. Both are activating an event without permission, be it a key to fit into the ignition switch or a mitogen to stimulate DNA synthesis. Normal control has been bypassed and an important event starts without its normal regulation.

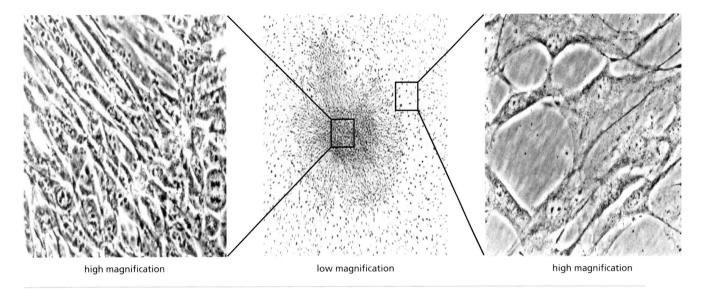

high magnification   low magnification   high magnification

FIGURE 6.9 DNA isolated from a human tumor can transform mouse cells in culture, inducing them to lose normal growth control. The central image shows a transformed culture at low magnification; the pile of cells in the middle is a focus of overgrowth. The higher-magnification images on either side show both transformed (left) and untransformed (right) cells. This experiment showed that human tumors contain oncogenes that can transform normal, cultured cells into cancerous cells, even when the DNA comes from a different species. (From M. Perucho et al. [1981] Cell 27:467–476.)

A normal c-*myc* gene can, however, become oncogenic by a different route. There is a different mutagenic bird virus that causes c-*myc* to be overexpressed in certain white blood cells. This overexpression can result from either of two genetic changes: (1) addition of extra copies of the c-*myc* gene, so-called gene amplification, or (2) placing a strong promoter from a viral gene next to a single c-*myc* gene (FIGURE 6.10). In either case c-Myc is overexpressed, providing hyperactivity and therefore carcinogenic action. In this case the tumor virus is not introducing an oncogene; instead, it is working as a mutagen that changes the expression level of a proto-oncogene, forcing the cell into an oncogenic transformation. Cancer

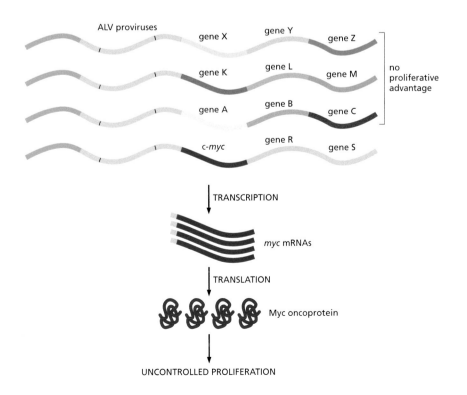

FIGURE 6.10 Viral DNA insertion can cause overexpression of nearby host genes. Avian leukosis virus (ALV) is a tumor virus that infects birds and can insert its DNA into the DNA of host cell chromosomes. The sites of insertion are fairly random, so the provirus DNA can enter host DNA almost anywhere. Here, ALV DNA enters in several places: near clusters of genes called X, Y, and Z or K, L, and M, etc. By chance, the provirus DNA might integrate into a host chromosome right next to c-*myc*. Because the viral promoters are strong, any gene that lies near the viral sequences will be induced to express at a high level. For many genes this result is unimportant, but for a gene involved in controlling cell cycle behavior, like c-*myc*, it can be highly significant. The resulting synthesis of many copies of c-*myc* RNA and the subsequent synthesis of lots of Myc protein result in excess cell proliferation. (Adapted from S.J. Flint, L.W. Enquist, R.M. Krug et al. [2000] Principles of Virology. Washington, DC: ASM Press.)

SIDE QUESTION 6.3

Describe two ways by which a virus that integrates its DNA into a host cell chromosome could cause a host cell gene to be expressed more strongly.

viruses that contain an oncogene are called acutely transforming viruses. They are more carcinogenic then viruses that are simply mutagenic, but either kind of virus can achieve oncogenesis.

In the face of these examples, it becomes surprising that there aren't a large number of human tumor viruses. While there are not many, there are enough to be important in human cancer. Epstein–Barr virus (EBV) is associated with Burkitt's lymphoma; human papilloma virus (HPV) is a causal factor for cervical and penile cancers; and there are several more examples, as mentioned in Chapter 5. All in all, approximately 15% of human cancers are now thought to be virus-related, and viruses are acknowledged as important human carcinogens, although most human-infecting viruses (measles, flu, mumps, cold viruses, etc.) are *not* carcinogenic. When work with animal tumor viruses was first taking off in the 1970s and 1980s, human tumor viruses were almost unknown. However, the knowledge about carcinogenesis obtained from study of animal pathogens was critically important in developing our current understanding of cancer. Dozens of oncogenes were discovered and their corresponding proto-oncogenes were identified in the animal host for a given tumor virus. The methods of molecular biology then allowed scientists to seek and find analogous proto-oncogenes in human cells. Subsequent work has shown that these are the very genes that are mutated by chemical or radiation-induced carcinogenesis to form the gain-of-function alleles that drive many cases of human cancer. This information has been essential for progress in understanding both the normal and abnormal regulation of human cell cycle progression.

PURSUING THE FUNCTIONS OF ONCOGENES AND THEIR PRODUCTS

The identification of oncogenes was an important step in the growing body of knowledge about cancer, but simple identification of a DNA sequence was not enough information to understand the role that a particular gene might play in carcinogenesis. To take the next step, scientists and doctors needed to learn the cellular functions of oncogenes, proto-oncogenes, and their protein products. Only then could they piece together the causes for the dramatic changes in cell behavior that gave rise to cancer. Fortunately, progress in several technologies important for understanding cells allowed cancer scholars to get this information.

The ability to sequence and compare different pieces of DNA led to a wealth of information about oncogenes and proto-oncogenes.

One important example of the power of DNA sequence information for understanding cancer is seen in the characterization of a human oncogene called *erbB*, originally identified as the transforming gene in a virus–induced leukemia of mice. DNA sequencing showed that this oncogene was a mutant form of a membrane protein you have previously met, the epidermal growth factor receptor, EGF-R. The DNA sequence of EGF-R and some biochemical study of its properties had demonstrated that the wild-type protein is a membrane-bound, protein tyrosine kinase with the ability, like Src, to transfer phosphate from ATP to tyrosine in a specific set of cytoplasmic proteins, including itself (FIGURE 6.11). Indeed, as more was learned about the pathways controlling cell growth and division, many of the key enzymes turned out to be protein tyrosine kinases, like Src and EGF-R.

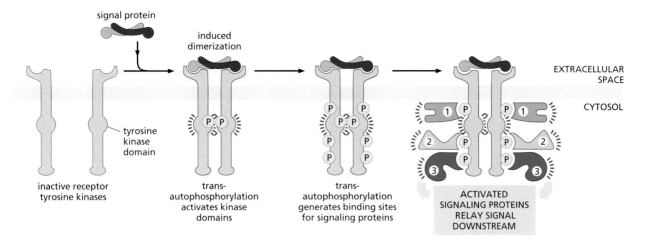

FIGURE 6.11 Activation of protein tyrosine kinase activity by binding a signaling protein to its receptor. Several signal receptors are transmembrane proteins that become protein tyrosine kinases when activated. The cytoplasmic part of these proteins contains a kinase domain, but on its own, the enzyme is inactive. However, the receptor can bind a corresponding signaling protein, such as epidermal growth factor (EGF), that is outside of the cell. The cytoplasmic domains of the receptor, such as EGF-R, will now become active protein tyrosine kinases and phosphorylate each other. The enzyme that activates EGF-R is EGF-R itself, but this happens only when two copies of the receptor protein have come close enough to phosphorylate each other. EGF binding to its receptor dimerizes EGF-R, so the two parts phosphorylate each other, making a new cytoplasmic structure. This new structure binds multiple additional proteins (1, 2, and 3) and activates them, and thus it can transmit signals to the rest of the cell. (Adapted from Alberts B, Johnson AD, Lewis J et al. [2015] Molecular Biology of the Cell, 6th ed. Garland Science.)

Several oncogenic forms of EGF-R were found, but two were particularly informative. In one case, the mutation that made the gene for EGF-R oncogenic was a change in the gene's promoter, which caused an overexpression of the gene product, just like the example of overexpressed Myc described earlier. Cells that produce too much EGF-R become supersensitive to even small amounts of EGF, so they enter the cell cycle when they should not. Another oncogenic form of EGF-R is a shortened allele of the gene, which leads to a gene product that lacks the extracellular domain (FIGURE 6.12). This truncated protein dimerizes, as if it were always interacting with EGF; it gives signals for the cell to grow and divide, regardless of the amount of stimulatory signal, clearly a cancer-promoting action.

As often happens in science, the answer to one question defines new questions whose answers are needed to approach a deeper understanding of nature. The obvious next questions for students of cancer were, What proteins did these protein tyrosine kinases phosphorylate, and how did the phosphorylation-dependent changes in the structure of those proteins lead a cell to initiate DNA replication and cell division?

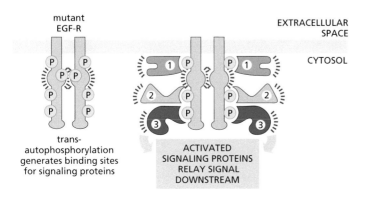

FIGURE 6.12 One oncogenic form of EGF-R lacks its extracellular domain. One oncogenic allele of this important signaling protein produces an EGF-R that is truncated, so it lacks the extracellular, receptor-binding domain. (Compare EGF-R in this diagram with the receptor in Figure 6.11.) This mutant form is always dimerized and thus is always phosphorylated, so it signals cells to progress through the cell cycle without any stimulation from EGF.

Pathways by which normal and cancerous cells control their growth and division cycles.

We can now describe a series of proto-oncogenes whose products lie in information-processing pathways that begin with EGF-R, or closely related growth factor receptors, and end with the activation of genes whose products are transcription factors that increase, or up-regulate, the synthesis of proteins required for a cell's progression toward mitosis. A key player in the EGF-R pathway is the protein **Ras**, the product of a proto-oncogene initially identified as a viral oncogene in mouse. Later, Ras was found to be the protein encoded by DNA isolated from human cancer cells that both Weinberg and Wigler found to have transforming activity for cells in culture. Ras is a representative of an important class of proteins that helps to control cellular events, so it deserves careful description.

Ras and other proteins like it are small proteins that bind GTP, a molecule similar to ATP, but rather than A (adenine), GTP is built on guanine, the base we have called G. Such a protein is called a **G-protein** (FIGURE 6.13A). When Ras binds GTP, its structure changes so it will now bind to and activate other proteins, such as a protein kinase (FIGURE 6.13B). For such a signal to be meaningful, however, it must be short-lived; any signal loses its meaning if it is given continuously. To achieve this turning off, Ras is itself an enzyme that cuts the terminal phosphate from GTP, turning it into GDP and releasing inorganic phosphate into solution. When Ras has GDP bound, it loses its activating structure, so it stops sending signals. Thus, when Ras binds GTP, Ras-dependent processes are turned on, but Ras spontaneously cuts one phosphate from the bound GTP and turns itself off.

The signal transduction pathway that begins with EGF-R uses the activated receptor to get GTP onto Ras, thereby stimulating it to activate other factors, like downstream protein kinases. These kinases send a signal that activates the appropriate transcription factors, but the signal from Ras is short-lived. Not only does it turn itself off, but it can be turned off even faster by increasing its GTPase activity. A protein that does this to Ras is called a **GTPase-activating protein** (**GAP**). A GAP conveys an inhibitory signal to Ras. Conversely, a factor that stimulates the rate at which

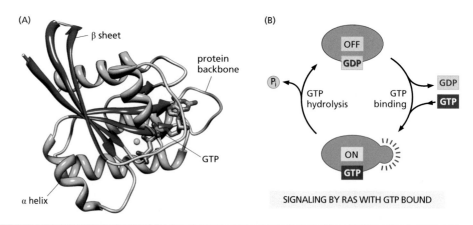

(A)
β sheet
protein backbone
GTP
α helix

(B)
OFF
GDP
P_i
GTP hydrolysis
GTP binding
GDP
GTP
ON
GTP
SIGNALING BY RAS WITH GTP BOUND

FIGURE 6.13 The small G-protein Ras and its cycle of activation and inactivation by GTP binding and hydrolysis. (A) The fold of the Ras protein is shown with helices in turquoise and sheets in purple. GTP (red) is bound to the pocket in Ras protein. (B) The cycle of Ras activation and inactivation. Ras or a similar small G-protein (orange) can bind GTP from the cytosol. When it does so, the G-protein takes on a shape that can send a signal. However, Ras is a GTPase, so it hydrolyzes the bound GTP, releasing a phosphate molecule (P_i for inorganic phosphate), leaving only GDP bound. This form of the protein is inactive and sends no signals. To reactivate the protein, GDP must come off and a new molecule of GTP must come on. The phrase signaling by Ras with GTP bound means that this form of the protein can bind to and activate other proteins, such as a protein kinase. (A, courtesy of ElaineMeng, CC BY-SA 3.0, via Wikimedia Commons.)

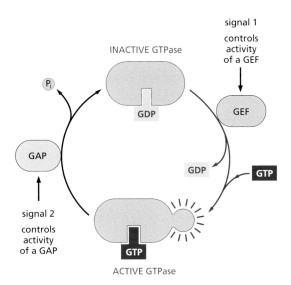

FIGURE 6.14 **The amount of active Ras is controlled by proteins that regulate the fraction of Ras that has GTP bound.** As described in Figure 6.13, Ras is a GTPase that hydrolyzes bound GTP and turns itself off. To be active, it must release that bound GDP and replace it with GTP from solution. The speed of the latter process is increased by a GTP exchange factor (GEF). Thus, a GEF promotes Ras signaling. A GTPase activating protein (GAP) increases Ras' natural GTPase activity, which turns Ras signaling off. The activities of GEF and GAP control the relative amounts of Ras-GTP and Ras-GDP, and thereby the amount of signal that comes from Ras.

Ras exchanges GTP from solution for a bound GDP works as an excitatory signal to Ras. Such a factor (usually a protein) is called a **GTP exchange factor** (**GEF**) (FIGURE 6.14). This kind of regulation of small GTP-binding proteins is a common form of control that cells use in several signaling pathways. One reason for its popularity is that it allows a single protein, such as Ras, to be a point at which signals from different pathways converge, so their relative strengths can be evaluated; some signals activate Ras whereas others inhibit it (Figure 6.14), thereby contributing to a cell's ability to balance information received from multiple inputs.

Because of its flexibility in signal processing, Ras is a key molecule in the pathway by which EGF-R speaks to the rest of the cell. When EGF-R is activated by binding EGF, its latent protein tyrosine kinase activity is turned on. First, it phosphorylates itself, thereby changing the structure of its cytoplasmic domain. This altered structure leads it to bind and activate a pair of proteins that work as GEFs for Ras, turning it on (FIGURE 6.15). While Ras is activated, it binds a different protein kinase called Raf and activates it. This kinase phosphorylates yet another protein kinase, activating it, which in turn activates yet another protein kinase, our old friend MAP kinase from Figure 4.17. The substrates of MAP kinase are transcription factors that regulate the rate of mRNA synthesis from genes that encode proteins that will help to turn on DNA replication. If you are interested in learning more about this important pathway for cellular signal transduction, watch MOVIE 6.1.

The pathways controlling cell division are of great importance in understanding cancer.

The EGF-R signaling pathway might seem complicated, but it is worthwhile to study the relevant text and figures because changes in this pathway and others like it are essential components of cancerous transformation for a great many human cancers. The Ras protein alone is a key oncogene in many cancers. Tumors of the pancreas, thyroid, and colon include an oncogenic form of Ras in 90%, ~57%, and ~45% of cases, respectively. Once you understand how Ras works, the issue of how Ras-dependent pathways are up-regulated during a cancerous transformation is remarkably simple. One such carcinogenic mutation changes Ras into a structure that always retains its activating shape. Another kind of mutation, a particular loss-of-function mutation, causes Ras to lose its ability to hydrolyze GTP, even when stimulated by a GAP. In this case, a loss of function (such

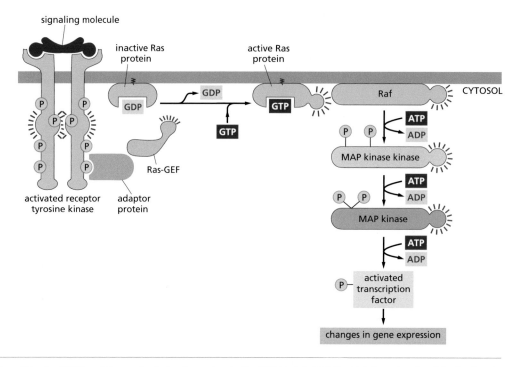

FIGURE 6.15 The activation of Ras by EGF-R and its results. A signaling molecule, like EGF, binds to EGF-R, which then phosphorylates itself and makes a binding site for other proteins. One of these (adaptor protein, shown in pink) activates a GEF for Ras (shown in turquoise), which causes Ras to lose bound GDP, bind GTP from solution, and become active. Activated Ras binds to and activates a protein kinase called Raf, which sets off a chain of kinase activations that leads to the activation of MAP kinase. The substrates of MAP kinase are transcription factors that regulate the rate of mRNA synthesis from genes that encode proteins that will help to turn on DNA replication. Note that Ras and Raf are shown up against the plasma membrane. They are tethered there to keep them close to membrane-bound receptors and make signaling efficient.

that the protein is no longer a GTPase) leads Ras to send its stimulatory signals, regardless of the signals it receives. Either of these mutations will give Ras a gain of function (it is now always signaling), making it oncogenic. (Note that there can be other loss-of-function mutations in the *ras* gene that are not oncogenic; for example, one that kills its ability to bind GTP. Such mutations are not of interest here, because they would not be oncogenic; they would lead Ras to lose its ability to send signals, which would probably lead the cell to die.)

As you work to understand the EGF-R signaling pathway, you may ask, Why is it so complicated? It is rarely easy to know why cells do things the way they do, but the answer for this pathway seems to be the idea sketched in Sidebar 4.2, called Signaling pathways and networks. Several of the steps along the way from EGF-R to the regulation of transcription are not simply unidirectional steps; they are nodes in a network of information processing. The plasma membrane includes receptors for multiple signaling molecules. Each of these receptors can initiate a signaling pathway like the one that starts with EGF-R. Some of the resulting pathways converge on Ras, as shown in FIGURE 6.16. Others go directly to their own regulators of gene expression or to other cellular functions. Figure 6.16 illustrates many aspects of information processing by cells. It contains some proteins you have not encountered and which will not be treated in this book. You should not try to memorize this figure, but instead look at the pathways and get a feeling for the complexity of inputs a cell can respond to. The figure is presented here because it shows how Ras is affected by streams of information that flow from several receptor molecules.

The complexity of cell signaling is, however, significantly greater than is shown in Figure 6.16. An activated protein in any one of these signaling

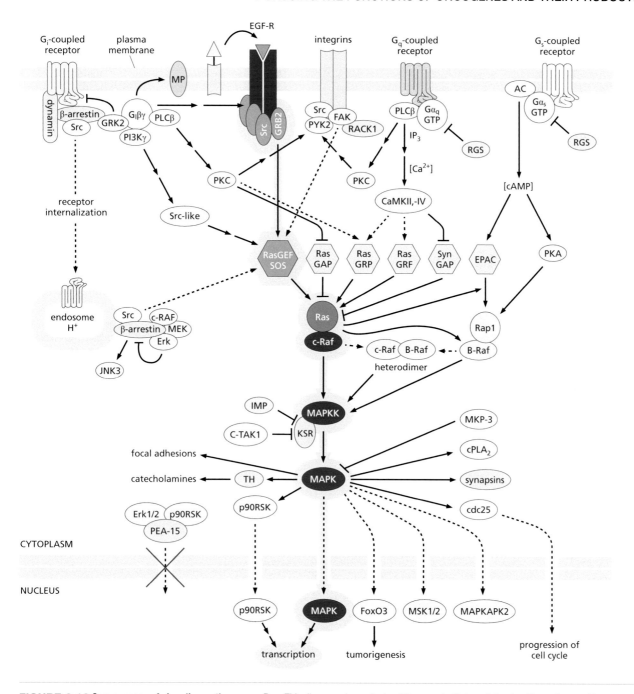

FIGURE 6.16 Convergence of signaling pathways on Ras. This diagram gives a taste of the complexity in cellular signaling pathways. Here you see several membrane-associated receptors, each of which can bind a different signaling molecule and thereby receive different information from outside the cell. One of these receptors is EGF-R, described in Figure 6.15. Components in the pathway that leads from EGF-R are highlighted, so you can trace the pathway you know amid this forest of interactions. Activating signals from each receptor are indicated by arrows that lead to multiple signaling proteins, showing the weblike character of the way cells process information inputs. At three of the receptors, you can find the Src protein, a reflection of its participation in multiple kinds of signaling. In the middle of the diagram, you see Ras with excitatory inputs ($\rightarrow$) and inhibitory inputs ($\dashv$) coming from different receptors. Thus, Ras is a node in this web of information processing: a place where signals from different receptors can be compared and evaluated. Many of these pathways convey information to transcription factors (inside the nucleus); others activate cytoplasmic processes. Clearly, cellular responses to signals are complex! (Illustration reproduced courtesy of Cell Signaling Technology, Inc.)

pathways, such as a protein kinase, can have more than one substrate, so it can turn on (or off) additional enzymes and thus set in motion processes that will affect additional aspects of cell behavior. This phenomenon leads to pathway branching, as diagrammed for MAP kinase (MAPK) in Figure 6.16. Branching is found (but not shown in the figure) for many other components of the pathways in this diagram. Ras offers another

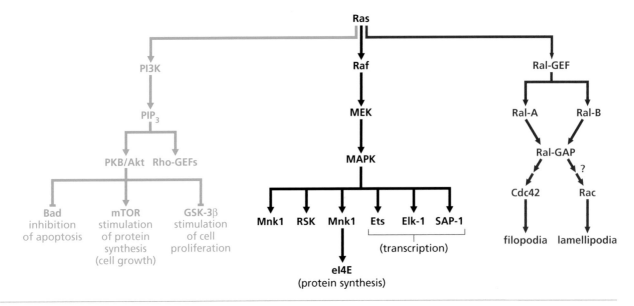

FIGURE 6.17 Branching of a signaling pathway at Ras. In this pathway, the central figure is Ras. The black arrows coming directly down from Ras indicate the pathway that works through MAP kinase (MAPK), which can phosphorylate several different proteins, some of which are transcription factors. However, Ras can activate two other pathways, showing how signaling can branch. The branch on the left is a membrane-associated pathway that helps to regulate cell growth (blue). The branch on the right affects actin microfilaments, the cytoskeletal fibers that impact cell shape and motility (pink). It activates cell movement. Ras is a hub in this net of information flow.

example of signal branching, and given its importance in cancerous trans-formation, Ras branches are worth looking at in more detail. Ras can activate several proteins in addition to the one kinase described earlier and diagrammed in Figure 6.16. For example, it can alter the activity of a kinase that phosphorylates membrane lipids, called PI3K (FIGURE 6.17, left branch). When PI3K is activated, it produces second messengers that act on several pathways. One of these, GSK-3β, up-regulates proteins required for cell proliferation. Another, mTOR, activates protein synthesis, which supports cell growth. Yet another inhibits the pathway that might lead a cell to commit suicide. Ras can also activate a GEF, known as Ral-GEF, that activates another small G-protein, thereby helping to up-regulate the actin cytoskeleton (Figure 6.17, right branch). These actin-dependent processes are involved in cell motility, which is of great importance for metastasis. In summary, the pathways from cellular receptors branch and converge, so signals can influence multiple cellular processes. The flow of information is not really down a pathway but into a web of information processing reactions.

In spite of all this complexity, a cell can distinguish signals from different receptors, showing that it can keep track of the multiple threads in the information web and respond to them in different ways. As an example of different cellular responses to related but different receptors, watch MOVIE 6.2, which shows the same liver cells responding in different ways to two quite similar growth factors. As an example of signals from multi-ple receptors acting to oppose one another, remember contact inhibition of growth, the behavior in which normal cells stop growing when they are interacting with neighbors on all sides. There is a flow of information from such contacts into the pathways that regulate the onset of DNA syn-thesis, so contact inhibition makes a normal cell rather insensitive to the action of a growth stimulator, like EGF. This inhibition works through cell membrane molecules that sense cell–cell contact and send signals that silence the effects of the pathways previously described. The stimulation of EGF-R to start DNA replication is usually inhibited so long as cell–cell contact is maintained. This scenario suggests, though, that a loss-of-function mutation in this contact-sensing pathway would inactivate this silencing. With this signal silenced, cells would continue to initiate DNA synthesis, even when confluent. This kind of mutation is the subject of the next chapter, but the complexity of Figure 6.16 shows how a multiplicity of mutations can be involved in cancerous transformation.

SIDE QUESTION 6.4

Why do oncogenic mutations in growth-stimulating pathways involve gain-of-function mutations, while oncogenic mutations in growth-repressing pathways involve loss-of-function mutations?

THE DISCOVERY OF ONCOGENES THROUGH ANALYSIS OF DNA SEQUENCE

In the late 1970s, methods for sequencing DNA became sufficiently efficient that the comparison of genes from viruses and their hosts was a straightforward if laborious task. Scientists could even sequence parts of host chromosomes in their own laboratories, though it was not until the beginning of the twenty-first century that sequencing became comparatively fast and cheap, leading to the frequent sequencing of entire human genomes. The knowledge that flowed from DNA sequences of viruses and their hosts led to a remarkable period of discovery for biomedical scientists interested in oncogenesis. This progress came both from the study of viruses and from a close look at the genomes of cancer patients.

Examination of animal tumor viruses has led to the discovery of many human oncogenes.

As the work to understand oncogenes and their functions went forward in laboratories all over the world, a number of previously unrelated growth-stimulating factors and pathways began to fit together into rational patterns. Continued work on animal tumor viruses led to the identification of additional proteins that influenced the growth and development of certain animal blood cells; this signaling worked through receptors analogous to EGF-R. Some of these newly discovered animal cell receptors are represented in the extra pathways shown in Figure 6.16. You may say, however, that you don't care about cancer in mice or chickens; you want to know the genes that are important in human cancer. It turns out that virtually all of the oncogenes identified in animal tumor viruses are oncogenes in humans, even though they rarely arise in our bodies through viral infection (TABLE 6.1).

Human oncogenes can be formed by the rearrangement of chromosomes.

The distribution of genes on chromosomes can change in any of several ways. These rearrangements are not common, but they do occur. One way is through **translocation**, where two chromosomes bump into one another in such a way that double-strand breaks form on both chromosomes, and whole segments of these chromosomes exchange. Some translocations are benign, because the breaks in both chromosomes occurred between genes. This kind of DNA exchange leaves both the structural genes and their regulatory elements unchanged; genes have simply been moved from one chromosome to another, which usually has little effect on organism phenotype. However, chromosomal breaks can occur in places that do matter. A gene may be broken by the translocation, altering it so the gene product is nonfunctional. Alternatively, the double-strand breaks may occur between the regulatory elements and the structural gene they control. In this case, a gene that used to be expressed under the control of one promoter is now placed under different control; such a change can alter the level of gene expression. Just as a virus can transform a proto-oncogene into an oncogene by mutation or

Table 6.1 Animal virus-associated oncogenes that have been linked to human cancers

Name of virus	Species	Oncogene	Type of oncoprotein	Homologous oncogene found in human tumors
Rous sarcoma	chicken	*src*	non-receptor TK	colon carcinoma[a]
Abelson leukemia	mouse	*abl*	non-receptor TK	CML
avian erythroblastosis	chicken	*erbB*	receptor TK	gastric, lung, breast[b]
McDonough feline sarcoma	cat	*fms*	receptor TK	AML[c]
H-Z feline sarcoma	cat	*kit*	receptor TK[d]	gastrointestinal stromal
murine sarcoma 3611	mouse	*raf*	Ser/Thr kinase[e]	bladder carcinoma
simian sarcoma	monkey	*sis*	PDGF	many types[f]
Harvey sarcoma	mouse/rat	H-*ras*[g]	small G protein	bladder carcinoma
Kirsten sarcoma	mouse/rat	K-*ras*[g]	small G protein	many types
avian erythroblastosis	chicken	*erbA*	nuclear receptor[h]	liver, kidney, pituitary
avian myeloblastosis	chicken	*ets*	transcription factor	leukemia[i]
avian myelocytomatosis	chicken	*myc*[j]	transcription factor	many types
reticuloendotheliosis	turkey	*rel*[k]	transcription factor	lymphoma

Abbreviations: TK, tyrosine protein kinase; CML, chronic myelogenous leukemia; AML, acute myelogenous leukemia; PDGF, platelet-derived growth factor.
[a]Mutant forms were found in a small number of these tumors. [b]Receptor for EGF; a related protein is overexpressed in 30% of breast cancers. [c]Fms, the receptor for colony-stimulating factor (CSF-1), is found in mutant form in a small number of leukemias; a related protein is frequently mutated in these leukemias. [d]Receptor for stem cell factor. [e]The closely related B-Raf protein is mutant in the majority of melanomas. [f]Protein is overexpressed in many types of tumors. [g]The related N-*ras* gene is found in mutant form in a variety of human tumors. [h]Receptor for thyroid hormone. [i]Twenty-seven related transcription factors are encoded in the human genome. This protein is overexpressed in many types of tumors; others are involved in chromosomal translocations. [j]The related N-*myc* gene is overexpressed in pediatric neuroblastomas and small-cell lung carcinomas. [k]Rel is a member of a family of proteins that work as human transcription factors; these are constitutively activated in a wide range of human tumors.
Adapted in part from Butel J (2000) *Carcinogenesis* 21:405–426, and Cooper GM (1995) Oncogenes, 2nd ed. Jones and Bartlett.

by a change in level of expression, so also a translocation can lead to oncogenic changes.

Before scientists and doctors had the tools to understand chromosomal translocation at a molecular level, scholars studying one kind of leukemia found an odd-looking chromosome in the hyperproliferating white blood cells of this cancer. The leukemic cells from several patients with the same disease all contained one chromosome that was abnormally short. This chromosome was called the **Philadelphia chromosome** in honor of the city where it was first spotted. We now know that the Philadelphia chromosome is the result of translocating a small amount of DNA from chromosome 9 in exchange for a longer piece from chromosome 22 (FIGURE 6.18A). The unfortunate result of this exchange is that a gene that encodes a protein tyrosine kinase is brought right next to the strong promoter of a completely unrelated gene (FIGURE 6.18B). The gene for this protein tyrosine kinase has a DNA sequence very similar to an oncogene named *abl* from a mouse leukemia virus. Because of the similarity in DNA sequence, the gene on human chromosome 9 that encodes this protein tyrosine kinase is called c-*abl*. It is considered a proto-oncogene, and the translocation is oncogenic because it puts c-*abl* under the control of a strong promoter from a gene on chromosome 22, resulting in the overexpression of the Abl kinase, an oncogenic change. Many people with this kind of leukemia, which is called chronic myelogenous leukemia (CML), have the Philadelphia chromosome. Evidence

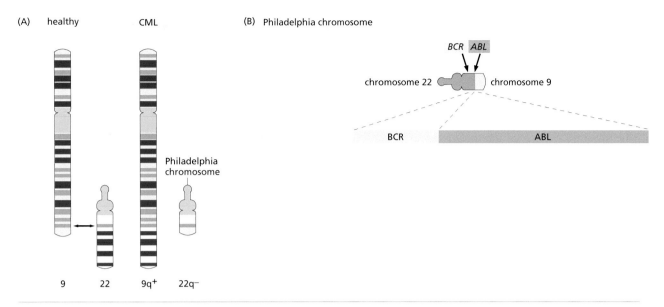

FIGURE 6.18 Identification of the Philadelphia chromosome. (A) Comparison of chromosomes 9 and 22 from a healthy human (on the left) and from a patient with the kind of blood cancer called chronic myelogenous leukemia (CML) (on the right). The chromosomes on the right, labeled 9q⁺ and 22q⁻, have undergone a translocation, so the normal chromosome 9 has been extended by the addition of DNA from the normal chromosome 22. Chromosome 22 has been shortened, having received only a short bit from chromosome 9, including the end of that chromosome. Note that the stripes on these diagrams represent visible bands that can be made on chromosomes by certain stains. The shortened chromosome 22 is called the Philadelphia chromosome. Its altered length led to its being spotted by light microscopy some years ago, when the study of chromosome structure was still comparatively crude. (B) What is left of the normal chromosome 22 is shown in blue. The gene just to the left of the translocation break is called *BCR*; not much is yet known about its normal function, but it has a strong promoter. The gray portion indicates DNA from chromosome 9, and the gene just to the right of the break is called *ABL*, which encodes a protein tyrosine kinase. The translocation brings this kinase under the regulation of the strong promoter from *BCR*, greatly increasing the level at which it is expressed. *ABL* is a proto-oncogene in the sense that its product is part of normal cell cycle control. When it is up-regulated by this translocation, it becomes an oncogene and contributes to the pathology of CML. (Adapted from A.S. Advani and A.M. Pendergast [2002] Leuk. Res. 26:713–720.)

discussed in Chapter 8 makes it clear that overexpression of the kinase encoded by c-*abl* is a driving mutation in the development of CML.

Translocations of chromosome arms are responsible for other human cancers. An analogous event has been spotted as part of the cancerous transformation in Burkitt's lymphoma. Once again, careful examination of chromosome structure in the cancerous cells of patients with Burkitt's lymphoma identified a translocation, this time between chromosomes 8 and 14. Molecular analysis of the regions near the breakpoints of the two chromosomes identified DNA encoding the important transcription factor Myc, which we have discussed previously. In Burkitt's lymphoma, the cause for overexpression is again a translocation-mediated up-regulation in the expression of a proto-oncogene, increasing protein levels and driving cells into growth and division cycles when no natural signals are telling them to do so.

THE MANY GENETIC CHANGES NEEDED FOR CANCEROUS TRANSFORMATIONS THAT LEAD TO MALIGNANCY

Our understanding of the genes important for human cancer has been aided significantly by scientific work on animals that appear to bear little similarity to humans. Studies on the development of worms, fruit flies, and fish have all led to the discovery of genes and gene products that participate in controlling growth and development in humans. Apparently, some of the pathways that regulate cell behavior are so basic that they are found in a wide range of multicellular animals. Here, we needn't be concerned

The weighing of multiple inputs that cells can do is a little like the consideration of factors you do when making a decision. Do you see a cell's abilities as intelligence, or do you think something more is involved in the workings of a human mind? If you think there are additional aspects to intelligence, what are they?

with the details of how these pathways work and interact with one another. However, you must realize that the subject of cellular information processing has become almost unbelievably intricate. This complexity endows cells with the ability to receive multiple inputs and weigh them to make choices that are valuable for the survival or well-being of the animal as a whole. Cells make balanced choices, which leads cell biologists to think that cell behavior is almost intelligent. Bearing this complexity in mind, we can now look at some of the factors other than cell cycle regulation that are important for cancerous transformation.

Some mutations that advance cancerous transformation affect cellular motility.

The most dangerous property of a cancerous cell, beyond inappropriate growth, is its capacity for metastasis. The importance of this behavior for cancer malignancy was emphasized in Chapters 1–4, but now you are in a position to understand several of the mutations that contribute directly to this aspect of altered cell behavior. These mutations promote cancerous transformation, even though they have nothing to do with a loss of control on cell division. The genes involved are not really proto-oncogenes, but their mutation to a hyperactive form makes them important agents of cancerous progression.

For a cell to leave the place where it was born by mitosis/cytokinesis and to wander through the body, it must loosen its connections with neighbors and become actively motile. These properties make the cell **invasive**; that is, able to move away from its site of birth. To go beyond that and become capable of real metastatic growth, the cell must also be able to get both into and out of blood or lymphatic vessels, allowing it to establish new sites of growth in locations distant from its site of birth. Mutations in several pathways of normal cellular control contribute to the acquisition of these properties. One is gain of function in a small, GTP-binding protein called Rac1, which is much like Ras. (You can find Rac on the right arm of the pathways for information processing by Ras in Figure 6.17.) As was true for Ras, the GTP-bound form of Rac1 is the active form, and Rac1 is a GTPase that slowly hydrolyzes bound GTP, thereby turning itself off. In spite of these similarities to Ras, Rac1 is encoded by a distinct gene, and its binding partners (the GEFs and GAPs that activate and inhibit it) are different from those of Ras. Indeed, we now know that there are more than 100 Ras-like proteins, of which Rac1 is a single example.

While Rac1 is in its activated form, thanks to GTP binding, it interacts with a set of proteins that control the behavior of the actin cytoskeleton. Activated Rac1 helps to reorganize actin into networks of microfilaments just under the plasma membrane (FIGURE 6.19). These push on the membrane, causing it to bulge outward, forming a lamellipodium (platelike foot in Latin). This is one aspect of activating cell motility, a key characteristic of metastatic cells. A second aspect of up-regulating cell motility results from the action of yet another small GTP-binding protein called Rho, which helps to polarize the cell, giving it a directionality for migration. This small GTP-binding protein helps to organizes Rac1, so it activates the actin cytoskeleton in a single direction. Gain-of-function mutations that make Rac1 always active are cancer-promoting, because they push the cell toward the dangerous behavior of invasiveness.

Metastatic behavior in carcinomas requires manipulation of the ECM.

An additional factor in becoming metastatic is the activation of extracellular proteinases, that is, enzymes that can cleave the bonds between

actin staining   actin staining

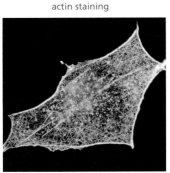 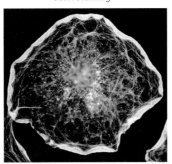

(A) QUIESCENT CELL   (B) Rac ACTIVATION

FIGURE 6.19 **The action of Rac1 on a cell's distribution of actin.** (A) Light micrograph of a cultured mammalian cell stained with fluorescent antibodies to actin, the protein subunit of microfilaments. There is some staining immediately inside the plasma membrane. A few cytoplasmic fibers are also stained, and there are indications of fibrous webs elsewhere in the cytoplasm. (B) A similar cell in which Rac1 has been activated. The amount of fluorescence beneath the plasma membrane is markedly increased, and a scalloped appearance of the cells edges indicates the presence of so-called ruffles, which are results of an active cytoskeleton. These actin microfilaments are pushing on the membrane in the way that is necessary for cell motility. For this cell to move, however, the pushing activity must become polarized, meaning that it must take on a direction, so it will cause the cell to move in a directed way. (From A. Hall [1998] Science 279:509–514. With permission from AAAS.)

amino acids and thereby chew up proteins. Normal cells use limited and highly controlled extracellular proteolysis to modify the extracellular matrix (ECM) in which they live. Metastatic cells amplify this behavior and modify their environments in ways that enhance their cancerous behavior. One example of this behavior is to secrete an activator of a normally inactive extracellular proteinase that is widespread in blood and lymph, the protein called plasminogen. Cells that are becoming metastatic commonly secrete plasminogen activator, which turns plasminogen into **plasmin**, an active extracellular proteinase. Active plasmin weakens the ECM and makes it easier for cells to migrate through it.

An additional way in which metastatic cells achieve the same goal is through secretion and activation of the **matrix metalloproteinases (MMPs)** mentioned in Chapter 5. Hyperactivating extracellular MMPs is an effective way to loosen the web of extracellular fibers that helps to stabilize tissue shape; the result makes it easier for cells to migrate and even to pass through the basal lamina. However, MMPs have naturally occurring inhibitors. The extent of MMP activity in normal circumstances results from a balance between enzyme secretion/activity and the concentrations of these inhibitors. This balance is used by cells to help shape and modify materials in the ECM that surrounds them. However, just as cells can influence the character of their ECM, changes in the composition and texture of the ECM have an impact on cell behavior. In the case of cancer cells, these changes can be quite extensive. For example, the cells of a carcinoma, which by definition arise from an epithelium, can encounter modified ECM. They then begin to look more like the cells of embryonic connective tissue, called **mesenchymal cells**. This is one aspect of the changes in cell shape that a pathologist looks for when assessing whether the cells of a neoplasm are dysplastic or anaplastic (Chapter 2). An **epithelial to mesenchymal transition** (EMT) affects cell shape and motility, as well as the genes that the cancer cells express. Just how changes in the ECM promote or facilitate EMT is not well understood, but the transition often accompanies the progression from simple invasiveness to a truly metastatic state.

The hyperactivation of extracellular proteinases is also important for migrating cells to be able to enter and leave a blood vessel, which is important for their travel to distant tissues where they can set up sites of metastatic growth (FIGURE 6.20). Like the boundary between an epithelium and the underlying connective tissue, the edges of blood vessels are surrounded by a basal lamina. If a metastatic cell can enter a blood vessel, a process called **intravasation**, it can take advantage of blood flow for rapid transport to new sites in the body. Thus, the extracellular proteinases that facilitate local cell movement are also important for long-distance migration. At some distant location in the body, the same

The development of mesenchymal properties in adult epithelial cells introduces the idea that carcinogenesis is the return of a mature cell to an embryonic state. Do you support this idea? Why or why not? What experiments could you do to test the validity of this idea?

SIDE QUESTION 6.6

FIGURE 6.20 The many processes that a cell must accomplish to establish a site of metastatic growth. A transformed cell that forms a clone of hyperproliferating cells must change further to become metastatic. It must loosen the bonds that normally keep it anchored to its neighbors. It must increase its level of motility, so it can wander from the site of its birth. It must secrete proteases that will loosen not only its connections with neighbors but also the surrounding extracellular matrix. It must cross through the basal lamina that separates all epithelia from nearby connective tissue and then migrate through that tissue and cross yet another basal lamina to enter a blood vessel or a lymphatic vessel. The migrating cells must now survive in their new environment as they are carried by fluid flow. In some cases the tumor cells become surrounded by tiny blood cells, called platelets, which help to protect them. This cluster of cells must then adhere to the wall of the vessel through which it has been traveling, so the individual cells can pass out through the wall of the vessel and enter the nearby tissue. Finally, the cells must be able to adapt to their new environment, so they can grow and divide in the new location, establishing a clone of cells at a distant growth site. (Adapted from Cotran, R.S. Kumar, V., and Robbins, S.L. [1994] Pathological Basis of Cancer, 5th ed. Philadelphia, W.B Saunders. Co.)

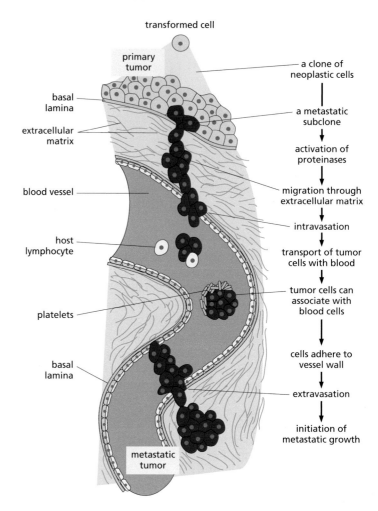

proteinases come into play both for the cell's leaving a blood vessel, the process of **extravasation**, and for invading nearby tissue. Note, however, that metastatic cells can develop an additional mechanism to facilitate their migration.

Transformation to a metastatic state often includes activating genes that encode membrane receptors for proteins in the basal lamina and in the walls of blood vessels, where these proteins help to hold the cells of the vessel together. Expression of these receptors on the surface of a metastatic cell facilitates the passage of that cell through both fibrous layers and the walls of blood vessel, with or without the action of MMPs.

Matrix metalloproteinases also contribute to increased blood vessel formation, another factor in the success of cancerous growth.

Extracellular proteinases can promote the formation of new blood vessels. Blood vessel formation, known as **angiogenesis**, is highly controlled. This is necessary to assure that the circulation of blood is appropriately balanced among all the cells and tissues of the body. For a dynamic view of this process, watch MOVIE 6.3. If a tumor is to grow more than about a millimeter in size, it needs to increase its local blood supply. Angiogenesis is therefore a key part of getting the body to support cancerous growth (FIGURE 6.21A). MMPs provide one of the mechanisms for inducing local angiogenesis, but there are other factors that contribute. Important among these are secreted proteins that serve as growth factors and attractants for the cells that line nearby blood vessels. Collectively, these are called

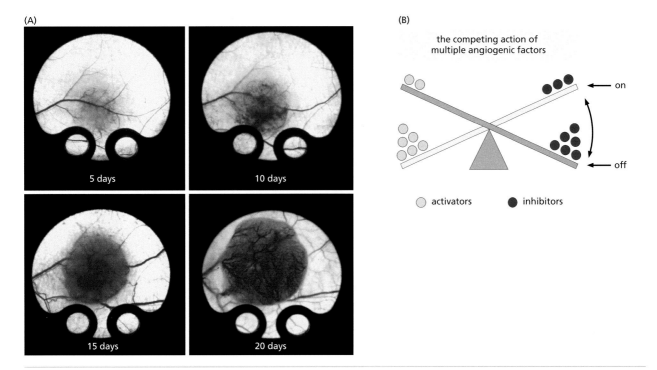

FIGURE 6.21 Tumors can induce angiogenesis to feed their many cells. (A) Human colorectal carcinoma cells were implanted under the skin of a mouse, where they could be imaged through a window inserted into the mouse's skin (light circle in the black background). The four images show the progress of angiogenesis over 20 days, during which the human tumor secreted angiogenic factors that stimulated mouse vasculature to grow into the foreign tissue. Although the mouse vasculature is responding to angiogenic signals from the implanted tumor, the new blood vessels are commonly disorganized and not as effective as they would be in normal tissue. (B) Angiogenesis can be either stimulated or inhibited by multiple protein factors. These factors signal to receptors on the membranes of cells in the walls of blood vessels. The seesaw represents the competition between activators and inhibitors of angiogenesis. Tumors, like normal tissues, can secrete both kinds of factors in an effort to shape the surrounding blood supply and provide themselves with the nourishment they need to thrive and grow. (A, from M. Leunig et al., Cancer Res. 52:6553–6560, 1992.)

angiogenic factors, and there are lots of them. Some enhance angiogenesis, while others inhibit it (FIGURE 6.21B). The up-regulation of enhancers of angiogenesis is yet another oncogenic phenomenon that depends on the up-regulation of gene function. Like the factors that promote excess cell motility, angiogenic factors are important contributors to the progression of cells to a state of malignancy.

Tumors use the stimulation and inhibition of angiogenesis to thrive.

The ability to stimulate angiogenesis is essential for a tumor to grow beyond a tiny size. Even when a tumor is only about a millimeter in diameter, the rate at which oxygen and nutrients can diffuse to the cells at the tumor's center is not fast enough to keep those cells growing well. Increased circulation of blood to the tumor allows its continued growth. The development of new blood vessels enhances a tumor's growth, even though the resulting web of blood vessels is commonly less well organized than the vasculature of normal tissue.

The vasculature that carries blood around our bodies plays an additional and important role in cancer, because blood vessels provide the most common pathways for metastatic cell migration. It follows that the pathways of blood flow influence the directions that metastatic cells will commonly travel. Lung cancers frequently metastasize to bones and both brain and spine, one of the reasons that lung cancer is a dangerous disease. Breast cancers commonly metastasize to those tissues and also to lung and liver. Colon and pancreatic cancers often metastasize to the liver. These migration targets are due in part to the way blood flows from the site of primary tumor formation to other parts of the body, but they also depend on the compatibility of cells from one tissue with the environment in another tissue. This is a complicated subject that is not yet well understood. Some oncologists believe that understanding the compatibility of one cell type with the environments in other tissues will be an essential step in understanding and limiting metastatic growth.

CHANGES THAT CONTRIBUTE TO THE IMMORTALITY OF CANCER CELLS

You have seen that cells in culture show contact inhibition of growth, but there are additional controls on growth that help to keep normal cells from hyperproliferating. In the 1970s, when many cell biologists began growing different kinds of cells in culture, Leonard Hayflick of the Wistar Institute in Philadelphia noticed that cells taken from an animal's body and grown in culture (primary cell cultures) would grow and divide at most about 70 times; then they stopped their cell cycles, arresting in the G1 phase. Even if they were well-fed with a medium that contained all necessary growth factors, and they were not contact-inhibited for growth, these cells simply would not initiate DNA synthesis. Hayflick and others found that cells taken from older animals would undergo fewer cycles in culture than cells from young animals, suggesting that primary cells could somehow count the number of times they reproduced. It seemed that they were allowed only a given number of divisions before they would cycle no more.

The process of DNA replication sets natural limits on cell proliferation.

For a long time, the phenomenon of limited proliferation was a mystery. We now know that the Hayflick effect is true for most cells taken from an animal's body, and there is a good reason for it. The limit to a cell's total number of growth and division cycles depends on details of the way DNA polymerase works to replicate the two strands of a DNA duplex, details that were not previously mentioned but will be described here briefly.

Each strand of DNA is directional, based on the orientation of its component nucleotides. This orientation is established at the time the nucleotides are added during DNA synthesis. The two strands of DNA in a double helix point in opposite directions. This is actually a geometrical requirement for the bases on opposite strands to bond well with their complementary bases. This anti-parallel arrangement of the two strands of normal DNA means that DNA polymerase runs in opposite directions on the two strands of a double helix.

When DNA synthesis starts at a replication origin, the polymerase on one strand can move smoothly in one direction from the site of initiation. On the complementary strand, the polymerase moves smoothly in the opposite direction (FIGURE 6.22). But to replicate the complementary strands on the other side of the origin, DNA polymerase has to start, run a short distance, and then fall off. As the replication fork moves farther along, the polymerase on the strand that is replicating backwards will start again, run a short distance and again fall off. The smoothly replicating strand is called the leading strand, and the faltering one is called the lagging strand (Figure 6.22). The details of how all these short pieces of DNA are put together to make a continuous strand are not important here. What does matter is that when the replication forks reach the ends of a chromosome, the lagging strands cannot be completely replicated.

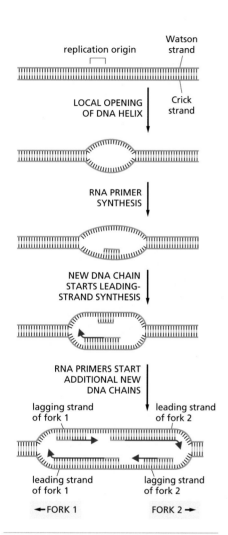

FIGURE 6.22 The two strands of a DNA duplex are anti-parallel, so the DNA polymerases that bind them move in opposite directions. DNA replication begins with a local unwinding of the double helix, which produces a loop where the enzymes for DNA replication can bind. Replication starts with the synthesis of an RNA primer (blue), first on one strand and then on the other. The primer helps DNA polymerase to start DNA synthesis (red arrows). For each fork, there is a strand of DNA pointing in each direction. One direction can be replicated continuously (this is called the leading strand), but the other direction requires repeated initiations of replication, because DNA polymerase moves in the direction back from the fork. This DNA strand is called the lagging strand. (Adapted from Alberts B, Johnson AD, Lewis J et al. [2015] Molecular Biology of the Cell, 6th ed. Garland Science.)

The consequence of incomplete replication is that with every round of DNA replication, each chromosome gets a little shorter at both ends. After multiple rounds of DNA replication (about 70 in human cells), the chromosomes have shortened enough that their systems for sensing DNA damage are activated. Cell cycle progression is blocked by the DNA damage checkpoint, accounting for the Hayflick effect. Note that this behavior in a cancer cell would limit its proliferation and stop the production of new cancer cells.

Certain mutations can override the Hayflick effect.

You can see from the preceding discussion that complete transformation to malignancy requires elimination of the Hayflick effect. The transformed cell must overcome this natural cessation of cell cycling, so it won't simply stop growing. There are several different mutations that can eliminate the Hayflick effect and make a cell immortal in the sense that it can keep on growing and dividing without limit. The most common of these is a gain-of-function mutation in an enzyme called **telomerase**.

To keep chromosomes stable and to allow cells to function, even as the chromosome arms are getting a little shorter with every round of DNA replication, every chromosome is capped at both ends by special DNA sequences called **telomeres**. These end regions contain no genes and their nucleotide sequences are repetitive. As a result, they can bind special proteins to make a structure that protects the rest of the chromosome from nucleases that chew DNA from its ends. This arrangement is sufficient to maintain chromosome integrity through the life of normal somatic cells, even though the chromosomes lose a bit of telomere length with every round of DNA replication. Stem cells, however, may have to divide more times than the number set by the Hayflick effect, so they have a special mechanism to prevent their chromosomes from shortening.

Stem cells, like those at the bottom layer of skin or in the crypts in intestinal epithelium, as well as the sperm-forming cells in a man's testes, have to divide many times to produce all the cells that the body needs for health and fertility. To maintain this ability, stem cells express a group of genes that is silent in most of our body's cells. These genes encode the multisubunit enzyme known as telomerase. Telomerase is a DNA polymerase, but rather than using an existing DNA strand as the template to define its choices of which nucleotides to add, it uses an RNA molecule that is continually bound to the protein. The sequence of this RNA instructs the polymerase to add the same six-base sequence, over and over again (FIGURE 6.23). Telomerase is a special example of a reverse transcriptase, since it is using an RNA sequence to define the sequence of the DNA it makes. It works during S phase at both ends of every chromosome and solves the problem of telomere shortening. Thus, stem cells are immortal in the sense that they can keep on dividing without limit, as long as they are properly fed and stimulated by growth factors. The genes for telomerase are not normally expressed in somatic cells. However, they can be turned on by a mutation that up-regulates their expression. This mutation makes the cell that carries it immortal in the same sense that stem cells are immortal. This is also the sense in which cancer cells are immortal. The gain-of-function mutation that brings about telomerase expression in somatic cells is another important part of cancerous transformation.

Blocking apoptosis is another key factor in a cancer cell's immortality.

You know from Chapter 4 that cells with DNA damage will often undergo apoptosis, a behavior that eliminates the danger that such a heavily

Could a tumor grow from transformed cells that lacked telomerase? Why or why not? What constraints would the lack of this enzyme put on tumor growth?

SIDE QUESTION 6.7

FIGURE 6.23 Elongation of DNA through the action of telomerase. The two orange bars represent the two strands of DNA in a normal chromosome. The letters represent the base sequences at a chromosome's end, where telomere sequences help to protect the rest of the chromosome from degradation. In the top diagram, the lagging strand is shorter than the leading strand because of the problem with DNA replication on the lagging strand. This problem is solved by the enzyme telomerase (green), which includes a bound piece of single-stranded RNA (light blue). Telomerase can elongate the leading strand (sequence in red), allowing space for DNA polymerase to bind and elongate the lagging strand (sequence in green). In normal stem cells, this system is regulated by a balance between telomerase activity and the action of other enzymes (not shown) that chew away the ends of DNA, keeping telomeres from getting too long. (Adapted from Alberts B, Johnson AD, Lewis J et al. [2015] Molecular Biology of the Cell, 6th ed. Garland Science.)

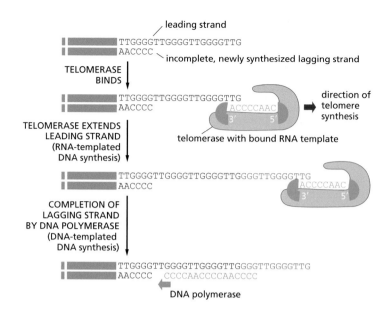

mutated cell might proliferate. The multiple mutations and genetic instability of cancer cells should, therefore, lead to immediate cell death. A fully transformed cancer cell can avoid this limitation to its proliferation by acquiring gain-of-function mutations in genes that encode proteins that inhibit apoptosis. A normal cell is always maintaining a balance between factors that would induce apoptosis and those that would inhibit it. Mutation can upset this balance in several ways, one of which is a gain of function in apoptosis inhibition. This behavior is analogous to the other cancer-inducing genes described in this chapter; they too are gains in function. Another way to inhibit apoptosis is to lose activity in a factor that enhances the process. In Chapter 7, we will discuss many of the genes whose loss of function promotes cancerous transformation. These are known as tumor suppressors.

SUMMARY

Oncogenes are genes whose expression drives inappropriate cell division. Many oncogenes have been discovered through the careful study of viruses that promote cancer in experimental animals. Some of these viruses carry an oncogene added to their own essential genetic makeup. Following infection by such a virus, the oncogene can be expressed, causing the infected cell to be transformed in shape and behavior. It becomes much like cells derived from any cancer. Study of oncogenic viruses has led to the realization that oncogenes are actually mutated forms of normal host cell genes; they are oncogenic alleles of a gene whose normal function is in the regulation of cell cycle progression. The normal form of such a gene is called a proto-oncogene, and it serves as a part of the pathways that control cell growth and division. A proto-oncogene can be turned into an oncogene by mutations that either make its product more active or increase the expression of the normal product. Either of these kinds of mutation can also be caused by the carcinogenic factors discussed in Chapter 5, for example, ionizing radiation and carcinogenic chemicals.

Many proto-oncogenes and their oncogenic alleles are parts of the regulatory pathways that control a cell's responses to mitogenic factors. Examples include molecules that signal a cell to initiate DNA replication and enter mitosis. There are, however, many genes besides oncogenes whose up-regulation can be part of cancerous transformation: for example, genes that encode proteins capable of degrading the fibers of the ECM (extracellular matrix), inducing the proliferation and expansion of nearby blood vessels (angiogenesis), inhibiting a cell's ability to induce programmed cell death (apoptosis), or inhibiting a cell's ability to maintain chromosome length at the telomeres.

Oncogenesis depends on many aspects of cell behavior, and there are many genes whose increased activity will promote cancerous transformation. The number of features a cell must acquire to be truly cancerous is fairly large, which is the reason that numerous mutations are required to effect cancerous transformation. From the diversity of genes whose gain of function contributes to cancerous progression, we are learning what a multifaceted disease cancer really is.

ESSENTIAL CONCEPTS

* Viruses are tiny pathogens made from a nucleic acid and some proteins. They reproduce by infecting a host cell and using host cell functions to make many copies of their own molecules, allowing the construction of many virions (virus particles), which can go on to infect other cells.

* There are viruses that infect bacteria. Some bacterial viruses can integrate their DNA into the chromosome of their host cell and later bring it back out again.

* Certain specialized viruses can transmit cancer from one animal to another, as was first shown by experiments with chicken leukemia and then with chicken sarcoma. Subsequently, many viruses have been found to cause cancer in laboratory animals, and a few viruses cause cancer in people. These are all called tumor viruses.

* Some tumor viruses use DNA as the material for their genes, while others use RNA.

* Infection by a tumor virus changes the behavior of normal cells growing in culture. The infected cells behave like cancer cells. They lose contact inhibition of growth and can pile up on top of one another. The discovery of this effect encouraged the belief that the use of cultured cells could provide a good experimental model for cancer in humans.

* The DNA of tumor viruses was found to encode some proteins that alter the host's cell cycle behavior by binding to and inactivating known regulators of cell cycle progression, such as the Rb protein.

* Transformation of cultured cells by DNA tumor viruses depends on the integration of viral DNA into a host cell chromosome. This chromosomal integration is what makes the cancerous phenotype inheritable in somatic cells.

* The problem of how RNA tumor viruses might transform host cells was solved with the discovery that RNA tumor viruses contain an enzyme that copies RNA into DNA. This reverse transcriptase allows them to make a copy of their genome that will integrate into the host cell chromosome, just as the genome of a DNA tumor virus does.

* Mutations in tumor viruses identified genes that caused host cell transformation. (The site of a mutation that inactivated the virus's transforming ability identified the transforming gene.) The transforming gene in the famous Rous sarcoma virus (RSV) encodes a protein kinase called Src. Src transfers phosphate to the amino acid tyrosine, analogous to other kinases that have been identified as regulators of cell reproduction.

* Studies using viral nucleic acids as probes demonstrated not only that viral genomes are present in the chromosomes of transformed cells but also that host cells contained genes that are very similar to the viral oncogenes that transform them.

* Hypotheses about carcinogenesis can now be formulated around the fact that our cells contain many proto-oncogenes (regulators of cell cycle progression). An oncogene is a mutated allele of a proto-oncogene that increases the function of the gene product, inducing that cell to misbehave in a cancerous way. This idea was confirmed by isolating a

cell-transforming piece of DNA from a chemically induced tumor. The relevant oncogene was a hyperactive form of a gene called *ras*, whose protein product, Ras normally participates in cell cycle regulation.

- Tumor viruses can be oncogenic in two ways: they can bring oncogenes into a host cell, or they can serve as mutagens by integrating into the host cell genome in such a way that they cause a mutation that up-regulates the activity of a proto-oncogene.

- Some human cancers are caused by oncogenic viruses; for example, human papilloma virus (HPV) promotes cervical and penile cancer, while hepatitis B virus (HBV) and hepatitis C virus (HCV) promote liver cancer. Taken all together, viruses probably account for about 15% of all human cancers.

- Several well-known oncogenes are mutant alleles of genes that encode important parts of the pathways that regulate normal cell growth and division. For example, such mutations are found in epidermal growth factor receptor (EGR-F), in Ras, and in other proteins from the pathways that turn on the expression of genes required for DNA replication.

- These growth control pathways are actually parts of networks that control many aspects of cell behavior. Many of the genes encoding proteins in these networks are proto-oncogenes.

- Not all cancer-inducing mutations affect cell growth and division. There are also mutations that increase the activity of extracellular proteinases, which help to regulate the condition of the matrix surrounding cells. These changes are important for metastatic behavior. Other oncogenic mutations affect the way cells induce the development of nearby blood vessels (angiogenesis).

- Certain oncogenic mutations endow cells with immortality by turning on the synthesis of telomerase, an enzyme that prevents a cell's chromosomes from shortening with every round of DNA replication. Others block cells from turning on apoptosis. Clearly, cancerous transformation is a complex, multi-step process.

KEY TERMS

angiogenesis	invasive	Ras
bacteriophage	matrix metalloproteinase	retrovirus
cell transformation	mesenchymal cells	reverse transcriptase
contact inhibition of growth	MMP	Src
DNA hybridization	oncogene	T antigen
epithelial to mesenchymal transition	oncogenesis	telomerase
extravasation	oncogenic virus	telomere
G-protein	papilloma virus	transfection
GAP	phage	translocation
GEF	Philadelphia chromosome	tumor virus
hybridize	plasmin	virus
intravasation	polyoma virus	
	proto-oncogene	

FURTHER READINGS

Temin HM (1975) Nobel Lecture: The DNA Provirus Hypothesis. http://www.nobelprize.org/nobel_prizes/medicine/laureates/1975/temin-lecture.html

Baltimore D (1975) Nobel Lecture: Viruses, Polymerases and Cancer. http://www.nobelprize.org/nobel_prizes/medicine/laureates/1975/baltimore-lecture.html

Varmus HE (1989) Nobel Lecture: Retroviruses and Oncogenes I. http://www.nobelprize.org/nobel_prizes/medicine/laureates/1989/varmus-lecture.html

Bishop JM (1989) Nobel Lecture: Genetic Lesions in Cancer Cells http://www.nobelprize.org/nobel_prizes/medicine/laureates/1989/bishop-lecture.html

Egeblad M & Werb Z (2002) New functions for the matrix metalloproteinases in cancer progression. *Nat Rev Cancer* 2:161–174.

Blackburn EH (2009) Nobel Lecture: Telomeres and Telomerase: The Means to the End. http://www.nobelprize.org/nobel_prizes/medicine/laureates/2009/blackburn-lecture.html

Weinberg R (1998) *One Renegade Cell: How Cancer Begins.* Basic Books.
    A general treatment for nonscientists about the biology of cancer.

Weinberg R (2007) Tumor viruses. In The Biology of Cancer. Garland Science.

Pecorino L (2012) Molecular Biology of Cancer: Mechanisms, Targets, and Therapeutics, 3rd ed. Oxford University Press.

QUESTIONS FOR FURTHER THOUGHT

1. What are the essential differences between a virus that promotes the formation of cancer and one that does not?

2. You have found a virus that promotes the formation of tumors in the mammary glands of mice. You want to know whether this virus is an acutely transforming virus (i.e., one that carried an oncogene) or one that is simply a mutagen. To answer this question, you infect 25 female mice with the virus and get 20 mice with breast tumors. You have a probe that lets you see the viral DNA, even when it is integrated into host cell chromosomes, so you can map the sites at which it integrated into the hosts' chromosomes. In 16 of the 20 mice with tumors, the viral DNA has integrated in a region that is very similarly placed: on one particular chromosome and at a particular position on that chromosome. Does this evidence lead you to think your virus is acutely transforming or not? Why?

3. You have invented a broad-spectrum inhibitor for matrix metalloproteinases (MMPs). Because of the importance of this kind of enzyme in metastasis, you propose that all people should take your inhibitor every day, like a vitamin pill. This way, even if a cancer should start to arise in their bodies, its metastatic behavior would be blocked, and the cancer would not become malignant. A group of oncologists has argued that your proposal is a bad idea. What reasons do you suppose they are giving to argue against the daily administration of your inhibitor? If you decide they are right, what alternative strategy could you propose whereby your drug might actually be useful for limiting the damage a cancer can do?

4. You went back to the lab and have now invented a drug that blocks telomerase. Think about the previous question concerning the MMP inhibitor and explain whether the same logic you used there can be applied here as well.

5. Figure 6.22 diagrams several stages in the process of DNA replication, starting with the separation of the strands that occurs at an origin of replication, the making of RNA primers that help DNA polymerase initiate replication, the propagation of the two replication forks in opposite directions, accompanied by elongation of the new strands of DNA on each of the two old strands. What is not depicted is how the lagging strand goes beyond making one short piece of DNA by running back from the fork. Think about what must happen to the lagging strand to accomplish complete replication. Draw diagrams that show what DNA polymerase must do to make all necessary complementary sequence while replicating the lagging strand. What additional enzyme activity might be needed to allow complete lagging strand replication with continuous sugar-phosphate backbones running all the way along the DNA double helices that are produced?

ANSWERS TO SIDE QUESTIONS

6.1. Any aspect of cancerous transformation that relates to cell growth and division should be available for study. Even some early aspects of invasiveness, such as increased motility, can be seen and characterized in culture. Some aspects of metastasis, on the other hand, would be much more difficult to study, since there isn't other tissue around, through which and to which cells can migrate.

6.2. The genome of an RNA tumor virus must contain a nucleotide sequence that encodes reverse transcriptase. It should, therefore, be possible for these sequences to be translated right after infection, making the enzyme necessary to synthesize DNA from the viral RNA. This strategy would accomplish the same goal as is achieved by having the enzyme present in the virion at the time of infection. The fact that all known retroviruses contain the reverse transcriptase enzyme suggests that this is a much more efficient way for the virus to organize its life cycle.

6.3. One way is for the viral DNA to integrate in such a way that it puts a strong viral promoter just upstream from a host cell gene. If this happens, the host gene will be regulated by the viral promoter, leading to overexpression of the host gene. Another way is for the virus to integrate into the host gene's promoter, mutating it to become stronger.

6.4. Since oncogenesis is a process that leads to an increase in cell growth and division, oncogenic mutations must achieve this goal. For cancerous transformation, growth-stimulating factors must be up-regulated and growth-inhibiting factors must be down-regulated.

6.5. This question asks for an opinion, and there can be many legitimate views on this issue. The biggest difference I see between a cell's decisions, based on multiple inputs, and a person's choice of action is the importance of history. People have memories about what they did previously and how it worked; they can let experience guide them in finding the best decision. Stated another way, they can learn from experience. Cells, on the other hand, will always respond to a given set of inputs in the same way, because their decisions are programmed by their genes, gene products, and the chemistry by which all these molecules interact. This is a little like instinctive behavior in an insect or lower vertebrate.

6.6. This is an interesting idea, and in essence the observation itself is support for the concept. If you are a skeptic (often a useful point of view), you could say that a single observation of embryonic appearance or behavior doesn't mean that cancer cells are really returning to the embryonic state. This is certainly a legitimate criticism of the suggestion. The interesting part of this question is, How could you obtain more evidence on the issue? There are many genes expressed in embryos that are not expressed in adults. A good way to get more data on the issue would be to identify as many of those genes as possible and then ask whether cancer cells are making mRNA from those genes at a level higher than is found in normal adult cells.

6.7. A tumor could certainly start from a neoplastic growth of cells that lacked telomerase activity. In fact, it could grow quite large. If you imagine that, prior to its cancerous transformation, a cell had already gone through 40 of its 70 possible divisions, it would still have about 30 divisions to go before reaching its Hayflick limit. Thirty doublings of a single cell, neglecting cell death, would lead to about 1 billion cells, a mass about 10 cm (4 in.) on a side. That tumor would be big enough to require angiogenic support to survive.

Tumor suppressors and their roles in resisting cancerous transformation

7

CHAPTER 7

Tumor suppressors are just what their name implies: genes or gene products that resist cancerous progression. If a tumor suppressor is inactivated by mutation, the risk of cancer increases. Just as with oncogenes, tumor suppressors come in several classes. Some are like the retinoblastoma protein, Rb, which normally inhibits the start of S phase, as described in previous chapters. Indeed, there are many tumor suppressors that hold back a key cell cycle event until some necessary task has been done right. When any of these cell cycle regulators is absent or is present with reduced function, the event it controls can occur too soon or too often, behavior that is cancer-like. Other tumor suppressors participate in the cellular processes that prevent or correct mistakes; for example, the mechanisms that detect DNA damage and repair it. Still other tumor suppressors are involved in a cell's ultimate mechanism for damage control: programmed cell death, also called apoptosis. These genes are important because their absence renders a cell unable to kill itself when it should for the sake of the organism as a whole. In this chapter we explore all these functions and the molecules responsible for them, using descriptions of their action as ways to deepen your understanding of how cells go astray during cancerous transformation. The chapter ends with a description of processes that are now thought of as hallmarks of cancer, that is, properties a cell must acquire to complete the transformation from normal to cancerous.

DNA REPAIR AS A CANCER-INHIBITING PROCESS

TUMOR SUPPRESSION BY REGULATORS OF CELL CYCLE PROGRESSION

CANCER-INDUCING MUTATIONS AND APOPTOSIS

IT TAKES MULTIPLE MUTATIONS TO COMPLETE A CANCEROUS TRANSFORMATION

LEARNING GOALS

1. Describe how a cell repairs different kinds of DNA damage and how those repairs help to prevent cancer.

2. Discuss how the cell cycle is normally controlled by protein kinases.

3. Explain how checkpoints use information about the unsuccessful completion of an essential cellular process, like DNA repair or mitotic spindle formation, to prevent the cell cycle from going forward.

4. Recount why the transcription factor p53 is of critical importance for cancer prevention and relate its actions to the checkpoints that are responsible for quality control and DNA damage correction in the cell cycle.

5. Compare the several ways in which mutations can inactivate tumor suppressors and contrast the behaviors of the mutant proteins with those of normal tumor suppressors.

6. Describe how the loss of heterozygosity can contribute to the oncogenic action of a tumor suppressor.

7. Compare the functions of tumor suppressors and proto-oncogenes.

8 List seven hallmarks of cancer.

DNA REPAIR AS A CANCER-INHIBITING PROCESS

You have seen in previous chapters how certain mutations—for example, those that increase the frequency of cell division or that enhance a cell's ability to become metastatic—are cancer-promoting. It follows that processes that reduce mutation frequency or correct mutations will help to prevent cancerous transformation. This logic is equally valid for processes that detect and repair DNA damage and those that correct mistakes in DNA replication. Although most mutations are benign, given the organization of the human genome, unlucky mutations that damage an important gene will increase cancer risk. Fortunately, our cells have several mechanisms for both sensing DNA damage and repairing it. Indeed, a significant fraction of all our genes encode components of various DNA repair processes. It has been estimated that these repair processes, summed over all the cells in a human body, correct something like 30,000 errors in DNA every day! These repair functions are therefore critically important for the body's ability to resist cancerous transformation.

Some mutation-causing DNA damage can be detected through the resulting changes in DNA shape.

If damage has changed one nucleotide in a segment of double-stranded DNA, the altered nucleotide will commonly fail to complement the undamaged nucleotide on the opposite DNA strand. As described in Chapter 4, this kind of damage will lead to trouble when the modified DNA is replicated during the next S phase of the cell cycle: one of the new DNA duplexes will be correct, but the other will contain an altered sequence (FIGURE 7.1). We can now look more closely at these kinds of damage and what a cell can do about them.

SIDE QUESTION 7.1
Diffusion is driven by the motions of molecules that give rise to what we call heat. The higher the temperature of a material, the faster its molecules are moving. In what ways is diffusion of a molecule along a polymer, such as DNA, different from diffusion of a molecule in three dimensions?

Some DNA damage can be detected by a system that surveys the surface shape of DNA's double helix. If damage has altered a base so it is no longer complementary to the base it faces, the altered base is no longer held in place by base pairing. It can swing out, making a bump on the otherwise smoothly helical surface of a DNA duplex (FIGURE 7.2). This perturbation can be detected by enzymes that monitor DNA structure. Some of these enzymes bind the surface of a DNA duplex and move along it by diffusion (random motions that are driven by heat energy). The cell contains many copies of these enzymes, bound at random places all over the DNA, so all your genes are being scanned continuously for places where the double helices are not smooth. When every base in a DNA duplex is complementary to the base on the opposite strand, enzyme motion is unimpeded and the protein moves rapidly along the DNA. When DNA has a bump in it (Figure 7.2B), the diffusive motion stalls and the enzyme dwells exactly where its activity is needed to initiate DNA repair. This kind of DNA repair is called **mismatch repair**, because the solution to the problem is to replace the damaged base with one that matches (is complementary to) the nucleotide on the opposite DNA strand (FIGURE 7.3).

Any mutation that reduces the functions necessary for detecting or fixing damaged DNA increases the likelihood of cancer.

The importance of mismatch repair for cancer prevention is demonstrated by the fact that there is a clear correlation between a high frequency of

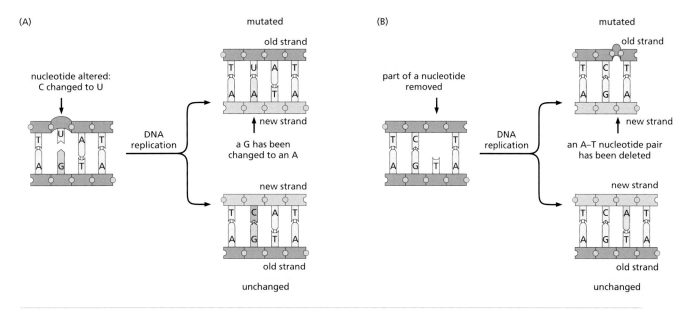

(A)   nucleotide altered:
C changed to U

mutated

old strand

DNA replication

new strand

a G has been changed to an A

new strand

old strand

unchanged

(B)   part of a nucleotide removed

mutated

old strand

DNA replication

new strand

an A–T nucleotide pair has been deleted

new strand

old strand

unchanged

FIGURE 7.1 Two examples of DNA damage and the ways in which it can lead to mutation if not repaired. (A) The nucleotide C, which normally pairs with G, has been modified by an outside influence, such as ROS, knocking off a nitrogen atom and its associated hydrogens, which converts the C into a U. The structure of U is complementary with A, not G; this is the base pairing found in RNA. After DNA replication in this example, one DNA duplex is normal, but the other is mutant at this position. (B) Radiation or ROS has knocked the base, A, off from the sugar–phosphate backbone of one DNA strand. If DNA replication occurs before this damage is repaired, one duplex will be okay but the other will be mutated by the removal of a base pair. These examples show why it is important to repair DNA damage before the cell goes through S phase. (Adapted from Alberts B, Johnson AD, Lewis J et al. [2015] Molecular Biology of the Cell, 6th ed. Garland Science.)

colon cancer and loss-of-function mutations in any of the genes that encode the enzymes required for this kind of repair. This correlation was found by studying families in which colon cancer was common. In such families, mutations that reduced the effectiveness of mismatch repair are passed from parent to child, so successive generations carry those mutations and are less able than other people to repair this kind of DNA damage. Apparently, the environment in the lower intestine makes these epithelial cells particularly susceptible to this kind of DNA damage. A person lacking enzymes for mismatch repair tends to accumulate mutations in his/her colonic epithelium, which increases the likelihood of colon cancer.

Another form of inheritable cancer that results from defective DNA repair is called **xeroderma pigmentosum** (**XP**). (The name of this cancer

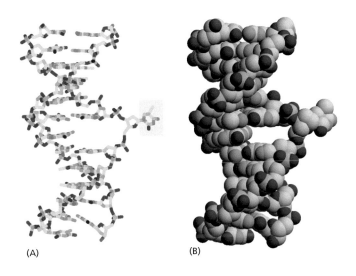

(A)   (B)

FIGURE 7.2 Some kinds of DNA damage alter the surface shape of a DNA molecule. (A) A stick model for what DNA with an unpaired nucleotide might look like, for example, when a C has been changed to a U (highlighted in yellow), which cannot pair with the G on the opposite strand of the duplex. (B) A space-filling model of the same alteration, built from spheres whose sizes represent the sizes of the atoms in DNA. Both of these representations show how this kind of damage has led to a swinging out of one or two bases, making a lump on the DNA surface. This protrusion will slow the diffusion of repair enzymes, allowing them to initiate DNA repair. (Adapted from Alberts B, Johnson AD, Lewis J et al. [2015] Molecular Biology of the Cell, 6th ed. Garland Science.)

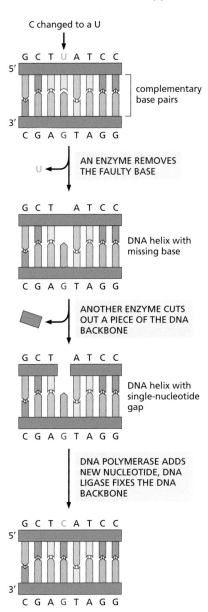

FIGURE 7.3 Mismatch repair replaces a damaged base, using information from the complementary nucleotide on the opposite DNA strand. In the example of the C being changed to a U, the faulty nucleotide must first be cut out. One enzyme cuts the bond between the base, U, and the sugar in the DNA backbone, so the bulge is gone. Then another enzyme cuts the DNA backbone in two places and removes the sugar–phosphate portion of the faulty nucleotide, leaving a gap big enough to allow the insertion of the correct nucleotide, chosen by complementarity to the nucleotide that is opposite on the DNA duplex. Finally, DNA ligase links the new nucleotide with the sugar–phosphate backbone on either side. (Adapted from Alberts B, Johnson AD, Lewis J et al. [2015] Molecular Biology of the Cell, 6th ed. Garland Science.)

derives from the Latin words that describe the most obvious phenotype of the disease: parchmentlike skin with unusual coloration, like a multitude of freckles.) XP results from a defect in one or more of the eight enzymes necessary to repair the kinds of DNA damage that are commonly caused by the UV radiation in sunlight. Either an abnormal bond between Ts and/or Cs that are adjacent along a single DNA strand, as diagrammed in Figure 5.7, or a missing base, as diagrammed in Figure 7.1B, can be repaired by the following process. A significant stretch of nucleotides is removed from the damaged DNA and replaced with nucleotides complementary to the bases on the undamaged DNA strand. Thereafter, replication can occur without introducing mutations. This process is called **nucleotide excision repair** (FIGURE 7.4). After the stretch of nucleotides has been removed, undamaged DNA is made by a DNA polymerase, using the undamaged strand as a template. This strategy assures that the new piece of DNA contains nucleotides that are exactly complementary to those in the undamaged strand.

An inability to carry out nucleotide excision repair, as in people who have XP, results in extreme sensitivity to UV light and a high incidence of skin cancers. People with XP develop skin cancers at rates as much as 1000-fold higher than people with functional alleles of all the relevant genes. Indeed, every protein involved in nucleotide excision repair is a tumor suppressor; their disappearance or inactivation through mutation seriously increases cancer risk. Children who carry any of these mutations are at special risk, so some generous people have donated money to fund a summer camp that provides youngsters with XP some of the pleasures that most children can enjoy (**SIDEBAR 7.1**).

Repairing single- and double-strand breaks in DNA

Some kinds of DNA damage are even more severe than those described above; for example, actual breaks in the DNA strands. If ionizing radiation, such as X-rays or β particles from radioactive decay, makes a direct hit on a DNA molecule, it can add so much energy to the molecule that the sugar–phosphate backbone of one or both DNA strands breaks. Alternatively, ROS generated by radiation damage to water can cause chemical reactions that lead a DNA backbone to break. If only one strand is severed, the other strand will maintain the physical continuity of the double helix; DNA ligase can reattach the ends of the broken backbone, making double-stranded DNA whole again. This is called single-strand break repair, and cells do it very well.

If both strands of DNA are broken, the problem is much harder to fix. The most straightforward process for double-strand break repair is simply to reattach the two broken ends. This is possible because the structures generated by a double-strand break are unique. You know from Chapter 6 that the ends of normal chromosomes are capped by telomeres, specialized structures that contain specific DNA sequences and bind several telomere-specific proteins. Telomeres are not free ends floating in the nucleus. However, the DNA ends created by a double-strand break are free, and they are quickly recognized as damage. The nucleus always contains many copies of a protein called Ku that will bind these free ends and recruit other proteins to the sites, making structures that are sticky at the molecular level (**FIGURE 7.5**). This stickiness means that if the two ends bump into one another, as a result of their diffusing with normal thermal motion, they will bind and initiate repair. The previously mentioned enzyme, DNA ligase, is then able to reconnect the broken backbones, completing the repair. This process is called **nonhomologous end join-**

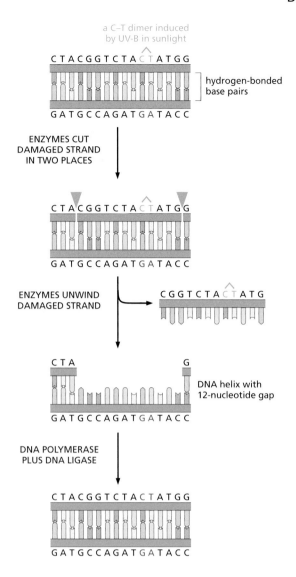

FIGURE 7.4 Nucleotide excision repair can fix damage done to DNA by UV light. Energy absorbed from the ultraviolet components of sunlight has caused two adjacent bases (C and T) to form a dimer. The abnormal bond between these adjacent bases will block their ability to base-pair properly during the next round of DNA replication. To fix this problem, the bonded nucleotides must be cut out from the DNA. The DNA strand that contains the damage is recognized as faulty, and enzymes then make cuts in the backbone of that strand on either side of the damage (shown by blue triangles). A different enzyme then unwinds that region of the double helix and releases the damaged piece into solution, where it is degraded. Now DNA polymerase inserts fresh nucleotides, making a DNA piece that is complementary to the undamaged strand in this region. Finally, DNA ligase links the new piece of DNA with the old parts of the previously damaged strand, and the DNA is as good as new. (Adapted from Alberts B, Johnson AD, Lewis J et al. [2015] Molecular Biology of the Cell, 6th ed. Garland Science.)

ing (**NHEJ**), where nonhomologous refers to the fact that the sequences put back together are not similar and that no similar DNA sequence from another piece of DNA was used to correct any errors that might have occurred when the two broken ends were simply stuck back together. The value of NHEJ is that it reassembles the chromosome, so it now has telomeres at both ends and a kinetochore (spindle attachment site) in between. These features assure that the repaired chromosome will be stable during interphase and able to segregate at mitosis. The limitation of NHEJ is that any changes in DNA sequence that occurred, either when

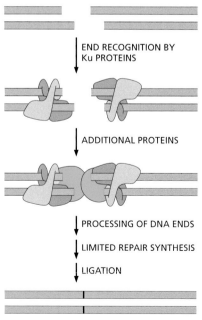

FIGURE 7.5 Nonhomologous end joining repairs a double-strand break by simply rejoining the broken ends. An event that breaks both strands of DNA generates structures that never occur in a normal cell: free ends of a DNA duplex with no telomeric structures to protect them. These sites bind Ku proteins, which are already available in the nucleus, waiting for a chance to do their job. Once Ku has bound to the DNA ends, it then binds additional proteins, making sticky sites that can glue the two sides of the DNA break together, initiating repair. Additional proteins now bind in place, making possible the chemistry needed to link the two pieces of double-stranded DNA. Note that the repaired DNA has generally suffered a deletion of nucleotides. (Adapted from Alberts B, Johnson AD, Lewis J et al. [2015] Molecular Biology of the Cell, 6th ed. Garland Science.)

SIDEBAR 7.1

A CLOSER LOOK: Camp Sundown.

Camp Sundown was established through the energy and caring of the parents of a child born with xeroderma pigmentosum (XP). The camp has thrived thanks to the donations from many generous people, and it is now the principal activity of the Xeroderma Pigmentosum Society. The mission of the camp is to give young people who are coping with this disease the kind of summer camp experience they cannot normally enjoy, because of their sensitivity to sunlight. The solution is simple: organize most camp activities between dusk and dawn (Figure 1). This camp is a beautiful example of what can be done to improve the quality of life of people with a special, cancer-related need. Analogous efforts that help with the quality of life for people dealing with other cancerous conditions at different stages of cancer progression will be described in Chapter 12.

Figure 1 The nighttime activities at Camp Sundown for children with XP. (Courtesy of Camp Sundown.)

the break was made or during its repair, will persist. If these changes happen to fall between genes, they will probably be of no consequence. However, if one of the changes is within a gene, it could mean the production of a faulty gene product. Thus, NHEJ is better than leaving the broken chromosome alone, but it is likely to have introduced mutations, which may make trouble for the cell in the future.

There is, however, another process for repair of double-strand DNA breaks that can be flawless; it is called **homology-directed repair** (**HDR**). If a cell is far enough along in its growth and division cycle to have finished DNA replication by the time a double-strand break occurs, the nucleus contains two identical copies of every chromosome. The broken DNA duplex and the identical unbroken duplex are the products of DNA replication, so they are coupled to one another by the cohesins that were added at the time of DNA synthesis (see Chapter 4). The undamaged DNA can therefore be used to guide DNA repair in the damaged double helix (FIGURE 7.6). Although this process is complicated, it is remarkably faithful and usually results in flawless repair. Thereafter, it is as if the damage never happened.

The steps in this kind of double-strand break repair are worth looking at, because they show the trouble a cell will take to keep its DNA in good shape. The cell first cuts back one strand of the DNA double helix on

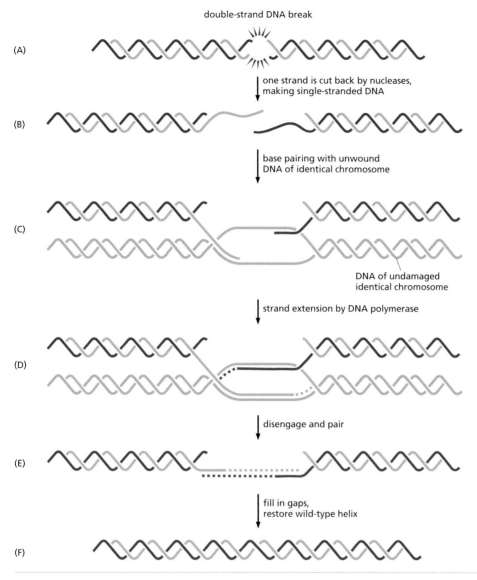

double-strand DNA break

(A)

one strand is cut back by nucleases,
making single-stranded DNA

(B)

base pairing with unwound
DNA of identical chromosome

(C)

DNA of undamaged
identical chromosome

strand extension by DNA polymerase

(D)

disengage and pair

(E)

fill in gaps,
restore wild-type helix

(F)

FIGURE 7.6 **Homology-directed repair uses undamaged DNA to guide the repair of a double-strand break.** (A) A DNA duplex has suffered a double-strand break. The break may have destroyed some DNA structure, such as knocking off one or more nucleotides. In this case, simply fusing the two ends back together would lead to an incorrect DNA sequence. Cells can perform error-free DNA repair by means of a multi-step process. (B) First, they use nucleases to cut back one strand of the DNA on both sides of the break, making the two DNA ends single-stranded over a short distance. (C) Now the cell can take advantage of sequence information from an undamaged DNA duplex that carries the same genes, for example, a sister DNA duplex, generated when the cell went through S phase. The nucleotide sequences of each single-stranded piece are exactly complementary to a particular region on one strand on the identical, undamaged DNA. The single-stranded pieces can therefore form perfectly complementary base pairs with the undamaged DNA. (D) The resulting arrangement resembles structures that form during DNA synthesis, so the enzymes that would normally synthesize DNA can extend the broken ends, using the undamaged DNA as a template to guide the selection of nucleotides at each position. (E) After some new DNA has been made, the two DNA duplexes can disengage from one another without the broken DNA falling apart into pieces, thanks to the recently made sequences, which are complementary. (F) Following a little more DNA synthesis of complementary sequences at the free ends, all missing DNA has been replaced. Now, DNA ligase can join the free ends (red to red and blue to blue), whereupon the red and blue strands can twist around one another and re-form normal, double-helical DNA whose sequence is perfectly repaired. (Adapted from Alberts B, Johnson AD, Lewis J et al. [2015] Molecular Biology of the Cell, 6th ed. Garland Science.)

either side of the break, making single-stranded ends on the two pieces to be joined. This is accomplished by always-present enzymes called nucleases that chew away one strand of any DNA duplex that has a free end. These segments of single-stranded DNA now serve as binding platforms for several kinds of protein that were again waiting around in the nucleus for a time when their functions would be useful. Three proteins, called the MRN complex, are among the first to get involved. Proteins like the MRN complex and the repair enzymes that associate with it are the tools that make double-strand break repair possible. The genes that encode these proteins are all tumor suppressor genes.

DNA damage and repair present different problems at different stages of the cell cycle.

The S phase of the cell cycle is a time of particular concern for a cell's faithful maintenance of its genome. DNA damage can result from a malfunction of DNA polymerase, which might introduce the wrong nucleotide or lead a replication fork to stall. Problems can also result from insufficient amounts of one or more nucleotides, causing stress to the replication machinery. Moreover, DNA damage from external factors, as described above, can occur at any time in the cell cycle, and during DNA replication, when segments of each duplex are unwound, DNA is particularly susceptible to damage. Over and above these issues, the mode of repair a cell will use can differ at different stages of the cell cycle. NHEJ can work before, during, or after DNA replication, but the more complex process of HDR will work in different ways, depending on the cell cycle stage at which damage is being repaired. Before S phase there is no identical DNA duplex to guide repair, so HDR, as depicted in Figure 7.6, cannot work from an identical DNA duplex. One of the beauties of our body's design, however, is that almost all our cells are diploid, meaning they contain homologous chromosomes, the two sets that came from mom and dad. Double-strand break repair can therefore take advantage of an undamaged piece of DNA whose sequence may not be identical to the damaged duplex, but at least it encodes the very same gene products. By using the homologous chromosome, as mentioned in Chapter 4, the DNA repair enzymes can match up the broken chromosome with a similar region on the homologous, unbroken chromosome and then use the undamaged sequence to guide the choice of nucleotides to fill in any missing sequences on the damaged DNA, as diagrammed in Figure 7.6. There are, however, two important differences: (1) The broken chromosome and its homologue are not bound together by cohesins, the way the products of replication are, so they have to find each other in the nucleus. This takes time and may be inaccurate. (2) When the homologous chromosome is used to define the nucleotide sequences that will be added during repair, the result is novel: a chromosome whose sequence comes partly from mom and partly from dad. The key point, though, is that the result is usually a repaired chromosome that will do the job.

You may wonder when a cell will use NHEJ versus HDR to fix a double-strand DNA break. NHEJ can start almost immediately, but if there is a DNA duplex around that can guide the more accurate kind of repair, the latter process displaces the components of NHEJ. In a sense, the processes compete to get the job done. As a result, the DNA is sure to be repaired, but exactly how repair is done depends on the phase of the cell cycle and what is available to the cell at the time damage occurred. HDR is slow but accurate, so it is preferred, if it can be done. For this reason, however, cells need ways to allow time for HDR, even if it means delaying progression of normal cell cycle events. It turns out that cells have ways to make this delay happen, and they are important in tumor suppression.

Several of the genes needed for DNA repair are named in TABLE 7.1, along with the functions they carry out and the diseases that occur when they fail. Two of these are ***BRCA1*** and ***BRCA2*** (pronounced braka 1 and braka 2), genes whose protein products participate in HDR. You may rec-

SIDE QUESTION 7.2

Why is it better for the cell to join the two ends of a broken chromosome by NHEJ, rather than simply putting telomeres onto the free ends made by a double-strand DNA break?

SIDE QUESTION 7.3

Think about the event diagrammed in Figure 7.6, where the single-stranded DNA from the end of a broken chromosome invades the region of the nearby DNA duplex at a site where the sequences are either identical or complementary. What do you think is really going on? How does the single-stranded DNA from the damaged chromosome worm its way into the undamaged double helix? (Hint: Think about thermal motion.)

Table 7.1 Inherited Diseases That Involve Defects in DNA Repair

Gene or Condition	Phenotype	Enzyme or Process Affected
MSH2,3,6, MLH1, PMS2	Colon cancer	Mismatch repair
Xeroderma pigmentosum (XP)	Skin cancer, UV sensitivity	Nucleotide excision repair
Cockayne syndrome	UV sensitivity	Coupling of nucleotide excision repair to transcription
Xeroderma pigmentosum variant	UV sensitivity, skin cancer	Translesion synthesis by DNA polymerase
BRCA1	Breast, ovarian, and prostate cancer	Repair by homologous recombination
BRCA2	Breast and ovarian cancer	Repair by homologous recombination
Werner syndrome	Premature aging, cancer at several sites, genome instability	Accessory exonuclease and DNA helicase
Bloom syndrome	Cancer, stunted growth, genome instability	DNA helicase for recombination
Fanconi anemia	Congenital abnormalities, leukemia, genome instability	DNA interstrand cross-link repair
46BR	Hypersensitivity to DNA-damaging agents, genome instability	DNA ligase I

Adapted from Table 5.2 in Alberts B, Johnson AD, Lewis J et al. [2008] Molecular Biology of the Cell, 5th ed. Garland Science.)

ognize these protein names, because they are the products of two genes identified as important for certain kinds of breast and ovarian cancers that run in families.

The mechanisms for DNA repair are complicated, so it is no surprise that there are still gaps in our understanding of exactly how they work. Although many of the proteins important for DNA repair have been identified, the exact function of each protein and how they work together is not always clear. Many of the important genes were first discovered in bacteria or yeast cells, which also face the problems of DNA damage. Gene identification and study is far easier in unicellular organisms than in humans, so progress in understanding gene function in these cells has been quite fast. It is striking that the DNA repair processes are remarkably similar in all organisms studied so far. In some cases, however, DNA repair molecules have been identified because they are the products of genes known from the fact that their loss-of-function increases the likelihood of cancer or a related disease (Table 7.1). The details of DNA repair are under intense study in many labs because of the importance of this process for several aspects of human health.

Proteins involved in many different aspect of DNA damage repair are tumor suppressors.

Because there are many gene products involved in a cell's multiple responses to DNA damage, it is useful to sort these molecules into categories. The first category includes proteins that sense DNA damage and initiate repair. We will call these first responders. Secondary responders are proteins that signal to the cell that damage has occurred and that repair is underway. Effectors are proteins that either contribute to DNA repair or affect an ongoing process, like the cell cycle, which must halt and allow time for the repair job to be done right. Some effectors even initiate apoptosis (FIGURE 7.7).

Proteins that signal the presence of DNA damage are also tumor suppressors.

The secondary responders become activated either by binding to the DNA break or by being activated indirectly by the break. One of these proteins, called ATR, binds directly to the single-stranded DNA ends that are made

FIGURE 7.7 DNA damage initiates several response pathways that help the cell to carry out repair and/or initiate alternative programs. Either DNA damage or cellular stress from a problem with DNA replication will initiate a signaling pathway somewhat like the pathways that lead from growth factor receptors to intracellular responses. The first responders sense the problem and initiate the first events of repair, such as digesting one DNA strand on each DNA duplex near the site of the double-strand break. These events activate secondary responders, which include the protein kinases ATM and ATR. These kinases turn on the activity of multiple proteins to accomplish the many functions that are necessary for successful DNA repair: activating some of the repair enzymes themselves, delaying cell cycle progression to allow time for repair, and, if that should fail, inducing apoptosis to assure that a cell with damaged DNA does not go on to reproduce.

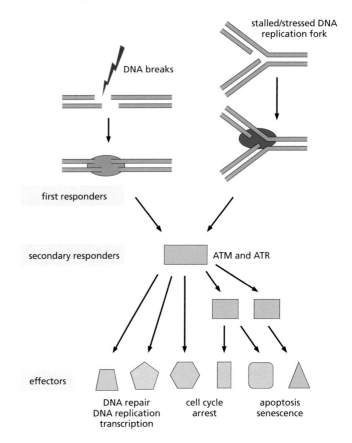

by the first responders as they initiate repair. Another secondary responder, called ATM, is distributed throughout the nucleus; it is activated by a double-strand break both to participate in repair and to signal that repair is going on. ATM and ATR are protein kinases that activate additional proteins, including additional protein kinases. Together, they initiate pathways for information transfer that is much like the pathways we saw in the previous chapter for passing information from receptors on the plasma membrane to processes occurring inside the cell. However, the signals that emanate from broken DNA tell the cell *not* to divide (Figure 7.7). Cell division with faulty DNA will increase the number of mutated cells, and DNA repair takes time, so slowing the cell cycle to delay division and allow time for repair is a useful strategy. The secondary responders, like ATM and ATR, are just as important for high-fidelity DNA repair as the first responders and effectors, so they too are tumor suppressors.

ATM gets its name from a human genetic disease called ataxia telangiectasia, a syndrome that was recognized by its symptoms long before there was knowledge about the mutations that caused it. Ataxia telangiectasia is a rare disease, commonly identified in late childhood by a jerky and unsteady gait. Mental development stops shortly thereafter, and leukemia is common. These symptoms are the result of fairly widespread cell death, which is probably a result of genome instability. The gene for ATM was mapped to a specific chromosomal region by the study of inheritance patterns in families afflicted by the disease. Once the position of the gene was known, scientists were able to identify the DNA that contained the gene and then determine the sequence of that DNA. Knowledge of the gene's DNA sequence identified the gene product as a protein kinase, and further work showed that this kinase is a part of a normal cell's response to DNA damage. The product of the ATR gene (ataxia telangiectasia-related) is a similar protein kinase, first identified in several laboratory organisms as important for sensing and fixing DNA damage. Subsequently, information from DNA sequencing identified a similar gene in humans.

The secondary responders to DNA damage, like ATM and ATR, can activate proteins not directly related to DNA repair that change the cell's control of its growth and division cycle. A pair of protein kinases, called Chk1 and Chk2 (pronounced check 1 and check 2), are turned on by the secondary responders. Together with ATM and ATR, they phosphorylate additional proteins, leading to a block in normal cell cycle progression. The logic behind this block has led to the concept of **checkpoints** in the cell cycle, which are processes that hold up cell cycle progression until an essential process, like DNA repair, has been properly completed. ATM, ATR, Chk1, and Chk2 are all parts of the **DNA damage checkpoint**. This is a set of functions that prevent some aspect of cell cycle progression as long as DNA has not yet been repaired. The DNA damage checkpoint is an excellent example of cellular intelligence.

TUMOR SUPPRESSION BY REGULATORS OF CELL CYCLE PROGRESSION

Patients who have developed any of several kinds of cancer, such as leukemia or lymphoma, commonly carry a loss-of-function allele of ATM or ATR. If either enzyme is inactivated by mutation, the patient is especially sensitive to X-rays and gamma rays, and s/he shows a general instability in chromosome number. However, there is a protein substrate of these kinases that is an even more important tumor suppressor than the secondary responders; this is the protein called **p53**.

p53 is one of the most important of all known tumor suppressors.

The name p53 comes from the protein's molecular weight (53,000 times the weight of a hydrogen atom), which was all that was known about the protein when it was first discovered. p53 was identified quite early in modern cancer research because it is involved in several aspects of cancerous transformation. For example, p53 is one of the proteins that binds the T antigens. (Recall from Chapter 6 that the T antigens are the products of tumor-related genes expressed in cells transformed by SV40 and some other DNA tumor viruses.) p53 was also identified as the product of a gene mutated to loss of function in several kinds of human cancer, so it appeared to be important for cancer biology in general. Just what the protein did was, however, mysterious. Gradually, studies in numerous labs have turned up significant amounts of information about p53. The more scientists have learned about the protein, the more they have realized its striking importance for helping cells to avoid cancerous transformation.

We now know that p53 is a transcription factor that regulates the expression of mRNAs that encode several proteins important in cell cycle control. For example, active p53 up-regulates the synthesis of an mRNA that encodes a small protein called p21, which inhibits the initiation of mitosis (FIGURE 7.8). p21 blocks the activity of a protein kinase that is essential for a cell's entry into mitosis, so the cell becomes stuck in G2 phase. Likewise, when cells are responding to DNA damage during the G1 phase of the cell cycle, p53 turns on the synthesis of other proteins that work like Rb to prevent cells from going into S phase, assuring that damaged DNA does not get replicated.

The regulation of p53 by secondary responders to DNA damage works by a surprising mechanism, so it deserves description. Activated ATM and

SIDE QUESTION 7.4

Would you expect NHEJ to be more common in the G1 or G2 phase of the cell cycle? Why?

FIGURE 7.8 DNA damage induces the synthesis of proteins that block cell cycle progression. The secondary responders to DNA damage turn on p53, a transcription factor, which activates the synthesis of several mRNAs. One of these encodes the protein p21, which is an inhibitor of mitosis. Thus, cells with DNA damage are inhibited from dividing and making two of themselves. (Recall from Sidebar 4.2 that the symbol ⊣ means inhibition.) If cells are in G1 at the time of DNA damage, p53 maintains the activities of Rb and other proteins that block the cell's entry into S phase, even if growth factors are urging the cell to initiate DNA replication. (Adapted from Alberts B, Johnson AD, Lewis J et al. [2015] Molecular Biology of the Cell, 6th ed. Garland Science.)

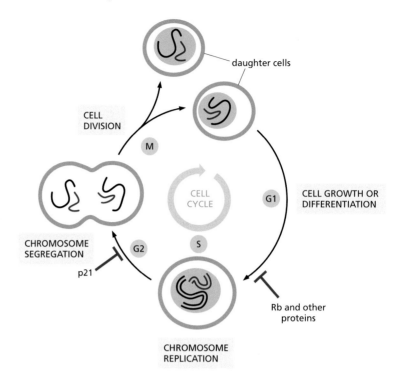

ATR work with Chk1 and Chk2 to increase the concentration of p53. You might expect that this would be accomplished by increasing the rate at which the protein is synthesized, but in fact, p53 is made at a high rate all the time. Its concentration is normally kept low by attaching it to another protein, called Mdm2, which tags p53 for degradation by one of the most powerful proteinases in the cell (FIGURE 7.9). Phosphorylation of p53 by Chk1 and Chk2 kinases causes it to separate from Mdm2, which immediately reduces the rate at which it is degraded. As a result, its concentration rises quickly, allowing it to be further phosphorylated and activated. Now it is an extremely active transcription factor, which can produce lots of mRNA, and thus p21, in a short time (Figure 7.9).

The importance of p53 in helping cells deal with DNA damage is shown be the fact that its action as a transcription factor also turns on genes that participate in DNA repair. These different factor activities would be sufficient to make p53 an important tumor suppressor, but in addition, this transcription factor can also activate apoptosis. Not all cells with activated p53 kill themselves; many simply arrest in the cell cycle. How a cell chooses cell cycle arrest versus apoptosis is still a subject of intense research. Nonetheless, the value of both these strategies for cancer prevention is clear: cell cycle arrest gives time to repair DNA, but if this job fails, then the cell is not fit to produce progeny. It must die for the good of the organism of a whole.

p53 function is lost or blocked in about half of all cancers.

The importance of p53 as a tumor suppressor is demonstrated by the frequency with which it is not expressed or is expressed in a non-functional form in humans and other organisms with cancer. For example, mice in which one or both genes for p53 have been mutated to loss of function show significant increases in early deaths from cancer (FIGURE 7.10). Additional evidence comes from an example in nature. Elephants have more cells in their bodies than humans, and they live a long time. These features should make elephants more prone to cancer than humans, but

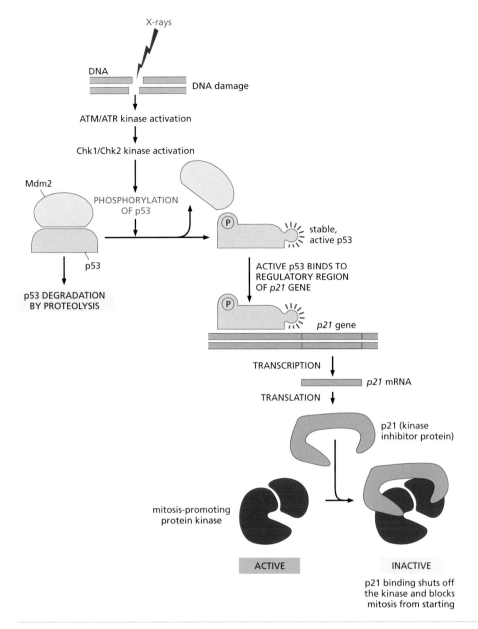

FIGURE 7.9 The mechanism by which DNA damage blocks mitosis. A double-strand break in DNA activates the ATM/ATR kinases, which activate other kinases (Chk1 and Chk2), leading to the phosphorylation and activation of p53. p53 is normally bound to the inhibitory protein Mdm2, so part of p53 activation is accomplished by inducing its release from this inhibition. The other part is the phosphorylation of p53 itself, which makes it fully active. When in this condition, p53 works as a transcription factor that induces the expression of the gene for p21, an inhibitor of a mitosis-promoting protein kinase. (Adapted from Alberts B, Johnson AD, Lewis J et al. [2015] Molecular Biology of the Cell, 6th ed. Garland Science.)

data show they are not. One possible reason for this situation is that an elephant's genome includes 40 copies of genes for p53, whereas human have only 2.

p53 is also an important factor in a cell's response to the shortening of telomeres, a kind of DNA damage. (For a refresher on telomeres, see Chapter 6.) In response to short telomeres, p53 turns on cellular senescence, which means that cells arrest in G1 and will never again enter S phase and divide. (This is the Hayflick effect, described in Chapter 6.) p53 is also part of the system that recognizes when a cell has too little oxygen; again it turns on the expression of mRNAs that encode proteins that modify cellular behavior, allowing the lack of oxygen to be tolerated. These diverse properties have given p53 the nickname guardian of the

FIGURE 7.10 **Evidence for the importance of p53 in avoiding cancer.** The percentage of mice still alive (vertical axis) is plotted versus the ages of mice (horizontal axis) for three different genotypes: (green) wild type or p53$^{+/+}$, meaning mice with two good p53 genes; (blue) p53$^{+/-}$, meaning that the mice were heterozygous, containing one good and one bad p53 gene; and (red) p53$^{-/-}$, meaning mice with no good p53 gene. Mice with all these genotypes survived normally during embryonic life, but those without functional p53 developed leukemias and sarcomas at rates far faster than normal, leading to their premature deaths. (Adapted from T. Jacks et al., Curr. Biol. 4:1–7, 1994.)

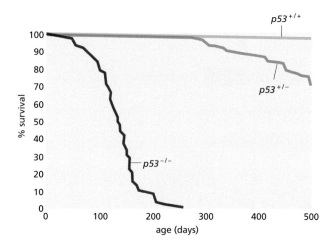

genome. The medical significance of p53 is reflected in the fact that about 5000 scientific papers are published each year by a virtual army of scientists who are trying to understand exactly how it works and how its functions are controlled.

Several aspects of cell cycle progression are controlled by checkpoints that are tumor suppressors.

We have already seen that the cell cycle control proteins activated by DNA damage, Rb and p21, are tumor suppressors. However, the integrity of DNA is so important for a cell's health that DNA-related factors other than damage can all feed into the regulation of cell cycle progression, helping to assure that cells don't divide when their genome is in trouble. For example, incomplete replication of DNA activates a checkpoint that blocks the onset of mitosis, helping to assure that a cell doesn't try to divide until there are complete genomes for both daughter cells. Problems with DNA replication, which can results from insufficient concentrations of the necessary nucleotides or from a defective copy of DNA polymerase, collectively called replication stress, will also arrest cell cycle progression. The processes by which these checkpoints work are all based on inhibition of a particular class of protein kinase that is important for cell cycle regulation, the **cyclin-dependent kinases** (**Cdks**). These enzymes get their name from the fact that their enzyme activity requires a protein subunit called a **cyclin**. One aspect of Cdk control is that cyclin is made by protein synthesis and then degraded by proteolysis as soon as its associated kinase activity is no longer needed. As a result, the concentration of the regulatory protein cycles up and down with time in the cell cycle, the property that gave this protein its name. The control of Cdk activity is a fascinating subject, but interestingly, no one has found an allele of a Cdk that is an oncogene or a tumor suppressor. Proper Cdk control is essential for proper cell cycle progression, so regulators of Cdk are oncogenes and tumor suppressors, but the absence or hyperactivity of Cdk itself is probably lethal. Mutations that might cause such changes are therefore not carcinogenic. Additional aspects of Cdk regulation are described in **SIDEBAR 7.2**.

Like DNA replication, chromosome segregation must be done right if a cell is to maintain the integrity of its genetic information and pass on a complete and accurate genome to its progeny. If a chromosome is poorly attached to mitotic spindle microtubules, it will probably fail to segregate properly during anaphase, when the duplicated chromosomes separate into two sets that move to opposite ends of the spindle. The results of faulty chromosome attachment are likely to be daughter cells that are

SIDE QUESTION 7.5

Imagine a mutant form of a cyclin that cannot be degraded by proteolysis. After it has done its job of activating a Cdk, this form of cyclin will not disappear the way a wild-type cyclin does. Would you expect this mutation to be carcinogenic? Why or why not?

A CLOSER LOOK: Control of cell cycle progression.

Cell cycle progression is under the control of cyclin-dependent protein kinases (Cdks). The activity of each of these enzymes increases at some point in the cell cycle, then rapidly decreases. Studies in which these enzyme activities were blocked have demonstrated that the transient activities of Cdks are necessary for the cell cycle to proceed at certain key transitions, such as the onset of DNA replication or mitosis. This phenomenology led cell biologists to be deeply interested in Cdks and the ways their activities are controlled. The necessary information has come from studies on cultured mammalian cells, on temperature-sensitive mutants of yeasts and fruit flies, and on the properties of specific proteins isolated by biochemical methods from the embryos of simple animals, like starfish, clams, sea urchins, and frogs. The information now available is a powerful testimony to the power of basic research for solving hard problems in biomedical science. It also demonstrates the fundamental similarity of cellular processes in a wide range of organisms

As mentioned in the text, each Cdk is built from two different proteins: an enzymatic subunit, which is a protein kinase, and a regulatory subunit, which has no

catalytic activity but must be bound to the catalytic protein for that enzyme to become active (**Figure 1A**). Each of these proteins is encoded by its own gene. The amount of catalytic subunit is approximately constant throughout the cell cycle, but its enzymatic activity is variable. This is because the regulatory subunit is initially absent, meaning that there is no enzyme activity at all. As the regulatory subunit is synthesized and increases in concentration, it binds to the catalytic subunit, which is necessary but not sufficient for enzyme activity. The activity of the catalytic subunit is also controlled by other factors that keep it inhibited until some important cell cycle event has been completed. For example, there are two different protein kinases that can phosphorylate the catalytic subunit of each Cdk. One of these phosphorylation events is essential for activity, while the other inhibits it. There is also a phosphatase that can remove the inhibitory phosphate, which then allows Cdk activation. This last step is commonly under checkpoint control. For example, the phosphatase that activates the M-Cdk is inhibited by incomplete DNA replication, which is one of the controls that keeps cells from entering mitosis before DNA replication is complete.

The many factors that regulate Cdk activity are under three kinds of control: (1) cell size, because the parent cell should not initiate DNA synthesis and head towards division until there are enough cytoplasmic components; (2) concentration of growth factors, which are necessary for the cell to enter S phase; and (3) DNA replication and repair, because a cell should not divide until all its genes have been accurately duplicated and any detectable mistakes have been corrected. When all these controls are satisfied, regulatory factors now allow the final activation of Cdk's catalytic subunit, whereupon the cell can finish its growth cycle and divide.

In human cells there are several Cdks controlling the cell cycle. Activation of any one of them leads to the phosphorylation of multiple proteins whose activities, in turn, have an impact on progression to the next cell cycle stage (**Figure 1B**). The activity of each Cdk has an additional consequence, which is of great importance to cell cycle progression. The active Cdk phosphorylates its own regulatory subunit, marking it for degradation by proteinases. As a result, shortly after a given Cdk becomes active, it is turned off by the disappearance of its regulatory subunit. The reactivation

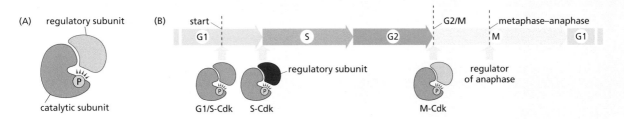

Figure 1 Diagrams of cell cycle-controlling kinases and their functions in regulating cell cycle progression. (A) A cell cycle-controlling kinase, called a cyclin-dependent kinase (Cdk); it is active only when it is bound to its regulatory subunit, cyclin. There are, however, additional changes in this kinase that also regulate its activity. For example, it must be phosphorylated by a different protein kinase to become active, as indicated by the yellow P. (B) Three different Cdks are important at different stages in the cell cycle: G1/S-Cdk, S-Cdk, and M-Cdk. These different Cdks regulate different cell cycle advances: G1/S-Cdk helps a cell get ready for S phase; S-Cdk is required for S phase to start; and M-Cdk is needed to get a cell into mitosis. The regulatory protein for each of these kinases is made at the time it is needed. When each Cdk is fully active and has done its job, the corresponding regulatory subunit is degraded by proteolysis, so that Cdk cannot be active again until more regulator is made. (Adapted from Alberts B, Johnson AD, Lewis J et al. [2015] Molecular Biology of the Cell, 6th ed. Garland Science.)

of this Cdk is thereby prevented until more regulatory subunit has been made. Thus, the activity of each Cdk goes up for only a rather short time, linked to the concentration of its regulatory subunit and to the other activation processes. Once the process controlled by this Cdk has occurred, the cell cannot turn backward in the cell cycle, because the enzyme important for that stage of cell cycle progression is no longer active.

The cyclic increase and decrease in the concentration of Cdk regulators was the first thing known about the control of these enzymes. This behavior is what led to the regulatory subunit being called a cyclin, which is, in turn, why the enzymes are called cyclin-dependent kinases. The control of cell cycle progression by fluctuating cyclin concentrations and Cdk activities was the first significant glimpse into how cells regulate their division cycle.

More recently, our understanding of cell cycle control has matured. Scientists have come to realize that there are two features of this system that make it remarkably successful for the control of cell reproduction. First, there are multiple cyclins, for example, one for timing events in G1, one to regulate the onset of S phase, and one or more to regulate mitosis (Figure 1B). These different regulatory subunits can endow the catalytic subunit (the one that works as a protein kinase) with different substrate specificities— that is, different proteins will be

phosphorylated once the kinase is active. This strategy allows essentially the same protein kinase to phosphorylate different proteins as needed for different cell cycle events. When this kind of specificity is combined with the existence of multiple catalytic subunits, the range of substrate specificities is large, and lots of different processes can be controlled by the same or similar kinases.

The second success-promoting feature of the system is that it can respond to a range of factors that might arrest its activity at different cell cycle stages. You know already that there are inhibitors of S phase onset, like Rb, and inhibitors of mitosis, like p21. p21 and some other proteins block cell

cycle progression by direct inhibition of the mitosis-specific Cdk (M-Cdk), as indicated in **Figure 2** and Figure 7.9. Such inhibitors are activated if DNA repair is done badly. Other proteins with a similar effect are activated if the cell is not big enough. Negative regulators of the cell cycle provide several checkpoints that arrest progression until numerous essential jobs have all been done right (Figure 2). The regulation of Cdks is obviously complex, but these mechanisms are worth understanding because the inhibitors of Cdks, like p21, and other negative regulators of cell cycle progression are all tumor suppressors (Movie 7.1). Loss-of-function alleles of the genes that encode them are found in several human cancers.

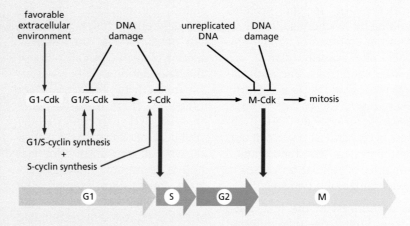

Figure 2 Cell cycle events that experience checkpoint control. A favorable extracellular environment is essential for a cell to complete G1 and initiate S phase. This environment must include sufficient nutrients and growth factors to let the cell thrive. In such an environment, a G1/S-phase cyclin is made, allowing the cell to initiate DNA synthesis. If DNA damage is detected, secondary responders feed back negatively on the activity of this Cdk and prevent the onset of S phase. Likewise, in S phase itself and at the entry into M phase, DNA damage will inhibit progression into the next cell cycle stage, ensuring that the cell maintains its genetic health as it grows and divides. (Adapted from Alberts B, Johnson AD, Lewis J et al. [2015] Molecular Biology of the Cell, 6th ed. Garland Science.)

defective in their complement of chromosomes (**FIGURE 7.11**). Just as there are checkpoints that hold up cell cycle progression until DNA replication and repair are complete, there is a checkpoint that prevents the completion of mitosis until the spindle is properly assembled. The key event here is the onset of chromosome segregation, because that is the moment when a cell makes a critical decision. Once the chromosomes have separated, the cell is on its way to becoming two cells, each of which will return to interphase and begin a new cell cycle. The **spindle assembly checkpoint** (**SAC**) is a negative regulator of chromosome segregation;

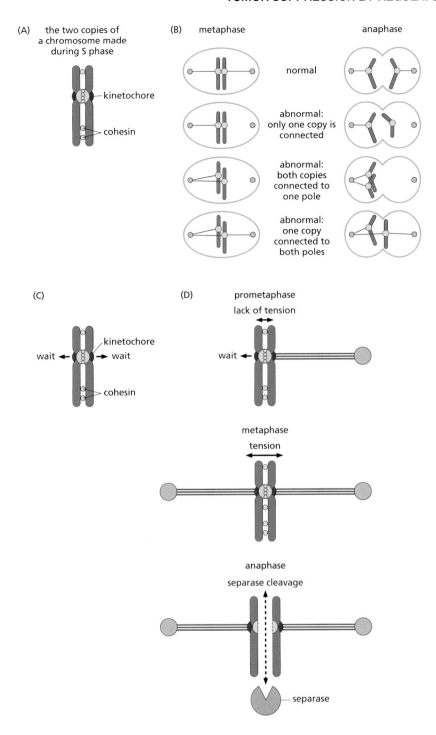

(A) the two copies of a chromosome made during S phase

(B) metaphase anaphase

normal

abnormal: only one copy is connected

abnormal: both copies connected to one pole

abnormal: one copy connected to both poles

(C)

wait

wait

(D) prometaphase
lack of tension

wait

metaphase
tension

anaphase
separase cleavage

separase

FIGURE 7.11 Mistakes that mitotic spindles can make and the logic of the spindle assembly checkpoint. (A) A mitotic chromosome contains two copies of each DNA duplex. Cohesins hold these copies together, and each copy has a kinetochore for attachment to mitotic spindle fibers. (B) Four possible patterns of connection between a chromosome and spindle microtubules. A normal connection (top) is bipolar, which will allow normal chromosome segregation. The other figures show three ways in which spindle attachments can go wrong. (C) Spindle assembly checkpoint (SAC). The kinetochores of a chromosome with no microtubules attached emit a signal that tells the cell to wait before starting chromosome segregation. (D) Mistaken attachments that fail to pull the two kinetochores in opposite directions do not turn this signal off. Once proper attachment has been achieved, the chromosome is under tension, which turns off the signal to wait. Once all chromosomes have reached this configuration, the signal to wait before starting anaphase is no longer being given. Chromosome segregation is now started by the activation of separase, a proteinase that cuts the cohesin links between the two copies of each chromosome, allowing them to move toward the pole to which they are attached. (From Musacchio A & Salmon ED [2007] *Nat Rev Mol Cell Biol* 8:379–393. With permission from Springer Nature.)

it blocks the separation of sister chromosomes until the last duplicated chromosome has become properly attached to the spindle.

The spindle assembly checkpoint works in a way that is logically simple, although its chemistry is still not fully understood. The onset of chromosome segregation is inhibited by any chromosome that is not yet properly attached to the spindle (Figure 7.11B). To achieve this inhibition, the kinetochores of unattached chromosomes emit a signal that tells the cell to wait before starting anaphase (Figure 7.11C). When each chromosome becomes properly attached to the spindle, its emission of this inhibitory signal is turned off. Thus, after all the chromosomes are properly attached, no kinetochore is emitting the wait signal, so the strength of the signal dies down and segregation can start. Proper attachment means that the

The presence of abnormal chromosome numbers in cancer cells was recognized early in the twentieth century by a very astute biologist, Theodor Boveri. He proposed that inaccurate mitosis, which leads to aneuploidy, was the cause of cancer. From what you now know, do you think he was right? Why or why not?

kinetochores on each of the two identical copies of a chromosome are attached to microtubules associated with opposite spindle poles (the poles at the two ends of the spindle). Under these conditions, a metaphase chromosome is being pulled by the spindle microtubules in two directions (toward both spindle poles). Somehow, the binding of microtubules to kinetochores and the tension that results turn off emission of the signal that inhibits segregation.

The spindle assembly checkpoint relies on proteins that participate in the attachment of chromosomes to spindle microtubules. Several of these are tumor suppressors. Their normal function is to help mitosis be accurate. As you will now expect, the function of these factors is inactivated by mutation in several human cancers. Without proper checkpoint function, mitosis proceeds before the cell is ready to divide, leading to an inaccurate distribution of chromosomes between daughter cells. The result is a cell division in which neither daughter gets a proper genome, a kind of genetic instability. This condition helps to generate the **aneuploidy** (improper chromosome number) that is characteristic of many cancer cells (see Chapter 4).

Some important progress in cancer research has come from understanding inherited tumor suppressors.

As mentioned previously, a loss-of-function mutation in a tumor suppressor can lead to a predisposition toward cancer that is inherited within a family. Colon cancer supplies an important example. Even though most cases of this disease occur without help from unfortunate inheritance, about 5% of cases occur in families where this cancer is common. The best studied of these conditions is called **familial adenomatous polyposis** (**FAP**), where familial implies that the disease is inherited and adenomatous polyposis means that people with this condition get many intestinal polyps (FIGURE 7.12). These neoplasms can progress to adenomas (benign tumors in epithelial tissues) and thence to intestinal cancer, so this genetic condition poses a serious medical problem.

FAP has been found in some families where a great deal is known about both the relationships among family members and whether or not polyposis was seen. Mormon people living in Utah have done a remarkable job of preserving family histories, so it is comparatively easy for them to trace relationships through several generations, as indicated in FIGURE 7.13. All individuals in this figure represented by a dark symbol had symptoms of FAP, and most of them developed colon cancer. By following the trait along with other genetic traits whose chromosomal locations were known, cancer scholars were able to locate the gene mutated in FAP to a specific

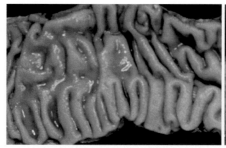

FIGURE 7.12 **The inside surface of a normal colon versus a colon from a person with familial adenomatous polyposis (FAP).** The image on the left shows the folds that are characteristic of the inner surface of a normal human colon. The image on the right shows a similar tissue that is covered with polyps as a result of inherited FAP. (Courtesy of Andrew Wyllie and Mark Arends.)

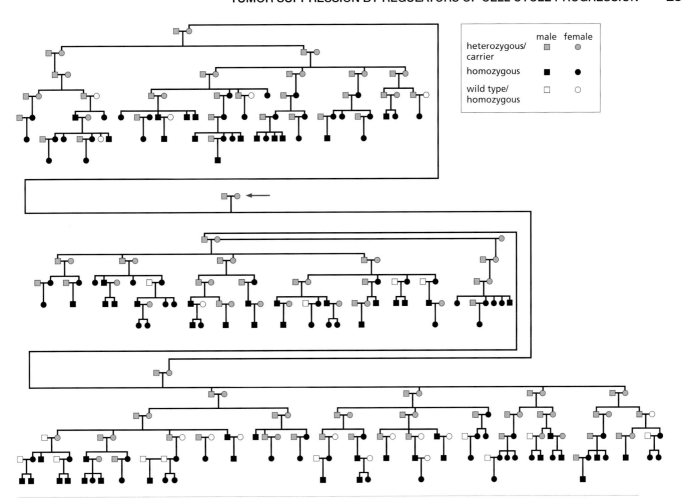

FIGURE 7.13 A tree showing relationships among an extended family in which FAP was common. The black shapes (circles and squares) represent people who were homozygous for a disease-causing allele of FAP and therefore came down with the disease. Gray shapes indicate carriers, meaning people who were heterozygous for the condition and therefore showed no symptoms. White shapes indicate people who had only wild-type alleles of the FAP gene. The red arrow points to the oldest members of this family, both of whom were heterozygous for FAP. They produced two daughters, each of whom married a man who was also heterozygous for the condition. From there you can trace the patterns of marriage and child bearing, allowing you to follow the emergence of disease several generations later. Note that in the middle of the diagram, one heterozygous man married two heterozygous women and had children by each. Patterns of inheritance like this were of great importance for the ability of cancer scholars to localize, or map, the chromosomal location of the gene responsible for FAP and then identify the DNA sequences of both wild-type and mutant alleles of this gene. (Courtesy of R.L. White, M.F. Leppert, and R.W. Burt.)

part of human chromosome 5. This information allowed them to identify the DNA that was mutated in this genetic disease, a gene now called
adenomatous polyposis coli (*APC*). It was then possible to learn the
DNA sequences of both the wild-type and mutant alleles of this gene.
This information has been of central importance in learning the function of the product of the *APC* gene and thus the way it serves as a tumor
suppressor.

Information about *APC* has allowed a deeper understanding of how colon cancer develops.

Although only a small fraction of all colon cancer cases result from an
inherited mutation in *APC*, the discovery and study of this gene has been
helpful in working out the pathways by which normal colon cells become
cancerous. FIGURE 7.14 diagrams an intestinal crypt, the place where new
colonic epithelial cells are born, so they can differentiate and move out
onto the wall of the intestine and participate in the absorption of nourishment and/or water from food recently eaten, as described in Chapter 4

FIGURE 7.14 An intestinal crypt and the differences in cell behavior that result from having either wild-type or mutant alleles of *APC*.
To review the organization of intestine, go back to Figure 4.11 and the related text. At the bottom of the crypt, there are stem cells that divide asymmetrically to produce one new stem cell and a cell that will differentiate. The latter cells receive signals from the stromal cells (red) that are just outside the crypt. These signals keep the normal cells (left) dividing as transit amplifying cells, thanks to the activity of internal, signal transduction protein called β-catenin. The resulting divisions increase the number of cells that can differentiate and line the wall of the intestine with nutrient-absorbing cells. Normally, as shown on the left, the strength of the division-inducing signal from the stromal cells weakens as the cells get farther up in the crypt wall, so signal transduction is no longer activated, β-catenin signaling is turned off by APC, and the cells stop dividing. This allows them to differentiate. In the absence of a functional APC protein, as a result of a loss-of-function mutation, the β-catenin signal is not turned off, so the cells keep on dividing and fail to differentiate, as shown on the right. This excess cell divisions gives rise to the cells that make a polyp, and the polyp cells are predisposed to change further, which can allow them to become real cancer cells. (Adapted from M. van de Wetering et al. [2002] Cell 111:251–263.)

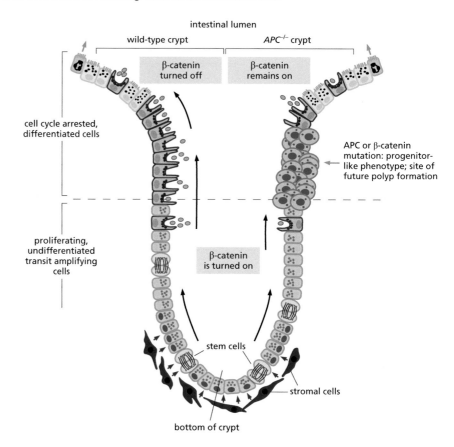

and Figure 4.11. Intestinal polyps arise as a result of improper controls in the cells that are part of this developmental pathway. The products of stem cell division at the base of the crypt would normally go through a few more cell divisions as **transit amplifying cells** and then differentiate into either absorptive cells or goblet cells, which produce the mucus that protects the intestine's surface from digestive enzymes in the gut lumen. Polyps arise when cells persist for too long in the transit amplifying state, continuing through cell cycles that produce more of themselves than is appropriate for the maintenance of intestinal structure (Figure 7.14, right side). This inappropriate behavior is common when the APC protein is missing or defective. We now know that APC protein is a negative regulator of a protein called β-catenin, which combines with other proteins to make a powerful transcription factor that induces the synthesis of proteins that keep these cells dividing.

Under normal circumstances, β-catenin is up-regulated by signals from cells called stromal cells, which lie just outside the bottom of the crypt. As the newly formed epithelial cells move up the wall of the crypt, pushed along by the cell divisions going on behind them, they get farther away from the stromal cells, so the signal that up-regulates β-catenin weakens. As this signal decreases, the action of APC protein becomes strong enough to down-regulate β-catenin and the cells stop dividing. If APC protein is missing or nonfunctional, then β-catenin is not down-regulated, and it continues to induce cell division. This means that the cells far up in the crypt don't leave the transit amplifying state; they make too many of themselves and produce a polyp. This behavior explains both why APC is a tumor suppressor and why its action is specific to cells of the intestine.

Only a small percentage of the people who develop colon cancer do so because they inherited a mutant allele of the *APC* gene. Likewise, only a small fraction of women who come down with breast cancer do so because they inherited a mutant allele of *BRCA1* or *BRCA2*. Inherited mutations that promote cancer have, however, been very important in our growing understanding of cancer. This is because the methods for mapping genes, together with modern methods for cloning DNA and determining its sequence, have often yielded enough basic information about one particular gene and the way its product contributes to cancerous transformation to provide an opening wedge in the complexity of a particular piece of cancer biology. This is a striking example of how medical science often proceeds: It is not necessarily by working on the most obvious medical problems that one gets the insights that lead to the most significant progress. This aspect of biomedical science has been hard for legislators to understand, so they continue to demand more work on what they see as the big problems of the day, when in reality the most important progress often comes from research on far more subtle issues.

CANCER-INDUCING MUTATIONS AND APOPTOSIS

You have previously read about the process of apoptosis, or programmed cell death. It was described as a natural process that most cells can turn on when the death of that cell would be advantageous for the organism as a whole. One example of a situation where apoptosis is appropriate is a cell that has accumulated enough DNA damage that it is likely to contain oncogenic mutations. It would certainly be advantageous for the organism hosting that mutated cell if the cell simply died, rather than reproducing. It follows that a loss of the ability to induce apoptosis can enhance the development of cancer and is therefore carcinogenic.

The pathway for apoptotic cell death is predictably complicated, but we can describe it as a result of two antagonistic activities, one that promotes cell death and one that inhibits it. The promotion of death can come from within the cell, as in the case of DNA damage, or from outside the cell, for example, when viruses have infected a cell and white blood cells come to help kill the infected cell before it can release large numbers of virions that would infect other cells. For current purposes, we will look only at the apoptotic pathway that is initiated from within the cell.

For a cancer to be successful, it must avoid programmed cell death.

The principal intracellular signal for apoptotic death is **cytochrome *c***, a soluble protein that normally resides in mitochondria where it participates in the making of ATP. Normally, cytochrome *c* is confined to the space between the outer and inner mitochondrial membranes. However, cells make proteins that form controllable channels in the outer mitochondrial membrane. When these channels are closed, cytochrome *c* stays in the mitochondria, but when they are open, cytochrome *c* can leak into the cytoplasm, where it is normally not found (FIGURE 7.15). Once in the cytoplasm, cytochrome *c* binds to a different protein, and together they form a large, star-shaped protein complex. This complex binds to and activates a cytoplasmic protein-digesting enzyme. This proteinase, called caspase 9, is highly specific in the proteins it will cut. It acts only on other proteinases, which are also called **caspases**, but they are given different numbers: 7, 6, 3, etc. These caspases all have a remarkable property. When they are whole, as when they have just come off a ribosome, they are inactive. In this condition they are called procaspases. Caspase 9 cleaves off a part from each of these procaspases that has been inhibiting its proteinase activity. Such cleavages open up active sites, so each procaspase becomes an active proteinase and starts to chew up the proteins

SIDE QUESTION 7.7

Imagine that you are a member of a committee charged with spending $5 million on cancer research. Another member of your committee, who wants to make sure that the money will be well-spent, insists that it be used for research on the cancer that kills the most American men: lung cancer. Would you argue with him that some of the money should be spent on a rarer condition, such as a cancer of the pancreas that runs in families? If so, what arguments and examples would you use to defend your point of view? If not, why not?

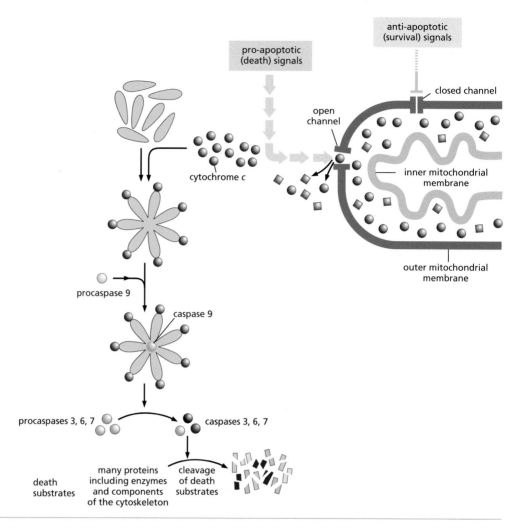

FIGURE 7.15 Apoptosis can be activated by the release of cytochrome *c* from mitochondria. Channels in the outer membrane of mitochondria can be opened by death signals, releasing cytochrome *c* into the cytoplasm. There, the protein combines with a different protein that is already present (yellow) to make a complex that binds caspase 9 (green) and turns it into an active proteinase. This proteolytic complex now activates multiple additional caspases, which in turn digest many cellular proteins, leading to cell death. Proteins like Bcl2, channel inhibitor that serves as an anti-apoptotic protein, are potentially oncogenic because their hyperactivation keeps the mitochondrial channels closed when they should open to release cytochrome *c* and initiate cell death. (Adapted from Weinberg R [2013] The Biology of Cancer, Second Edition. Garland: New York.)

of the cell. Some caspases also turn on nucleases, so RNA and DNA too become degraded. Caspases also affect the structure of the plasma membrane, a change that alerts white blood cells that their phagocytic activity will be needed. In the end, the cell collapses on itself and is then suitable for phagocytosis by white blood cells. (See Figure 4.25, the accompanying text, and Movie 4.4 for reminders of what this process looks like.) All parts of the apoptotic cell are ultimately degraded into small molecules that can be recycled for use by other cells. The release of cytochrome *c* from mitochondria is like a death sentence, and the caspases are the executioners. Thus, the factors that control the permeability of the channel in the mitochondrial outer membrane are important regulators of cell death.

Proteins that help to induce apoptosis are tumor suppressors.

A cell's ability to initiate apoptosis is an important part in the body's collection of processes that guard against cancer. Thanks to apoptosis, cells with excessive DNA damage will commonly die rather than proliferate. It is therefore not surprising that, among the genes known to promote cancerous transformation, we find some whose products regulate the release of cytochrome *c* from mitochondria. The protein **Bcl2** helps to keep the mitochondrial membrane channels closed; it inhibits cytochrome *c* release and is therefore anti-apoptotic. The hyperactivation of the *Bcl2* gene, either by a gain-of-function mutation or by overexpression, blocks the cell's ability to induce apoptosis. In this way, a normal guard against cancerous growth is foiled. The genes that encode Bcl2 and other proteins like it are proto-oncogenes because a gain-of-function mutation in their DNA promotes cancer. There is, however, a family of proteins called **BAX** that antagonizes the activity of Bcl2 by promoting the release of cytochrome *c*; these proteins can enhance apoptosis. Anything that down-regulates the function of a BAX protein deprives the cell of the tumor-suppressing action of apoptosis. These proteins are therefore tumor suppressors. Mutant alleles of both Bcl2-like and BAX-like proteins are commonly found in human cancers.

p53, the previously described guardian of the genome, has the ability to induce apoptosis. It achieves this goal through its action as a transcription factor, up-regulating the expression of pro-apoptotic proteins and down-regulating the expression of anti-apoptotic proteins. These functions define another way in which loss of function in p53 is tumorigenic. Without it, cells in the course of cancerous transformation are usually unable to kill themselves, regardless of the mutational state of individual pro- and anti-apoptotic factors.

We are now in a position to understand a nice example of the way cells use control by negative feedback to avoid overdoing a particular response (Sidebar 4.2, Signaling pathways and networks). You will recall the protein c-Myc, which you previously saw as an inducer of DNA replication. Like p53, c-Myc is also an inducer of apoptosis. Experiments on cultured cells growing in the lab have shown that when c-Myc is overexpressed in cells that have reached confluence (when they should show contact inhibition of growth), apoptosis is induced. This behavior is yet another safeguard against cancerous behavior, and it reaffirms why the loss of programmed cell death is an important step in cancerous transformation.

How can there be both oncogenes and tumor suppressors in the very same biological pathway that leads to apoptosis?

SIDE QUESTION 7.8

IT TAKES MULTIPLE MUTATIONS TO COMPLETE A CANCEROUS TRANSFORMATION

Loss-of-function mutations in tumor suppressors have been described as recessive, and human cells are diploid, so why doesn't the normal function provided by the wild-type copy of a tumor suppressor gene make the mutation of the other copy irrelevant? Stated another way, aren't loss-of-function mutations in tumor suppressors invisible in the phenotypes of cells that carry them? In fact, loss-of-function mutations of tumor suppressors *are* recessive, so this is an excellent question. Its answer takes us into another aspect of the processes that lead to the cancerous transformation of cells.

Loss of heterozygosity can lead recessive mutations to show up in cellular phenotypes.

A chance mutation that inactivates a tumor suppressor will commonly generate a cell that functions perfectly well, thanks to the wild-type copy of that gene, present on the homologous chromosome. Unfortunately, there are processes that can lead to a **loss of heterozygosity**, so a recessive allele (for example, the nonfunctional tumor suppressor) now shows up in the way a cell behaves. One route to loss of heterozygosity is defective mitosis. Imagine that a diploid cell has a wild-type allele of a tumor suppressor on mom's copy of chromosome 14 but a mutant copy on the chromosome 14 contributed by dad. This cell will behave normally until it goes through a faulty mitosis, in which the chromosome 14 carrying the wild-type allele gets lost. One of the resulting daughter cells will now have only one chromosome 14, and with bad luck it can be the copy that carries the mutant allele. This cell has lost heterozygosity for an important gene, so it now displays a phenotype that was previously masked by the wild-type gene. This cell is now on its way through a cancerous transformation.

Another route to loss of heterozygosity is through chemical modification of the promotor that controls expression of the wild-type copy of the tumor-suppressing gene. Much of the DNA in our cells is chemically altered by a process in which a single carbon atom attached to three hydrogens is added to one of the nucleotides. This simple change, called **methylation**, can alter the way in which DNA interacts with important proteins, such as transcription factors. Methylation can therefore change the level at which a gene is expressed. If chance methylation happens to down-regulate the expression of a wild-type tumor suppressor, and the other allele is already mutated to loss of function, the cell will behave as if it is homozygous for the nonfunctional allele, and the tumor-promoting phenotype will appear.

We have already seen that cells are well designed to guard against errors, so you can argue that events leading to a loss of heterozygosity will happen so rarely that they are unimportant. Chromosome loss is rare, thanks to the intrinsic accuracy of mitosis and the action of the spindle assembly checkpoint. Moreover, methylation of DNA is usually well controlled. The point that cells are well protected from making errors is right; the likelihood of such events is small. Sadly, however, the rate of cellular errors is not zero. Mistakes do happen during mitosis, and DNA does become methylated in disadvantageous ways. If a heterozygous cell goes through multiple cycles of growth and division, there is a chance of such events with every cell cycle. We have previously seen that tumor promoters work by increasing the frequency of cell division. Tumor promotors increase the chances of cancerous transformation by providing more chances for unlikely events to occur. The fact that mitotic errors do happen is shown by the abnormal distribution of chromosomes that is often found in cancer cells (FIGURE 7.16). These remarkable images were made by using fluorescent dyes coupled to pieces of DNA whose sequences match multiple places along each human chromosome. The same dye was coupled to all the DNA pieces from one chromosome, and a different dye was used for each chromosome, allowing the scientist to paint each chromosome with a chosen color, which is visible in a fluorescence microscope. The chromosome array in Figure 7.16A shows how nicely this method works to display a normal complement of chromosomes. Figure 7.16B shows how badly scrambled the chromosomes of a cancer cell can become. It is clear that even unlikely events can contribute to cancerous transformation and particularly to the progression from early cancerous transformation to a more dangerous metastatic condition.

Research has shown that there are actually multiple ways in which cells can lose heterozygosity. One of these involves an unusual event that might occur any time after S phase, in which the strands of DNA in one chromosome bump into the strands of DNA in the homologous chromosome. The similarity of their DNA sequences can lead them to line up and exchange strands, so both copies of this chromosome become mixtures

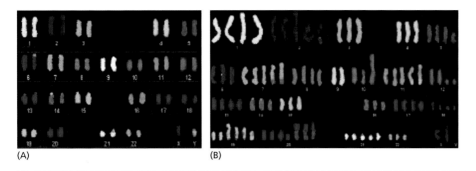

(A)   (B)

FIGURE 7.16 Chromosome composition of normal and cancerous human cells. Normal and cancerous cells were grown in cell culture and then treated with a chemical that blocked formation of the mitotic spindle. These cells completed DNA replication and went into mitosis in the sense that they condensed their chromosomes in preparation for division, but without a spindle, they were arrested in cell cycle progression by the spindle assembly checkpoint and couldn't segregate their chromosomes. (A) The chromosomes of a normal human cell were painted by binding short pieces of DNA whose sequences were complementary to successive sequences along each chromosome. These short pieces were marked with fluorescent dyes, one color for each chromosome. Note that the image of each chromosome was cut out from the original micrograph and moved to lie next to its homologue(s). Pictures like these, which show all the chromosomes from a single cell, are called karyotypes. (*Karyos* is the Greek word for nucleus.) (B) An analogous karyotype of chromosomes isolated from a cancer cell, showing how badly abnormal the chromosome content of a cancer cell can become as a result of faulty mitosis and perhaps other problems in cell cycle behavior. (Courtesy Wael Abdel-Rahman and Paul Edwards.)

of DNA from mom and dad (FIGURE 7.17). This may sound weird, but this kind of DNA exchange is in fact a normal process called recombination that occurs quite frequently in cells that are differentiating to become either sperm or egg cells. Such events occurring in a mitotic cell cycle can lead to a loss of heterozygosity, analogous to what happens when a chromosome is lost or a promoter is inactivated by methylation. There are also ways in which a gene can be converted from one allele to another, for example, by DNA polymerase jumping from one chromosome to its homologue as it is working during S phase. None of these events is common, but such things can and do occur. Inescapable evidence for this statement is found in studies of the chromosomes of cancer cells (FIGURE 7.18). By use of tools that paint each chromosome with one and only one fluorescent dye, it becomes obvious that the chromosomes in cancer cells can be abnormal in multiple ways. Chromosomes can break and join

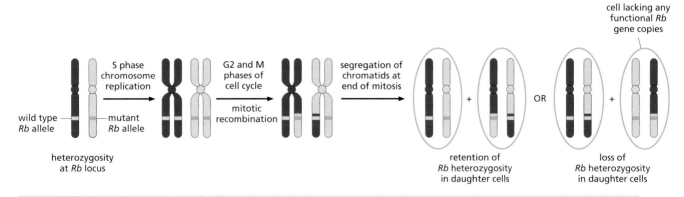

FIGURE 7.17 Chromosome recombination during mitosis can lead to a loss of heterozygosity. A loss-of-function mutation at the *Rb* gene is used as an example to show how an unusual event during a mitotic cell cycle can lead to a cell that is homozygous for a damaging, recessive trait. Two homologous chromosomes that contain the *Rb* gene are shown at left. The genotype of this individual is heterozygous for the *Rb* gene—that is, she possesses one wild-type and one mutant allele of this gene. During S phase, all the chromosomes are replicated, making two copies of both the wild-type and the mutant alleles of *Rb*. If the two homologous chromosomes are near one another, they may happen to exchange strands (an event called recombination, which is unusual during somatic cell division). Now, both chromosomes have some DNA from each of the two homologous chromosomes. When these chromosomes segregate at mitosis, there are two possible sets of division products. One set produces a cell with no wild-type copy of the Rb gene. Heterozygosity has been lost, and now the phenotype of the recessive allele will be expressed. (Adapted from Weinberg R [2013] The Biology of Cancer, Second Edition. Garland: New York.)

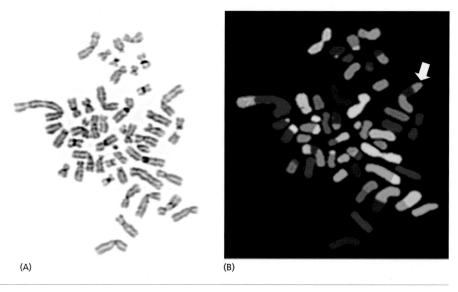

(A)   (B)

FIGURE 7.18 Chromosomes from a breast cancer cell. (A) This light micrograph shows all the chromosomes of a cancer cell blocked in mid-mitosis by a drug that prevented spindle formation. This image is similar to the ones shown in Figure 7.16, but here the chromosomes have not been cut out and arranged in pairs; they are shown as they appeared in a light microscope. Note, however, that in this cancer cell there are 48 chromosomes, not the normal 46, showing that this cell was aneuploid. (B) The same chromosomes are imaged in a fluorescence microscope after they have been been painted with small probes that are complementary to sequences along each chromosome, marking each normal chromosome with one dye molecule. Every chromosome should be only one color, but the chromosomes of this cancer cell are clearly abnormal. Some chromosomes have exchanged strands, as in Figure 7.17, making improper junctions between chromosome pieces. The white arrow points to a chromosome that has done this trick twice! Both aneuploidy and this sort of inappropriate exchange of chromosome arms, which is called translocation, are examples of the genetic instability characteristic of cancer cells. (Courtesy of Joanne Davidson and Paul Edwards.)

SIDE QUESTION 7.9

Do you see any connection between the ability of a cell to use identical or homologous DNA to aid in homology-guided repair of a double-strand break and the fact that cells sometimes carry out damaging translocations of DNA from one chromosome to another? Explain.

incorrectly, forming hybrids that are never found in normal cells. In conjunction with an original loss-of-function mutation, these kinds of events will lead to a cell that displays the mutant phenotype because it lacks the wild-type allele of an important tumor suppressor.

As described in Chapter 5, if one parent is heterozygous for a tumor suppressor and passes this trait on to the next generation, both parent and child need encounter only a single mutation that inactivates the other copy of that gene for the phenotype of the recessive allele to be expressed in some somatic cell (look back at Figure 5.16). The mutation that allows the phenotype to appear could either be a loss of function mutation in the wild type copy or a loss of heterozygosity; both will result in equivalent phenotypes. Thus, members of a family who carry a loss-of-function allele of a tumor suppressor are always one step closer to forming a cancerous cell than others in the population who have two wild type copies of that gene.

Although single mutations can be oncogenic, it takes multiple mutations to effect cancerous transformation.

You are now in a position to fully appreciate an important feature of the progression from normal to cancerous phenotype. Cancerous transformation does not happen in a single step. The process might start with an oncogenic mutation in a single somatic cell. This event might occur through a gain-of-function mutation in a proto-oncogene brought about by a change in DNA sequence, a translocation, or even through infection by a tumor virus. Oncogenic alleles of the gene for the epidermal growth factor receptor (EGF-R) or of *ras* (see Chapter 6) are good examples of such changes, because they promote excess cell division. Since oncogenic mutations produce a gain of function, they are dominant; no second

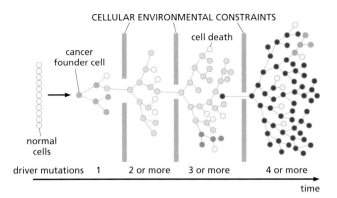

FIGURE 7.19 **Multiple mutations are required for cancerous transformation.** One normal cell experiences a mutation, symbolized by its turning blue. This cell divides to produce blue progeny, but another mutation alters one of these daughter cells to produce one daughter that dies (gray) and another that divides well (yellow). This cell with two mutations may now divide more frequently than normal, increasing its chances of acquiring additional mutations. Many of the new mutations increase the likelihood of cell death (gray), but one new mutation (green) increases the cell's tendency to divide, and the process continues. Eventually, after four or more mutations (red), the cells are far along the pathway to a cancerous state. Note that many cells may acquire mutations that do not enhance their proliferation. Some of these mutations will be irrelevant, and some may lead those cells to die. Only certain kinds of mutations convey proliferative advantage and immortality. These are the mutations that are potentially lethal to their host. (Adapted from M. Greaves [2010] Semin. Cancer Biol. 20:65–70.)

change is required for them to alter cell behavior. The result might be formation of a clone of cells that don't belong (a neoplasm).

Mutations that initiate cancerous transformation and promote excessive cell division are among the most important factors in the subsequent behavior of that clone of cells; these are often called **driver mutations**. Any mutation that promotes excess cell division or compromises the fidelity of inheritance can be a driver mutation. So also are changes that promote metastatic behavior. In a particular set of oncogenic events, the second mutation might inactivate a tumor suppressor but this mutation is recessive, and from a single recessive mutation there is not yet an associated change in phenotype. A subsequent loss of heterozygosity could, however, bring out that phenotype too (FIGURE 7.19). If the loss-of-function mutation in a tumor suppressor, followed by a loss of heterozygosity, made the cell less able to take good care of its DNA—for example, through a loss of p53 or a checkpoint gene for either mitosis or DNA repair—the likelihood of further genetic change is increased. This cell is on its way to cancerous behavior.

From what you have previously read, it is clear that cells have many guards against cancerous misbehavior, but through multiple mutations, enough changes can occur to bring this unwanted event to pass. Note that some oncogenic mutations will lead the genome to become progressively less stable. A result of this change is that mutations are likely to occur at a faster rate. Some of the mutations that happen will play little or even no role in cancerous progression. These are called **passenger mutations**, because they really are simply coming along for the ride as these cells grow and divide. In the context of these considerations, a more realistic scenario for carcinogenesis is diagrammed in FIGURE 7.20, where different symbolism is used to portray driver mutations and other events that lead to the formation of a breast cancer. Note, however, that with a loss

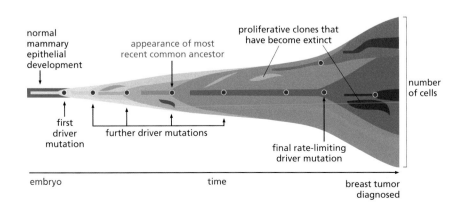

FIGURE 7.20 **A depiction of how driver mutations are thought to cause cancerous progression over long periods of time, before enough proliferating cells are produced to be detected as a tumor.** This diagram is a more realistic portrayal of events in the development of a breast cancer, including the numerous driver mutations that can happen and contribute to cancerous progression. The different colors that fan out from the central axis of the figure imply new mutant cells that increase in number but eventually die, as displayed by the disappearance of that color as time passes. This behavior could result from the fact that the mutation that occurred induced such a high level of genetic instability that the daughter cells could not survive. Many cell divisions and multiple carcinogenic mutations are likely to have occurred before there are enough cells to make a tumor that can be identified by normal screening methods (Chapter 2) and then diagnosed. (Adapted from S. Nik-Zainal et al. [2012] Cell 149:994–1007.)

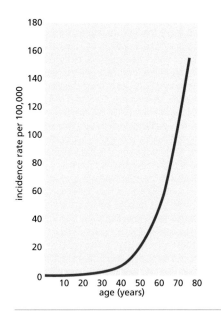

FIGURE 7.21 The incidence of cancer as a function of age. Although there are certainly cases in which young people get cancer and even die from it, the frequency of cancer is much higher in older people than in the young, as this graph indicates. Here, you see the frequency as a function of age for newly diagnosed cases of colon cancer in women during one year in England and Wales. The way in which this curve gets steeper with age suggests that somewhere between five and eight cancer-causing mutations are needed for a cancer to develop. These data are therefore consistent with the research that suggests multiple mutations are needed for cancer to develop. (Data from Muir C et al. [1987] Cancer Incidence in Five Continents, Vol. V. International Agency for Research on Cancer.)

SIDE QUESTION 7.10

Human populations are now living longer than in the past, partly as a result of improved nutrition, partly due to improved workplace safety, and partly thanks to modern medicine. What do you think this trend does to the frequency with which people die of cancer now, as opposed to 100 years ago? Explain.

of fidelity in DNA replication and repair or with increased errors in mitosis, cells are ever more likely to acquire lethal mutations, whereupon they simply die out. Other cells in the proliferating clones will accumulate passenger mutations that have nothing to do with cancerous transformation. Some cells, however, acquire by chance all the changes they need to become truly cancerous.

Robert Weinberg, the scientist who first showed that an oncogene from a human bladder tumor could transform a cultured mouse cell, has written insightfully about the multiple changes that are required for cancer to form in humans. In the year 2000, he defined six changes required for cancerous transformation and called these the hallmarks of cancer. They are (1) sustained cell proliferation, even when growth signals are not present; (2) the evasion of tumor suppressors that would normally limit cell growth and division; (3) the ability to resist cell death by apoptosis; (4) the acquisition of cellular immortality, for example, through the up-regulation of telomerase; (5) the ability to induce angiogenesis, so any large clone of cells that forms can obtain an adequate blood supply; and (6) the ability to invade neighboring tissues and engage in metastatic growth. During our examination of oncogenes and tumor suppressors in this and previous chapters, you have encountered some of the genes involved in all these cancerous behaviors.

Time is a significant factor in the likelihood that cancer will occur.

Given that the emergence of cancer requires many mutations in a single population of cells, it is not surprising that the disease is more common in people who have lived a long time. Data on the frequency with which cancer is identified as a function of the age of the patient makes this point clearly (FIGURE 7.21). The curve in this graph is far from the straight line that one would expect if a single event, occurring with constant probability, were responsible for cancer. If two events were required, the curve should look like half a parabola. If three events were needed, the curve would rise even more steeply. The shape of the curve in Figure 7.24 is consistent with the idea that five or more independent events are needed for cancer to appear. However, this statement presumes that the likelihood of a cancer-causing event stays constant. From what is shown above about carcinogenic mutations, it appears that some of them increase genetic instability. These changes would be likely to increase the frequency of mutations and may be partly responsible for the shape of the curve in Figure 7.21.

Genetic instability allows a population of cancer cells to evolve resistance to many forms of treatment.

In a fully developed cancer, it is common for cells to have chromosomes that are significantly altered from normal, both in number and in the composition of each chromosome (Figures 7.16 and 7.18). The genetic instability that gives rise to such conditions is a key reason why cells in the process of cancerous transformation can become metastatic, evade the body's immune system, grow in parts of the body where they don't belong, and become immortal by turning on telomerase. Once a cancer is established, genetic instability means that its cells are no longer identical to the cells from which they initially grew. This is good news in the sense that it suggests they may have become sufficiently different from normal that they should present novel, identifiable characteristics that could be exploited to kill them without harming our normal cells. Modern, scientific cancer treatments try to build on this idea.

Genetic instability is also bad news because it results in cells that are sufficiently unstable that they will continue to change as they grow and divide. A treatment that works to kill many cancer cells one day may not work a few weeks later, because the targeted cells have changed. Physicians who are trying to treat cancer are well aware of this changeability in the disease they confront.

With a little thought, one can see that the time-dependent changes of cancer cells are really a form of evolution (Figure 7.20). The mutations that allow a given cell to divide more rapidly, to survive when their parental cell would have initiated apoptosis, to overcome constraints like the Hayflick effect (Chapter 6), and to migrate to novel places in the body are all traits that support an increase in the number of these cells. This is analogous to evolution, which allows some organisms to thrive in a given environment while other die off. The problem for cancer cells, however, is that by evolving to be superb at the task of growing and dividing, they foul their own nest, meaning that they alter their environment (the person with the cancer) in ways that may ultimately destroy it. If the cancer patient dies, the cancer cells die with him or her. The parallels and differences between a cancer cell and a newly evolving microorganism are interesting to think about. The ability of cancer cells to evolve is one of the greatest challenges that doctors face in finding a real cancer cure. It means that a single treatment is unlikely to effect a cure unless it is applied early, before genetic instability has become advanced. Indeed, the need to cope with cellular variability is at the heart of many of the cancer treatments described in Chapter 8.

You have now looked at enough tumor suppressors and oncogenes that it is appropriate to assemble them in a symbolic way that should help you see relationships among multiple carcinogenic factors. FIGURE 7.22 provides such a view and may help you understand the complexity of carcinogenesis. A diagram like this makes it clear why multiple mutations are necessary to inactivate all the pathways that help to regulate cell behavior. While viewing Figure 7.22, you can imagine the many functions that must be compromised to allow carcinogenesis. This view emphasizes the importance of genetic instability, whether it comes from inabilities to repair DNA, inaccuracies in mitosis, or failures in checkpoint pathways. Increased genetic instability is considered a seventh hallmark of cancerous progression.

SUMMARY

Cells contain many functions that act to prevent cancerous transformation. These functions depend on genes (and the proteins they encode) called tumor suppressors. This name is appropriate because the absence or inactivation of such genes allows the emergence of cancerous behavior. During cancerous transformation, at least some of these functions are lost as a result of mutation. The functions important for preventing cancerous transformation include high-quality repair of damaged DNA, strict regulation of cell cycle progression, accurate mitosis, and the ability to induce apoptosis. Loss of any of these functions, through mutation of the relevant genes, promotes the likelihood of cancer. Many aspects of a cell's normal behavior help to prevent the formation of cancer, so several mutations are required to inactivate all the relevant cellular controls. The acquisition of multiple mutations in a single cell is unlikely, so it takes time. This is why cancer is more common in older people than in the young. It is also the reason that so much time must pass between exposure to a carcinogen and the appearance of cancer, as described in Chapter 5. At

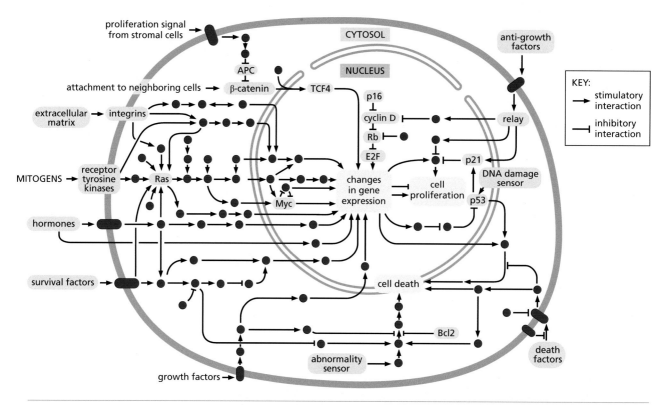

FIGURE 7.22 Multiple mutations are necessary to inactivate all the pathways that help to regulate cell behavior and prevent cancerous transformation. This diagram of a cell and its nucleus displays many of the control pathways that have previously been discussed. At the left, you see mitogens acting on receptors, like EGF-R, that have protein tyrosine kinase activity. If you follow the arrows going to the right, you see Ras, followed by symbols that represent the steps necessary for activation of the Myc transcription factor, which will alter gene expression. Below this pathway you can see hormones, such as growth hormones, acting in similar ways. Below that, you see both survival factors, which have not yet been discussed, and growth factors acting on their receptors, initiating pathways that lead through multiple steps to interact both with the Ras pathway and with other pathways that influence gene expression. Continuing counterclockwise, you see an abnormality sensor, which could be any of the checkpoints discussed in this chapter. Then you find Bcl2, which is on the pathway that controls cell death. Here the diagram also shows a receptor for death factors, which will be described after you have learned more about the immune system in Chapter 9. Continuing counterclockwise, you see p53, which is linked to DNA damage sensors, the factors we called primary and secondary responders to DNA damage. You also see p21 acting to inhibit cell proliferation. This pathway is parallel to one that includes a cyclin, which controls a Cdk, and Rb, which inhibits the onset of DNA replication. Going farther counterclockwise, you see the APC interacting with β-catenin, as it does in cells that are about to differentiate into intestinal epithelial cells. This cluster of factors is receiving signals through a receptor that senses factors secreted by the stromal cells at the bottom of an intestinal crypt, and it acts on gene expression through a transcription factor called TCF4. Finally, you come around to attachments to neighboring cells and to interactions with the extracellular matrix (ECM), which binds to membrane-bound receptors called integrins, which we have talked about but not named. Don't try to memorize this diagram, but study it with care. It will help you develop your comprehension of both the complexity of cellular responses to signals and why multiple mutations are needed to inactivate all the controls that regulate cell behavior in general and cell division in particular.

least seven important changes in cell behavior must occur for cancer to emerge: an increase in cellular growth rate, a loss of sensitivity to growth-suppressing signals, a loss of the ability to induce apoptosis, immortality (meaning a loss of normal regulation on the number of cycles a cell can go through), up-regulation of the ability to induce a blood supply, acquisition of the ability to metastasize, and an increase in genetic instability. These are hallmarks of cancerous transformation.

ESSENTIAL CONCEPTS

- Errors in DNA structure are unavoidable because DNA replication is not perfectly accurate and because our environment includes DNA-damaging factors, such as radiation and genotoxic chemicals.

- Our cells contain mechanisms that are efficient in identifying errors in DNA structure and repairing them. These mechanisms include mismatch repair and nucleotide excision repair.

- Cells can also repair both single- and double-strand breaks in DNA backbones. Single-strand break repair is quick and accurate, but the simplest mechanism for double-strand break repair (nonhomologous end joining) commonly introduces mutations. Double-strand break repair that uses identical or homologous DNA to select the correct nucleotides for accurate repair can be very effective, but it is complicated and takes time.

- The importance of DNA repair for human health is demonstrated by the fact that people who acquire or inherit mutations in the genes that encode the relevant proteins are prone to various kinds of cancers.

- Normal cells respond to DNA damage in two ways: by recruiting the enzymes that can make the necessary repairs and by sending signals that activate other cellular processes, such as stalling cell cycle progression.

- Many of the enzymes needed for double-strand break repair are tumor suppressors. Some, such as *BRCA1* and *BRCA2*, were initially discovered as genes whose loss of function leads to an increase in the incidence of a particular cancer, such as carcinoma of the breast.

- Some of the proteins that alert the cell to the presence of DNA damage are also tumor suppressors, for example, ATM and ATR. These tumor suppressors are protein kinases responsible for activating both other kinases and p53, a transcription factor of great importance for preventing cancer, as shown by the fact that it is mutated to loss of function in many human cancers.

- When p53 is turned on by the signals from DNA damage, it initiates the transcription of genes whose protein products help to stall the cell cycle, allowing time for DNA repair.

- Activated p53 achieves cell cycle arrest by inducing the synthesis of proteins that inhibit cyclin-dependent kinases (Cdks). Inhibition can occur at any of several times in the cell cycle, thanks to the different inhibitors induced by p53 and the multiple roles of Cdks at different times in the cell cycle. All of these cell cycle arrests are called checkpoints. Many checkpoint proteins are tumor suppressors.

- p53 can also up-regulate the synthesis of proteins that induce apoptosis. Both cell cycle arrest and the induction of apoptosis are strategies to prevent cells with damaged DNA from going on to divide.

- Once DNA is properly replicated, checked for damage, and repaired (if necessary), the cell can proceed into mitosis. Here again, high precision is required to prevent changes in the cell's genome. Mitosis is accurate but not perfect. Its mistakes can lead to an inappropriate number of chromosomes per cell (aneuploidy), which is yet another mark of genetic instability and a common property of cancer cells.

- Mitosis too has checkpoints that work to assure its accuracy. The relevant proteins delay the onset of chromosome segregation until all the chromosomes are properly attached to the microtubules of the mitotic spindle. The source of this checkpoint is a signal sent from unattached kinetochores that blocks the segregation process. Once all chromosomes are properly attached, this segregation-blocking signal becomes silent, and mitosis can continue. Several proteins of the mitotic checkpoint are tumor suppressors.

- Some tumor suppressors, like adenomatous polyposis coli (APC), are more important in one particular tissue, such as the intestine, than others. This is a result of tissue-specific processes for cellular renewal and repair.

- Mutations in tumor suppressor genes are commonly recessive, so they have no effect in our diploid cells unless they are accompanied by an additional genetic change that leads to a loss of heterozygosity in the relevant gene. Generally, this occurs as a result of another mutation, such as a change in the structure of the wild-type gene. It can also occur by a modification of DNA that down-regulates the strength of the promoter of the normal gene or by any of several more complicated changes in the structure of relevant DNA. These are rare events, but they happen sufficiently frequently that tumor suppressor mutations can eventually appear in the cellular phenotype, advancing cancerous transformation.

- A cell's ability to induce apoptosis is an important tumor-suppressing activity. The onset of apoptosis can be blocked by either gain-of-function mutations in some genes or loss-of-function mutations in other genes. Programmed cell death is part of a cell's cancer-preventing system, because many of the changes that accompany cancerous transformation would normally induce apoptosis and prevent that cell from proliferating. If a cell is unable to turn on apoptosis, it can reproduce, complete an oncogenic transformation, and progress to cancer.

- Recognition of all the processes that defend a cell against genomic instability (checkpoints that recognize DNA damage or mitotic errors and functions that correct these errors) and all the processes that keep the cell from dividing too frequently (controls on Cdk activity and on the onset of apoptosis) makes one appreciate the significant number of cellular processes that must go wrong for cancer to emerge.

- The number of changes needed for a complete cancerous transformation is around seven, and getting so many mutations in a single cell is unlikely. This is why cancer is so much more common in older people than in the young. It also explains the long time that is commonly seen between exposure to a carcinogen and the emergence of cancer.

- In light of all this information about tumor suppressors and oncogenes, we can identify seven properties of a normal cell that must change for it to become cancerous: its growth rate must increase, it must lose sensitivity to growth-suppressing signals, it must lose the ability to induce apoptosis, it must acquire immortality through preventing chromosomes from shortening, it must be able to induce a blood supply, it must acquire the ability to metastasize, and it must become genetically unstable. These are thought of as the hallmarks of cancer.

KEY TERMS

adenomatous polyposis coli (*APC*)
aneuploidy
BAX
Bcl-2
BRCA1 and *BRCA2*
caspase
checkpoint
cyclin
cyclin-dependent kinase (Cdk)

cytochrome *c*
DNA damage checkpoint
driver mutation
familial adenomatous polyposis (FAP)
homology-directed repair (HDR)
loss of heterozygosity
methylation
mismatch repair

nonhomologous end joining (NHEJ)
nucleotide excision repair
p53
passenger mutation
spindle assembly checkpoint (SAC)
transit amplifying cell
tumor suppressor
xeroderma pigmentosum (XP)

FURTHER READINGS

Pecorino L (2012) Molecular Biology of Cancer: Mechanisms, Targets, and Therapeutics, 3rd ed. Oxford University Press.

Hanahan D & Weinberg RA (2000) The Hallmarks of Cancer. *Cell* 100:57–70.

Davies K & White M (1996) Breakthrough: The Race to Find the Breast Cancer Gene. Wiley.

Go to the Website for the American Cancer Society (cancer.org) or the National Cancer Institute (cancer.gov) and type in the words tumor suppressors as a search term.

Sullivan K, Gallant-Behm CL, Henry RE et al. (2012) The 53 circuit board. *Biochim Biophys Acta, Rev Cancer* 1825:229–244.

QUESTIONS FOR FURTHER THOUGHT

1. Why is the transmission of cancer risk within families almost entirely due to mutations that inactivate tumor suppressors rather than mutations that generate oncogenes?

2. When normal cells leave the cell cycle and differentiate, they activate control pathways that are analogous to the pathways we have seen regulating cell proliferation. Even some of the same molecules are involved, like cyclins and Cdks and a variety of transcription factors. Would you expect there to be oncogenic mutations in the pathways that lead to differentiation? If so, how might they work?

3. We have seen that a mutation that helps to prevent a cell from inducing apoptosis can be oncogenic. Can you imagine a mutation in the apoptotic pathway that would help to prevent cancer?

4. You have learned several hallmarks of cancer. For each one, identify a gene whose mutation could help to induce this particular hallmark.

5. Imagine that you have invented a gene-based cancer therapy in which you use a virus to insert a wild-type copy of a tumor suppressor gene into cancer cells, allowing the patient to overcome the effects of a loss-of-function mutations in that gene. (Recall from Chapter 6 the ways that tumor viruses can insert DNA into a host's chromosomes.) What are the principal problems you will have to overcome to make your therapy useful in curing people with cancer?

6. Now consider trying to use gene therapy to help cure a patient who has a gain-of-function mutation that has made an oncogene. Is this situation likely to be more or less successful than working to treat a mutated tumor suppressor? Why?

7. Compare the transformation that leads to a cancer cell with the emergence of a newly evolving microorganism, such as a bacterium or a virus that is now causing a rapidly spreading infection.

ANSWERS TO SIDE QUESTIONS

7.1. Diffusion is a result of the molecular motions that give rise to heat. Because all molecules are jiggling around, thanks to heat energy, they bump into each other, bounce off, and go on to bump into another neighbor. Thus, they wander slowly from wherever they are to places nearby. This is diffusion. It is much faster in gases than in liquids, and it is faster in liquids than in solids, but it occurs in all phases of matter. When molecular motions are confined to one dimension, as when a protein is bound to DNA, the molecule will diffuse back and forth on that polymer, again driven by thermal motions but confined in its motions by being bound to a linear polymer. The only difference between free diffusion and this one-dimensional diffusion is that the latter is constrained by polymer binding.

7.2. If a double-strand break in DNA led to the establishment of new telomeres, the newly made ends of DNA would be protected from further damage by DNA-digesting enzymes, but there would still be two problems: (1) Any gene that happened to span the place where the break occurred would now be broken in two, so only one part of it would be expressed. That change is a serious mutation in the damaged gene. (2) Only one of the two chromosome pieces would contain the region of DNA that can attach to mitotic spindle fibers. That piece would be segregated faithfully at mitosis, but the other piece of DNA would be lost. This would result in the loss of many genes and might well be lethal.

7.3. The key idea here is that the same thermal motions that cause diffusion also cause all molecules to vibrate. The structures you see for proteins and DNA, for example, Figure 7.2, are drawings of the average positions of the component atoms, but every atom is jiggling around those average positions all the time. Thus, a stretch of single-stranded DNA is wiggling like a little eel, and the nearby stretches of double-stranded DNA are also wiggling, including very short times of partial separation. The wiggling single strand can sometimes get in between the complementary strands of a DNA duplex, and if it finds a sequence of nucleotides to which

it is complementary, it will bind there and initiate the kinds of events that are diagrammed.

7.4. NHEJ is more common in the G1 cell cycle stage than in G2, because in G1 there is no already-made second copy of the DNA to facilitate homology-guided repair. For HDR to occur in G1, the broken chromosome must come near its homologous chromosome in order to receive any guiding information. That process requires the diffusive motions of the broken chromosome pieces and their homologous partner until they happen to meet. Such diffusion takes time, so during G1, NHEJ can often proceed to completion before the homologous chromosome is found.

7.5. A nondegradable cyclin would lead to continued activity of the Cdk with which it is associated. This is a gain-of-function mutation, which suggests it might be an oncogene. However, when you think about the consequences of continued Cdk activity, you can see that it is far more likely to lead to a confused cell cycle in which events fail to occur in the proper order. Such a mutation is more likely to be lethal than oncogenic.

7.6. From what you have learned about oncogenic mutations and the loss of function of tumor suppressors, which can all occur by local changes in DNA sequence rather than aneuploidy, it seems unlikely that Boveri was right. Interestingly, however, two features of cancerous progression, not yet discussed in this book, keep Boveri's ideas alive: (1) Aneuploidy is almost certainly an important factor in the progression and evolution of cancer, because it is both a result of genetic instability and a cause of massive genetic errors. (2) Scientists looking at several different cell types have noted that a failure in the fidelity of mitosis, which causes aneuploidy, also leads to a significant increase in the frequency of other kinds of mutations, such as errors from faulty DNA repair. These observations suggest that Boveri was partially right in the sense that aneuploidy is a driver for cancerous progression.

7.7. There is no right answer to this question, but it raises interesting and important issues. Investing money in research on the most dangerous cancer sounds like a good idea. Experience has shown, however, that lung cancer is very hard to treat well, so there is no obvious way to spend those dollars wisely. Money invested in a cancer of the pancreas that runs in families holds the promise of identifying a gene or genes that are important for pancreatic cancer in general, and this too is a reasonably common disease and difficult to treat. Experiences with retinoblastoma, FAP, and breast cancer all show that analysis of the hereditary factors in a cancer can identify genes that are also mutated in nonhereditary forms of the same cancer, which in turn can lead to a deeper understanding of that disease and the possibility of effective therapies. There is, therefore, an argument for a significant investment in the rare, heritable cancer.

7.8. The control of apoptosis is complicated. It involves some proteins that promote apoptosis and others that inhibit it. In the former category are components of channels in the mitochondrial membrane that can leak cytochrome c, as well as proteins that up-regulate the leaking process. A loss-of-function mutation in any of these genes will inhibit the onset of apoptosis, which promotes cancerous progression. As for the latter category, other parts of this regulatory pathway inhibit the onset of cell death; for example, proteins that tend to keep the channels closed so cytochrome c cannot leak out. Gain-of-function mutations in any of these genes lead to a mutant cell in which cell death is less likely, and therefore cancer can again thrive.

7.9. At first sight, there does not appear to be a connection, because the association of identical or homologous chromosomes on the basis of identical or similar nucleotide sequences should facilitate an exchange of chromosome arms that is completely benign; it involves simply swapping the location of the very same or very similar DNA. Thinking about the issue a bit more deeply, though, one can see that regions on different chromosomes that contain completely different genes might have a run of nucleotides with very similar sequences, even though the genes have completely different functions. Such regions of sequence similarity might provide a way for a chance chromosome break to lead to a translocation rather than a repair. Thus there is a similarity, and this may be part of the mechanism that leads to rather frequent translocations between certain chromosomal segments, such as the process that forms the Philadelphia chromosome in progression to chronic myelogenous leukemia.

7.10. Since people are now living longer, the average age of the human population is older than it was in the past. Given the strong dependence of cancer frequency on age, this implies that a larger fraction of the population is now likely to be coming down with cancer. This is indeed the case, but improved methods for early diagnosis and treatment are keeping the frequency of cancer deaths in check. Data relevant to this situation will be presented in Chapter 12.

Medical treatments for cancer

8

Many diseases can be left untreated, and the body's natural defenses will take care of them. Viral infections are often like this, in part because our immune systems are so effective at eliminating pathogens that it can recognize as nonself. Likewise, a benign tumor may sometimes be left alone, at least for a while, and there will be no ill effects. If a tumor is malignant, however, it is very unusual for it to get better on its own. There have been rare cases in which a widespread cancer that seemed life-threatening has simply gone away. These events, called **spontaneous remissions**, are not yet understood; they are so rare they have been very hard to study. Indeed, they occur so infrequently that simply hoping for spontaneous remission would be a foolish way to treat cancer. Doing nothing about a tumor that is or might be malignant incurs the risk that continued tumor growth and probable metastases are likely to prove fatal. Moreover, the farther the rogue cells have progressed in cancerous transformation, the more likely they are to become genetically unstable and therefore harder to treat. Most oncologists agree, therefore, that the best treatment plan is to act quickly with protocols that are likely to kill as many tumor cells as possible. Remember, though, that by the time a tumor is only half an inch in diameter, it already contains about a billion cells, so there is a tremendous number of these unwanted objects to eliminate.

In this chapter, you will read about the cancer treatments that medical science has been using for some years. These include surgical removal of the tumor; treatment with ionizing radiation, which can damage cells so much that they die; and administration of chemicals that are toxic to growing cells, a class of treatments called chemotherapies. A few kinds of novel chemotherapies, recently identified by scientific medicine and fully tested in the clinic, are also described, as are some of the cancer treatments recommended by alternative approaches, such as traditional medicines. There are, of course, novel scientific cancer treatments being invented all the time, and some of these are covered in Chapter 10. However, most cancer patients are currently treated by the tried-and-true methods with which clinicians have extensive experience.

DIAGNOSTIC METHODS SET THE STAGE FOR CANCER TREATMENT

SURGERY

RADIATION TREATMENT

CHEMOTHERAPY

NEW CHEMOTHERAPIES FROM BASIC CANCER RESEARCH

ALTERNATIVE MEDICINE AND CANCER

LEARNING GOALS

1. Describe what surgery can and cannot do for removing cancer.
2. Specify the strengths and limitations of ionizing radiation for cancer therapy.
3. Describe the ways in which effective radiation therapy depends on high-quality medical imaging.

4. Explain in one sentence why chemotherapy is likely to provide the best treatment approach in the case of metastatic cancer.

5. Detail the limitations of chemotherapies and the origins of their side effects.

6. Describe the factors driving current efforts to improve cancer treatments.

7. List the ways in which alternative therapies may interact with the approaches and treatments developed by scientific medicine.

DIAGNOSTIC METHODS SET THE STAGE FOR CANCER TREATMENT

Like the assignment of a cancer prognosis, the choice of which cancer treatment(s) to apply depends heavily on the condition of the tumor(s) at the time of diagnosis. The TNM (tumor, nodes, metastasis) cancer staging system was described in Chapter 2; you can refresh your memory of it by looking back at Table 2.2. In modern cancer medicine, however, the many ways in which doctors can now study biopsied tissue have greatly enriched the information available for making a sound decision about the best modes of treatment. These data too are important for choosing the lines of therapy to take.

Scientific advances have extended the value of microscopic examination of tissue biopsies.

The dyes hematoxylin (for nucleic acids) and eosin (for proteins) have long been used to give doctors a good overview of tissue structure. Pathologists are doctors trained to recognize tissues, specific cell types, and their diseases using these dyes. A biopsy that contains cancerous tissues will reveal several aspects of tumor development when slices from that tissue are stained with these dyes and imaged in a good light microscope (FIGURE 8.1A). More recently, doctors have begun using antibodies, coupled to specific dyes, to see whether a particular antigen is present in the tumor. FIGURE 8.1B shows a sample of cancerous breast tissue, analogous to the one in Figure 8.1A, but now stained with an antibody to the estrogen receptor. Many of the cells in this biopsy show strong staining with this antibody (note the dark brown nuclei). This means that the cells of this tumor are expressing the estrogen receptor and can therefore respond to estrogen in the patient's blood. Such tumors are excellent candidates for treatment with chemicals that decrease the body's estrogen levels or that block estrogen signaling. Either kind of treatment will reduce the growth-stimulating action of this hormone on the cancer cells, causing them to reduce their growth and in some cases to die. In this example, information from the biopsy can help to define what is likely to be a successful treatment.

Antibodies can also be used to see whether the product of a particular oncogene is present in the tumor's cells. Some breast cancers are stimulated to grow by excessive response to epidermal growth factor. This can be a result of the tumor cells making too much epidermal growth factor receptor (EGF-R). When this membrane protein is overexpressed, it can be detected by antibodies coupled to a visible stain (FIGURE 8.1C). The presence of this protein in such quantity indicates that this tumor might be treatable by an antibody that interferes with this receptor's binding to its growth factor, as described later in this chapter.

Pathologists can also use the kind of DNA probes discussed in Chapters 6 and 7 to ask whether a particular gene is present in extra copies. One kind of oncogenic mutation is an increase in the number of copies of a key gene,

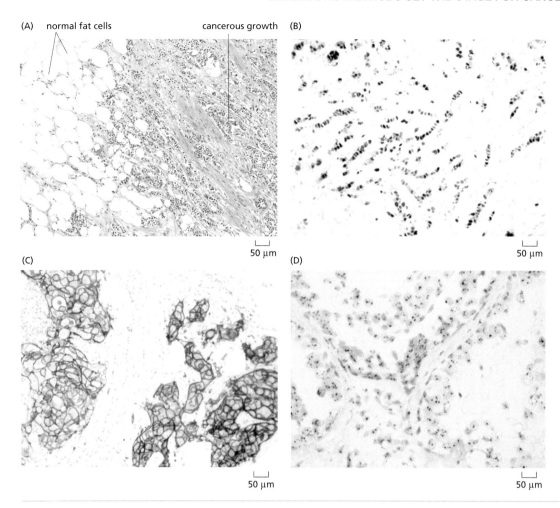

FIGURE 8.1 Biopsy tissues stained in various ways to bring out different aspects of tumor structure and cancer cell condition. (A) A high-grade breast cancer stained with hematoxylin and eosin. On the upper left, the white ovals are droplets of milk fat made by normal breast cells. On the right, the dark blue dots are the nuclei of cancer cells stained by hematoxylin. The pink is largely connective tissue proteins of the tumor's stroma, stained by eosin. The image shows the large number of cancer cell nuclei that have been produced by excess cell divisions. (B) A breast cancer biopsy stained with antibodies to the estrogen receptor. The nuclei of cells expressing this receptor have been turned brown by the stain. Unstained regions of the picture show mostly connective tissue stroma. The presence of the estrogen receptor means that this tumor could be treated by depriving it of estrogen. Methods to do this are described later in the chapter. (C) A breast cancer biopsy stained brown with an antibody to a receptor for epidermal growth factor (EGF). The presence of this receptor means these cancer cells are being stimulated to grow by EGF. Therefore, an antibody against this receptor might reduce their rate of growth. (D) A breast cancer biopsy treated with a DNA probe for the gene that encodes an important growth factor receptor. The cells whose nuclei show dark spots contain multiple copies of this gene. Gene amplification is an example of a mutation that can cause cells to grow and divide too frequently. The image implies that treatments blocking the action of this growth factor should be effective in reducing tumor growth. (Images courtesy of Dr. Alar Enno, Liverpool Hospital.)

for example, the one that encodes EGF-R. Extra copies of this DNA will lead to overexpression of its gene product, making the cells supersensitive to normal levels of this growth factor. The presence of such an oncogenic mutation can be detected by using a labeled DNA probe that reveals the extra copies of the gene with a light microscope (FIGURE 8.1D).

All of these images reflect ways in which modern biomedical science is extending the power of biopsies to inform doctors about the specifics of a particular disease and therefore the kinds of treatments that are likely to be effective.

A doctor's approach to cancer treatment is also influenced by the tumor's stage at the time of diagnosis.

A localized tumor (one that has been scored T = 0 or 1 and N = 0) is an excellent candidate for successful removal by surgery. If the edges of the tumor are well defined, so there is no indication of invasive behavior by the tumor cells, then surgery can probably remove all the tumor cells in a single operation. Higher values of T imply a bigger primary tumor, which may complicate treatment by surgery, but they carry no implication of metastasis. Values of N greater than zero indicate that tumor cells have moved out from the primary tumor into nearby lymph nodes, even if M = 0. In this case, there is danger of metastatic growth. A treatment either different from or in addition to surgery is likely to be worthwhile, in spite of the resulting increased stress on the patient. If metastasis has begun, then chemotherapy is almost always part of the recommended treatment. Thus, some comparatively straightforward observations by well-trained oncologists and pathologists can help to define appropriate ways for treatment to proceed.

SURGERY

If a tumor can be located before it has metastasized, it can often be removed completely. Skillful surgery is impressively effective, causing minimal trauma to the patient as all the cancer cells are cut away, what a doctor calls **extirpated**. Complete extirpation of a tumor is an ideal treatment. The chances for such an operation are increased by early diagnosis and by getting medical attention from an experienced surgeon who works in a good hospital. An example is the identification of an early breast cancer, such as a ductal carcinoma *in situ* (DCIS) (FIGURE 8.2). DCIS is commonly treated with a **lumpectomy**, meaning that a surgeon removes the apparently abnormal tissue and little or none of the surrounding tissues. Such an operation uses local anesthetic, requires only a small **incision** (that is, a cut in the skin), and it can be performed in an **outpatient** set-

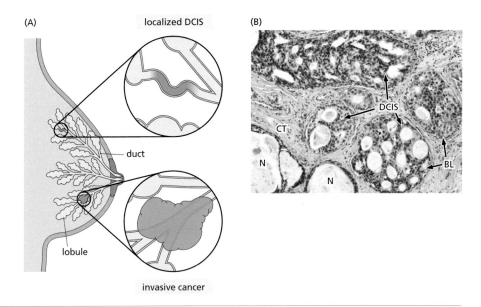

FIGURE 8.2 Anatomy of breast tissue and a common site of breast cancer formation. (A) Schematic of a human breast showing the lobules where milk is produced and their connections via ducts to the nipple. In the upper inset, one duct is magnified to show the formation of ductal carcinoma *in situ* (DCIS). This is not a serious cancer. It has not metastasized, and the abnormal cells are still confined to the duct where they were born. The lower inset shows a later stage in cancerous progression. Cancer cells have migrated through the basal lamina surrounding the duct and are invading the nearby tissue. (B) Light micrograph of a slice cut from a breast biopsy in which the connective tissue (CT) is more abundant than normal (often the case for breast cancers), and three ducts contain DCIS (arrows). These cells are overgrowing, which clogs ducts that should permit the flow of milk. However, these overgrowing cells have not passed through the basal lamina (BL). This confinement of the proliferating cells is the criterion that defines this neoplasm as cancer *in situ*. Compare the structure of these ducts and their luminal contents with those in the two ducts labeled N for normal. (Image courtesy of KGH, CC BY-SA 3.0.)

CASE STUDY: Rosemary's cancer treatment

Rosemary got good advice about treatment for her recently identified breast cancer. As she described it, "The pathologist's full report stated that I had an invasive, ductal carcinoma that was moderately differentiated [for an explanation of what this means, see Figure 8.2]. I got two opinions from surgeons, both very good people, and their views were essentially the same. As a result, there was really no choice; I simply had to go through with an operation that would remove the tumor [lumpectomy] and sample some nearby lymph nodes. Nowadays, of course, there is the improved procedure of sampling the sentinel nodes [the few nodes that are positioned so they are most likely to have been invaded by wandering cancer cells; http://www.cancer.gov/cancer-topics/factsheet/Therapy/sentinel-node-biopsy], but in the early 1990s they took lots of nodes, just to get a good sampling. My surgeon removed the tumor very successfully, but he also took 17 nodes, which can, of course, lead to problems after the operation, like **lymphedema** [swelling due to reduced drainage of the lymphatic system]. I was lucky on this score, because I had only mild lymphedema, which I was able to treat myself by using an elastic sleeve to help prevent swelling. In fact, until recently I have always worn a sleeve like this when I have gone on an airplane.

After the operation, the hospital wanted to send me right home, but my husband was away, and our son was still only in high school, so I insisted on staying in the hospital that night. A few days after I got home, I got the pathologist's report on tissue samples taken during surgery. All the lymph nodes were clear, but the tumor itself was less encouraging. It wasn't too big, about 1 inch in diameter, but its margins were not as 'clean' as the doctors would have liked. [A smooth margin to a tumor is good news, because it implies a boundary between the cancer and surrounding tissue. 'Clean' is doctor's jargon for a smooth cancer surface.] The good news was that tests on the tumor cells showed that they required estrogen to grow, so I was a good candidate for the therapies that deprived breast cells of that essential growth stimulus."

ting, meaning that no overnight hospital stay is needed. DCIS is vastly different from a condition in which there are tumors all over the body, demonstrating widespread metastatic growth, which has a very poor prognosis. With good protocols for cancer screening, the latter situation is now unusual for breast and some other cancers. However, tumors in some other organs, such as the pancreas or the ovaries, are frequently not found until they are malignant, and even breast cancer is sometimes not identified as early as one would like. Fortunately, however, prompt and rigorous medical treatment can bring such situations to a happy outcome (SIDEBAR 8.1).

The structure of a tumor sometimes provides inherent complications for surgery.

Surgery commonly requires an anesthetic of some kind to reduce pain. The anesthetic can be local for a small operation, but for a big tumor or one that is within the **viscera** (the body's internal organs), general anesthesia is commonly used, meaning that the patient is given a drug that makes him/her unconscious. A local anesthetic is almost always tolerated well, but being put to sleep by drugs can be hard on some patients—particularly the elderly or people with heart problems or other systemic difficulties.

Another problem with cancer surgery is that for a tumor to be detected, it must be fairly big (FIGURE 8.3). If the tumor is several inches in diameter at the time of diagnosis, removing it may damage or displace the surrounding tissues. Moreover, a tumor that is big enough to be detected and removed by surgery is always more than just a mass of cancer cells. It contains **stroma**, a word that refers to both the **connective tissue** that supports the physical arrangement of the tumor cells and the blood

SIDE QUESTION 8.1

If the margins of Rosemary's tumor had been clean, what would that have implied to the doctors in terms of the possibility that metastasis had already occurred?

SIDE QUESTION 8.2

If you were the surgeon responsible for removing a large tumor from a small child, like Scott, what kinds of medical imaging would you want to have done before you started? Explain your choices and give a justification for each method you propose, bearing in mind that every procedure costs money and that the big machines for looking inside a person's body can be quite frightening for a child.

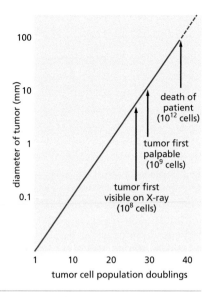

FIGURE 8.3 **Graph of the number of times a tumor cell has doubled versus the tumor's size (on a logarithmic scale).** With each round of cell division, the number of cells increases by a factor of two. Twenty divisions can make 1,048,576 cells (which is a little over a million, or 10^6). 10^8 is 100 times that many, 10^9 is a billion, and 10^{12} is a trillion. A tumor is visible by X-ray when it is between ¼ and ½ inch in diameter. Palpable means that you can feel it, for example, as a lump in a breast exam. To be palpable, a tumor must be almost one-half inch in diameter. The data on this graph indicate the striking facts that there is a long period of growth before detection is likely, and if the cancer cells are growing without constraint, there is a comparatively short period of growth between detection and lethality. (Adapted from Alberts B, Johnson AD, Lewis J et al [2015] Molecular Biology of the Cell, 6th ed. Garland Science.)

vessels that bring nourishment and blood cells to the tumor (FIGURE 8.4). Stroma is made from and by normal cells that migrate from elsewhere in the body into the region of the tumor. These normal cells from the patient's body behave much as they would in response to a wound, as described in Chapter 4 and shown in Figures 4.15 and 4.16. The cells that line nearby blood vessels migrate into the region of the tumor and form new blood vessels. When connective tissue cells join the tumor, they make and secrete the same fibers that lie beneath and between normal body cells. In some cases, tumors develop a large amount of fibrous tissue. The cells, fibers, and blood vessels of stroma make the tumor somewhat like a piece of regular tissue. Indeed, stroma helps to create an environment that can support tumor cell growth. Because this environment may differ from what is going on outside the tumor, it is called the tumor **microenvironment**, which can be either stimulatory or inhibitory for tumor growth,

The structure and behavior of a tumor are the result of conversations between the cancer cells themselves, which doctors call the **parenchymal cells** of the tumor, and non-cancerous cells that have migrated into the tumor to become tumor stroma, surrounding and pervading the parenchyma and helping to create the microenvironment (Figure 8.4). Thus, when a tumor is extirpated, blood vessels must be cut, and normal connective tissue cells are removed along with the tumor parenchyma (the cancer cells). Neighboring tissues may become damaged as the tumor is removed, leading to wounds that the patient's body must repair after the operation. Sometimes this results in a prolonged period of postoperative recovery. When a cancer can truly be removed, however, the discomfort and inconvenience of the operation are almost certainly worthwhile (SIDEBAR 8.2).

Surgery can also be used to reduce a tumor's size or to remove a tissue in which tumor formation is likely.

If a tumor is very large at the time of diagnosis, the cells in its interior may be so far from a good blood supply that they are now dead or dormant and are not likely to respond to treatments that act on growing and dividing cells. In this case, surgery to reduce tumor mass by cutting out a large fraction of the growth may be a good treatment option. For this type of surgery, there is no effort to remove all the tumor parenchyma but simply to reduce the **tumor burden**, that is, the number of tumor cells in the

FIGURE 8.4 **A tumor is composed of cancer cells (parenchyma) and normal cells that make up the tumor's stroma.** Cancer cells (pink with red nuclei) make up the parenchyma of the tumor, but normal cells from the patient's body gather as they might in healing a wound. The normal cells interact with the parenchymal cells and form the tumor's stroma. Endothelial cells that line blood vessels migrate into the tumor and form new blood vessels that will provide nourishment for the parenchyma. Fibroblasts migrate into the area and make new extracellular matrix. White blood cells, such as lymphocytes and macrophages, may enter the tumor and help to form the tumor's microenvironment. (Adapted from Xu X, Farach-Carson MC & Jia X [2014] Biotechnol Adv 32:1256–1268.)

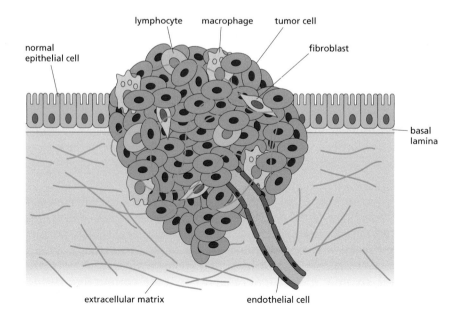

CASE STUDY: Scott's cancer surgery

After Mary's three-year-old son, Scott, had received the diagnosis of a Wilms tumor, based on a CT scan, the kidney specialist told the family what was coming. As Mary described it, "Scott would go into surgery the next day, and it would be a long operation. The problem was that the tumor had already gotten big, so he would be on the table for hours. In fact, Scott's surgery took about 5 hours. The good news was that the tumor was still encapsulated [meaning that all the tumor tissue was in one place, encased in a layer of connective tissue], so the chances of metastasis were small. The bad news was that the tumor was huge relative to the body of the child. They had to move a lot of his organs to get out both the tumor and the affected kidney. Then they tried to put everything else back in the right place. While they were at it, they also put a catheter into one of his abdominal veins [a catheter is a tube that connects an internal compartment with the outside of the body], so Scott could have chemotherapy without having to get injections.

Scott had problems from the surgery, because the surgeons had to move so many organs around. The result was some scar tissue on his intestines, and after the surgery, his bowels didn't get working properly. About a week after the first surgery, they did another operation to help clean things up. This seemed to work, but in the end he had to have even one more surgery before his bowels really got going, letting him digest food properly. So Scott spent quite a long time in the hospital. Fortunately, they were really nice about it, and I was able to sleep in the same room with him.

Scott was really good about taking all this in stride, but there was one time while he was still in the hospital when he began getting very frustrated with all the treatments, and he started hitting me. Just then, a great big male nurse happened to come into the room and saw what was going on. He was a kind of gentle giant, so the kids really respected him. He took Scott's face in his great big hands and he said, right up close, 'Scott, this is not your mom's fault.' Scott really seemed to take that in, even though he wasn't even four years old. After that, he was easier with me, though he had to put up with quite a bit of morphine to help with the pain from his multiple operations."

body, to the point that the patient will be more comfortable and other treatments may be more effective.

Another use of surgery in cancer medicine is to reduce the likelihood that a cancer will form. For individuals in families with a history of cancer, such as women carrying a mutated allele of the *BRCA1* gene (for a reminder about *BRCA1*, see Table 7.1 and associated text), surgery can be used to remove breast tissue when the patient is in early middle age, before any tumors have been identified. This kind of operation is called a **prophylactic mastectomy**. The logic behind such action is based on the likelihood that a tumor would probably form in the coming years if the tissue were left in place. Some women with a family history of breast cancer will choose this route but will postpone surgery until after they have borne children and nursed them. In this case they are taking advantage of the fact that cancer risk increases with age, which allows them the experience of motherhood, followed by an action that may protect them from the difficulties of breast cancer. The same strategy can be applied in the prevention of ovarian cancer, which is also more likely in women with a mutant *BRCA* gene. Now the operation requires removing a woman's entire reproductive system in an operation called an ovarian **hysterectomy**. In the case of a genetic tendency toward testicular cancer, an analogous operation to remove a man's testes is uncommon, because it eliminates sexual drive as well as fertility. Preemptive surgery for people with a high likelihood of colon cancer is even more problematic, because one needs intestinal function to digest food. The removal of even a part of the intestine can lead to postoperative complications, and benign removal of the entire lower intestine is impossible. For this reason, individuals whose family history of colon cancer indicates a high risk of colon cancer will frequently be screened by colonoscopy, and everyone older than 50 should receive such screening occasionally.

Sometimes, metastasis has already occurred by the time of surgery.

Figure 8.3 makes clear the large amount of cancer cell growth that has gone on before a tumor is likely to attract medical attention. This is particularly true for tumors that form in organs of the abdominal viscera, such as the pancreas, liver, or ovaries. Doctors are therefore on the lookout for indications of metastatic behavior from the time of initial diagnosis. As previously described, the presence of tumor cells in nearby lymph nodes, as determined at the time of cancer diagnosis, is one important indication of metastatic activity. If cancer cells were seen in nearby lymph nodes, the surgeon must decide how much tissue to remove from around the tumor. Sometimes, cancerous cells have migrated out from the site of first growth, the so-called **primary tumor**, and initiated small tumors nearby. Such tumors are often called **local metastases**, although this is something of a misnomer, since metastatic growths are, by definition, distant growths. Such clusters of cancer cells are often too small for the surgeon to spot, so it is all too easy for local metastases to be left behind during tumor extirpation. On the other hand, removing a significant amount of normal tissue, in an effort to get both the primary tumor and any local metastases that might be present, will cause greater damage to the patient. Thus, a balance is required. One way in which cancer surgeons are trying to improve their effectiveness is by using stains that are tumor cell specific to make the cancer cells more visible. With this proce-

---

SIDEBAR 8.3

CASE STUDY: Woods' cancer surgery

Woods described his experiences with surgery as follows: "In August of 2012 I had my first surgery. They had the information from the endoscopy, so my urologist anesthetized me and went to work to try to clean out the cancer. He used a method that again went up the urethra [the tube that empties the bladder through the penis], so they didn't have to cut me open; it's called transurethral resection of a bladder tumor, or TURBT. It was slow work, and he could only keep me under for a given amount of time, so after a while he had to quit and leave about 10% behind. All in all, though, that operation was a success. In October, they went back in and did the operation again. They had hoped to get the bladder really clear that time, but the 10% left behind had grown up, so there was a lot of tumor to deal with. Again they couldn't get

it all, and the fact it had grown back so fast made it clear that this cancer was not good news.

The surgeon told me that I should really have a third surgery, and this time, they wouldn't just try to clean me out, they'd take out the whole bladder and prostate gland at the same time. When you don't have a bladder, they can fix you up so you have a bag to store urine. It's a nuisance, but it works pretty well. Losing my prostate would mean that my sex life was over, but in fact, over the last few years I had been pretty celibate, so that wasn't going to be such a big change. Nonetheless, all that news threw me for a bit of a loop, and I said I needed to think about it. I meditated on it, and I also went to see my general practitioner. She was not only my doctor, she was a real source of support and caring. I talked it over with her,

and she said this really is the path you need to take if you want to stay alive. So I decided to go ahead with it.

This surgery was a bigger deal. I was in the hospital for 10 days, and it took me a while to get over it. Of course the doctor tried to paint the whole situation as positively as he could. I asked him, is this the kind of cancer where I'll be dead in 6 months, and he said no, no, not at all, but he sure wasn't very specific. He knew the news wasn't good, because they did biopsies on the tissues that were removed, and they did find bladder cancer cells in my prostate, which meant it was already metastatic. I didn't look up the stats about the future for my condition, but a month later the man who did my surgery told me that I was now considered 'medically noncurable.' This was just a fact."

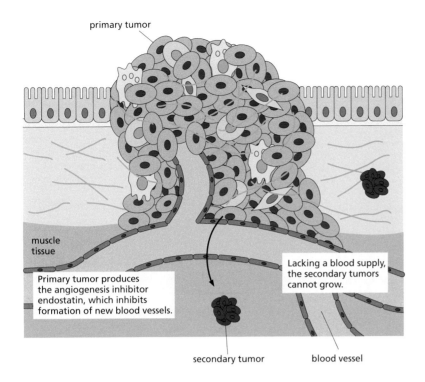

primary tumor

muscle
tissue

Primary tumor produces
the angiogenesis inhibitor
endostatin, which inhibits
formation of new blood vessels.

Lacking a blood supply,
the secondary tumors
cannot grow.

secondary tumor   blood vessel

FIGURE 8.5 Big tumors stimulate the growth of blood vessels into their parenchyma and inhibit the formation of vessels into small tumors nearby. Proteins like endostatin block the growth of blood vessels; they are called angiogenesis inhibitors. Angiogenesis inhibitors are made and secreted by some tumors, which prevents nearby tumors from getting the blood supply they need to grow well. The removal of a big tumor can relieve this inhibition and allow smaller tumors nearby to become vascularized and grow.

dure, even small, local metastases can often be seen and removed without taking any more normal tissue than necessary.

Skillful surgeons are also using a range of clever methods to reduce the trauma to the patient that is associated with anesthetics and incisions. For example, Woods had a constructive experience with cancer surgery, defined by the fact that his tumor was primarily in his bladder (SIDEBAR 8.3).

Many big tumors have the remarkable ability to secrete proteins called **angiogenesis inhibitors** (for example, the protein endostatin) that interfere with the growth of blood vessels toward smaller tumors that lie around them (FIGURE 8.5). In a strange way, this inhibition makes sense. The big tumor would probably not have become so large if the nourishment it needed was shared with neighboring tumors. The unfortunate result of this situation, though, is that the surgical removal of a big tumor means that the growth of blood vessels to small, nearby tumors is no longer inhibited. Now the small tumors, which the surgeon couldn't see, stimulate angiogenesis and acquire a good supply of blood vessels. As their level of nourishment improves, their speed of growth may increase. Consequently, an apparently successful surgery may provide only temporary control of a cancer that had already spread locally. This situation suggests that cancer should often be treated by more than just surgery. Fortunately, additional treatments are available.

RADIATION TREATMENT

X-rays can be used to kill cancer cells, doing a job analogous to surgery but without the patient having to be cut open and subjected to the resultant trauma. You will remember, however, that we have previously met X-rays as tools for cancer diagnosis (Chapter 2) and as risk factors for carcinogenic mutations (Chapter 5). The use of X-rays to *treat* cancer, therefore, needs explanation. The difference between X-rays for diagnosis versus those for therapy is a matter of dose. A **radiation oncologist** knows what X-ray dose is needed to form an image and how much will induce sufficient cell damage to promote tumor shrinkage.

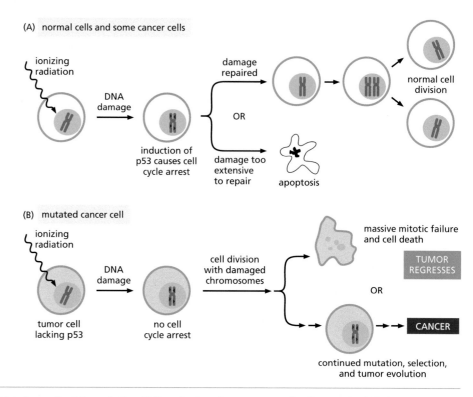

FIGURE 8.6 Cellular responses to high doses of ionizing radiation. (A) Normal cells and some cancer cells will respond to DNA damage by initiating DNA repair and up-regulating p53, which arrests cell cycle progression (as described in Chapter 7). A stalled cell cycle allows time for DNA repair, but if that fails, p53 will induce apoptotic cell death. (B) If p53 activity has been lost through mutation, cells can still be killed by radiation, due to extensive damage to DNA and other macromolecules, but it is also possible that the mutations induced in some irradiated cancer cells will enhance the cell's genetic instability and cancerous progression. (Adapted from Alberts B, Johnson AD, Lewis J et al [2015] Molecular Biology of the Cell, 6th ed. Garland Science.)

When many X-rays pass through a given piece of tissue, all kinds of macromolecules become damaged. When this damage is sufficiently severe, the irradiated cells, including both normal and cancer cells, will die. In times past, doctors thought that tumor cell death was simply because high levels of radiation killed the cells, but we now know that radiation-induced cell death is commonly a result of the fact that DNA damage can activate the apoptotic pathway. Recall that p53, a transcription factor discussed in Chapter 7, is an important part of the pathway that links radiation damage to apoptosis. When p53 is still active in the parenchymal cells of a tumor, X-rays or other kinds of radiation can readily induce significant amounts of cell death (FIGURE 8.6). Recognition of this pathway has allowed radiation oncologists to greatly reduce the X-ray dose they use in cancer treatment. The irradiation of tumors used to be so extensive that patients experienced burns. Now, much lower doses are given, with treatments repeated every few days over a longer period of time. This regimen yields the same or greater damage to cancer cells but with less harm to the patient. If, however, p53 activity has been lost through mutation, it takes more radiation to kill the cancer cells, and the greater dose is likely to mean more damage to the body's normal cells. The mutagenic action of this radiation may lead to trouble for those normal cells in the future.

Properly administered radiation can cause tumors to shrink significantly.

Radiation treatment has the advantage that it can kill the cells in tiny tumors that might have escaped notice during surgery. It can also be aimed

CASE STUDY: Woods' follow-up treatment

Woods, the patient with bladder cancer in Sidebar 8.3, gave this account of the treatment he received as follow-up to a radical cystectomy: "After I learned that I was not curable, I did go to the Web and look up the stats on my condition. I used information from both this country and Great Britain, and the sites agreed. Since I had stage 4 bladder cancer, the chances were that I had 9–12 months to live. But in fact, it had already been two years by the time I read that, so I figured I was doing pretty well. Maybe I'm just too mean to die! As I have been going along, though, there have been problems. A while ago I developed a blood clot in my leg. I had a tumor growing in my groin that was inhibiting circulation, and probably that led to the clot. I didn't go to the emergency room, because I had an appointment with my regular doctor in just a few days. She looked at things and put me into the hospital immediately. Fortunately, I had brought my laptop and its electri-

cal hookup, so I was able to do stuff and amuse myself. That really made a difference.

First they did a bunch of things to clear up the clot. I was on two drugs for a while, and I'm still on one of them to reduce the chances of its happening again. Then they decided it was time for me to go for radiation treatment to help keep those leg tumors as small as possible. That should help reduce the chances of another clot. The radiation treatment they gave me was pretty extensive. I had a treatment every day, five days per week, for four weeks, but it wasn't bad at all. It didn't take that long, and all you had to do was lie on a table; it's like they are taking a picture of you. It was the everyday aspect of it that was tough. Travel was not so easy for me, so that took a big effort, but the fact is that after the treatment was done, I really felt better. I had a whole lot more energy, and I was able to get a bunch of things done that really needed doing."

at likely sites of metastasis to kill tumor cells that might be there too. Thus, radiation can be an effective treatment for cancer in its own right. It can also be an important companion to surgery, used either before an operation to shrink the tumor or after an operation to damage any tumor cells that surgery missed. When a second form of treatment accompanies the primary treatment, doctors call it an **adjuvant therapy**. As an example of adjuvant therapy, Woods had radiation after his surgery to slow the growth of his several metastatic growths (SIDEBAR 8.4).

A localized tumor can be treated effectively with several kinds of radiation.

One good strategy for maximizing damage to a localized tumor, while minimizing damage to other body cells, is to focus the radiation and bring the irradiating beam in from several angles, so the cancer cells are much more affected than the healthy cells around them. This idea can be illustrated with a tumor that has formed in an internal organ, for example, the pancreas. First, the doctor localizes the tumor with all possible accuracy, using a CT scan or MRI. This is often done with a device to hold the relevant part of the patient's body in exactly the same position, whenever that is wanted. The radiation oncologist then uses a special radiation source to produce a narrow beam of radiation and aims it exactly at the tumor. This will, of course, hit all the cells along the line of the beam, those in the tumor and the normal cells situated both before and after the tumor. However, the orientation of the beam can be changed, again and again. If every beam is aimed at the tumor, its cells are blasted many times, but if each successive beam passes through a different part of the rest of the body, the dose to the normal cells is only a fraction of the dose to the tumor (FIGURE 8.7). When this procedure is done right, the normal cells are likely to be able to repair any DNA damage they suffer, whereas the tumor cells cannot.

Similar procedures can also be carried out with beams of gamma rays emitted by radioactive cobalt, in a device known as a Gamma Knife®

(A)

(B)

(C)

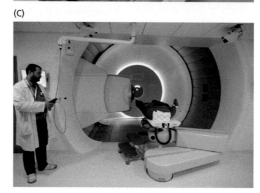

FIGURE 8.7 Instruments to irradiate tumors with focused beams of radiation from multiple angles. Properly aimed beams of ionizing radiation can deliver a lethal dose to the cells of a tumor without doing as much damage to the surrounding, normal cells. (A) A Gamma Knife® is a source of gamma rays that can be rotated around the patient to come at the tumor from multiple angles. (B) A CyberKnife® uses a linear electron accelerator to generate beams of high-energy X-rays. Again, by aiming the beam from multiple angles, the radiation oncologist can minimizing damage to surrounding tissues. (C) A device that generates a beam of protons for the same purpose. (A, courtesy of Marmama University, Turkey; B, courtesy of Accuray Incorporated; C, from Petr David Josek/Associated Press.)

SIDE QUESTION 8.3

You are the radiation oncologist responsible for irradiating a cancer that has been imaged by magnetic resonance imaging (MRI) in a 67-year-old man's prostate gland. How are you going to do it? (1) Do you think X-rays would be appropriate, given that the prostate is near the testes, and it may be hard not to irradiate those organs as well as the prostate? (2) You have lead shielding that you can use to protect the parts of your patient's body that you don't want to irradiate. How would you distribute that shielding to achieve the best effect?

(Figure 8.7A). Recall that some forms of naturally occurring atoms contain unstable atomic nuclei that spontaneously rearrange and emit some kind of radioactivity. (For a review of radioactivity, see Chapter 5.) The nucleus of one isotope of cobalt (cobalt-60, also called ^{60}Co) rearranges and emits both a beta ray (a high-energy electron) and a gamma ray (a high-energy X-ray). The latter is so energetic that it can penetrate bone and pass through a significant thickness of soft tissue. This kind of radiation has been found particularly successful for treating inoperable tumors localized to one or a few positions in a patient's brain. Medical engineers have also invented a device called a CyberKnife® that uses beams of high-energy X-rays for analogous irradiation of tumors anywhere in the body (Figure 8.7B). Yet another radiation treatment uses a beam of protons, which are even more damaging to cells than X-rays and gamma rays. When the tumor is localized near the surface of the skin, this kind of radiation can deliver a damaging treatment to the tumor with only modest harm to the cells between the tumor and the radiation source. The cells beyond the tumor are essentially unaffected, because the radiation is absorbed by the tumor before it reaches them.

The benefits available from these kinds of treatments are a result of significant engineering that has gone into both the sources of radiation (for example, forming a narrow beam that can be aimed with accuracy) and the physical structure that rotates the radiation source (making sure that the radiation is always aimed where treatment is needed). In some circumstances, these kinds of radiation offer an excellent treatment option. A significant downside of these methods, however, is their cost. High-technology radiation therapy is a good example of an improvement in cancer treatment that has been achieved through modern science and engineering, but it also serves as a poignant example of why health care costs keep rising.

Another way to irradiate tumor cells is through carefully placed radioactive materials.

As seen with the Gamma Knife®, ionizing radiation from radioactive decay can serve the same apoptosis-inducing role as X-rays. A successful example of this approach is for treatment of tumors in the thyroid gland. This organ produces thyroid hormone, which is the only molecule in the body that includes the element iodine as one of its atoms. If a doctor administers radioactive iodine by mouth to a person with thyroid cancer, the element is taken up by the thyroid gland and incorporated into precursors of thyroid hormone, which are made and stored in the thyroid gland. The rest of the radioactive iodine is eliminated from the body in urine. Now thyroid hormone-producing cells have brought in radioactive atoms whose emissions are quite sharply localized to the tissue where the tumor resides. This procedure is called **radioiodine therapy**.

The radioactive iodine used for tumors in the thyroid gland is ^{131}I, an isotope whose unstable nuclei emit beta rays. Each beta ray is sufficiently energetic to cause double-strand breaks in DNA (see Chapter 5), so cells situated close to the incorporated iodine will experience significant radiation damage. The beta rays interact so strongly with water and other molecules in cells that they do not travel very far from their site of origin. As a result, the cellular damage done by this radiation is quite localized. Moreover, the radioactivity of ^{131}I is relatively short-lived; half of all the radioactive atoms incorporated into the thyroid will have emitted their radiation in just 8 days. After 16 days, only a quarter of the original radioactivity is left. After a month, only about 7% of the iodine will still be emitting radiation, so the treatment is self-terminating within a reasonable period of time. Most of the radiation experienced by the patient will have hit thyroid cells and will thus have irradiated the cancer. However, experience with this treatment makes it clear that the beta rays from ^{131}I are sufficiently energetic that they can travel some distance out of the thyroid gland and irradiate nearby tissues as well; there they can cause mutations in noncancerous cells. Thus, the treatment is not perfect, but it is a good treatment option and is commonly used for treatment of thyroid cancer.

Interestingly, there is an effort among some clinically oriented cancer scientists to harness the killing power of radiation from ^{131}I to antibodies that react with surface proteins of other kinds of cancer cells. In principle, this approach should allow radioiodine treatment of any cancer cell, not just the ones that make thyroid hormone. This is an example of efforts to bring new technologies to bear in cancer treatment, which is the subject of Chapter 10.

Radioactivity has also been used successfully to treat tumors that are situated in places where surgical removal of the tumor is difficult. Tumors of the prostate gland are now quite commonly treated by **brachytherapy**, the surgical implantation of a small pellet of a radioactive element, like the ^{131}I used for thyroid tumors. This or other radioactive elements can provide a moderate level of radiation for weeks, an effective treatment for slow-growing tumors. An alternative form of brachytherapy uses a very strong emitter of radioactivity, such as iridium-192 (^{192}Ir). This element emits both beta rays and gamma rays, so it is very destructive to the cells near the radiation source. A highly radioactive sample of this isotope can be inserted into the tumor on a needle for a matter of minutes. The treatment is then repeated every day for a few days. This kind of

SIDE QUESTION 8.4

There is a radioactive form of sulfur that emits a beta ray quite like the one emitted by ^{131}I. Most proteins have sulfur in them, so giving radioactive sulfur to a patient would get cell-killing radiation right into the proteins of all tumor cells. Do you think this would be a good way to administer radiation to a tumor while leaving the rest of the body unharmed? Why or why not?

SIDE QUESTION 8.5

In Side Question 8.3, you were asked to consider the best way to use X-rays for treating prostate cancer. Now that you know about brachytherapy, you have an alternative treatment to consider. Think again about the treatment of that 67-year-old man diagnosed with prostate cancer and decide whether you would prefer using implanted radioactivity or X-rays. What are the reasons for your choice?

high-dose brachytherapy is useful for treating fast-growing tumors, where rapid action is called for.

Some tumors are not well localized, so treating them with radiation requires special procedures.

In blood cancers, like leukemia and lymphoma, the rogue cells are not well localized. In these cases the treatment of choice is sometimes radiation of the blood-forming cells, the stem cells that are giving rise to the cancerous leukocytes. Blood-forming stem cells are localized mostly in the marrow of a few large bones. The trouble with such treatment is that the radiation dose required is high enough to kill not only the cancer cells but also the stem cells that produce normal blood cells. For the patient to survive, it is necessary to transplant bone marrow from a nonirradiated donor into the irradiated patient. One way to do this is to take bone marrow from the patient prior to irradiation, and then use those cells to re-introduce nonirradiated blood-forming stem cells after irradiation (FIGURE 8.8). The virtue of this approach, which is called an **autologous transplant**, is that there is no danger that the injected cells will be identified as nonself by the patient's immune system and rejected, which means killed by the patient's own lymphocytes, as described in Chapter 9. The limitation of the approach is that the injected material may contain cancer cells that have not been irradiated and may, therefore, reestablish the cancer following the killing by radiation of all the cancer cells that were left in the bone marrow.

An alternative approach is to find a generous person who is willing to donate some bone marrow to the irradiated patient, thereby helping to reestablish a healthy blood-forming system after bone marrow irradiation. This approach is called an **allogeneic transplant** (allogeneic means transported to its current position from somewhere else). In an allogeneic transplant, the reestablishment of cancer is much less likely, but unless the donor has an immune system that is accurately matched to that of the recipient, the stem cells in the transplanted marrow will be killed by the recipient's immune system. (There is more information about antibodies and tissue rejection in Chapter 9, which covers the immune system.)

Many transplants of both kinds, autologous and allogeneic, are now done all over the world, but they require both a skilled medical team and the right donor, commonly a close relative, whose immune system is sufficiently similar to that of the patient that the injected bone marrow cells are not rejected by the recipient. Moreover, bone marrow transplants are expensive: $50,000–100,000 for an autologous transplant and $100,000–200,000 for an allogeneic transplant. The high cost puts this treatment out of reach for many.

SIDE QUESTION 8.6

Your 14-year-old brother has been diagnosed with leukemia. The doctors are recommending irradiation of his bone marrow and a bone marrow transplant. You know that the options involve either an autologous transplant or an allogeneic transplant of bone marrow. For the latter option, you are a prime candidate to be the donor. What factors would you discuss with your brother in helping to decide the best treatment for his condition?

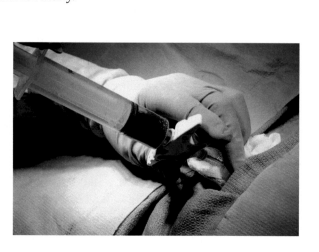

FIGURE 8.8 Preparation for irradiating bone marrow to kill cancerous blood-forming cells. Cells from bone marrow, including stem cells that can divide and differentiate into blood cells (see Figure 4.15), are withdrawn from the hip bone, or some other major bone, using a rather large needle. The large needle size ensures that the viscous marrow can be drawn into the syringe without damage to the cells it contains. This material is stored in a sterile bag and refrigerated. After the cancer patient's bone marrow has been irradiated with a dose of radiation sufficient to kill the cancerous blood cells, the previously harvested bone marrow is reintroduced into the patient by simple intravenous injection. Ideally, the stem cells in the transplanted material enter the patient's bone marrow and resume their function of making new, healthy blood cells. (From Photographer's Mate 2nd Class Chad McNeeley, Navy News Service, 021204-N-0696M-180.)

CHEMOTHERAPY

If a tumor is widespread, either as a result of metastasis or because this cancer developed from a tissue like blood, which doesn't form a cell mass, then surgery is likely to be useless, and even radiation may not be a good treatment choice. In these cases, one wants a therapy that can circulate through the whole body and influence cancer cells, wherever they may be. Certain drugs fulfill this need, so they are commonly used to treat metastatic cancer. They are also commonly used as adjuvant therapies in conjunction with surgery and/or radiation, addressing the possibility that metastasis has already occurred, even though it has not been detected. The good news about the use of medicinal drugs is that virtually every cancer cell is exposed to them. The bad news is that most of the normal cells in the body get the same treatment; the ones that are naturally growing and dividing can be badly affected by it. In an ideal world, cancer scholars would have found chemicals that are toxic only to cancer cells and are benign for all normal cells, but this goal has so far been achieved with only limited success. Most cancer drug treatments, or **chemotherapies**, have side effects as a result of their action on the body's normal cells.

Many of the chemicals used to treat cancer have been chosen because they are more damaging to cells that are growing and dividing than to other cells.

Some cancer chemotherapies interact with DNA. In some cases they damage DNA by forming adducts (chemical complexes with nucleotides); in other cases they either block DNA synthesis or introduce enough errors during DNA replication to promote apoptosis. Thus, although radiation and chemotherapy appear very different, they have one thing in common. Both work best when the tumor cells still have an effective system for programmed cell death.

Two examples of drugs that are commonly used for DNA-directed chemotherapy are **cisplatin** and **carboplatin**. Both contain a platinum atom bound to other atoms that help the platinum to get into a cell and react with its essential internal components (FIGURE 8.9). These drugs are quite reactive, so they damage proteins, RNA, and DNA. The latter damage is the most important, because these reactions can cross-link the two strands of a DNA duplex and/or form adducts with the nucleotides. As a result, they disrupt accurate DNA replication; thereafter, apoptosis is the cell's most likely fate. Treatments with these kinds of chemicals are more damaging to dividing cells because DNA is most sensitive to chemical damage when it is replicating.

Actinomycin D is another commonly used chemotherapeutic agent. This chemical, which is produced by some bacteria, binds DNA and blocks the transcription of active genes into RNA. Drugs like carboplatin and actinomycin D can certainly kill many cancer cells, but they kill other dividing cells as well. They cannot discriminate between DNA in normal versus cancerous cells, hence their side effects. In SIDEBAR 8.5, Mary describes the chemotherapy given to her son Scott, including actinomycin D.

There are numerous drugs that affect DNA replication. Some resemble the nucleotides from which DNA is made. Because they are similar to normal nucleotides, DNA polymerase will incorporate them into DNA during replication. However, their shapes are not quite right, so the resulting new strands of DNA cannot base-pair properly, and subsequent rounds of DNA synthesis continually introduce mutations. An example is the drug **fluorouracil**, abbreviated **FU**. This chemical strongly resembles the base T, but

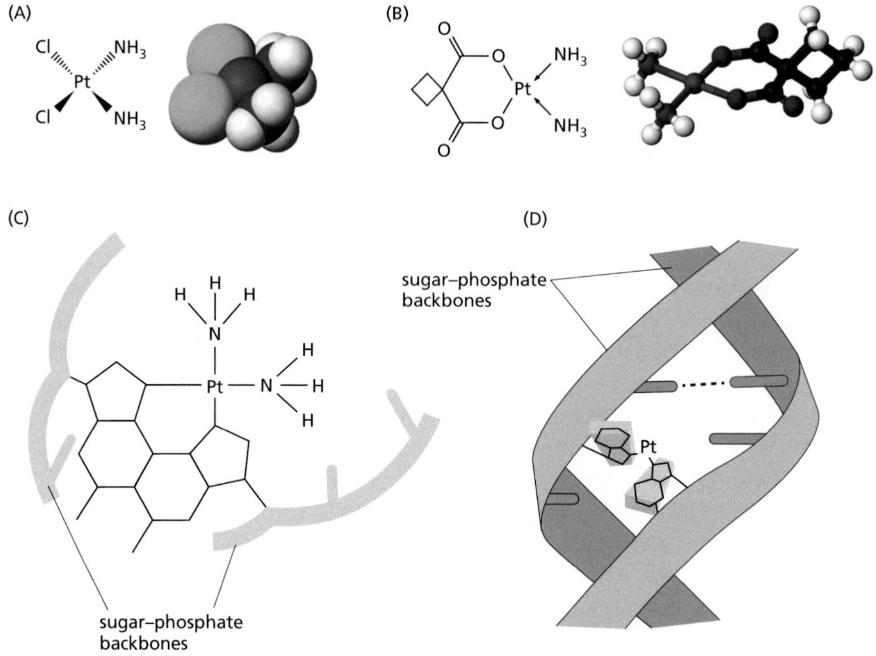

FIGURE 8.9 Two forms of platinum (Pt) used in chemotherapy and their reactions with DNA. (A) Two representations of cisplatin; Pt and the blue sphere in the center of the model (right) stand for an atom of platinum, Cl (green) indicates chlorine, N's and the blue spheres farther to the right are nitrogens; H's (silver) are hydrogens. The triangular lines imply bonds that go to atoms lying above or below the plane of the diagram. (B) The structure of carboplatin; O's and red spheres on the right stand for oxygens. Each line represents a chemical bond, and the unnamed vertices between lines are sites for carbon atoms.

the fluoro part of its name means it includes a fluorine atom, which makes it different from T. As a result, future DNA replication is inaccurate and mutations occur, again leading to apoptosis. Other drugs interfere with DNA synthesis by blocking the synthesis of one or more of the small molecules required to make new DNA. Methotrexate is one example; it blocks or retards DNA synthesis, causing the DNA synthesis checkpoints to stall

CASE STUDY: Scott's chemotherapy

Scott is the boy whose kidney tumor was first diagnosed by sonography and then by a CT scan. The tumor was large, but its removal by conventional surgery was successful. His mother Mary described the situation after surgery like this: "After Scott's surgeries, the doctors put him on two kinds of chemotherapy. The first was Actinomycin D. We were at Children's Hospital for the first round of chemo, but after that, he stayed home for most of his treatments. The doctor thought it might be a good thing for Scott to take his chemo at home, and it sure was a lot easier for us all. Someone came up to the house to show me how to do it, which was actually pretty easy. [Mary worked in a biology lab, so she was familiar with this kind of technical work.] A pharmacist delivered the chemo every day we needed it. We had already been trained in the hospital on how to take care of the Scott's catheter, so Scott's

sister Emily and I did that at home. There were two different cleaning solutions, to make sure that it didn't get infected: the 'whitey' was alcohol and the 'brownie' was a disinfectant. Emily was great about it. The actinomycin made Scott feel really sick, but that feeling passed. Afterwards, he had vincristine [an inhibitor of cell division], which gave him sore joints, and he lost his hair. The chemo lasted from April to October.

Scott also went to the Hope Clinic, a place for cancer survivors to get additional care. In this clinic there was a dentist, a psychologist, and a bone doctor, someone to look at just about everything and see how the kids did after chemo. This was important, because those early years are times when many parts of the body are still developing. Fortunately, it looks as if Scott is really OK."

cells in their growth and division cycle. At a minimum, blocked cells are no longer increasing in number; in the best case, they initiate apoptosis and die.

There are also several drugs that interfere with the formation and function of the mitotic spindle, the cell's machinery for segregating chromosomes (see Chapter 4). **Vincristine** and **vinblastine** come from periwinkles, flowering plants that are often used to make colorful flower beds in shady places. These drugs bind tubulin, the protein that makes up microtubules. Once tubulin is in a complex with these drugs, it cannot polymerize to make microtubules, so the mitotic spindle cannot form. Now cells fail to divide, and again they often initiate apoptosis, although why they do so is still a matter of study in labs that do basic cancer research. Also under study is why vincristine and its relatives are more active against leukemias than most other kinds of cancer.

An even more effective group of chemotherapeutic drugs for blocking cell division is the **taxanes**. These compounds, originally identified and isolated from bark of the Western yew tree, can now be synthesized in chemistry labs, yielding drugs that bind to microtubules and make them unusually stable. Under these conditions, mitosis again fails (FIGURE 8.10), bringing on cell death. As before, dividing cells are killed preferentially. An unexpected value of the taxanes is that their ability to kill cells does not seem to depend on functional p53, although the reason for this action is not yet known.

SIDE QUESTION 8.7

Why do chemotherapies focus so much on blocking DNA synthesis? Why not block either RNA or protein synthesis, since these processes are also required for cells to live?

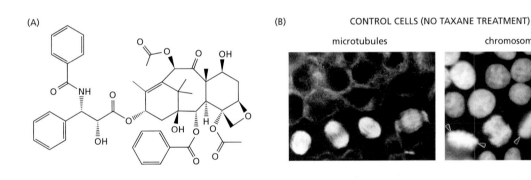

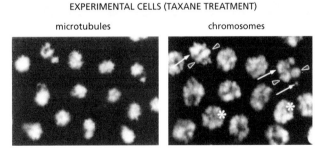

FIGURE 8.10 Structure and action of taxanes. (A) The chemical structure of paclitaxel, one of the taxanes that is widely used for cancer chemotherapy. Although this molecule is big, it is sufficiently hydrophobic that it will cross the plasma membrane and enter living cells. Once there, it binds tightly to microtubules, making them unnaturally stable. (B) Upper images show cultured cells that serve as controls; they have not been treated with paclitaxel. On the left, the cells are stained to show the microtubules of the mitotic spindle. On the right, the same cells are stained to show the DNA. By comparing the pictures, you can match up the cells and see both microtubules and DNA in the mitotic cells. Lower images show similar cells in a culture treated with paclitaxel. No mitotic spindles form, but all the cells contain small, starlike arrays of microtubules (left). On the right, the same cells are stained to show the DNA. Two structural features of these cells are abnormal: (1) all the cells have condensed chromosomes, showing that they are stuck in mitosis, and (2) none of the cells shows a proper mitotic arrangement of chromosomes. Instead, brightly stained DNA is clustered around a central point from which microtubules are growing. Some cells show an approximation of a metaphase plate (open arrowheads), but in others, the chromosomes are simply clustered (arrows). (B, images from Jordan MA, Wendell K, Gardiner S et al [1996] *Cancer Res* 56:816–825 With permission from American Association for Cancer Research.)

The most effective way to use chemotherapies is a research subject in itself.

All these treatments sound quite straightforward when described simply, as in the previous sections, but there is a significant art to delivering cancer chemotherapies so they will do maximal damage to cancer cells and minimal harm to the patient's normal cells. For a few cancers, such as acute lymphocytic leukemia (ALL) in children, doctors have identified such effective treatment regimens that a true recovery is quite common: something like 80% of patients overall. When originally devised, the treatments used were very harsh, following the idea that cancer therapies will be the most successful if they don't allow the rogue cells time to develop insensitivities to the drugs being used. More recently, this approach is being modified to milder treatments in an effort to find regimens that are still effective but not so uncomfortable for the patient. This kind of progress represents a significant success in cancer treatment, but it is counterbalanced by the more limited successes of chemotherapies in other kinds of cancer, such as lung and kidney cancers. In the latter cancers, chemotherapies usually retard progression of the disease, and under fortunate circumstances they can achieve remission, but it is uncommon for them to effect a cure.

Research that determines the effective concentration of a given drug has commonly been carried out on experimental animals before doctors measure the drug's effects on people. A group of experimental animals, for example mice, will be given the drug at a range of doses to determine both the dose that is lethal and the minimal dose that has a beneficial effect. Because there is natural variation among animals (as among people), a meaningful measurement of these doses requires experimentation on multiple animals, say 100 in each group. Such experiments are not trivial, and they are undertaken only when there is a good indication that there will be a medical benefit from the results. From such experiments, the scientist will thereby learn the minimal dose that is effective on 50% of all animals tested (ED_{50}) and the dose that is lethal to 50% of all animals treated (LD_{50}). The ratio of lethal dose to effective dose, LD_{50}/ED_{50}, is called the **therapeutic index** of the drug. Such information is vitally important for clinicians who are treating people, but humans and mice often respond quite differently to a given drug. It is therefore necessary to repeat such experiments in patients. This is an example of clinical research. In such research, one can use information gleaned from experiments with mice or other animals, scaled up by the ratio of human-to-animal body weights, to get close to the right doses for people quite quickly. Rather than identifying a dose that is lethal to people, doctors usually measure the highest dose that a patient can tolerate without getting sicker. This is called the maximal tolerated dose or the toxic dose (TD). Thus, research on human patients determines the therapeutic index as the ratio of toxic dose to effective dose, TD_{50}/ED_{50} (FIGURE 8.11). The higher this ratio, the more likely a drug is to be effective without causing distress to patients. With a well-established and often-used chemotherapeutic agent, doctors know the conditions that are likely to be effective but to cause minimal stress for the patient (SIDEBAR 8.6).

Effective chemotherapy is based on two core principles.

The first widely held principle of chemotherapy is to use more than one drug at a time. For example, carboplatin is administered to interfere with DNA replication and simultaneously a taxane is given to interfere with cell division. This kind of double whammy is almost invariably more effective than a higher dose of either chemical alone, both in killing tumor

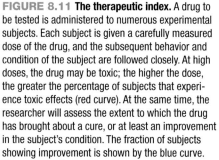
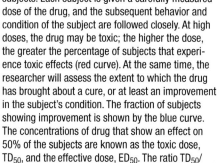

FIGURE 8.11 The therapeutic index. A drug to be tested is administered to numerous experimental subjects. Each subject is given a carefully measured dose of the drug, and the subsequent behavior and condition of the subject are followed closely. At high doses, the drug may be toxic; the higher the dose, the greater the percentage of subjects that experience toxic effects (red curve). At the same time, the researcher will assess the extent to which the drug has brought about a cure, or at least an improvement in the subject's condition. The fraction of subjects showing improvement is shown by the blue curve. The concentrations of drug that show an effect on 50% of the subjects are known as the toxic dose, TD_{50}, and the effective dose, ED_{50}. The ratio TD_{50}/ED_{50} is the therapeutic index.

CASE STUDY: Rob's chemotherapy

Once Rob's lung cancer was diagnosed, he was referred to the Columbia University Medical Center, where he met with a team that included a radiation oncologist. These doctors were not very explicit about prognosis, but it was clear that they expected Rob to live for only a few months. To slow the cancer down, however, they administered several bouts of radiation therapy to Rob's head, where the cancer seemed to be well settled into his brain and growing quite fast. The oncologist also prescribed chemo-therapy with a taxane. The radiation began almost immediately, the chemotherapy shortly thereafter. It was administered to him as an outpatient by intravenous injections that lasted a couple of hours. This combination of therapies significantly slowed the growth of his cancer and was probably responsible for the fact that he lived another seven months, giving him the chance to do many things that he really cared about.

cells and in achieving a prolonged effect. A likely explanation for this result is that some cells in a tumor may already have mutated so they are at least partially resistant to one of the drugs. By using two or more different drugs at the same time, a physician increases the chance that most, or maybe even all, of the cells in the tumor will be killed. Some drug combinations are now so commonly used that they are referred to by acronyms, like ABVD, CHOP, FOLFIRINOX, and TIP. Letters in the acronym represent the several drugs used to attack multiple cellular processes at the same time. For a fuller description of these combination therapies, see TABLE 8.1.

The second core principle of effective chemotherapy is to employ cycles of treatment and rest. A set of drugs is administered for a period of days in an effort to poison as many tumor cells as possible. Then drug treatment is stopped, and the patient's body is given a chance to recover. This allows normal cells, like white blood cells, to increase in number, reducing the chances of infection. When this treatment pattern works well, it

Table 8.1 Drugs Used in Combination Therapies

Drug combination name:	ABVD	FOLFIRINOX	CHOP	TIP
Components and their functions:	Doxorubicin, also called **A**driamycin® slips into DNA between the nucleotides and interferes with DNA replication	**Fol**inic acid interferes with synthesis of natural nucleotides; used with fluorouracil	**C**yclophosphamide forms adducts on DNA and blocks faithful DNA replication	Paclitaxel, a form of **T**axol®, a taxane, that stabilizes microtubules and blocks normal mitosis
	Bleomycin, a large molecule isolated from a bacterium, causes double-strand breaks in DNA	**F**luorouracil (FU) forms an analogue of the DNA base thymine (T); when used with folinic acid, FU is incorporated into DNA, fails to pair normally with adenine (A), and thereby introduces mutations	Doxorubicin (**h**ydroxydauno-mycin), with a different name, is described under ABVD	**I**fosfamide reacts with DNA and makes adducts, blocking accurate nucleotide pairing and introducing mutations
	Vinblastine (Velbe®), isolated from a plant, blocks the formation of the mitotic spindle	**Iri**notecan inhibits an enzyme needed for DNA to unwind and thus blocks DNA replication	Vincristine (**O**ncovin®) is similar to vinblastine	Cisplatin (which contains **p**latinum) is similar to oxaliplatin; it interferes with accurate DNA replication
	Dacarbazine forms adducts on DNA and blocks faithful DNA replication	**Ox**aliplatin, a reactive, platinum-containing compound, reacts with many biological molecules, including DNA, probably acting to block its accurate replication	**P**rednisolone, an artificial steroid, mimics natural steroids and helps to block inflammation	
Cancers commonly treated:	Hodgkin's lymphoma	Pancreatic cancer	Non-Hodgkin's lymphoma	Many, including testicular cancer

can reduce the number of tumor cells by huge amounts in a relatively short period of time. For example, the initial treatment might kill 99% of the tumor cells, a very effective first step. However, a tumor that is big enough to attract attention, say an inch or more in diameter, contains as many as 10 billion cells. Even if only 1% of these cells survive the treatment, that means there are still 100 million tumor cells in the patient, far too many to leave untreated. The next treatment may kill another 99% of the tumor cells, bringing their number down to around a million cells. If the treatment continues to work well, multiple repeats can ultimately remove the tumor altogether, but such procedures are tough on the patient, given their toxic effects on nontumor cells.

Chemotherapies are powerful and useful treatments, but they are certainly not ideal.

One problem with the kinds of chemotherapies described in the previous section is that we don't yet really understand how they work. Although each of these treatments was designed to kill dividing cells, many cells in a big tumor are dormant, meaning they have stopped cycling, probably because of inadequate vasculature in the tumor where they live. In spite of this, many such cells are killed by chemotherapies, and at the moment, doctors don't know why.

A second and even more important point is that the side effects of chemotherapies can be very unpleasant. With many of the chemotherapies, side effects seem to be inescapable. You know several examples of tissues in which continued cell growth and division is a normal and essential event, for example, the intestine. When these cell divisions are blocked or impaired by cancer-fighting drugs, the intestine doesn't work properly, leading to nausea, vomiting, and general abdominal discomfort. Likewise, the skin cells responsible for hair growth are part of a tissue that is normally rich in cell divisions. Blocking skin cell cycles leads existing hair to fall out, and it takes a while for the remaining epithelial cells to recover well enough to replace the hair that has been lost. Blocking cell division also leads to a drop in the number of circulating white blood cells, which means the patient is unusually susceptible to infection. It also means that red blood cells are not made at a normal rate, yet old red blood cells are continuously removed from circulation by the spleen. The resulting loss of blood cells promotes anemia (a low red blood cell count), which leads to fatigue and lack of stamina. Because of the widespread occurrence of these kinds of side effects, doctors have sought additional treatments that make them less unpleasant or dangerous. Drugs have been found that greatly reduce nausea. Antibiotics can help prevent infection, and both diet and drugs can treat fatigue and anemia. Hair loss, however, continues to be a problem with some kinds of chemotherapy.

Yet another problem with the cycles of chemotherapy and rest, as described, is that the periods between treatments allow the tumor cells to grow back, at least partially, while the rest of the body is recovering. This means that some of the ground gained by chemotherapy is lost during recovery from the treatment. Clearly, the biomedical community would like to find chemicals and treatments that improve on this situation.

NEW CHEMOTHERAPIES FROM BASIC CANCER RESEARCH

The ideal cancer drug would kill, damage, or block the reproduction of all cancer cells but leave all normal cells unharmed. Medical science has made some significant progress toward this goal. Although there is not

Table 8.2 Naming Conventions for Anti-Cancer Drugs

Biological process or molecule class	Stem	Example drug
Angiogenesis inhibitors	-anib	Semax*anib*
Protein kinase inhibitors	-tinib	Ima*tinib*
Mitotic inhibitors; tubulin binders	-bulin	Mivo*bulin*
Cyclin-dependent kinase inhibitors	-ciclib	Seli*ciclib*
Multi-drug resistance inhibitors	-spodar	Val*spodar*
Vascular growth factor inhibitors	-bermin	Tel*bermin*
Epidermal growth factor inhibitors	-dermin	Muro*dermin*
Proteinase inhibitors	-zomib	Borta*zomib*
Monoclonal antibodies	-mab	Imciro*mab*
Aromatase inhibitors	-mestane	Plo*mestane*
Platinum derivatives	-platin	Cis*platin*
Antineoplastic antibiotics, daunorubicin type	-rubicin	Eso*rubicin*
Alkylating agents	-sulfan	Bu*sulfan*
Taxane derivatives	-taxel	Pacli*taxel*
Uracil derivatives	-uracil	Fluoro*uracil*
Vinca alkaloids	vin-	*Vin*epidine

yet a magic bullet to kill all cancer cells, there have been some important advances, particularly in the treatment of a few specific cancers. These forward strides in the treatment of some cancers open windows of hope for others. Here, we discuss a few well-established examples of this progress, but the newest cancer treatments, those still in development or undergoing clinical tests, are described in Chapter 10.

As medical science has identified more and more drugs for cancer treatment, the naming of these drugs has become a complex subject. Many different biological processes have been targeted in the effort to retard cancer cell growth. A drug company that discovers a useful drug often gives that chemical a patented name, which is company-specific, and this pattern leads to a bewildering array of names with little meaning. To help minimize confusion, doctors have developed a scheme by which to assign drug names, based on the function that is being affected. In this book we will sometimes mention the drug name given by a company, but commonly we will use the generic names, as described in TABLE 8.2.

A powerful treatment for chronic myeloid leukemia was discovered through insightful laboratory work.

One drug that comes encouragingly close to the ideal cancer chemotherapeutic agent is the chemical called **Gleevec**®, whose generic name is **imatinib** (pronounced im-ă′-tinib, FIGURE 8.12A). It was invented by a remarkable mixture of knowledge and experimentation. In Chapter 6 you learned that chronic myeloid leukemia (CML) is commonly associated with a translocation of DNA from chromosome 9 to chromosome 22, which makes the abnormal Philadelphia chromosome. The resulting up-regulation of a modified version of the gene called *abl* provides an oncogenic version of the normal, cell cycle-regulating protein tyrosine kinase Abl. Scientists realized that inactivating this abnormal protein kinase might inhibit inappropriate cell division, so they studied the kinase carefully. It was purified, and its structure was determined in atomic detail.

FIGURE 8.12 **The structure of Gleevec®, also known as imatinib, and its interaction with a mutant protein kinase.** (A) The chemical structure of imatinib. As you have previously learned, lines indicate chemical bonds and vertices are the sites of carbon atoms, unless the letter N is present, which means nitrogen instead of carbon. H is hydrogen and O is oxygen. This molecule serves as a mimic for ATP; it fits into the ATP-binding cleft of the oncogenic form of the Abl protein tyrosine kinase and prevents the entry of ATP, which is necessary for the kinase to work. (B) A diagram showing the backbone of the Abl oncoprotein (blue) with imatinib (green) slotting into its active site, showing where the drug binds. (Adapted from Alberts B, Johnson AD, Lewis J et al [2015] Molecular Biology of the Cell, 6th ed. Garland Science.)

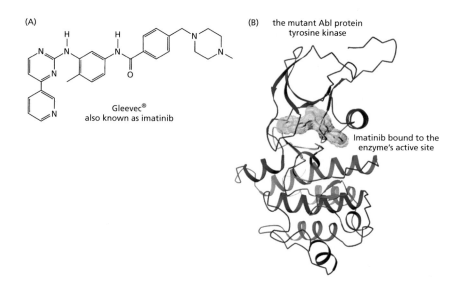

(A)

Gleevec®
also known as imatinib

(B)   the mutant Abl protein
tyrosine kinase

Imatinib bound to the
enzyme's active site

Imatinib was designed to bind this mutant protein tyrosine kinase and inactivate it. Starting with a process that screened thousands of possible protein kinase inhibitors, basic scientists found one that worked moderately well on this oncogenic enzyme. Scientists working in a drug company then used the three-dimensional structure of that drug and the atomic structure of the oncogenic protein tyrosine kinase to modify the initial inhibitor so it would fit better into the pocket that served as this enzyme's active site. Remarkably, this process worked well; the chemical they finally identified can bind tightly to the active site of the mutant protein and block its ability to bind ATP (FIGURE 8.12B). When the inhibitor is present, the mutant enzyme no longer works as a protein kinase. A patient with CML whose cancer cells have developed the Philadelphia chromosome can take imatinib and turn off the activity of the mutant enzyme that is causing white blood cells to grow and divide too frequently. This treatment often controls the cancer and can lead to the disappearance of most symptoms.

Imatinib isn't really a cure, because cancer cells are still present in the patient and are still mutant. Fortunately, however, their tendency to divide too frequently is blocked by the treatment. Unfortunately, some patients experience a gradual loss in the effectiveness of the treatment. This is in part a result of additional genetic mutations that alter the structure of the oncogenic protein kinase so it is no longer well inhibited by the drug. A second mechanism for emerging drug resistance is the up-regulation of a membrane protein that cells can make to pump unknown chemicals out of their cytoplasm. This up-regulation works well against many chemotherapeutic agents, so the protein responsible is called the **multi-drug resistance** transporter. Fortunately, there are also drugs that antagonize the multi-drug resistance transporter, such as valspodar, mentioned in Table 8.2, but here too, mutations can arise that reduce the effectiveness of that inhibition. In yet another path to drug resistance, cells can up-regulate the action of enzymes in their endoplasmic reticulum (ER), like the ones that make indirect carcinogens into truly carcinogenic molecules. The resulting chemistry can sometimes inactivate a drug by modifying its structure. Such changes are clear but painful examples of the ability of cancer cells to evolve; the ones that mutate by chance to drug insensitivity are the ones that now can thrive and reestablish disease.

In spite of the limitations of chemotherapeutic agents, it would be wonderful if drugs similar to imatinib could be found for all cancers. Unfortunately, very few cancers have such a well-defined genetic origin as CML, which is commonly initiated by formation of the Philadelphia chromosome. Nonetheless, imatinib has recently been found to have beneficial effects in some cancers other than the one for which it was designed. Doctors simply tried it on their patients with other kinds of cancer and found that it helped them. Given the saga of how imatinib was first made, this is surprising. However, you know from Chapters 6 and 7 that there are several protein tyrosine kinases involved in regulation of the cell cycle, so perhaps imatinib works by interfering with one or more of these related enzymes. This suggestion is supported by the fact that doctors are now using other protein tyrosine kinase inhibitors as effective drugs for the treatment of a range of cancers. In addition, inhibitors of the receptors for growth signals are a promising, relatively new target for drug intervention in cancer cell growth, a subject covered in Chapter 10.

Some chemotherapeutic agents for breast cancer are based on our growing understanding of estrogen.

Several types of cells in the mammary gland grow and differentiate under the control of sex hormones. One of the most important of these hormones is estrogen. Many of the cells in a woman's breast require some level of estrogen stimulation, not only to divide and differentiate but simply to survive; without this hormone, these breast cells induce apoptosis. Estrogen provides survival signals for these cells that keep the process of apoptosis turned off. Likewise, many breast tumors cells need estrogen to survive. Thus, treatments that block a tumor's access to estrogen are likely to cause the tumor cells to die and the tumor to shrink.

Tamoxifen is a laboratory-synthesized molecule that bears a slight resemblance to estrogen (FIGURE 8.13). Once inside a cell, tamoxifen is

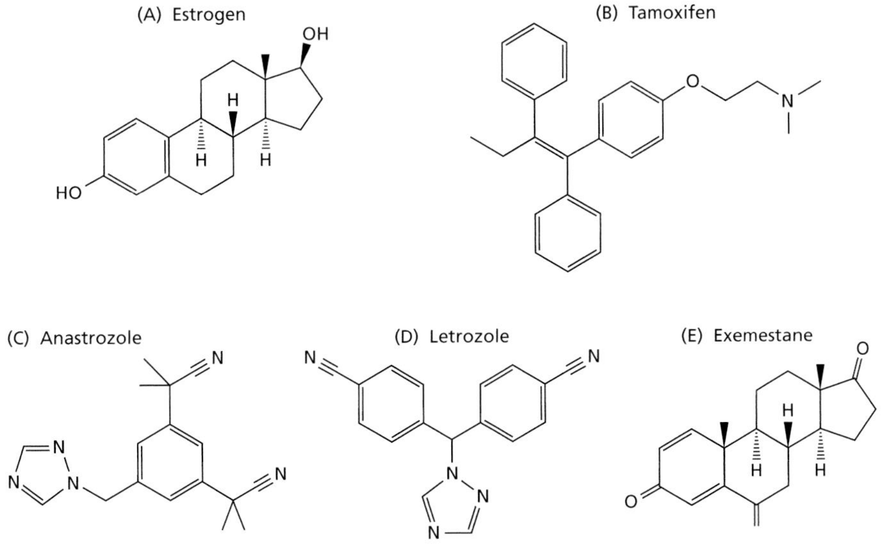

FIGURE 8.13 **Analogues of estrogen that inhibit either its action or its synthesis.** (A) The structure of estrogen. The pairs of lines imply what a biochemist calls a double bond between the carbons at those vertices, and the triangular lines imply bonds that go to atoms that are above or below the plane of the diagram. (B) The structure of tamoxifen. Although these two molecules look quite different, tamoxifen is a very effective anti-estrogen, probably because its structure is modified by the cells that take it in. Tamoxifen binds to the estrogen receptor and prevents normal estrogen from doing so. Although tamoxifen binds the estrogen receptor, it does not activate it, so the drug blocks the survival action of estrogen on cells. (C) Anastrozole, (D) letrozole, and (E) exemestane are three aromatase inhibitors that block one of the pathways for synthesis of estrogen.

altered by cellular enzymes to be more similar to estrogen; this modified tamoxifen interacts strongly with the estrogen receptor. Even in its modified form, however, tamoxifen is still sufficiently different from estrogen that it does not stimulate the estrogen receptor. Thus, it not only fails to support cell health and stimulate cell division, it also inhibits real estrogen from doing that job. When tamoxifen is interacting with the estrogen receptor, estrogen itself cannot bind the receptor, so the drug acts like an anti-estrogen that makes the body's estrogen ineffective. You can think of this as a process like jamming a radio signal. If a receiver is tuned to pick up signals that will give important information, but someone sends in meaningless signals that are received but carry no information, the meaningful signals cannot be heard.

The cells of many breast tumors continue to make the estrogen receptor (see, for example, Figure 8.1). When such tumors are treated with tamoxifen, many of their parenchymal cells will die, and the tumor will shrink in size, becoming less dangerous, simply because the cancer cells are unable to sense the estrogen signals that are circulating in the patient's body. For many patients, tamoxifen is very effective, and for some, the treatment can even lead to complete remission. Because it is working through a pathway that is specific to a subset of the body's cells, its side effects are milder than those of drugs like cisplatin or a taxane, which interact with essential components of all dividing cells.

For some women, however, tamoxifen is ineffective, even if their breast tumor cells require estrogen for survival. Fortunately, tamoxifen is only one example of a class of drugs called selective estrogen receptor modulators. Other compounds from this group can sometimes replace tamoxifen and be more effective. In addition, there is an alternative that can do the job even better in some cases. Doctors have found chemicals that block one of the pathways by which the body normally produces estrogen; these chemicals inhibit **aromatase**, a key enzyme in this pathway, so they are aptly called **aromatase inhibitors**. The generic names for three drugs with this function are anastrozole, letrozole, and exemestane (Figure 8.13). Aromatase inhibitors are very effective when used on the right patients. However, they are successful only in women who have passed through menopause. At earlier times in a woman's life, there are regulatory pathways that make sure enough estrogen is being made. If aromatase is inhibited, the body compensates by up-regulating other pathways for estrogen synthesis, so the treatment doesn't work. In postmenopausal women, however, the ovaries are no longer making estrogen, so this hormone comes from low levels of synthesis in other tissues. In this circumstance, an aromatase inhibitor will drive the body's estrogen levels to a very low level. Cells whose survival depends on estrogen will then die. This therapy is commonly used after an operation to remove a breast lump or radiation to shrink such a lump, as a way of assuring that any tumor cells that escaped treatment will be unable to form a new tumor elsewhere. Note that breast cancer becomes more likely with age, so the requirement that a patient has passed through menopause is not very restrictive. Aromatase inhibitors are now successfully used on many older women who have had a lumpectomy.

There are, however, some side effects from aromatase inhibitors, even though this chemotherapeutic agent is nicely focused on the cells of interest. Some women experience joint or muscle pains, which probably result from the fact that estrogen has effects on cells other than those in the reproductive system. When the body makes essentially no estrogen, there can be discomfort, but the effects are mild compared with the nausea, anemia, and hair loss that are common with more generally acting chemotherapies.

Treatments that reduce estrogen's effects don't work on some breast cancers.

Breast tumors whose cells have accumulated many mutations often do not respond to either tamoxifen or aromatase inhibitors. Such cancers emerge because the high level of genetic instability that is characteristic of cancer can lead to the emergence of breast-derived cells that have lost their dependence on estrogen. This is a special case of a common pattern in cancer progression. As the parenchymal cells of the tumor become more mutated, they become less and less like the cell from which they originated. We first saw this pattern in Chapter 2, where pathologists used the transition from dysplastic to anaplastic as a way to assess the danger of a tumor by microscopic examination of a biopsy. Fortunately, if breast tumor cells have lost their dependence on estrogen, one can sometimes turn to another powerful, scientifically conceived drug called **Herceptin**®, also known as **trastuzumab** (pronounced tras-tü′-zü-mab). This is a commercially available antibody that binds to one of the receptors for epidermal growth factor (EGF) in the plasma membrane of breast cells (FIGURE 8.14). The binding of trastuzumab turns the EGF signaling system off, reducing the growth stimulus that these cancer cells receive. Trastuzumab works very well for a special class of breast cancers that overproduce the EGF receptor (Figure 8.1C). It does not work well on other breast cancers, and it is ineffective against most other kinds of cancer. However, for cancers that do respond to it, trastuzumab is almost a miracle drug. Indeed, it has recently been found to be highly effective for a small group of patients whose lung cancer cells overexpress EGF-R. Unfortunately, a positive response to trastuzumab treatment occurs in less than half the breast cancer patients whose cells express the relevant receptor. The reasons for this limited range of effectiveness are not yet well understood, but other treatments that work along similar lines are now being developed. These are described in Chapter 10, after the nature of antibodies has been discussed more thoroughly in Chapter 9. For a personal description of one person's responses to these treatments for breast cancer, see SIDEBAR 8.7.

Hormone-directed therapies also provide chemotherapeutic agents for prostate cancer.

In prostate cancer, as in breast cancer, the parenchymal cells of the tumor are commonly dependent on continuous stimulation by a steroid hormone for growth and even for survival. The relevant hormones are collectively called androgens; the two most important of these are **testosterone** and dihydrotestosterone. The patterns of prostate cancer cell behavior in response to androgens and the chemicals that antagonize androgen action are much the same as those described above for estrogen in women. The ways in which medical science can deal with them are also similar. For example, there are drugs, like flutamine, that bind to androgen receptors and interfere with the stimulatory effects of testosterone, a mechanism analogous that that of tamoxifen. Other drugs, like leuprolide, interfere with the synthesis of testosterone by the testes and thereby reduce the body's concentration of this steroid hormone, making it harder for prostate-derived tumor cells to grow and survive. Leuprolide treatment is sometimes referred to as chemical castration, because the resulting reduction in testosterone levels is similar to what would happen if the man's testes were removed. As with breast cancer, these treatments become ineffective if the tumor cells have mutated to the point they no longer require hormonal stimulation to thrive; then the physician must apply conventional chemotherapies or radiation, as described previously.

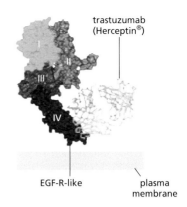

FIGURE 8.14 Trastuzumab binds to a growth factor receptor like EGF-R and blocks its activity. Growth factor receptors protrude from the plasma membrane, so they can bind the molecule that stimulates them in the extracellular medium. The four structural domains of the receptor are marked with Roman numerals and different colors. Trastuzumab is an antibody that binds this receptor and prevents it from sending signals to the cell's interior. (Adapted from Alberts B, Johnson AD, Lewis J et al [2015] Molecular Biology of the Cell, 6th ed. Garland Science.)

Compare and contrast the emergence of drug resistance in tumors to the selective breeding of plants and animals.

SIDE QUESTION 8.8

SIDEBAR 8.7

CASE STUDY: Rosemary's chemotherapy

Once Rosemary's surgery was over, she went to see an oncologist in a really good teaching hospital. Rosemary reported that doctor's recommendation and the opinion of another physician: "My oncologist said I might get some benefit out of chemotherapy with drugs designed to kill cancer cells, but probably not a lot. There was no evidence for metastasis, and the surgery might well have dealt with the problem. This doctor thought that the best treatment would not include conventional chemotherapies but simply tamoxifen. Overall, this doctor was a bit negative about chemotherapy, but I went to get a second opinion. The second physician said that chemotherapies designed to kill cancer cells might only increase my chances of long-term survival by 2% or 3%, but that if my cancer recurred in the future, it would be much harder to treat.

This second opinion changed the way I thought about things. I had tried to do a lot of reading, because that's how I normally get my information, but I have to say that all my efforts turned out to be pretty useless. Most of the books and articles I saw described all the different features of tumors, mentioning many different markers that distinguished different kinds of breast cancers, but when I ask my doctors about these things, it turned out that knowing which markers were expressed by my tumor didn't make any difference. At that time (the mid-1990s), there was no information to say that a tumor carrying a particular marker responded better to one treatment than another, so even though tests for these markers would tell me more about my tumor, it wouldn't make any difference to my treatment.

At that time, and even now, there were two features of my breast tumor that really did matter. The first was, did the cells require estrogen to grow; the second was whether the tumor cells were expressing Her2/neu [the membrane-associated growth factor receptor diagrammed in Figure 8.14], which might mean that they would respond to Herceptin®. I already knew my tumor needed estrogen to grow, so tamoxifen was a treatment that should help, but I also had the test for Her2/neu. At that time, these tests were pretty unreliable. The first time it showed that my cancer cells did express this marker, but a second test indicated that they did not. Ultimately, my doctors didn't do anything about Herceptin®, and I came to feel this was right. In fact, at that time, there didn't seem to be much of a range of choices for treatments. Pretty much the same thing would have been done, almost no matter where I was being examined, so long as I was treated by reputable physicians.

I guess you would say that my treatment was conservative, because first I had surgery, then chemotherapy, then radiation, then tamoxifen, and finally an aromatase inhibitor. My first chemotherapy was a mixture of three drugs known as CMF [cyclophosphamide, which is another compound that forms adducts with DNA; methotrexate, which inhibits the synthesis of some nucleotides; and fluorouracil, which is a thymine analogue]. This lasted for about 6 months, and it was a big deal. My hair fell out on about day 17, which I found very upsetting. With these drugs there is essentially nothing you can do to prevent this happening, so I simply had to put up with it. I looked around to find wigs that would look more or less plausible, but they really are an unpleasant addition to life. They are hot and scratchy, and I was always worrying about their blowing away.

After I had recovered from chemotherapy and had my hair back, I took tamoxifen for five years. Here I was lucky, because I experienced absolutely no side effects. I didn't gain weight or have any other problems. When you read about tamoxifen, you see a lot of anxiety on the part of the patients, but I didn't have any troubles at all."

Large tumors need a blood supply, which opens up a novel approach to chemotherapy: blocking the formation of new blood vessels.

You have learned that tumors can grow large only if they stimulate the ingrowth of blood vessels, which will then provide their parenchymal cells with nourishment and oxygen (Figure 6.21 and associated text). The formation of new blood vessels is a result of angiogenesis. Blocking this process is a comparatively new chemotherapeutic strategy for cancer. Its discovery and testing in laboratory mice were spectacular successes; they served as an impressive example of the ways in which clever medical scientists can invent a novel approach to cancer medicine.

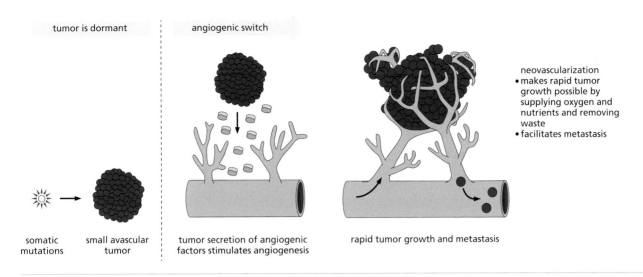

tumor is dormant   angiogenic switch

neovascularization
• makes rapid tumor growth possible by supplying oxygen and nutrients and removing waste
• facilitates metastasis

somatic   small avascular   tumor secretion of angiogenic   rapid tumor growth and metastasis
mutations   tumor   factors stimulates angiogenesis

FIGURE 8.15 **The onset of angiogenesis, which enhances tumor growth.** The cells in small tumors can get the oxygen and nourishment they need by diffusion of these substances from the surrounding tissues. However, once the tumor has reached a critical size (1–2 mm in diameter), either cell growth stalls for want of nourishment or some of the tumor cells mutate in such a way that they now secrete factors that stimulate angiogenesis. This change is called the angiogenic switch. The growth of new blood vessels is called neovascularization. It is required for continued tumor growth. (Adapted from Carmeliet P & Jain RK [2000] *Nature* 407:249–257, and from Bergers G & Benjamin LE [2003] *Nat Rev Cancer* 3:401–410. With permission from Springer Nature.)

Tumors induce angiogenesis by secreting proteins that promote the proliferation and migration of endothelial cells, that is, the cells that make the walls of blood vessels. The onset of this secretion is called the angiogenic switch. Its result is an increase in the number and extent of blood vessels running into and out of the tumor (FIGURE 8.15), which is essential for continued tumor growth. In a novel effort to block tumor growth, doctors treated cancerous mice with proteins that inhibited angiogenesis. As a result, their tumors shrank for want of an adequate blood supply. This discovery was rapidly brought to the clinic for treatment of humans. Although there has been some success with the method, progress has been much slower than was initially hoped. Nonetheless, some cases demonstrate that the treatment of patients with proteins that inhibit angiogenesis can slow or even reverse tumor growth. One can hope that this approach will be refined and made more effective, giving the oncologist a completely different method for treating metastatic cancer. For a personal account of response to this kind of treatment, see SIDEBAR 8.8.

SIDE QUESTION 8.9

Why would a surgeon want to shrink a tumor before operating on it? If she is going to be cutting the whole thing out anyway, what is the value of reducing the tumor's size before surgery?

The long-term effectiveness of all chemotherapies is limited by the accumulation of new mutations, so doctors are always seeking new targets for cancer treatment.

If a tumor is being treated with a taxane, the cells that are least affected by this drug are the ones most likely to be among the few that survive each round of chemotherapy. Thus, chemotherapy is actually a selection for cells that are not affected by the treatment. This idea underlies the strategy of using multiple drugs and harsh initial treatments. The more cells that are killed in the early stages of chemotherapy, the lower the chances that the treatment is selecting for survivors that are unaffected. However, the repeated emergence of drug resistance has provided strong motivation for doctors to identify new chemotherapies that act on completely different targets.

As an example of a novel drug target with cancer relevance, telomerase is an enzyme that helps to maintain the length of chromosomes so they

SIDEBAR 8.8

CASE STUDY: John's response to chemotherapy

John's kidney tumor, a renal carcinoma, was so big at the time of diagnosis that the surgeon didn't want to operate. Instead, the oncology team prescribed chemotherapy to shrink the tumor. Kidney tumors are notoriously insensitive to conventional chemotherapies, but the doctors on this case knew about drugs that interfered with the formation of blood vessels. They reasoned that these might make trouble for the tumor and cause it to shrink. Even though these chemicals were not the conventional, cell-killing drugs, John found that "the side effects from this treatment were harsh. I had hives, skin peeling off my fingers, and rashes; it was nasty." Between January and May this treatment made John's tumor shrink by about 30%, as seen in CT scans. The doctors finally took him off the chemotherapy because they were afraid that the drug was doing too much harm. The surgeon was still skeptical that an operation was going to work well, so John's case was brought before a tumor board, consisting of an oncologist, a radiation oncologist, a pathologist, and a surgeon. They looked at four CT scans taken over a series of months, and concluded that the operation was doable, though high risk. John heard his options and decided that the knife was his best choice. The surgeon wouldn't give odds for the success of his surgery, but he did say, "John, you could die in surgery." John's response was, "That's not my problem. I won't know; it will only be a problem for you and my friends who are still around."

The surgeon's skill and John's own physical strength stood him in good stead, so he came through very well. The tumor was still encapsulated, so it had not spread everywhere. However, it weighed over 10 lb, it surrounded one of the kidneys, and it was highly vascularized. This raised the worry that metastasis could have occurred through either lymphatic or blood vessels. The surgeon clearly did a great job, because there was no infection and no need for clean-up operations. Several nearby lymph nodes were taken out, along with the tumor and the kidney it surrounded, allowing the pathologist to do extensive tissue analyses. The tumor was indeed malignant, and some of the lymph nodes were positive for metastatic growth, others not. The situation was bad but not awful.

John spent only four days in the hospital at the time of his surgery. After that he was able to live at home, which he really preferred, even though postoperation recovery was slow. His weight dropped from about 180 to 139 lb, and it was hard for him to get going again. He wanted to avoid lying there and thinking, "Why me?," but that took oomph. His wife, Lee, was a great help here, because she is a can-do kind of person, and pretty soon, John got back on his feet.

Given that there was metastasis, John was put back on chemotherapy, even though the extent of metastasis had not been determined. His doctors did an MRI to make sure that the brain was clear, and the CT scans of his abdomen were a good check on the liver. They didn't do a PET scan to look for metastatic growths, because the insurance company wouldn't pay for it. Given that the cancer was already known to be metastatic, that made some kind of sense. The doctor's choice for chemo was again the one that interfered with blood vessel formation (sorafenib, which had been effective in shrinking the tumor before surgery). This treatment went on for several months, given as a pill, so John could take it at home. After another couple of months, the doctors switched him to sunitinib [Sutent®], which made him lose his hair, but not his sense of humor. This drug made him sleepy, but he found that if he took it in the evening, it just helped him sleep, while if he took it in the morning, he wasn't good for much that day. Sunitinib, like the previous drug, required a time of trial and error to get the dose right, and those were hard times, because John felt pretty poor when the doses were too high. Once they had a working dose that he could tolerate, though, he began to feel OK. He had regular checkups and quite a few CT scans to track the size of his lymph nodes. During this time, he regained his strength and was able to get back to work on a pretty regular basis.

After a few months, though, the CT scans made it clear that some of the metastasized tumors were growing again, so John joined a phase 3 clinical trial for the treatment of kidney tumors. (Phase 3 means that the treatment under study had already been shown not to be damaging to a patient's health, and it might be helpful, but data from more patients were required to know for sure.) This treatment used a mixture of drugs: erlotinib [Tarceva®], which was commonly used at that time for the treatment of non-small-cell lung cancer, and rapamycin [serolimus], which helps to block the cell cycle and suppress the immune system, thereby helping to prevent inflammation. The doctor's goal for this treatment was to stop the progression of John's cancer for 6 months.

John had nine tumors in different lymph nodes, and the doctors kept track of them with CT scans. Six responded well to this drug regimen; they actually shrank, and it looked as if they would disappear. Two continued to grow, albeit slowly, but then one took off and grew from 1 to 5 cm in 6 months. This growth was fast enough that John had to drop out of the clinical trial. Once a tumor has grown by 20%, doctors will generally conclude that the treatment under study is not working. John lasted in the program for 11 months, but then he had to find other options.

One possibility was another surgery, but the doctor who did the first job was skeptical. The tumor was located between the pancreas and a kidney artery, where it would be hard to get at. A different surgeon said he'd be glad to operate, but John really trusted the man who operated on him first. The fact that the second doctor was self-confident didn't cut much ice with John, who was leery of cocky people. Radiation was out of the question, and the number of useful drugs to try was getting small. John's spirits stayed very good, because he wanted to live the life he had and not worry about the life he would no longer be able to have. Sadly, after a couple of years of additional experimental drugs, John died. He passed away at home, surrounded by friends and family, making his life and death with cancer one that you can really admire.

don't shorten with successive rounds of cell division, an event that initiates a DNA damage response, as described in Figure 6.23 and associated text. The drug BIBR1532 is a reasonably specific inhibitor of telomerase. It works by inducing the enzyme to fall off the DNA, blocking telomere elongation. It is now being tested as a chemotherapeutic agent for cancer. Other methods for inhibiting this enzyme employ molecular methods that are more complicated and so will not be described here.

As another example, doctors are now targeting proteinases, the enzymes that break proteins down into their constituent amino acids. These enzymes are an important part of many cellular functions, including cell cycle control. A proteinase inhibitor called bortezomib is effective in slowing the growth of two blood cancers, multiple myeloma and mantle cell lymphoma. Because proteinases are essential components of cycle control, in normal cells as well as cancer cells, bortezomib is not as specific for cancer as one would like. The issue of cancer specificity is yet another motivation for much of the work described in Chapter 10.

ALTERNATIVE MEDICINE AND CANCER

There are many forms of health care other than the scientific medicine that has grown up in the Western world over the last 100 or so years. I am calling all of these approaches **alternative medicine**, though the terms **traditional medicine** and **complementary medicine** are also used. These kinds of medicine include a wide range of attitudes and practices towards human disease and its treatment. A complete description of them as they pertain to cancer therapy is an impossibly large task, and therefore the following description is necessarily superficial.

Most forms of alternative medicine derive from traditional medicines that have flourished for centuries all over the world.

The most highly developed traditional medicines are found in Eastern and Southern Asia. Countries in these parts of the world have been active sources of therapies for millennia, though there are also traditional medicines from many other parts of the world, including African and Native American cultures, to name but two. Cancer has been known for a very long time, so traditional medicines include recommendations for both its prevention and its treatment. Most of these traditions present points of view about life and well-being that are quite different from the more recently developed scientific medicine of the Western world. The traditions usually include prescriptions for foods and behaviors that will keep one healthy, rather than pills that will make a sick person well again. Such approaches commonly address the body as a whole, rather than focusing on a single part of the body that is failing at a particular time. For example, alternative medicines often recommend a vegetarian diet. They also suggest effective ways for reducing stress, such as yoga, and they urge abstaining from self-destructive behaviors, like smoking tobacco, drinking to excess, and overeating. Adoption of some aspects of these approaches to health can yield a real improvement in one's general well-being, but they are not designed to cure an already developed cancer. However, some people who have seen a scientifically trained doctor and been given a diagnosis of metastatic cancer with a correspondingly grim prognosis will turn to traditional medicines in a quest for therapies that might at least give them longer to live. While this is very understandable in human terms, it does not make good medical sense. The real benefits from these alternative approaches require living one's whole life in the recommended way. Asking a particular herb, mushroom, diet, or tea to deal with a metastatic cancer is unreasonable.

Many alternative therapies are chemically complex.

One valuable way to think about alternative therapies is based on a very simple realization: if one of them was really a cancer cure, cancer deaths would have been significantly reduced in the country that followed that medical tradition. Yet cancer mortality is about as high in the countries of Eastern and Southern Asia as anywhere else in the world. It follows that these remedies are unlikely to save a desperate situation. Some may be worth trying, but an important principle emerges. Herbs, teas, and mushrooms contain a very large number of chemicals (after all, every living thing is made of chemicals, mushrooms included). Taking a natural substance that has been chosen by traditional medicine to have strong effects on medical problems means that one is ingesting a complex mixture of chemicals, some of which may do you good and some of which may harm you.

Mushrooms offer a good example. They are a diverse group of complex fungi that include some species, like the Angel of Death (*Amanita phalloides*), which make several highly poisonous substances. Hallucinogenic mushrooms make substances like psilocybin, which affect the way nerve cells work and thus how people think. It is therefore no surprise that some mushrooms make chemicals that affect aspects of cell growth and division. If a given remedy based on mushrooms is really traditional, it will have been tried on many people before you, so the danger that it will do you serious harm is small. However, if you are also taking chemotherapeutic agents that have been prescribed by a scientifically trained doctor, eating substances recommended by alternative medicine puts you at risk of harmful interactions among the multiple chemicals you are now putting into your body. It is therefore wise to discuss any alternative therapy with an oncologist (**SIDEBAR 8.9**).

Medicinal mushrooms are a source of several biologically active substances that can serve in cancer therapy.

Methods for the extraction and purification of some medicinal chemicals from mushrooms are now well worked out. Several such molecules have now undergone clinical trials, including lentinan from the mushroom *Lentinus edodes* (commonly called shiitake) and schizophyllan from the mushroom *Schizophyllum commune* (**FIGURE 8.16**). Analogous substances from *Trametes versicolor* are known by their initials PSK and PSP. Additional

CASE STUDY: Rosemary's experience with alternative therapies

Rosemary, the breast cancer patient we encountered in Sidebar 8.7, looked into alternative therapies to treat herself. She reported, "I did do a little exploration of treatments recommended by alternative medicine. A friend of East Asian origin knew quite a bit about traditional Chinese medicine, and he got me some medicinal mushrooms, which I took for a while. At one point during chemotherapy, though, my oncologist noticed that my white blood cell count was dropping more than she expected, and she asked what other medicines I was taking. She then suggested I really ought not to mix therapies like this, because the mushrooms might be contributing to what was happening, and she didn't know how to interpret the data she was getting. I stopped then and didn't go back to alternative treatments."

compounds from other medicinal mushrooms that are thought to have anti-cancer properties, such as reishi, are part of traditional medicine in China, Japan, and Korea. These substances have been studied in preclinical models and will increasingly be submitted for clinical trials.

Several of the potentially beneficial chemicals from mushrooms are polysaccharides, that is, polymers of various sugar molecules (Figure 8.16B,D; see also Figure 3.15 and associated text). Some such substances seem to have a stimulatory effect on the patient's immune responses; they have been said to activate phagocytic cells (white blood cells that can engulf pathogens and other foreign materials), but just what activate means is

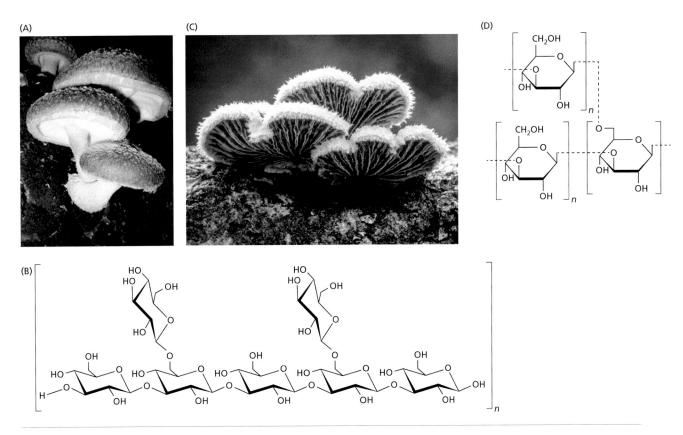

FIGURE 8.16 Two medicinal mushrooms and two medically active chemicals isolated from them. (A) Four examples of *Lentinus edodes*, also known as shiitake. This mushroom is the source of lentinan, whose chemical structure is shown in part B. The symbols representing the atoms of this large carbohydrate are all as previously described, but the brackets and the symbol *n* on the lower right means that this structure is repeated many times in each molecule of lentinan. When lentinan was administered to cancer patients, it prolonged their lives in some cases. (C) Three samples of *Schizophyllum commune*, the California mushroom from which schizophyllan, shown in part D, has been isolated. Again, the square brackets and the symbol *n* imply polymers of these sugarlike rings. (A, image courtesy of frankenstoen, CC BY 2.0; C, courtesy of Michael Wood.)

not clear. Some of these substances do appear to have anti-cancer activity in laboratory mice, and some mushroom-derived molecules have proceeded to clinical trials, initially in Japan and China and more recently in the United States. Lentinan has been successful in prolonging the survival of patients with gastric and colorectal cancer, and it is now approved as a drug in Japan. It is also considered a potentially valuable supplementary treatment in several other cancers. Schizophyllan seems useful for reducing symptoms in cases of recurrent, inoperable gastric cancer, as well as for increasing survival times of patients with head and neck cancers. Neither of these compounds induces significant side effects.

Some mushroom-derived polysaccharides have been shown to kill cancer cells growing in the laboratory. It may be that their described ability to stimulate the immune system derives in part from their action as poisons. At this time, no one knows exactly how they work, but these chemicals are interesting examples of an important idea. There may be less difference than one might think between the actions of alternative cancer therapies and scientifically developed ones. Both may provide poisons for dividing cells, thereby having a negative effect on cancer's growth and spread but also some effects on our normal cells. It is clear that further scientific work on the ways alternative therapies work could help us understand the complex comparison between traditional medicines and scientifically identified chemotherapeutic agents.

Another important observation from these studies is the apparent ability of several mushroom-derived molecules to reduce the side effects from radiotherapy and/or scientifically based chemotherapy. Such relief can be a blessing for people who are experiencing problems associated with treatments based on high doses of radiation or the cytotoxic drugs that scientific medicine has developed.

All people who prepare and sell medicines make money from their work, but traditional medicines are not subjected to government regulation.

The pharmaceutical companies that produce and sell drugs for use in scientific medicine are obviously motivated, at least in part, by the profits they can make in this enterprise. The charges made by these companies are explicit to the patient and any insurance companies involved in that patient's health care. High as these charges may be, one knows from the start what they will be, and one can evaluate how much a particular medical insurance program will cover. The costs of treatments with alternative medicines are also high, and they may not be covered by health insurance. As with all financial transactions, it is worth weighing costs versus benefits.

The Food and Drug Administration (FDA) of the United States has established strict rules about the tests a drug must pass before it can be approved for human treatment. As a result, there is usually reliable information available on the number of people who have already tried a drug, what it did for their condition, and the side effects to be expected. Thus, one can have a clear sense of a drug's likely value and how it stacks up against the associated cost and dangers. Most natural products, such as herbs, teas, and mushrooms, are not classified as drugs, so they are not subject to the same controls. There is no requirement that alternative medicines be evaluated in objective trials. This is one of the meanings of alternative medicine; it has both pluses and minuses. It allows the patient to get access to chemicals that would otherwise be unavailable, for example, the ones in medicinal mushrooms. Even though these are natural products, they are made of chemicals, just like you and me! As with the

SIDE QUESTION 8.10

A medicinal plant recommended by practitioners of traditional medicine contains many chemicals. A drug prescribed by a Western doctor consists almost entirely of a single chemical (the drug itself, plus whatever contaminants were not purified away during drug synthesis). Which of these treatments do you think is better? Give reasons for your answer.

drugs made by a pharmaceutical company, there is risk associated with putting these substances into your body. Because the treatments have usually not been subjected to objective testing, the risk of consuming them is probably not well defined, and it may be large. An evaluation of risk relative to possible benefits is simply not part of alternative medicinal thought.

Many Websites devoted to alternative medicine include testimonials that state without qualification the value of a particular treatment for a given condition. These statements claim that the health of some individuals improved dramatically as a result of taking this particular product. Some such statements may be true. Remember, though, about spontaneous remission (the disappearance of a cancer, simply on its own). There are also probably some cases where the initial diagnosis of cancer was wrong, so the apparent cure of cancer was not really a cure at all. All that a provider of an alternative medicine needs is one example of marked improvement while the patient was taking their medicine, and he can claim that he has proof that the medicine works. However, a few examples of health improvements are *not* proof; the patient might have gotten better anyway. The only way to be sure about the effectiveness of a treatment is to carry out the kind of objective testing that the FDA requires, which is exactly why they require it.

To evaluate a treatment for cancer, one needs data from studies that are carried out by unbiased investigators on large numbers of patients who are given either the substance to be tested or the treatment that is customarily given for this condition. In some cases, another group of patients is given a benign, irrelevant substance, like sugar, which is called a **placebo**. Patient progress is then monitored without either the patients or the observers knowing which treatment was given to each individual. These kinds of studies, aptly called **double-blind tests**, are time-consuming and expensive (SIDEBAR 8.10). Only a few traditional cancer remedies have undergone this kind of evaluation, and most that have been so tested have come up wanting. For example, East Asian medicine has traditionally recommended the cartilage (a connective tissue) from sharks as an effective treatment for cancer, even when the disease is well advanced. Shark cartilage as a cancer chemotherapeutic agent has now been evaluated with double-blind tests by reputable scientists at the National Institutes

A CLOSER LOOK: Constructing a double-blind test

Trials or experiments that follow a double-blind design are organized to minimize the chances that either the experimenters or the subjects of a test can influence the results. To achieve this, neither the experimenter nor the subject knows whether any one individual is receiving the experimental or the control therapy, hence the name for this kind of procedure.

Double-blind testing is an extremely important element of experimental design because it assures the objectivity of the data that result. In any test or experiment, the participants are likely to have an opinion or feeling about how the results should turn out. This is particularly true when a researcher and/or company have labored long and hard to bring a new

drug or innovative treatment far enough along in research that it can be tested on people; these scientists and businesspeople are likely to want very much to find out that the treatment or procedure works. Patients who are being treated by a researcher who wants a specific outcome are likely also to want that outcome or, if something has previously gone wrong, to want the experiment to fail. These very human reactions can lead to bias in data collection and analysis, and thus to incorrect results, which in the long run will be both painful and expensive. Thus, objective trials based on a double-blind test are at the heart of reliable evaluation for a drug, an idea, or a procedure.

SIDEBAR 8.10

of Health in the United States. These scientists found no value in this substance relative to a placebo. Currently, several studies are ongoing to provide reliable information about other remedies proposed by alternative medicine, as described in Chapters 11 and 12.

The negative tone of the above discussion motivates an important additional comment. It is well known by scientists that many of the drugs upon which modern medicine relies, like aspirin, were originally identified by the analysis of treatments recommended by traditional healers. Moreover, some very recent medical successes have resulted from a similar approach. Artemisinin, the most successful current treatment for malaria, was discovered by using scientific methods to identify the medically active chemical in the herb recommended by a traditional Chinese treatment for this disease. Some treatments for cancer from both East Asia and South Asia are now undergoing analogous scrutiny, and the results may be valuable. In India, the Ayurvedic tradition of health and healing has about 5000 years of experience in helping people to wellness, and some practitioners of these arts are now subjecting their cancer treatments to scientific testing in both the laboratory and the clinic. The results may provide valuable tools for future cancer treatments.

SUMMARY

The goal of cancer treatment is to eliminate all cancerous cells from the patient's body, either by cutting them out with a surgical operation or by killing them in place. Radiation is an effective way to kill cancer cells, but it will generally have a negative effect on the body's normal cells, producing side effects. Treating tumors that have already begun to spread through the body (that is, metastasized) is difficult, because both surgery and radiation treatment require that you locate a tumor to treat it. While medical imaging can help with this issue, a widely metastasized cancer must be treated in ways that affect cells, regardless of where they are. Chemotherapy can accomplish this goal, but since cancer cells derive from the cells of our own body, it is hard to find treatments that kill the unwanted cells without affecting one's normal cells too. As a result, most chemotherapies have side effects. The drugs chosen to kill cancer cells tend to damage normal cells as well as their targets.

The repertoire of well-tested cancer treatments is already quite large and reasonably effective. Surgery in combination with radiation and/or chemotherapy can eliminate a significant fraction of cancers, especially when they have been identified early. Radiation has been brought to a fine art in which the damaging rays are targeted quite precisely to the tumor cells, so harm to the rest of the body is kept to a minimum. There are numerous chemotherapeutic agents, some of which interfere with DNA replication and others with cell division. These can be used in various combinations, minimizing the chances of selecting for cancer cells that are resistant to one particular drug. At the same time, all these treatments are hard on the patient as a result of their effects on healthy, noncancerous cells. Thus, the motivation for invention of new and more specific treatments is strong. The specific protein kinase inhibitor, imatinib, and drugs that inhibit the action of sex steroid hormones, like estrogen and testosterone, offer encouraging examples, but more and better treatments are clearly needed. They are now sought continuously, both by targeted searches for drugs that block cellular pathways now known to be important for cancer and by scientific analyses of remedies suggested by alternative medicine. In addition, doctors are seeking ways to harness the patient's immune system and turn it against cancerous cells. Some of these advances will be described in Chapter 10 after you have been introduced to the immune system itself, the subject of Chapter 9.

Many cancer patients seek alternative treatments, either because the ones their doctors have prescribed are not working or because they would like to try all possible options. Alternatives to scientific medicine are available from traditions of healing that have been developed all over the world. The likelihood that these treatments will cure an advanced cancer is small, and one must be aware that interactions between traditional and scientific therapies can be a problem. Getting good medical advice about alternative treatments is important before you begin one, and it is always wise to think in terms of cost versus benefit.

ESSENTIAL CONCEPTS

- Using surgery to cut out a tumor is a simple and direct approach to eliminating a cancer, but given the possibility of small, undetected tumors lying in the neighborhood of the primary tumor, it is sometimes not as effective a cancer therapy as one would hope. It is therefore commonly used in combination with other therapies.

- Tumors contain several kinds of tissue: parenchymal cells (the cancer cells themselves) and stroma, which includes vasculature, blood cells, and connective tissue. Removing a tumor by surgery therefore cuts out more than simply cancer cells.

- Examining lymph nodes is an effective assay for evidence of metastatic behavior in the cancer under study, but lymph node examination is not perfect. The pathologist can miss wandering cancer cells, either through faulty diagnosis or because these cells did not happen to move to the nodes being examined.

- A sufficient dose of ionizing radiation can kill cells. The most common and effective route for this process is the use of X-rays to induce apoptosis, a cellular process that is initiated in response to DNA damage. This induction depends on the activation of p53 by DNA damage; cells lacking functional p53 can also be killed by radiation, but it takes a much larger dose.

- Carefully placed sources of radioactivity can sometimes deliver lethal radiation to cancer cells in a more focused way than X-irradiation. Focused delivery of radiation can be accomplished either by a radioactive isotope of a specific element that is used by a single cell type, for example, iodine for thyroid cells, or by the careful placement of radioactive atoms, for example, the insertion of a small chip of radioactive material into a cancerous prostate gland.

- Chemotherapeutic agents work by blocking some cellular process that is essential for cell growth, such as DNA replication or mitosis. Although cancer cells are sensitive to such treatments, so too are our normal, dividing cells. Such therapies therefore have side effects, which can be unpleasant for the patient and can compromise recovery.

- Common side effects of chemotherapeutic agents include nausea, gastric distress or diarrhea, fatigue and anemia, sensitivity to infection, and hair loss.

- Given the relatively high mutation rate of cancer cells, they will commonly produce progeny that are resistant to almost any chemotherapeutic agent unless essentially all the cancer cells can be killed before such mutations arise.

- Treatment by two or more drugs with different targets is better than treatment with a single drug, because the ability of the surviving cancer cells to evolve resistance to a double treatment is less than their ability to survive a single drug.

- Some cancer treatments work by affecting the tumor's stroma, for example, the vasculature that provides parenchymal cells with a blood supply.

- Basic cancer research is discovering new ways of treating cancer that target properties specific to particular cancers; for example, a cell surface receptor that is characteristic of a specific tumor. Such treatments can be selectively lethal for the cancer cells, thereby reducing the side effects that are common with chemotherapies.

- Molecules that regulate cell cycle progression, such as protein tyrosine kinases, are proving to be useful targets for drugs to block cancer cell growth, particularly in certain cancers.

- The cancer treatments discovered over many years in many countries around the world provide an alternative cancer medicine. Most of this kind of medicine is advice on how to live a healthier life and avoid cancer in the first place. Some of the effective treatments recommended by alternative medicines, such as those based on medicinal mushrooms, work through chemicals whose actions on our cells may not be very different from the chemotherapeutic agents discovered by scientific medicine.

KEY TERMS

actinomycin D	fluorouracil (FU)	radiation oncologist
adjuvant therapy	Gleevec®	radioiodine therapy
allogeneic transplant	Herceptin®	spontaneous remission
alternative medicine	hysterectomy	stroma
angiogenesis inhibitor	imatinib	tamoxifen
aromatase	incision	taxanes
aromatase inhibitor	local metastases	testosterone
autologous transplant	lumpectomy	therapeutic index
brachytherapy	lymphedema	traditional medicine
carboplatin	microenvironment	trastuzumab
chemotherapy	multi-drug resistance	tumor burden
cisplatin	outpatient	vinblastine
complementary medicine	parenchymal cells	vincristine
connective tissue	placebo	viscera
double-blind test	primary tumor	
extirpate	prophylactic mastectomy	

FURTHER READINGS

Mukherjee S (2010) The Emperor of All Maladies: A Biography of Cancer. Scribner. This is a very readable and interesting book written by an oncologist that talks about cancer treatments, cancer patients, and the history of cancer. A very good read.

Dollinger M, Rosenbaum EH, Tempero M & Mulvihill S (2002) Everyone's Guide to Cancer Therapy: How Cancer Is Diagnosed, Treated, and Managed Day to Day, 4th ed. Andrews McMeel Publishing.

For more information about radiation therapy, see Radiation Therapy and You: Support for People with Cancer at http://www.cancer.gov/publications/patient-education/radiationttherapy.pdf or RadiologyInfo.org: The Radiology Information Resource for Patients at http://www.radiologyinfo.org/index.cfm?bhcp=1.

For more information about tamoxifen, see Hormone Therapy for Breast Cancer at https://www.cancer.gov/types/breast/breast-hormone-therapy-fact-sheet.

For more information about Herceptin®, see HER2-Positive Breast Cancer: What Is It? At http://www.mayoclinic.com/health/breast-cancer/AN00495.

For an interesting account of the discovery of angiogenic factors as a possible cancer therapy, see Cancer Warrior Judah Folkman at http://www.pbs.org/wgbh/nova/body/folkman-cancer.html.

If you want information on the frequency of various kinds of cancer and the deaths that ensue, arranged both by state within the United States and by ethnicity, see the annual publication by the American Cancer Society called Cancer Facts and Figures at https://www.cancer.org/research/cancer-facts-statistics/all-cancer-facts-figures.html.

Servan-Schreiber D (2009) Anticancer: A New Way of Life. Viking. This describes in readable form many things you can do to minimize the chances of getting cancer and what to do about it if you do get it.

For descriptions of the Ayurvedic perspective on cancer prevention and treatment, see Widodo N, Takagi Y, Shrestha BG et al (2008) Selective killing of cancer cells by leaf extract of Ashwagandha: Components, activity and pathway analyses. *Cancer Lett* 262:37–47, and Balachandran P & Govindarajanin R (2005) Cancer—an ayurvedic perspective. *Pharmacol Res* 51:19–30.

QUESTIONS FOR FURTHER THOUGHT

1. Define on paper the properties of an ideal cancer therapy.

2. Take each of the conventional treatments prescribed by modern medicine and relate it to your ideal. Spell out both the good and the bad (or at least limited) features of that treatment.

3. Imagine that you have a compound that kills cells very effectively, so long as it gets through their plasma membranes. Think about ways in which you might be able to deliver that killer compound selectively to cancer cells, thereby making a really good chemotherapeutic agent.

ANSWERS TO SIDE QUESTIONS

1. A clean margin on a tumor refers to the smoothness of its edges. Smoothness implies that the cells of the tumor are contained and are not invasive, so they are unlikely to have metastasized.

2. Although ultrasound is the least expensive and least scary form of medical imaging for looking inside a person, the images it produces are not very sharp. Before operating, a surgeon usually wants the clearest possible idea of where a tumor lies, how large it is, and how much vasculature is present. A CT scan following the injection of an X-ray contrasting substance into the blood would provide this kind of information, as would an MRI. PET scanning would be inappropriate. On the basis of location of the tumor, the surgeon would probably choose either a CT scan or an MRI, not both. The magnet required for MRI is commonly thought to be more scary than the CT scanner, and MRI is more expensive, but X-rays are more likely to cause damage to DNA than the magnetic fields and radio waves of MRI.

3. Although the testes are near the prostate gland, and there is a likelihood that sperm-producing cells will receive X-irradiation in a procedure of this kind, this is of only modest concern with a 67-year-old man, because the chances that he will be siring children are very small. Nonetheless, before beginning the irradiation, you would probably want to distribute the lead shielding over all nearby parts of the patient's body, making a window through which the irradiating beam could pass, but meanwhile protecting as much of the rest of the body as possible.

4. Radioactive sulfur would get incorporated into proteins everywhere in the patient's body, wherever protein synthesis was taking place. Although this would lead to irradiation of the tumor cells, it would also mean a great deal of radiation exposure for all other cells in the body. Almost all cells make proteins continuously, providing new proteins to replace ones that are getting old and worn out. Thus, there would be little or no preference for tumor cells or even for growing cells. This approach would give the patient a high general level of irradiation with no specific attention to tumor cells, so it would be a bad choice for cancer therapy.

5. Brachytherapy has the advantage that it defines quite accurately the place in the body that will receive the most radiation: the volume that surrounds the implanted radioactive chip. If the chip can be placed directly in the prostate tumor, it will do the most damage to the cells you want to kill, and that sounds ideal. Brachytherapy has the disadvantage that when an implant is left in place it will continue to emit radiation for quite a while, albeit at ever-decreasing strength as the nuclei of the radioactive atoms decay. Moreover, some of the radiation will affect surrounding tissues, just as with X-irradiation of a tissue. Also, getting a well-measured dose to the prostate is harder this way than with X-rays. Nonetheless, the prospect of getting high-intensity radiation to the site of the tumor is a major consideration, so brachytherapy is probably the treatment method of choice.

6. This is a difficult question with no simple right answer. You know that an autologous bone marrow transplant would give your brother the risk of re-introducing untreated cancer cells, whereas an allogeneic transplant requires a donor, such as yourself. The extraction of bone marrow is not fun, and it would require you to be hospitalized, at least briefly. On the other hand, if your immune system was close enough to that of your brother, it would increase the chances that the new bone marrow would provide blood-forming stem cells without cancer cells. On the downside, an allogeneic transplant is more expensive. Are you willing? It might help your brother to survive.

7. The focus of cancer chemotherapy is on DNA synthesis, rather that RNA or protein synthesis, because the latter two are going on all the time, whether a cell is dividing or not. DNA synthesis is specific to cells that are committed to divide. Thus, a treatment that blocks DNA synthesis will affect cancer cells and only the somatic cells that are also dividing. This strategy allows damage to the tumor with a lesser chance of side effects.

8. The selection of drug-resistant strains of cancer cells is the result of killing the cells that do respond to the drug. Only those that are resistant can continue to grow and divide. This is an example of survival of the fittest, a principle of Darwinian evolution. Selective breeding is similar in the sense that the breeder is making decisions about what phenotype is fit (a cow that produces more milk, a horse that is stronger, a plant with better drought resistance, etc.). One key difference is that in selective breeding the breeder is making the choice directly, whereas with chemotherapy a doctor is defining an environment that makes the selective choice.

9. A smaller tumor is commonly easier to extirpate than a larger one, at least down to an inch or so in diameter. The likelihood of displacing or damaging nearby organs is less, the amount of stroma to be removed is less, and the trauma to the patient is reduced. Thus, if radiation or chemotherapy can make the tumor smaller, the chances of an effective and nondestructive surgical operation are increased.

10.   This question is interesting to think about, but there is no clearly right answer. When you are using a pure chemical, the chances are very good that its action will be reproducible. Moreover, it is likely that a doctor knows what its effects will be. A natural substance, like a medicinal mushroom or a brewed green tea, contains a large number of chemicals, some of which may do you good, most of which will be irrelevant, but some of which might harm you. One point of view says, Try the chemical zoo and see what happens. Another point of view says, Stick with drugs of known chemistry. In some cases there will be clear data that favor using the well-defined drug (for example, with imatinib, for which there is no equivalent in alternative medicine). In cases like advanced lung cancer, where no single drug has yet been found to have a dramatic positive effect, the decision is harder.

The immune system and its relationship to cancer

CHAPTER 9

The immune system is the sum of all the ways a human body protects itself from infection by other organisms: viruses, bacteria, protozoans, and even more complex pathogens, such as insects and worms. The essential feature of the immune system is its ability to identify things that don't belong in the body, be they molecules or other organisms, and prevent them from causing harm. Cancer poses a special problem for the immune system because the hyperproliferating cells that are transforming into a special kind of pathogen began their lives as normal parts of the body, cells that were accepted by the immune system as a part of self. To combat a cancer, the immune system must therefore learn to identify tumor cells as foreign.

The human immune system is best described in two segments, one based on our body's *innate* characteristics that help to stave off infection and one based on our ability to *adapt* to and deal with the presence of invaders. The first part of this system, called innate immunity, includes some straightforward defenses that are easy to understand, such as our skin and its role in keeping microorganisms out. Innate immunity also includes some more sophisticated defenses, such the action of certain white blood cells that can bind to foreign cells and kill them. The second part of the system, called adaptive immunity, is based on other kinds of white blood cells that make proteins capable of specific interactions with the current invader. The most remarkable aspect of adaptive immunity is the way by which it can generate these interacting molecules, called antibodies and T-cell receptors, in such a way that they react with foreign molecules but not with the molecules of our own cells.

The first goal of this chapter is to help you learn enough about both innate and adaptive immunity to understand the essential mechanisms by which they work. We will then build on this knowledge to describe both the challenges the immune system faces in eliminating a cancer and the reasons why many doctors believe that the immune system often identifies and eliminates tumors before they become medical problems. The material presented here will also allow you to understand some of the innovative treatments for cancer that are now under development by scientists who seek ways to gear up a patient's immune system and generate a strong anti-cancer action from within the patient's own body, a subject treated in Chapter 10.

INNATE IMMUNITY

ADAPTIVE IMMUNITY

HUMORAL ADAPTIVE IMMUNITY

CELLULAR ADAPTIVE IMMUNITY

THE IMMUNE SYSTEM AND PREVENTION OF CANCEROUS PROGRESSION

THE IMMUNE SYSTEM AND CANCER TREATMENT

LEARNING GOALS

1.   Outline the multiple components of innate immunity, and describe why each is valuable for your staying healthy.

2.   Describe the signals coming from stressed or pathogen-infected cells that stimulate white blood cells to kill the signaling cells and any pathogens they contain before damage or infection can spread.

3.   Describe the key cell types of the adaptive immune system and specify the basic functions of each.

4.   Diagram the steps of DNA rearrangement and RNA processing that allow B cells to make a wide range of antibody molecules.

5.   Diagram the ways by which the adaptive immune system avoids making antibodies or T-cell receptors to the cells and molecules of your own body.

6.   Give two examples of situations where the innate and adaptive immune systems work together.

7.   Summarize the evidence that the immune system is actively involved in controlling cancer.

8.   Describe two ways in which T cells can combat cancerous growth.

9.   Describe two ways by which cancer cells can inhibit the action of the immune system.

10.   Describe the strengths and limitations of the cancer treatments that use monoclonal antibodies to growth factor receptors as a strategy for limiting tumor growth.

INNATE IMMUNITY

Innate immunity, meaning the immunity you were born with, refers to the features of your body that help to protect you from disease without any learning or instruction. This kind of immunity is simply a result of the way your organs are built. FIGURE 9.1 diagrams several functions, sites, and mechanisms of innate immunity; it also displays the broad functions of **adaptive immunity** and the cells that accomplish these tasks. Features and properties of adaptive immunity are described later in this chapter.

Innate immunity is a fast and nonspecific system of responses to attempted or successful invasions by a pathogen. Within seconds, or up to a few hours of encountering a pathogen, innate immunity springs into action, because the cells and molecules of innate immunity are always ready. These responses are not specific to particular pathogens; the immune cells and molecules of the innate immune system attack any pathogen. In contrast, adaptive immunity can take several days or even weeks to become effective. This time is needed to prepare the immune molecules and cells that can target specific pathogens. Innate immunity is a consistent source of protection, but adaptive immunity changes quite dramatically in response to whatever new pathogens you encounter.

Covering epithelia have innate protective functions.

Several features of the human body enable us to fend off infections and the diseases they might cause. Skin offers valuable protection against most of the bacteria, viruses, and other parasites that inhabit the world around us. The importance of skin is demonstrated by the relative frequency with which we get an infection at places where skin is cut versus places where it is not. Our so-called inner skins are also valuable for the same purpose: the epithelia that line our throat, stomach, and intestine help to confine

conceptual organization of the immune system				
Innate Immunity			adaptive Immunity	
protection by epithelia	protection by molecules	protection by cells	antibody production	cell-mediated action
skin	RNases	neutrophils	B lymphocytes	T lymphocytes
throat	DNases	monocytes	antibody production (plasma cells)	T helper cells
stomach	proteases	macrophages	memory cells to help fight subsequent, similar infections	T killer cells
intestine	enzymes to cut polysaccharides	natural killer cells (NK cells)		T regulatory cells
lung	complement			
urethra				

FIGURE 9.1 **Components of the immune system.** Innate immunity provides rapid responses to infection by a pathogen. The protections offered by epithelia and pathogen-fighting molecules are always on, so there is no lag between an assault and protective action. The protection provided by cells of innate immunity offers a rapid backup to the immediate factors. Neutrophils, monocytes, and macrophages are all phagocytes that can ingest pathogens as well as host cells killed by pathogen attack. NK cells can kill some pathogens directly. All these cells can also cooperate with adaptive immunity. They are stimulated into action by antibodies bound to the surface of a pathogen. The adaptive immune response, on the other hand, takes time to develop. Its great strength is the specificity with which it can bring its actions to bear. B lymphocytes produce antibodies that react specifically with antigens on pathogens. A branch of this system provides long-lived cells (memory cells) that can mobilize quickly if there is a subsequent attack by the same or a similar pathogen. The cell-mediated action of T cells provides three important facets of adaptive immunity: highly specific killing of pathogenic cells, help for B cells in developing antibodies specific for particular antigens, and regulation of all immune activities in an effort to assure that the immune system does not act against cells and proteins of the host.

the large numbers of bacteria that live in the lumens of those tubes, preventing them from getting into our blood and internal cells. Moreover, inner epithelia commonly secrete materials that supplement their protective function. The epithelium of the stomach secretes acids that make an environment that is hostile to many of the microorganisms we might ingest. Most internal epithelia are coated with mucus, a viscous and slimy aqueous solution that contains polysaccharides with acid groups bound. This material adds to the protection provided by the cellular barriers. The problems that result when the linings of these tubes are no longer robust, for example, if one has an ulcer, are again evidence for the importance and success of epithelia in preventing infection.

Soluble molecules contribute to innate immunity.

Some skin cells make and secrete enzymes that degrade RNA and DNA. These nucleases account for part of our skin's ability to keep viruses from infecting our cells, because they can degrade the nucleic acids that viruses use as genomes (Chapter 6). Some body fluids, like tears, and some of the extracellular spaces in skin contain lysozyme, an enzyme that degrades the walls of bacteria. Lysozyme cuts a sugar polymer that gives bacterial walls much of their strength (see Chapter 3 and Figure 3.13) and thereby promotes cell lysis and death. Extracellular spaces also contain antibacterial peptides (short strings of amino acids that are either synthesized directly or cut from a protein made on a ribosome). These small molecules can induce a bacterial cell to burst and die.

White blood cells help fight against infection.

Inside our bodies, there are several types of cells whose job is to back up the protective activities of epithelia. These are our white blood cells, or **leukocytes** (*leuko* is the Greek word for white). Different types of leukocytes are important for innate defense against infection: **neutrophils**, which are usually the first to arrive at a site of infection, and monocytes, which can mature into **macrophages**, large cells that are specialized to engulf other cells and degrade them (FIGURE 9.2). All these cells are **phagocytes** (*phago* comes from the Greek word for to eat). Once the

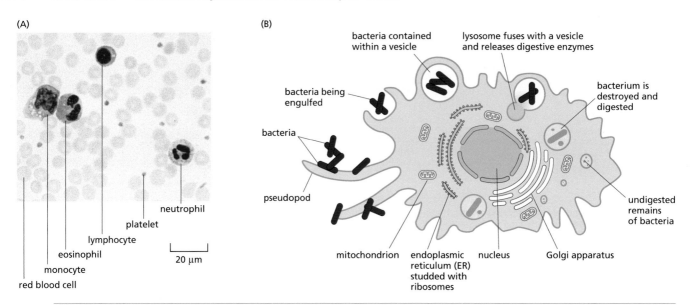

FIGURE 9.2 White blood cells are important parts of the immune system. (A) Light micrograph of blood, showing four kinds of white blood cells. (1) Monocytes can mature into macrophages, which are phagocytic cells important for innate immunity. (2) Neutrophils are another kind of phagocyte, usually the first to arrive in dealing with bacterial infections. (3) Lymphocytes are of two kinds: B and T cells. These cannot be distinguished in the light microscope, but all lymphocytes are important in providing acquired immunity. (4) Eosinophils get this name from their affinity for the dye eosin, which stains proteins. Their cytoplasm is filled with granules that contain enzymes, like RNases and proteinases. They are important in inflammation and in fighting large pathogens, such as worms. Most of the cells in this image are red blood cells (pale pink circles). Blood also contains numerous small objects called platelets, which are fragments of cells. Platelets play a key role in the formation of blood clots. (B) A macrophage phagocytosing bacteria. Multiple stages in this process are shown in different parts of this diagram. (Adapted from Alberts B, Johnson AD, Lewis J et al. [2015] Molecular Biology of the Cell, 6th ed. Garland Science.)

phagocytes reach an infection, they phagocytose both the bacteria that got through the skin and any of our own cells that the pathogens have already killed by secreting toxins. The resulting mixture of fluid and dead cells (most commonly neutrophils) is what we call pus.

The names of particular leukocytes are usually a reflection of what a microscopist could see many years ago, since these cells were identified and named long before their functions were understood. Neutrophils bind both acidic and basic dyes, so they are referred to as neutral. Monocytes have a single round nucleus. Macrophages are literally big eaters. They commonly form from monocytes in the presence of infecting microorganisms, but some macrophages inhabit permanent locations to help block infection, for example, in lung and liver. All these cell types can change their shape and become thin enough to pass through the walls of blood vessels (see Figure 1.12A), which is what allows them to get close to sites of infection. They phagocytose pathogens by extending their plasma membranes to surround the target cell, forming what is called a pseudopod, meaning a false foot (FIGURE 9.3). Then, they can draw the pathogen into a special compartment within their cytoplasm (Figure 9.2B). Once they have internalized the pathogen, they use several methods to kill it. For example, phagocytes make enzymes to generate ROS in the pathogen-containing compartment (for a review of ROS, see Chapter 5). For a cartoon representation of phagocytosis, see MOVIE 9.1. Thereafter, the pathogen-containing compartment fuses with cytoplasmic organelles called lysosomes that contain digestive enzymes (Figure 9.2); these degrade proteins, polysaccharides, lipids, and nucleic acids.

Several of the white blood cells shown in Figure 9.2A contribute to innate immunity, but **lymphocytes** are different. The get their name from the fact they frequently inhabit the lymphatic system, the lymph nodes and

bacterium   pseudopod   plasma membrane

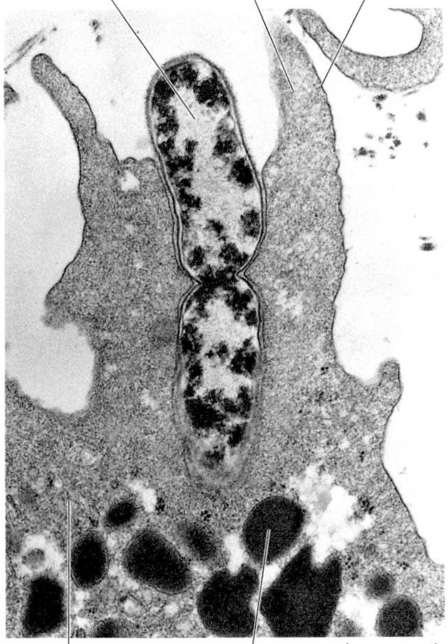

phagocytic white blood cell   cytoplasmic granule filled with digestive enzymes   1 μm

FIGURE 9.3 A phagocytic white blood cell engulfing a pathogen. A bacterium in the process of dividing is being surrounded by pseudopods (Greek for false feet) from a phagocyte. In this electron micrograph, you can see that the plasma membrane of the phagocyte is closely applied to the surface of the bacterium. This close fitting continues as the pathogen is brought into the cytoplasm of the phagocyte. The internalized vesicle of plasma membrane forms a boundary that isolates the pathogen from the cytoplasm of the phagocyte. The phagocyte contains membrane-bound granules filled with toxic substances, like digestive enzymes and enzymes to generate ROS. The membranes of these granules fuse with the membrane surrounding the pathogen, dumping the granule's contents onto the pathogen, which usually kills it quickly. (Courtesy of Dorothy F. Bainton, from Williams RC Jr & Fudenberg HH [1971] Phagocytic Mechanisms in Health and Disease. Intercontinental Medical Book Corporation.)

lymphatic vessels that drain extracellular fluids back into the blood (see Chapter 2, Figure 2.9, and Chapter 4, Figure 4.12). Lymphocytes are involved in adaptive immunity, so they will be described later in this chapter.

Cells of the innate immune system are often attracted to and activated by small molecules.

Leukocytes are drawn to a site of infection by several kinds of molecular signals. Bacteria secrete small molecules that serve as signals to cells of the innate immune system. These help to bring leukocytes into play, increasing the chances of combating infection effectively. Moreover, if your body's cells are damaged by a pathogen, or even by some mutations, molecules that would normally reside inside the cell can leak out into the extracellular space. These molecules become signals that attract phagocytes. The general word for such signals is chemoattractants (FIGURE 9.4).

One kind of chemoattractant is a class of small proteins called **cytokines** (Greek for cell movers). Cytokines are secreted by pathogen-infected cells, both to attract and to activate the leukocytes that will help fight infection. In other books about this subject or on the Web, you may also meet the word chemokine. This word refers to the subset of cytokines involved specifically in chemoattraction, versus cell activation. The word **lympho-kine** refers to cytokines that signal specifically between lymphocytes. Each of these signaling molecules has its own receptor, so even though the attractant molecules are secreted, and are therefore diffusing throughout the extracellular medium, there is specificity in which cells will respond. A cell must have the right receptor for a given cytokine before it will sense its presence and change behavior. This implies that each cell type in the immune system will have a particular set of membrane-bound receptors that help it to receive specific signals. The receptors on the surface of a leukocyte are one way scientists distinguish among them, since many of these cells look identical.

Think back to what you have previously learned about the signals that cells receive and interpret. What processes do you imagine the phagocytes use to sense the molecules secreted by cells that are in trouble?

SIDE QUESTION 9.2

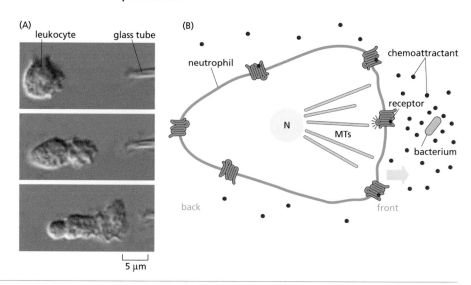

FIGURE 9.4 Several types of cells can perceive certain molecules in solution and migrate in the direction from which the molecules are coming. (A) Three images, taken minutes apart, of a cell that has receptors for the chemical being released from the tiny glass tube. The presence of the chemical induces the cell to move, and the direction of motion is toward the place where the concentration of the chemical is highest, a process called chemotaxis. (B) A diagram of chemotaxis in which a white blood cell (for example, a neutrophil) is sensing the presence of molecules secreted by a bacterium (a chemoattractant). Receptors on the surface of the neutrophil activate that cell's cytoskeleton to move the cell toward the highest concentration of the chemoattractant. This direction sensing is possible because the cell's receptors on the side exposed to the higher concentration have more chemoattractant bound than receptors on the opposite side of the cell. Chemotaxis is the mechanism that allows white blood cells to become polarized and move toward sites of infection, where their phagocytic abilities are needed. (N, nucleus; MTs, microtubules. The asymmetric positions of these organelles are aspects of the cell's becoming polarized, so it can move toward the bacterium.) (A, from Weiner OD, Servant G, Tranum-Jensen J et al [1999] *Nat Cell Biol* 1:75–81. With permission from Springer Nature.)

SIDE QUESTION 9.3

Figure 9.4B shows a diagram of a white blood cell next to a bacterium that is emitting chemoattractants. Those molecules (red dots) are shown bound to more of the receptors on one side of the white blood cell than on the other. Does this seem likely to be true? Why or why not? How might the white blood cells use this asymmetry of binding to define the direction in which it will move?

SIDE QUESTION 9.4

Do you think it is advantageous for an animal to kill its cells that are infected by viruses or other pathogens? Why? What would be the advantages and/or disadvantages of lysing a pathogen-infected cell in order to kill it?

The signals that a pathogen-infected cell sends are an important part of innate immunity. For example, some virus-infected cells make and insert into their plasma membranes a protein called the **Fas-receptor**. This protein interacts with the **Fas-ligand**, a protein exposed on the surface of certain leukocytes. When Fas-ligands interact with Fas-receptors, they induce death by apoptosis in the cell that expressed the Fas-receptor. Imagine how valuable it would be for our body's fight against cancer if doctors could find a way to make all of the parenchymal cells in a tumor express the Fas-ligand.

A less specific but important part of signaling by cells of the innate immune system is the secretion of a particular kind of cytokine by macrophages that are engaged in phagocytosis. These cytokines induce the phenomenon of inflammation: that is, the in-migration of other leukocytes and the expansion of local blood vessels to increase blood flow in the area. This causes swelling and reddening of the local tissue, a familiar aspect of local infection. Inflammation can contribute to the healing process, but if it becomes severe, it can be painful, and, if it persists for a long time, it can become harmful.

The innate immune system includes two ways to kill cells that are recognized as foreign.

One mechanism for killing foreign cells is a set of about 30 blood proteins called the **complement system**. Some of these proteins can bind the surfaces of foreign cells because their surfaces contain unusual macromolecules for which some complement proteins have affinity. The binding of complement proteins makes the foreign cells recognizable by phago-

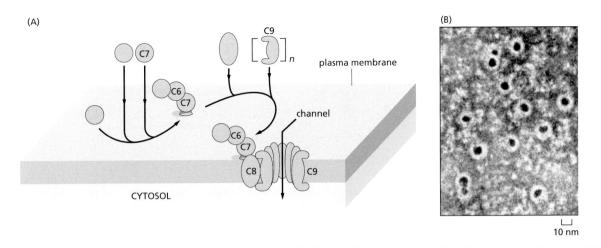

FIGURE 9.5 Proteins of the complement system bind to the surface of a pathogen and form a pore. The complement system is composed of soluble proteins (green) that reside in blood. Some of these proteins can bind to a pathogen's surface and then start activating each other. Activation is accomplished in most cases by proteolysis, a bit like the cascade of events that initiates apoptosis. (A) Proteins of the complement system, called C1, C2, and so forth, activate assembly of the protein C9, making a channel in the plasma membrane of the pathogen, which causes that cell to lyse and die. (B) Electron micrograph of pores made by components of the complement system that have already inserted into a plasma membrane. The assemblies of C9 proteins appear as white rings because the heavy metal used as a stain is excluded from the places where the proteins sit, a method called negative staining. (A, adapted from Alberts B, Johnson AD, Lewis J et al [2015] Molecular Biology of the Cell, 6th ed. Garland Science. B, from Bhakdi S, Hugo F & Tranum-Jensen J [1990] Blut 60:309–318. With permission from Springer.)

cytes. If a phagocyte doesn't come quickly, however, other components of the complement system join the initial binders and assemble into tiny cylinders that insert into the membrane of the invading cell. They form a pore big enough to drain cytoplasm into the extracellular space, killing the foreign cell (FIGURE 9.5).

A second cytotoxic component of innate immunity uses a special lymphocyte, called a **natural killer (NK) cell**. These lymphocytes have three ways of killing cells that don't belong in a healthy body: (1) The plasma membranes of NK cells expose the Fas-ligand, mentioned above, so if they touch a cell that is expressing the Fas-receptor, they will induce it to initiate apoptosis. (2) NK cells can secrete proteinases. (3) NK cells can also secrete a protein called **perforin**, which works like complement to assemble pores in the plasma membrane of its target cell. These pores allow both entry of the secreted proteinases and cytoplasmic leakage from the target cell, promoting apoptosis in the affected cell (FIGURE 9.6). Thus, NK cells work together with other components of innate immunity to provide a powerful defense against infecting cells, including bacteria. NK cells are also thought to be important in the body's control of cancer, as will be described later in this chapter.

A critically important issue for innate immunity is how NK cells can be sure they act only on cells that are potentially harmful.

One way NK cells recognize appropriate targets is based on the Fas-ligand/ Fas-receptor mechanism just described. A second way is that human cells under stress secrete proteins called **interferons** and **interleukins**. These proteins are a special kind of cytokine that attracts and activates NK cells. A third mechanism of activation is based on the adaptive immune response and will be described later in this chapter.

The proteinases secreted by NK cells are designed specifically to cleave and thereby to activate caspases, the cellular proteinases you met while learning about apoptosis (see Figure 7.15). Why is this an efficient way for an NK cell to kill the cell it has targeted?

SIDE QUESTION 9.5

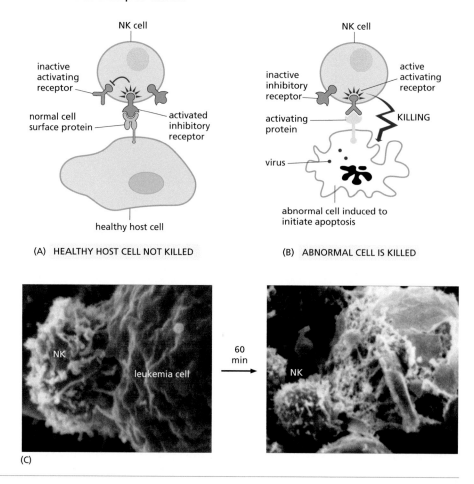

FIGURE 9.6 A natural killer (NK) cell can bind to the surface of another cell and kill it. (A) NK cells interact with other cells through surface receptors that recognize surface proteins on other cells. If a surface protein is identified as self, that cell is left alone. (B) Cells that display proteins identified as nonself lead the NK cell to initiate killing reactions. (C) Scanning electron micrographs of the close engagement between the plasma membranes of a killer cell and its target. Through this engagement and subsequent secretions, the small NK cell brings on the death of the much bigger cancer cell. At the right is a similar image, showing a later stage in the death of a leukemia cell, brought on by the action of an NK cell. (A and B, adapted from Alberts B, Johnson AD, Lewis J et al [2015] Molecular Biology of the Cell, 6th ed. Garland Science. C, from Herberman R [1985] Mechanisms of Cytotoxicity by NK Cells. Academic Press.)

ADAPTIVE IMMUNITY

The actions of the innate immune system are supported by an even more complex and effective system for defense against pathogens. Our adaptive immune system can make proteins that interact specifically with virtually any molecule, particle, or cell that doesn't belong in our bodies, be it a virus, a bacterium, or a cancer cell. Because the range of possible pathogens is very large, the number of molecules the adaptive immune system can make is even larger, a remarkable accomplishment. Moreover, the adaptive immune system must act *only* on foreign or abnormal molecules, not on any of the numerous molecules that are essential parts of our own bodies, making the task seem almost impossible. Nonetheless, our adaptive immune systems do exactly this and usually do it very well.

There are two key sets of molecules in adaptive immunity: **antibodies** and **T-cell receptors**. You have previously met antibodies in Chapters 1, 2, 6, and 8. Recall that an antibody is a soluble protein, made by our bodies, that binds specifically to one other molecule, which is called its **antigen**. T-cell receptors are also proteins, but they reside permanently in the membrane of lymphocytes called **T cells**. T-cell receptors, like antibodies, are designed to bind to one specific antigen. They use their binding to this antigen to generate a signal that induces the T cell to perform a specific job. There are three kinds of T cells, and the jobs they do are different, but all of these jobs help to kill the pathogens expressing that particular antigen. An essential feature of both antibodies and T-cell receptors is that the adaptive immune system has developed ways to make very large numbers of both kinds of molecules, many more than the number of genes in our genome. From these pools of molecules, adaptive immunity selects the ones that will be valuable in fighting a particular infection but will not have damaging effects on our own cells and molecules. The following sections describe how our bodies achieve these remarkable goals.

The key cells of the adaptive immune system are lymphocytes.

Both B and T cells, like red blood cells, are made in the bone marrow. In that tissue, a multipotent stem cell divides to produce cells that are committed to become lymphocytes, but they are not yet differentiated, so they are called lymphoid progenitor cells (Figure 4.12). These progenitors go on to initiate differentiation into the ancestors of specific kinds of lymphocytes, whereupon they either remain in the bone marrow and become B cells or migrate to the thymus gland where they differentiate into T cells (T stands for thymus) (FIGURE 9.7). **B cells** are the source of antibodies. Since most antibody molecules are soluble proteins, this branch of the adaptive immune system is called **humoral adaptive immunity** (*humor* is the Latin word for moisture), meaning the part of adaptive immunity that relies on soluble proteins. T cells accomplish the parts of adaptive immunity that involve lymphocytes interacting directly with unwanted cells, such as pathogens or cancer cells, so these parts are called **cellular adaptive immunity**.

Figure 9.7 distinguishes between two kinds of lymphoid organs: central and peripheral. The central lymphoid organs (bone marrow and thymus) are involved in making lymphocytes; the peripheral lymphoid organs

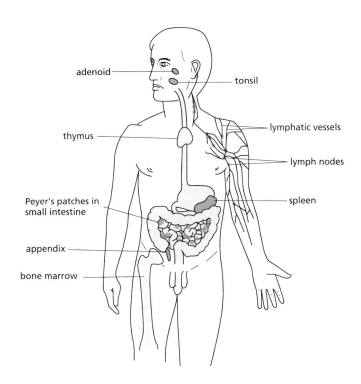

FIGURE 9.7 Relative positions of the many organs of the adaptive immune system. Bone marrow and the thymus (shown in yellow) are considered central lymphoid organs. All lymphocytes are born in the bone marrow. B cells continue their development there, whereas T cells migrate to the thymus to become mature. After maturation, both of these lymphocytes migrate to the many components of the peripheral lymphoid system (shown in blue), where they can interact with cells and molecules that don't belong in the body. The peripheral lymphoid organs include the lymph nodes, where metastatic cancer cells often go and can be found by a skillful pathologist, aiding in cancer diagnosis. Adenoids and tonsils are situated where they can interact with pathogens that may enter our bodies through the nose or mouth. The other blue structures help to identify problem cells and molecules that are already inside our bodies. The Peyer's patches in the small intestine are special because they must accept the large number and variety of bacteria that live in our intestines to help with digestion but reject other bacteria that might be harmful to us.

(lymph nodes, adenoids, tonsils, spleen, and Peyer's patches in the small intestine) are places where mature lymphocytes can interact with molecules and cells that don't belong in the body. Lymphocytes move from the central to the peripheral lymphoid organs and elsewhere in the body both with the flow of blood and by a slower migration through the lymphatic vessels, which are partially diagrammed on the right side of Figure 9.7. Remember, though, that lymphatic vessels, also called ducts, are in fact situated throughout the body, as diagrammed in Figure 2.9A.

HUMORAL ADAPTIVE IMMUNITY

Since many pathogens and foreign molecules reside in blood, lymph, and extracellular spaces, it is valuable to have one part of adaptive immunity make molecules that can diffuse in these liquids and encounter pathogens wherever they might be. This task is accomplished by antibodies. To know how they work, you must understand their structure and how they are made.

Antibodies contain both variable and constant regions.

Antibodies are soluble proteins made from four subunits, two big ones called heavy chains and two smaller ones called light chains, where chain means a sequence of amino acids (FIGURE 9.8A). There are several classes of antibody molecules, but all have essential features in common. Both their heavy and light chains contain regions with the very same amino acid sequence in all antibody molecules of that kind, so these are called the **constant region** (also called fragment constant, and abbreviated **Fc**). The other region of each antibody chain differs from one antibody molecule to another, so it is called the **variable region**; this region is responsible for the ability of an antibody to bind tightly to its own particular antigen, such as a protein or a chain of sugars on the surface of a bacterial cell. Only a few antibodies will have the right structure to bind tightly to any given antigen. Given that antigens can be of virtually any shape, there must be a huge variety of antibodies to provide at least one antibody that can bind to each of them. The ability to create this variety in antibody structure is at the heart of the adaptive immune response.

The term adaptive reflects the ability of the immune system to adapt its synthetic activities to the particular foreign antigens that are present at a given time. Thus, infection by a flu virus induces our bodies to make one group of antibodies, whereas a measles or streptococcal infection or a developing cancer cell will each bring on a different group of antibodies. Your adaptive immune system can make a large variety of antibodies and is thereby able to protect your body from almost any infection you encounter.

The portion of the antibody that reacts with its antigen, known as the antigen-binding site, is only a fraction of the antibody molecules as a whole (FIGURE 9.8B). It is made from the variable parts of both the heavy and light chains. Since there are two of each of these chains in every antibody molecule, there are two antigen-binding sites in each antibody (Figure 9.8A). The antigen-binding sites of an antibody usually react with only a small part of the molecule that is its antigen. This part of the antigen is called the **antigenic determinant**. The size of a single antigenic determinant is not rigidly defined, but when the antigen is a protein, it is usually between 8 and 11 amino acids. If the antigen is a polysaccharide, it can be one or a few sugars. The important idea is that whatever the structure of the antigenic determinant might be, the variable region of the right antibody will bind to it tightly. The immune system usually makes

SIDE QUESTION 9.6

Given the fraction of the surface of the antigen that is interacting with the variable part of the antibody, as shown in Figure 9.8B, how many different antigenic determinants do you think there could be on the antigen shown in this figure?

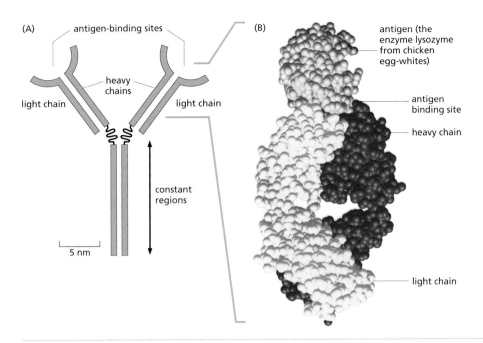

FIGURE 9.8 **The structure of an antibody molecule.** (A) Green lines indicate the four polypeptide chains that make a single antibody molecule. Two light chains and two heavy chains assemble to form one protein. Toward one end of the protein, there are two identical antigen-binding sites. These are the domains that interact with an antigen. Thus, antigen binding is defined in part by the structure of the heavy chains and in part by the structure of the light chains. The other part of an antibody, called the constant region, is the same for all antibody molecules of a given kind, regardless of the antigen they bind. This constancy allows all antibodies to interact with other parts of the immune system in the same way. (B) A molecular model of the variable region of an antibody interacting with its antigen. This part of a light chain (yellow) and of a heavy chain (purple) interact with the antigen (blue) to which they bind. In this model, the positions of all atoms in both antibody and antigen are shown as shaded spheres. Note that the size of the region on the antibody that interacts with the antigen is comparatively small. Likewise, the part of the antigen that interacts with the antibody, the antigenic determinant, is only a small portion of the antigen. (B, adapted from Amit AG, Mariuzza RA, Phillips SE & Poljak RJ [1986] Science 233:747–753. With permission from AAAS.)

multiple antibodies that can bind to any one antigen, but the ones that bind most tightly are the important ones in what happens next. For a more detailed description of the structure of an antibody molecule and the way it interacts with its antigenic determinant, watch MOVIE 9.2.

B lymphocytes differentiate to synthesize one and only one kind of antibody.

As B lymphocytes begin to differentiate in the bone marrow, they start to express DNA that encodes both the heavy and the light chains of a particular antibody. This antibody is first made in a form that can incorporate directly into the B cell's plasma membrane, with its antigen-binding part projecting into extracellular space (FIGURE 9.9). Shortly thereafter, this partially differentiated B cell is released from the bone marrow to wander through the body, both flowing with the blood and wandering through the lymphatic ducts and nodes. The antibody molecule projecting from its plasma membrane acts like the signal receptors you have met before. If the B cell encounters an antigen to which its antibody binds, the binding sets off a signaling cascade, inducing this B cell to proliferate. In short, the antigen works like a growth factor *specifically* for the B cells that make antibodies that will bind it, not for other B cells that would make irrelevant antibodies. Remarkable!

Each B cell can make one and only one form of antibody, so the large range of antibodies your body can make requires a large number of different B cells. When a specific antigen is present, perhaps as a result of

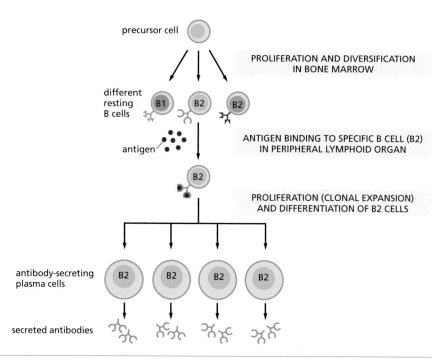

FIGURE 9.9 Stages in the development of a mature B cell. B cells are born in the bone marrow by division of a pluripotential stem cell to form a B-cell precursor. While still in the marrow, they go through an initial step of differentiation in which they define the structure of the variable part of the antibody they will make; this step leads to B-cell diversity. The antibody is first made on the rough ER, so it is transported to the cell surface and one part of it is exposed to the solution outside the cell. There, it works as a receptor that is exposed to all antigens present in the body. Any foreign antigen that binds to this antibody works as a mitogen (a stimulant of cell division) for that cell; the tighter the antigen binds, the more successfully it stimulates reproduction of that particular B cell (here the B2 cell, not the others), increasing the number of cells that make this particular antibody. Thereafter, the B cell differentiates into a plasma cell, an antibody-secreting cell that produces a secreted version of that antibody. The secreted antibodies can now help to fight the infection represented by that particular antigen. (Adapted from Alberts B, Johnson AD, Lewis J et al [2015] Molecular Biology of the Cell, 6th ed. Garland Science.)

a particular infection or a cancer cell with an abnormal surface protein, the B cells that can make the relevant antibodies not only divide multiple times, they also differentiate to become antibody factories. They develop large amounts of rough ER, the membranous organelle that binds ribosomes to make proteins for export from the cell (FIGURE 9.10). These differentiated B cells can secrete many soluble copies of the antibody they previously presented on their plasma membrane. This highly differentiated form of the original B cell is called a **plasma cell** or sometimes an **effector B cell**. These two names come from the facts that such cells were first discovered in blood plasma and are now known to be the B cells that effect an immunological response. Plasma cells devote most of their time and energy to the synthesis and secretion of the antibodies that can help to fight the infection or cancer at hand. The proliferation of particular B cells and their differentiation into plasma cells, followed by the synthesis of the right antibody molecules, is part of what is meant by mounting an immune response.

The key components of humoral adaptive immunity are antibodies.

The soluble antibodies secreted by plasma cells are called **immunoglobulins G** or **IgGs**. They can bind to their antigen wherever it might be and initiate the next steps in the action of the humoral arm of the adaptive immune system for coping with infection. If the antigen-binding sites of an IgG have bound to their antigen, for example, a molecule on the surface of a bacterium, the constant region (Fc) of the antibody is now

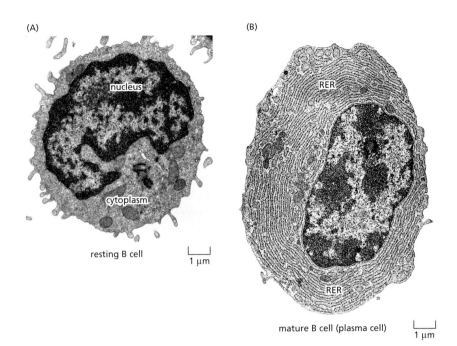

(A)

nucleus

cytoplasm

resting B cell

1 μm

(B)

RER

RER

mature B cell (plasma cell)

1 μm

FIGURE 9.10 Electron micrographs of B cells before and after maturation. (A) A resting B cell that has already differentiated to choose the anti-body it will make, but has not yet been stimulated by antigen binding to differentiate into a plasma cell. At this stage, B cells are characterized by a small and comparatively unspecialized cytoplasm. (B) A fully differentiated B cell is called either a plasma cell or an effector B cell. Nuclear size is unchanged, but the nucleus is producing large amounts of the mRNAs that encode the light and heavy chains of a particular antibody molecule. The enlarged cytoplasm is filled with rough endoplasmic reticulum (RER), so the cell can synthesize antibodies for secretion at a high rate. (A, courtesy of Dorothy Zucker-Franklin; B, courtesy of Carlo Grossi. A and B, from Greaves MF, Grossi CE, Marmont AM & Zucker-Franklin D [1988] Atlas of Blood Cells: Function and Pathology, 2nd ed. Lea & Febiger.)

projecting into extracellular space (FIGURE 9.11A). The Fc is not just a passive part of the IgG molecule. Once the variable part of an IgG has bound its antigen, the Fc takes on a shape that binds surface receptors on phagocytes and signals to them to engulf the antigen–antibody complex, including the entire structure to which the antibody is bound. Likewise, IgGs bound to the surface of a mammalian cells that presents foreign anti-gens on its surface will elicit the killing activity of NK cells (FIGURE 9.11B).

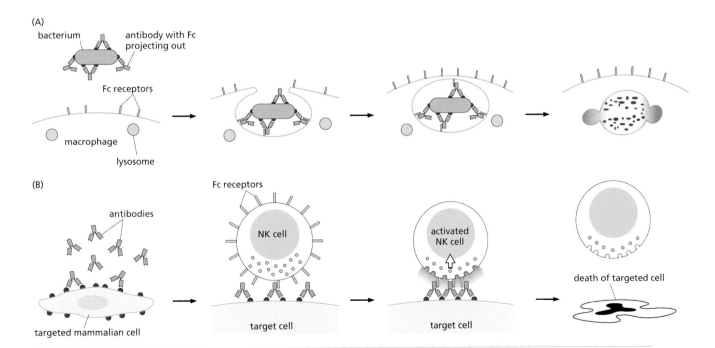

FIGURE 9.11 Coating a pathogen with antibody can lead to its elimination by cells of the innate immune response. (A) An invading bacterium presents antigens that have elicited the formation of antibodies (IgGs) by B cells. These antibodies (shown in blue) now coat the surface of the bacterium (shown in pink). The plasma membrane of a phagocytic cell, such as a macrophage, is covered with receptors that will bind to the constant regions (Fc) of any IgG molecule whose variable portion is binding to its antigen. This binding induces phagocytosis, which leads to the death of the pathogen in the cytoplasm of the phagocyte. (B) If a targeted mammalian cell displays antigens (shown in red) that are recognized by IgGs, this cell type too will become coated by IgGs. The Fc regions of these antibodies can bind to and activate an NK cell, leading to the death of the targeted cell.

SIDE QUESTION 9.7

The above description shows how antibodies that recognize antigens on foreign cells can lead to the death of the invader. What would be the consequence of a person making antibodies that recognized antigens on the surface of her own cells?

Thus, a virus, bacterium, or cancer cell that is floating in blood may at any time encounter IgG molecules that recognize (i.e., bind to) antigens on its surface. As a result, the antigen is now identified as unwanted, so it is phagocytosed and degraded or killed. These are examples of ways in which the adaptive and innate immune systems work together to eliminate pathogens.

The power of the humoral arm of the adaptive immune system derives in large part from the great range of antibodies it can make. Within this range there are likely to be one or a few antibodies that will bind to any foreign antigen that comes into any given person, either from a pathogen or from the surface of a mutated cancer cell. Indeed, the number of different antibodies that our B cells can make is far larger than the number of genes in the human genome. For years this posed a mystery for scholars of the immune system, but careful work has now revealed how the immune system achieves this remarkable job. We can start to understand the relevant processes by taking a closer look at antibody molecules themselves (FIGURE 9.12). The variable part of each antibody forms the antigen-binding site. This region of the antibody is called variable because different B cells produce antibodies with different amino acid sequences in these regions. Several subregions within the variable region are called **hypervariable regions**, because they show even more variation in amino acid sequence from one antibody to another than is found in the variable regions. All the antibodies produced by one B cell have the same amino acid sequence, so after the differentiation and clonal expansion of a given B cell into a population of plasma cells, there are now many cells making exactly the same antibody. Every other differentiated B cell that divides multiple times to form a clone of cells will make a different antibody. In a given immune response, there will commonly be multiple clones of plasma cells, all making different, useful antibodies, because most antigens contain multiple antigenic determinants. This description defines an important problem: how does B-cell differentiation lead to the generation of so many different genes for making all these different IgGs? A part of the solution to this problem is described in SIDEBAR 9.1.

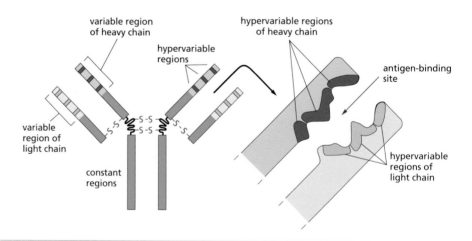

FIGURE 9.12 **Diagrams of the variable and hypervariable regions of an antibody.** The diagram on the left displays the organization of an antibody molecule: two identical light chains and two identical heavy chains are joined to form a single protein molecule by bonds between sulfur atoms (S) in each polypeptide chain. Here, the constant regions (Fc) of both heavy and light chains are shown in medium green, and the variable regions are lighter, with one exception: the hypervariable regions are shown as even darker bands. At the right is an expanded view of the variable parts of both heavy and light chains from one part of the dimeric antibody, showing how the antigen binding site of an antibody molecule is composed of all the variable and hypervariable regions of both chains. It is the specific structure of this region that gives each antibody the ability to bind tightly to only one antigen. The fact that the immune system can make a huge range of different variable and hypervariable structures accounts for the ability of antibodies to interact successfully and specifically with such a wide range of antigens.

A CLOSER LOOK: How the immune system can make so many different antibodies.

The structural variation in antibody molecules is generated by a process unique to cells of the adaptive immune system. There is a region in the DNA of every cell, called the immunoglobulin locus, which can be transcribed and translated to make antibody molecules. This region is expressed only in cells of the immune system, an example of the differentiation we have met in the development of several specialized cell types. The immunoglobulin locus contains genes that encode both the heavy and the light chains of antibodies (Figure 9.8). We can get the essential ideas about how these few genes can produce so many different antibody molecules by looking with a bit of detail at what happens to one of these genes during B-cell differentiation: a gene that encodes an antibody

light chain. Variability in the structure of the other proteins of the immune system, such as the antibody heavy chain, is achieved in similar ways.

The genes for antibody light chains can best be thought of as several adjacent regions separated by spacers (**Figure 1**). Near the left end of this gene, there is a cluster of about 40 different segments of DNA, each with a sequence that encodes an oligopeptide of the right size to be the variable part (V) of the final antibody light chain. All these sequences are different, so this part of the gene can be described as V1, V2, V3, etc. (**Figure 1**). The region near the gene's right end includes DNA that encodes the constant fragment (Fc) of this antibody subunit (C). Between the V regions and the constant region there are sev-

eral joining regions (J1, J2, J3, and so on). During the early stages of differentiation, when a B-cell precursor is still in the bone marrow but is committing to become a specific B cell, there is an unusual rearrangement of this DNA. The DNA between one of the V regions and one of the J regions loops out, is cut off, and is discarded, bringing two parts of the immunoglobulin locus that used to be far apart into immediate proximity (Figure 1). This DNA rearrangement is part of B-cell differentiation; after it has occurred, the B cell is on its way to having chosen the antibody it will make.

When a particular B cell starts to make a specific antibody, the mRNA that is transcribed starts with nucleotides that encode the single V region that is immediately next to a specific J region.

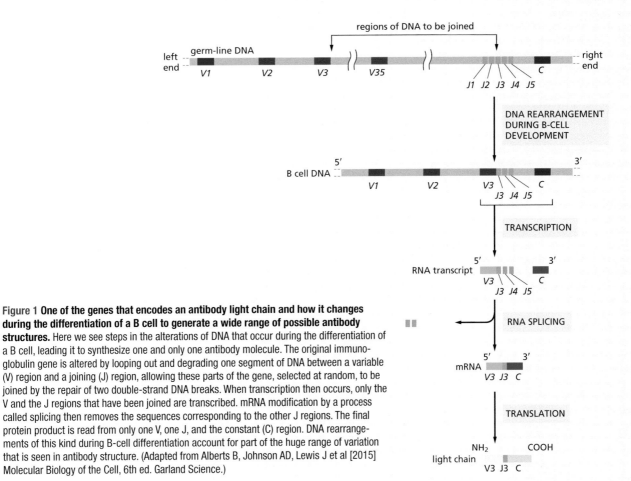

Figure 1 One of the genes that encodes an antibody light chain and how it changes during the differentiation of a B cell to generate a wide range of possible antibody structures. Here we see steps in the alterations of DNA that occur during the differentiation of a B cell, leading it to synthesize one and only one antibody molecule. The original immunoglobulin gene is altered by looping out and degrading one segment of DNA between a variable (V) region and a joining (J) region, allowing these parts of the gene, selected at random, to be joined by the repair of two double-strand DNA breaks. When transcription then occurs, only the V and the J regions that have been joined are transcribed. mRNA modification by a process called splicing then removes the sequences corresponding to the other J regions. The final protein product is read from only one V, one J, and the constant (C) region. DNA rearrangements of this kind during B-cell differentiation account for part of the huge range of variation that is seen in antibody structure. (Adapted from Alberts B, Johnson AD, Lewis J et al [2015] Molecular Biology of the Cell, 6th ed. Garland Science.)

None of the other V regions is transcribed into mRNA, so a choice has been made that means this antibody chain will include only one out of the ~40 possible V regions. Moreover, there is only one J region now adjacent to the chosen V region. If there were five possible J regions in the DNA, then the choice that puts a particular J adjacent to a particular V selects one of five possible J regions to contribute its sequence to the antibody light chain that will be made. The mechanism that will exclude from translation the other J sequences on the RNA that is transcribed from this modified DNA relies on a splicing of that RNA, which occurs after transcription is complete (Figure 1).

The key point here is that the differentiating B cell has gone through a process that has chosen one nucleotide sequence from among many possibilities. With 40 possible sequences of V region and five possible sequences of J region, there are 40 × 5 = 200 different light-chain sequences possible, based on which V and J regions were chosen during the looping and cutting of the DNA from the immunoglobulin locus. The choices of which V and J regions are used appear to be completely random, so roughly equal frequencies of all possible V and J combinations are used as multiple B cells initiate their differentiation.

Similar splicing events occur in the making of antibody heavy chains. The DNA that encodes an antibody heavy chain contains many variable regions, several joining regions, and a constant region, just like the DNA that encodes a light chain. In addition, however, the heavy chain includes D regions between its V and J regions; these too come in multiple copies that can be chosen at random during the events that produce a differentiated heavy chain gene. In short, a heavy chain gene includes about 30 V regions, 14 D regions, and four J regions, followed by one constant region. Work out for yourself how many different heavy chains this kind of B cell could make. Now when you consider the possible combinations of both heavy and light chains that can go into making one antibody, you will see that there are thousands of possibilities.

In reality, the range of antibody structures that B-cell differentiation can produce is significantly greater than is implied by the above description. When DNA is spliced out to bring a V region up against a J region, the process is not precise. The exact site where DNA is cut before it is joined back together (repairing two double-strand breaks!), can be different each time a B cell differentiation event occurs. Therefore, during the formation of the DNA that will be expressed to make a given antibody light chain, any two B cells that have looped out essentially the same DNA, bringing, say, V8 up against J5, will not make identical antibodies; the small differences in the position of the cut and splice sites give the resulting protein products slightly different amino acid sequences and therefore different structures. This is one aspect of forming the hypervariable regions found in the structures of both light and heavy chains in an IgG (Figure 9.12). A second aspect is that the immunoglobulin locus is more susceptible to random mutations than any other part of the genome, so both during normal B-cell differentiation and later during an infection, additional variation can come from changes in the DNA that is being manipulated, transcribed, and translated into antibodies. Add to this the fact that there are three different light chain genes, and all together these mechanisms allow a single person to make literally millions of different antibody molecules.

The adaptive immune system can turn itself off, but it remembers its previous responses.

The previous discussion and Sidebar 9.1 describe how our B cells can start to make antibodies that bind to a pathogen or cancer cell, but an equally important part of adaptive immunity is its ability to turn off. If we continued making antibodies to antigens that were no longer a threat, our blood would become clogged with useless proteins. For this reason, once the immune system has eliminated an antigen from your body, the relevant antibodies are slowly degraded, and the cells that made them die off with one important exception. A few special B cells, which were part of each B-cell clone that grew up in a given immune response, differentiate to become very long-lived. These long-lived B cells are called **memory cells** because they retain their partially differentiated state, with multiple copies of the antibody they make projecting through their plasma membranes to serve as antigen receptors.

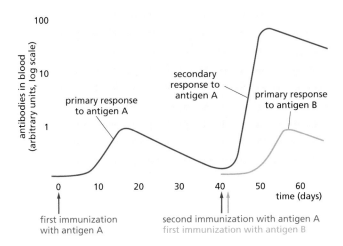

FIGURE 9.13 A graph comparing the amount of antibody in blood following initial exposure to an antigen versus subsequent exposure to the same antigen, when memory B cells are present. Both the speed of the immune response and its strength are greater when there has been prior exposure to an antigen. This phenomenon accounts for acquired immunity to many diseases through either exposure or immunization. There are, however, some pathogens that come along year after year, like colds, influenza, and malaria, to which the immune system fails to mount an effective secondary response. This unfortunate fact is a result of the pathogen's having learned how to vary its antigenicity, so your memory B cells don't do very much good at the time of the second exposure. (Adapted from Alberts B, Johnson AD, Lewis J et al [2015] Molecular Biology of the Cell, 6th ed. Garland Science.)

Memory B cells are stored in the peripheral lymphoid organs, including lymph nodes. There they remain dormant in the sense that they don't differentiate further. They are, however, able to respond quickly to a reappearance of the antigen to which their antibody binds. If at some later time you encounter a pathogen that expresses the same antigen, your memory B cells can leave dormancy and start replicating within about a day. Within a couple of days, your blood contains many relevant effector B cells, which make and secrete large amounts of the IgGs you need to fight the reappearing pathogen. The resulting immune response is both quicker and stronger than the initial immune response (FIGURE 9.13), meaning that your immune system can usually kill the pathogen or cancer cell before it makes you sick. The activities of memory B cells are the reason you will rarely get certain diseases twice. Measles, chicken pox, mumps, etc., leave enough memory B cells in your system to make a subsequent infection ineffective.

Memory B cells are also responsible for the ability to acquire immunity from a vaccine.

Any vaccine you get as a shot or pill, administered by a doctor, includes antigens from an important disease, like polio or smallpox. You mount an immune response to the antigens present in this small dose of foreign material that is usually inactivated in some way, so it won't make you sick. Since there was no virulent infection, this primary response dies down quickly, but the memory B cells you make during the response will persist and allow a fast and effective immune response to a subsequent exposure to the real pathogen. While this system is highly effective in protecting you from many pathogens, some pathogens, such as flu virus or human immunodeficiency virus (HIV), change the proteins on their surfaces so rapidly that previously generated memory B cells are of little use in combating a subsequent infection. The genetic instability of cancer cells presents the immune system with a similar challenge.

When functioning as it should, the adaptive immune system does not make antibodies to your own antigens.

The descriptions thus far have not addressed one of the most important parts of adaptive immunity: how does the system refrain from making antibodies that recognize antigens in your own body? This important issue can be understood in part by the behavior of B cells as they are differentiating. The rearrangements of DNA by which B cells define the antibody they will make (Sidebar 9.1) produce a gene whose product is a

Some pathogens can frequently change the antigens they present to their host's immune system. What are the advantages and/or disadvantages of this ability?

SIDE QUESTION 9.8

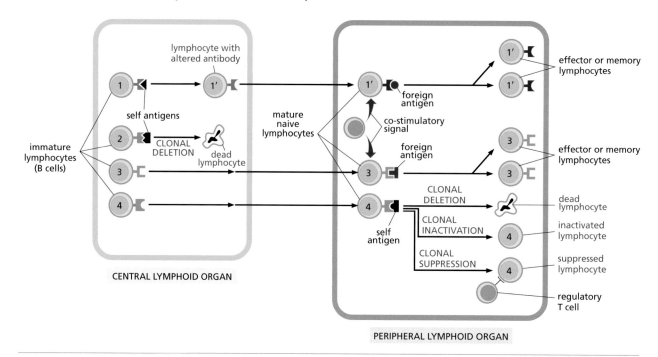

FIGURE 9.14 Editing B cells to eliminate those making antibodies that interact with self-antigens. In a central lymphoid organ, partially differentiated but still immature B cells, generated with randomly defined antibodies as surface receptors, will have different fates, depending on the antigen they bind. If the antibody presented on the cell surface (the B-cell's receptor) can bind strongly to a self-antigen that comes along while the B cell is still in a central lymphoid organ, apoptosis is induced, eliminating this B cell from the body's antibody repertoire, as indicated by cell 2. Weak binding by self-antigens is tolerated, as indicated in cell 1, but such cells increase the mutation rate at their immunoglobulin locus (by an unknown mechanism), allowing slightly different antibodies to be produced, some of which might bind well to a foreign antigen but not to self. This is called receptor editing. In peripheral lymphoid organs, B cells go through their final maturation to become either plasma cells or memory cells (cells 1 and 3). Interactions with T lymphocytes can, however, alter B-cell behavior in ways that suppress interactions with self-antigens, as indicated for cell 4 and described later in the chapter. (Adapted from Alberts B, Johnson AD, Lewis J et al [2015] Molecular Biology of the Cell, 6th ed. Garland Science.)

specific antibody that is first displayed as a membrane protein projecting into the extracellular space from the B-cell's surface. Some of these antibodies will, by chance, react with antigens in your own body. These differentiating B cells must be eliminated or inactivated before they mount an immune response against your own tissues.

After B cells have done their DNA rearrangements and have inserted an antibody into their plasma membranes, they are still in the bone marrow. Here, they are exposed to virtually every protein-derived antigen your body can make by a remarkable process that goes on in a different bone marrow cell type. The stromal cells of the marrow have the unique behavior of expressing small amounts of product from virtually every gene in your genome. Thus, while B cells are still in the bone marrow, they are exposed to almost all of your body's antigens. Any B cell in the bone marrow that displays a cell surface immunoglobulin that binds one of these antigens is now induced to initiate apoptosis. Thus, B cells that recognize self-antigens are usually killed before they can do harm. This process is called clonal deletion (FIGURE 9.14). Later, after the differentiating B cells have left the bone marrow and moved to a lymph node, they change their response to antigen binding; from then on, the binding of a molecule to their membrane-associated antibody works like a mitogen and a stimulus of differentiation, as described above. In reality, however, the avoidance of self-interacting antibodies is more complicated than this and involves interactions between B cells and a particular kind of T lymphocyte, so our next subject is T lymphocytes and how they work, both to carry out the cellular arm of adaptive immunity and to refine the ways that B cells behave.

CELLULAR ADAPTIVE IMMUNITY

Our adaptive immune system has another way to combat pathogens that is even more powerful than the humoral adaptive response for dealing with unwanted cells. Cellular adaptive immunity does not rely on antibodies but instead on analogous, highly variable protein molecules called **T-cell receptors** (**TCRs**). These molecules are proteins exposed on the plasma membranes of T lymphocytes, commonly known as T cells. Like antibodies, TCRs come in a huge range of structures, generated by manipulations of DNA and RNA analogous to the ones that generate antibody diversity. This structural variety is at the heart of the specificity shown by cellular adaptive immunity. As with antibody–antigen interactions, the binding of a TCR to the molecule it recognizes is highly specific, and when a tight bond it formed, it initiates a chain of events important for the elimination of that pathogen. T-cell responses are, however, more complex than B-cell responses, because there are several kinds of T cells that accomplish quite different functions. Thus, to understand cellular adaptive immunity, we must take things step by step. Only then will you able to appreciate the importance of cellular adaptive immunity and the ways in which the cellular and humoral adaptive responses cooperate to control both pathogens and cancers.

T cells are the principal players in cell-mediated responses of adaptive immunity.

Like B cells, T cells are initially formed in the bone marrow, but as part of their development, they migrate to the **thymus**, a central lymphoid organ situated where your neck meets your chest (FIGURES 9.7 and 9.15). T cells differentiate, specifying the structure of the TCR that they will make. All T cells are initially able to make an enormous range of different TCRs, which allows all T cells, collectively, to interact with a wide range of antigenic structures. However, as each particular T cell differentiates in the

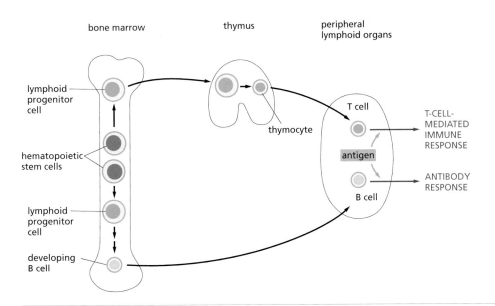

FIGURE 9.15 The migrations taken by different kinds of lymphocytes. All immune cells originate through the division of blood-forming stem cells in bone marrow. They begin life as lymphoid progenitor cells that could become either B or T cells. B cells begin to differentiate in the marrow, first going through the DNA rearrangements that generate antibody diversity and the culling that eliminates them if they make antibodies against self-antigens. Then they migrate to a peripheral lymphoid organ, such as a lymph node, for further differentiation as needed. T cells go directly to the thymus for their further steps of differentiation. (Adapted from Alberts B, Johnson AD, Lewis J et al [2015] Molecular Biology of the Cell, 6th ed. Garland Science.)

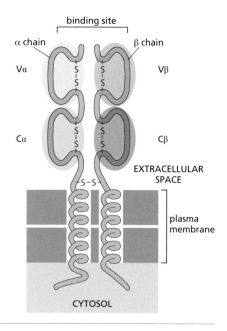

FIGURE 9.16 A T cell receptor (TCR). Like antibodies, TCRs are dimeric proteins. They are made from two polypeptides called α and β. As in antibodies, one part of the molecule is constant (C), having the same structure in all TCRs, analogous to the Fc regions of IgGs. The other part of the protein (V) is variable, resembling the variable and hypervariable regions of an IgG. T-cell receptors remain anchored in the plasma membrane of the cell that made them. The parts of the two polypeptides that project out of the cell include the variable part that will bind tightly to an antigenic determinant on any molecule that it recognizes. The constant part of the TCR is analogous to the receptor proteins for growth factors. It can convey the fact of the receptor's binding its antigen to the inside of the cell, initiating a cellular response.

SIDE QUESTION 9.9

The previous discussion states, in short, that our immune system generates a huge number of different B and T cells. Altogether, these cells make a wide range of pathogen-reacting proteins (antibodies and TCRs), thanks to the variable portions of these proteins. Before these immune cells have had a chance to interact with pathogens, the immune system kills all the lymphocytes making proteins that react with our own molecules. This sounds unbelievably wasteful! Why doesn't our immune system simply make antibodies and TCRs that will react with nonself?

thymus, it rearranges its DNA. This process is analogous to the formation of antibodies by B cells in the sense that each differentiated T cell can make one and only one TCR. However, TCRs are quite different from immunoglobulins. Right after their synthesis, they become part of the T cell's plasma membrane, and they stay there throughout the life of the T cell (FIGURE 9.16). Like the membrane-associated antibodies that are made in differentiating B cells, these proteins work as receptors for the antigen to which they can bind; they perform a signaling function to activate that particular T cell when the antigen it recognizes is present.

The generation of variability in T-cell receptors and the weeding out of T cells whose receptors recognize self-antigens are accomplished by processes analogous to those used for B cells, except T cells are screened for self-reactivity in the thymus. In this organ, the developing lymphocytes also make a choice about which one of three kinds of T cells to become. They can become **Helper T (Th) cells**, which assist B cells to differentiate or **Cytotoxic T (Tc) cells**, which are analogous to NK cells, in that they are specialized to kill other cells. Alternatively, they can become **Regulatory T (Treg) cells**, which differentiate to control several aspects of the cell-mediated immune response.

Cytotoxic T cells are equipped to kill cells that interact with their T-cell receptors.

Cytotoxic T cells resemble NK cells in that they contain cytoplasmic granules that help to lyse and kill any cell with which their TCRs interact (FIGURE 9.17A). These granules include both the pore-forming protein perforin (FIGURE 9.17B), and proteinases that can cleave inactive caspases, activating their proteinase activity and initiating apoptosis in the target cell. In resting Tc cells, these granules are dispersed throughout the cytoplasm (FIGURE 9.17C, upper panel), but when contact with a target cell is made, the granules congregate and are secreted to make pores in the plasma membrane of the target cell and to induce its death (Figure 9.17C, lower panel). The major difficulty faced by this kind of T cell is, however, identifying which cells are suitable targets and which are not. To explain how cytotoxic T cells do this, we must draw back and speak a little more about what happens when a cell is infected by a pathogen or is mutated to be different from self, as occurs during carcinogenesis.

All our cells contain enzymes that chew proteins into pieces and processes to place these pieces on the cell surface.

We have previously seen that cells contain proteinases, enzymes that degrade proteins. One of the most effective of these proteinases, called the proteasome, is designed to cut proteins into pieces, not digest them all the way down to single amino acids. The resulting **oligopeptides** (oligo means a few) are 7–9 amino acids long, just the right size to be an antigenic determinant. Given that there are 20 amino acids from which proteins are made, an oligopeptide containing nine amino acids is likely to be unique to the protein from which it came. This statement is supported by the following logic: there are 20 possible amino acids that might serve as the first amino acid of the oligopeptide, 20 possibilities for the second, 20 for the third, and so forth. The number of different possible oligopeptides nine amino acids long is therefore $20 \times 20 \times 20 \times 20 \times 20 \times 20 \times 20 \times 20 \times 20 = 5.12 \times 10^{11}$, which is about 500 billion. That's a lot of possibilities! Thus, an oligopeptide produced by this kind of proteolysis is likely to be a unique representative of the protein from which it comes. If it comes from one of your own proteins, its sequence will probably have

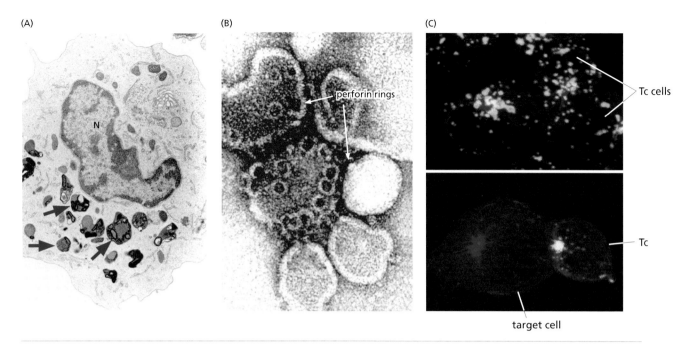

FIGURE 9.17 Structure and function of cytotoxic T lymphocytes. (A,) A low-magnification electron micrograph of a slice through a cytotoxic T lymphocyte (Tc) shows its nucleus (N) and its many cytoplasmic granules (red arrows). (B) Some of these granules contain perforin, a protein that can form pores in the plasma membrane of a target cell. Other granules contain proteases that can activate caspases in the target cell cytoplasm. Both of these functions are dangerous for the Tc, which is why they are packaged in granules, rather than floating free in the cytoplasm. (C) In a resting Tc cells, the granules are dispersed throughout the cytoplasm (upper image, green spots). Red staining in both images shows the positions of microtubules. When a Tc makes contact with a target cell, the granules gather at a place just beside the interface between the Tc and its prey (lower image, bright yellow spot). (A, from Podack ER & Dennert G [1983] *Nature* 302:442–445. With permission from Springer Nature. B, from Clark RH, Stinchcombe JC, Day A et al [2003] *Nat Immunol* 4:1111–1120. With permission from Springer Nature.)

been present at the time that differentiating B or T cells were being examined for reactions with self-antigens, so any B or T cell that makes an antibody or TCR that binds that amino acid sequence has previously been eliminated from the repertoire of lymphocytes. If, on the other hand, the oligopeptide comes from a virus that has infected your body or from a mutated protein made by a cancer cell, it is likely to be distinguishable from the oligopeptides that can be cut from your own normal proteins and is therefore identifiable as nonself, something that doesn't belong. If the immune system could see these oligopeptides, it would be able to identify any cell that displayed them as something that should be killed. This situation poses the question, How does the cell that has generated these oligopeptides get them to the cell surface, where antibodies and TCRs can interact with them?

These oligopeptides can be detected by the immune system because of a remarkable transfer machinery that carries them to the cell surface and displays them for other cells to interact with. This transfer and display is accomplished by a set of plasma membrane proteins called **major histocompatibility complex (MHC)** proteins. These proteins are made on the rough ER of all mammalian cells. From there, they migrate through the Golgi apparatus and are carried by vesicles to the plasma membrane. Once at the plasma membrane, they are positioned with one part exposed to the outside world and one part in the cytoplasm (FIGURE 9.18A). MHC proteins have a cleft on one surface that is designed to bind almost any oligopeptide about nine amino acids long (FIGURE 9.18B).

Binding of oligopeptides by MHC proteins occurs very soon after each copy of an MHC protein is made, because oligopeptides generated by proteolysis are concentrated the lumen of the ER by a pump in the ER membrane

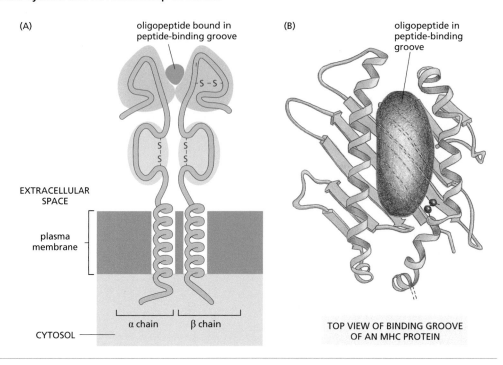

(A)

oligopeptide bound in
peptide-binding groove

(B)

oligopeptide in
peptide-binding
groove

EXTRACELLULAR
SPACE

plasma
membrane

α chain   β chain

CYTOSOL

TOP VIEW OF BINDING GROOVE
OF AN MHC PROTEIN

FIGURE 9.18 A major histocompatibility complex (MHC) protein and its binding to an oligopeptide. (A) MHC proteins span the plasma membrane, leaving one domain in the cytoplasm and one projecting into extracellular space. When MHC proteins are first synthesized on the rough ER, the domain that eventually will lie in the extracellular space projects into the lumen of the ER. (For a diagram of this behavior, see the positions of the MHC molecules in Figure 9.19.) When the MHC protein is in the plasma membrane, the bound oligopeptide (shown in pink) is positioned so TCRs and antibodies can interact with it. This is called antigen presentation. (B) A more detailed diagram of the outermost portion of the MHC backbone (shown in green) as it surrounds an oligopeptide (shown in pink). In this image, you are looking at the MHC from the direction that would be vertically downward in part A. Domains from the two subunits of the MHC protein form a pocket that fits nicely around the oligopeptide to bind it in place. Thus, if an oligopeptide created by proteolysis gets into the lumen of the ER, where this domain of the MHC protein is present, the peptide will bind in this cleft and then be conveyed to the cell surface as the MHC protein migrates to the plasma membrane. (Adapted from Alberts B, Johnson AD, Lewis J et al [2015] Molecular Biology of the Cell, 6th ed. Garland Science.)

SIDE QUESTION 9.10

The barrel-shaped proteinase in Figure 9.19 is called the proteasome. It degrades cellular proteins into oligopeptides that can be transported into the ER for binding to an MHC protein. At first, this may sound counterproductive. Why would our cells degrade their own proteins? With thought, though, one can see two advantages to the process. One is the display of self-proteins on the cell's surface by MHC proteins, as discussed. Another is that this system results in continuous degradation and replacement of our cell's proteins. Describe an analogy between this degradation and replacement process in cells and the process of preventive maintenance by a company.

(the transporter shown in FIGURE 9.19). Thus, when MHC proteins are conveyed to the plasma membrane, they carry a bound oligopeptide to the cell's surface and display it for any other cell or antibody to interact with (Figure 9.19). This is a somewhat complicated figure, but if you work through it carefully, remembering the cell biology of membrane traffic from Chapters 3 and 4, you will understand the process of oligopeptide display.

If a cell is infected by a virus, or if the cell is making mutant proteins that are no longer identical to its original self-proteins, some of the oligopeptides cut from these proteins will be presented by MHC proteins on the cell's surface. Once these oligopeptides are on the surface, one or a few of the body's Tc cells, which altogether have a wide range of TCRs in their membranes, will have the right shape to bind the oligopeptides generated from the viral or mutant protein and initiate the killing of that cell. Thus, even when a virus or other pathogen is inside a cell, it is still at risk from the adaptive immune system, thanks to the processes of oligopeptide display and T-cell receptor recognition. For a cartoon description of cell recognition and killing by a Tc, see MOVIE 9.3.

MHC proteins are the products of a genetic locus that varies from one person to another, so your own MHC proteins are different from your neighbor's, albeit they are similar to the MHC proteins of your close relatives. As such, MHC proteins are cellular markers of self versus nonself. Tc cells whose TCRs recognize your own MHC proteins are induced to initiate apoptosis early in their differentiation, so your repertoire of existing Tc cells will not bind your own MHC proteins, whether or not they are presenting oligopeptides from your own proteins. Either a foreign MHC protein, as would be encountered in the case of an allographic bone marrow transplant, or your own MHC protein displaying a foreign oligopeptide will induce Tc activity and result in cell death. Even cells displaying no MHC proteins are recognized as nonself and are subjects for elimination. Clearly, this mechanism makes our bodies a hostile environment for foreign tissues. This is a major reason that both blood transfusions and organ transplants must be done with great care to match the MHC proteins of the donor and recipient.

MHC proteins continually convey oligopeptides to the plasma membrane in all our cells.

The MHC protein transfer system does not discriminate between oligopeptides made by the proteolysis of pathogen or mutant proteins, and the native proteins from our own cells. Every cell is continuously breaking down its own proteins and transporting the resulting oligopeptides to the cell surface for immune cells to see. This process is also carried out by the stromal cells in bone marrow and by analogous cells in the thymus, so it is the mechanism by which our bodies inform the cells of the immune system which antibodies and T-cell receptors should *not* be made. B and T cells making receptors that bind self-antigenic determinants in bone marrow and thymus are induced to initiate apoptosis and are thereby pruned from the repertoire of immune cells. They are not part of the array of lymphocytes our bodies can use to fight infection or cancer.

Phagocytes can present antigens cut from pathogens they have ingested.

Macrophages and other cells that phagocytose pathogens also contain an MHC system to present antigens on their plasma membranes. Proteins of ingested pathogens are degraded into oligopeptides, but these foreign oligopeptides are carried to the surface of the macrophage by an MHC protein that differs from the one that conveys oligopeptides from the ER. The foreign oligopeptides, generated by proteolysis of a pathogen captured in an endocytic vesicle, become bound to this different MHC protein and are conveyed to the cell's plasma membrane, where they can again be seen by the proteins of the adaptive immune system. This presentation does not, however, activate Tc cells. Instead, it allows other kinds of T cells to interact with the antigenic determinants in ways that stimulate an immune response against the organism that was the source of the oligopeptides.

Dendritic cells are a kind of phagocytic cell that is critically important for adaptive immunity. Dendritic cells are numerous just beneath the skin and other epithelia; like macrophages and neutrophils, they also migrate through the body looking for evidence of infection or indications that a self-derived cell is abnormal, as in cancer. These phagocytes can engulf pathogens or mutant cells and kill them, just like a macrophage. Dendritic cells also use the different MHC protein to display the resulting oligopeptides on their surfaces, so they are not killed by Tc cells; instead, they activate a different kind of T cells. Dendritic cells also differ from macrophages in that they commonly migrate to lymph nodes, where they encounter lymphocytes and interact with them (FIGURE 9.20). This interaction activates any helper T (Th) cells whose receptors bind to the oligopeptides presented on the dendritic cell's surface. This activation stimulates the Th cells to interact with any B cells that recognize the same oligopeptide, be it from a pathogen or a mutated cell.

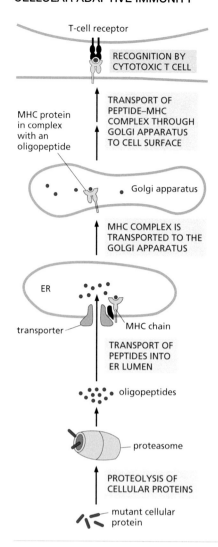

FIGURE 9.19 Diagram of the proteolysis of cellular proteins for display as oligopeptides on the cell surface by MHC proteins. All cellular proteins are subject to proteolysis by the proteasome, generating oligopeptides. These oligopeptides are transported into the lumen of the endoplasmic reticulum (ER), where they can bind to the peptide-binding pocket of an MHC receptor. The MHC protein is conveyed through the Golgi apparatus to the plasma membrane by the normal cellular processes of membrane traffic, described in Chapter 4. This transport brings oligopeptides cut from intracellular proteins to the plasma membrane and presents them at the cell surface for recognition by TCRs or antibodies.

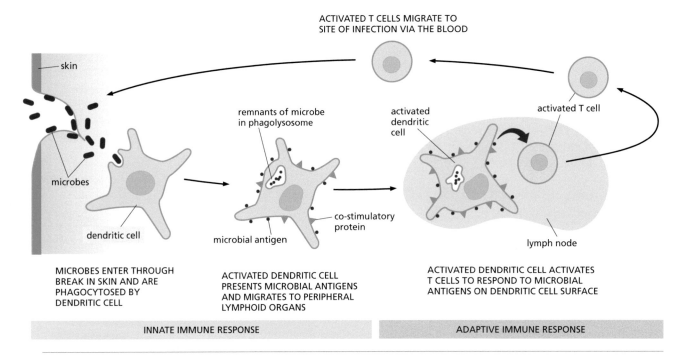

ACTIVATED T CELLS MIGRATE TO
SITE OF INFECTION VIA THE BLOOD

skin

remnants of microbe
in phagolysosome

activated
dendritic
cell

activated T cell

microbes

co-stimulatory
protein

dendritic cell

microbial antigen

lymph node

MICROBES ENTER THROUGH
BREAK IN SKIN AND ARE
PHAGOCYTOSED BY
DENDRITIC CELL

ACTIVATED DENDRITIC CELL
PRESENTS MICROBIAL ANTIGENS
AND MIGRATES TO PERIPHERAL
LYMPHOID ORGANS

ACTIVATED DENDRITIC CELL ACTIVATES
T CELLS TO RESPOND TO MICROBIAL
ANTIGENS ON DENDRITIC CELL SURFACE

INNATE IMMUNE RESPONSE

ADAPTIVE IMMUNE RESPONSE

FIGURE 9.20 The role played by dendritic cells in obtaining and presenting antigens to lymphocytes. Dendritic cells are phagocytes that are plentiful in the connective tissues just beneath the skin and other epithelia. They also migrate widely throughout the human body. They can arrive at the site of an infection, as shown here, or they can find antigens to internalize elsewhere. These cells are efficient at presenting the oligopeptides they have cut from the pathogens they have phagocytosed on their surface, so when they migrate into a lymph node, they display antigens to which both antibodies and TCRs are needed. Any T cell whose TCR binds that oligopeptide is now activated and can join the immune response. This is an example of cooperation between the innate and the adaptive arms of the immune system. (Adapted from Alberts B, Johnson AD, Lewis J et al [2015] Molecular Biology of the Cell, 6th ed. Garland Science.)

You have now learned about several kinds of phagocytic cells that contribute to the body's immune responses. Make a diagram of the similarities and differences among neutrophils, macrophages, and dendritic cells.

Th cells that have been stimulated by dendritic cells can interact with B cells as they interact with foreign antigens, contributing to the activation of those B cells that recognize truly foreign antigens. This process takes advantage of the proteolysis-based antigen presentation process. When a B cell has already defined the antibody it will make and has incorporated this antibody into its plasma membrane, the antibody can bind its antigen, as previously described. What was not previously mentioned is that this antigen–antibody complex is brought into the B cell by endocytosis (as in Figure 3.26), so it resides in a cytoplasmic vesicle called an endosome (FIGURE 9.21). In the endosome, the antigen is degraded by proteinases, just like the degradation of a pathogen's proteins in a macrophage. The resulting oligopeptides are now conveyed to the plasma membrane of the B cell by the second type of MHC protein. Oligopeptides from the original antigen, which are the right size to be single antigenic determinants, are now presented on the B-cell surface in such a way that they can interact with TCR receptors on Th cells. If the TCR on a particular Th cell recognizes this antigenic determinant, the interaction stimulates the Th cell to secrete cytokines that stimulate the B cell to complete its maturation into a plasma cell (Figure 9.21). This requirement for two cells that are both stimulated by the same antigen is an important part of the mechanism that reduces the chances of an immune response by a single B or T cell that happens to bind to a self-derived antigen.

B and T cells refrain from reacting with self-antigens thanks to additional steps of regulation.

The culling of self-reacting B and T cells does not completely explain how our bodies refrain from making antibodies and T-cell receptors that react

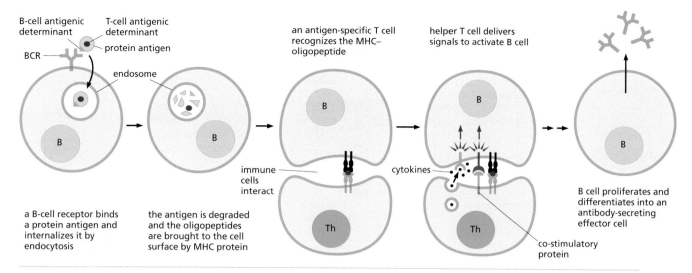

FIGURE 9.21 Helper T (Th) cells cooperate with B cells to assure that they mount a strong response to foreign antigens. The activation of immature B cells is more complex than previously described; it requires not only the binding of antigen to their membrane-bound antibody but also the action of a helper T cell that can bind the same antigen. This checking for nonself by two cells reduces the chances that a B cell will mount an immune response to a self-antigen. As diagrammed, the membrane-bound antibody on an immature B cell binds an antigen, which stimulates endocytosis of the antigen–antibody complex, followed by digestion of the antigen into oligopeptides, which can then be presented on the B cell's surface by MHC proteins. These presented peptides are examined by various Th cells, and if there is a Th that recognizes the oligopeptides as nonself, then that Th stimulates the B cell to complete its differentiation and become an effector B cell, also called a plasma cell.

with our own antigens. This situation has become clear through recent experiments that have identified immune cells in normal people that *do* recognize self-antigens, yet they do not result in an autoimmune response, that is, a mobilization of the immune system to kill the cells of our own bodies. One part of the explanation for this phenomenon depends on the fact that weak binding to a self-antigen is tolerated, probably because of regulatory T cells (Treg cells). These cells somehow suppress the activity of Tc and B cells that might otherwise mount an immune response against an antigen in our own body. A part of this inhibition probably comes from the ability of Treg cells to interact with B and Tc cells as they are becoming activated and suppress that activation process.

This description of adaptive immunity, complicated as it may seem, describes only a fraction of the intricacies of interactions among lymphocytes. The problem of being able to make antibodies and TCRs that will interact with any antigen that invades the body but not with any self-antigens is so difficult that our bodies invest heavily in making a system that works. To help you get a sense of the most important interactions, FIGURE 9.22 displays the relevant cells and the ways they contribute to adaptive immunity.

An additional mechanism by which your immune system suppresses self-cell destruction involves a group of signaling proteins called **immune checkpoints**. These are proteins on the surface of T cells that can interact with cells that might be targets for destruction and provide additional information about the character of the potential target cell. Their ability to bind or not bind helps the T cell know whether the potential target is normal and healthy, foreign, an infected cell, or a mutant. When these interactions were first discovered, they included only negative regulatory signals that were recognized to play an important part in the ability of the immune system to tolerate self-made proteins, that is, not mobilize an immune response against self. Recent work has shown that there are also stimulatory signals acting between immune cells and the surface

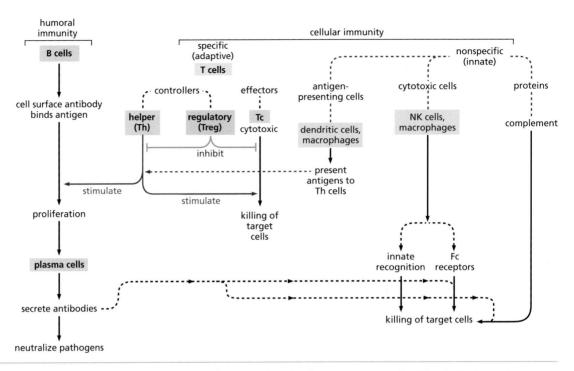

FIGURE 9.22 An overview of the multiple interactions among immune cells and pathogens or cancer cells. Although this diagram has many parts, you can trace the flow of interactions between cells of adaptive immunity and innate immunity to see how these parts of the immune system relate to one another when killing pathogens or cancer cells. For example, Treg cells can interact with B or Tc cells as they are becoming activated and suppress that activation process. (Adapted from Weinberg R [2013] The Biology of Cancer, 2nd ed. Garland Science.)

proteins of other cells, so the term immune checkpoints now refers to positive as well as negative regulatory signals. Immune checkpoints have become an extremely active area of work for cancer immunologists because current evidence shows that cancer cells manipulate immune checkpoints to suppress an immune response against themselves. It follows that blockers of this kind of inhibition should be useful in helping a cancer patient's own immune system kill cells in an already developed cancer. This approach to cancer therapy will be discussed more fully in Chapter 10, but to learn more about immune checkpoints in the normal function of an immune response, see SIDEBAR 9.2.

SIDEBAR 9.2

A CLOSER LOOK: Immune checkpoints.

The job of the adaptive immune system is to generate antibodies and T-cell receptors that recognize foreign antigens but leave self-antigens alone. Given the vast number of antigens in both categories, this is a very difficult task. The culling of both B and T cells that occurs during the early stages of their differentiation is an important first step in limiting immune reactions against self, but a few immune cells that pass through this selection make molecules that bind self-antigens. These cells present a threat to the tissues of one's own body. Indeed, some unfortunate people develop defects in their immune systems that begin killing some of their own cell types. Rheumatoid arthritis, Crohn's disease of the intestine, and systemic lupus are three of the diseases in which the immune system attacks self; this is called autoimmunity, and it can lead to significant pathologies.

Immune responses from these self-reacting antibodies and T-cell receptors are rare, thanks to the action of Treg cells and the inhibitory action of the protein–protein interactions of immune checkpoints. (Note that the word checkpoint is used here in a way that is quite different from the same word when you encountered it in cell cycle control. There, a checkpoint was a mechanism that evaluated the fidelity of a process and allowed continued cell cycle progression only after the job at hand had been done correctly, be it DNA repair or chromosome attachment to the spindle. Immune checkpoints are simply inhibitory or stimulatory interactions.)

The idea of immune checkpoints has emerged from an increased understanding of the interactions between immune cells and cells presenting antigens on their surfaces, be they macrophages, dendritic cells, or any other cell that is using MHC to put oligopeptides cut from its own proteins out for examination on the cell surface. Careful work by numerous scientists has shown that the interactions between T cells and antigen-presenting cells include many more factors than simply the MHC–oligopeptide complex and the TCR. The membrane of each of these cells contains numerous proteins that can interact with one or more components of the other cell's membrane, as diagrammed in **Figure 1**. Most of these membrane proteins have not been discussed in the text, though at the middle of this long list you can see the TCR interacting with an MHC–oligopeptide complex. The other protein pairs in this figure (depicted with antigen-presenting cells on the left and T cells on the right) are examples of protein pairs that are known to interact as the T cell is deciding what to do with the cell to which it is now bound. Most of these proteins are parts of the immune checkpoint system.

A cartoon of a typical immune checkpoint interaction is shown in **Figure 2**. If the T cell is a Tc, it will either initiate killing of the target cell or not. If it is a Th, it must be stimulated to participate in activating a B cell to produce antibodies or it will remain unstimulated. If it is a Treg, it will control the action of other T and B cells or not. These differences

are important for the proper functioning of the immune system, so it is no wonder that multiple factors are considered, just as we saw in cells responding to signals that made them decide whether or not to enter the cell cycle and replicate their DNA.

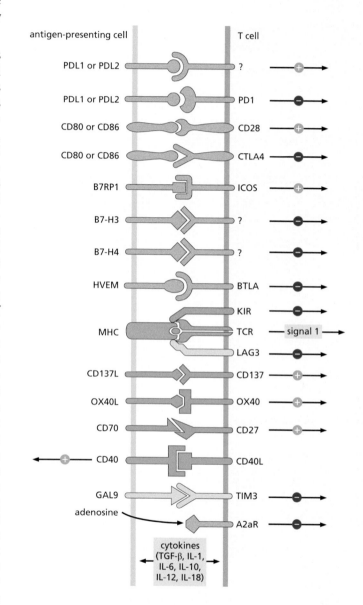

Figure 1 The protein pairs that can interact between a T cell and an antigen-presenting cell. At the center of this diagram is the MHC protein of the antigen-presenting cell with an oligopeptide bound. The MHC–oligopeptide complex is interacting with a T-cell receptor and two immediately related proteins that are part of the antigen recognition system of T cells. All other pairs of interacting membrane proteins are elements in the complex system of interactions between a T cell and an antigen-presenting cell. Protein pairs can be either excitatory (arrows with blue circles) or inhibitory (arrows with red circles). For example, the T-cell membrane protein CTLA4 is receiving an inhibitory signal, meaning that when it is activated, it will impede the killing action of a Tc. It is now well-documented that tumor cells often up-regulate the expression of CD80 and/or CD86, either of which can turn on the inhibitory activity of CTLA4. The use of antibodies to CTLA4 is now an FDA-approved treatment for malignant melanoma, because clinical trials have shown that this antibody blocks the inhibitory action of the tumor's CD80 or CD86, enhancing the ability of the patient's CTLs to kill tumor cells. Note that the pair labeled CD40 and CD40L sends signals in the opposite direction. This interaction is one of the ways in which T cells can modify the behavior of antigen-presenting cells. (From Pardoll DM [2012] *Nat Rev Cancer* 12:252–264. With permission from Springer Nature.)

SIDEBAR 9.2

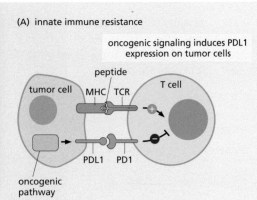

(A) innate immune resistance

oncogenic signaling induces PDL1 expression on tumor cells

peptide

tumor cell   MHC / TCR   T cell

PDL1   PD1

oncogenic pathway

Figure 2 Two possible mechanisms for the action of immune checkpoints on tumor cells. In the innate immune response, the tumor cell can express the membrane protein PDL1, which interacts with the T-cell membrane protein PD1 and sends an inhibitory signal to the T cell. (B) In an adaptive immune response, an oligopeptide from the tumor cell may initiate a T-cell response, but the subsequent expression of PDL1 in the tumor cell, as a result of up-regulating the relevant transcription factors, allows the tumor cell to inhibit further activation of the T cell and thereby block its cytotoxic action. (From Pardoll DM [2012] *Nat Rev Cancer* 12:252–264. With permission from Springer Nature.)

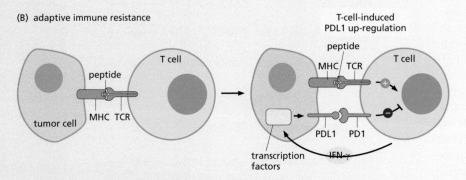

(B) adaptive immune resistance

T cell

peptide

tumor cell   MHC TCR

T-cell-induced PDL1 up-regulation

peptide

MHC / TCR   T cell

PDL1   PD1

transcription factors   IFN-γ

Although our knowledge of immune cell communication is still incomplete, scientists have learned enough about it to identify some of the immune checkpoints that are critical for proper functioning of the immune system. The membrane component called cytotoxic T lymphocyte-associated antigen 4 (**TCA4**) is an important component of the way an antigen-presenting cell inhibits an associated Tc from becoming activated. Likewise, the programmed cell death protein 1 (**PD1**) is an important part of the interactions that might lead an immune cell to induce apoptosis in its target cell. These membrane components, and others like them, are now in active use by clinicians who are trying to overcome the fact that tumors use the expression and cell-surface presentation of some immune-regulatory proteins as a mechanism to turn off the activity of T cells that might otherwise attack them and either kill them or limit their ability to grow and divide.

It is clear that there are multiple kinds of T cells and that they perform diverse functions. The T-cell differentiation that occurs in the thymus involves the regulated expression of different cell surface molecules in different T cells. Most of these molecules are called cluster of differentiation or CD, followed by a number. Thus, a more complete description of the immune system includes an account of the numerous CDs that are expressed on different T cells and a description of the different actions of these molecules on different T-cell types. For example, CD8 is characteristic of cytotoxic T cells, whereas CD4 is on the surface of helper T cells. For a summary of the processes sketched in this section, see MOVIE 9.4. For a more detailed understanding of the immune system than we are attempting here, one must master this language of cell surface proteins that help different T cells to identify and react with the various cells they must affect. These details are available in several of the recommended readings mentioned at the end of this chapter.

THE IMMUNE SYSTEM AND PREVENTION OF CANCEROUS PROGRESSION

Cancer develops from cells that were at one time self, so the antigens displayed on their surfaces were initially no different from the antigens presented on the surface of normal body cells. If the immune system is working properly, none of these cells will attract attention for all the reasons that the immune system learns to tolerate self-antigens. However, as cells move through cancerous progression, their DNA is altered by mutation, and they begin to make proteins that are not the same as the ones in normal cells. Moreover, genetic changes can reduce the expression of some molecules that would normally be present and increase the expression of others, including some that are not normally made in that cell type. These and other changes should alert the immune system that these cells are now different. We therefore have two questions to address: What changes are recognized by the immune system, leading it to eliminate cancerous or precancerous cells without medical intervention, and how does this system fail, allowing cancer to progress? These are both profound questions for which complete answers do not yet exist. Nonetheless, several points are clear and can be described.

There are three ways in which the immune system can help to prevent cancer. (1) There are oncogenic viruses whose presence can be detected by the same kinds of immune responses that protect us from any viral infection. If these are eliminated by B- and/or T-cell responses, their oncogenic effects can be prevented or limited. (2) Some pathogens work as tumor promoters, causing the irritation and inflammation that induces cell divisions and thereby increases cancer risk. Eliminating these pathogens through an immune response reduces their cancer-promoting effect. (3) As suggested previously, the immune system should be able to recognize a cancer cell as nonself, due to changes in the antigens it presents on its surface. If this recognition were to work, the action of cytotoxic T cells would kill the altered cells, limiting or eliminating cancer growth.

Considerable evidence supports the idea that the immune system normally limits cancerous progression.

Experiments that altered the immune system of laboratory mice have shown that loss of immune function leads to a significant increase in the incidence of cancer. In one example, mice were modified genetically so they could no longer make interferon γ, a cytokine important for communication among lymphocytes. These mice were 10–20-fold more susceptible than normal mice to the induction of tumors by a chemical carcinogen. In another experiment, mice were mutated to lose function in the gene encoding perforin, the protein used by cytotoxic T lymphocytes and NK cells to poke holes in target cells. These mice showed increases in their susceptibility to chemical carcinogenesis and in the frequency with which they developed tumors spontaneously. Similar results were obtained when mice were mutated in genes whose products help to generate the variable parts of both antibodies and T-cell receptors. All these results are consistent with the hypothesis that adaptive immunity normally helps to prevent tumor growth.

Innate immunity is also important in cancer prevention. NK cells can help to eliminate cancer through their recognition of alterations in a normal cell's surface. For example, some viruses suppress the ability of MHC protein to display peptides on the surface of the cell they infect, a subtle ruse by which they reduce the ability of cytotoxic T cells to recognize and kill

There are some cases in which an infection is not killed by an immune response, but the body's reaction to the pathogen leads to severe, persistent inflammation. Doctors have observed that cancer is increasingly likely when persistent inflammation is present. Here, the immune system is not fighting cancer; it is contributing to it! Do you see this as an example of carcinogenesis or tumor promotion?

SIDE QUESTION 9.12

off the infected cells. NK cells counteract this ruse by identifying cells with too little MHC protein on their surfaces. Likewise, some tumor cells down-regulate (decrease the amount of) the display of MHC protein. Again, reduced surface MHC protein activates NK cells. Strong evidence for the importance of this statement comes from studies of people who carry a mutation that blocks their formation of normal NK cells. These unfortunate people develop multiple infections from various herpes viruses and then multiple cancers at a strikingly early age, emphasizing the importance of NK cells in preventing cancer.

The view that both innate and adaptive immunity help to prevent cancer is supported by evidence from some current practices in human medicine. Doctors have developed many methods for transplanting organs from a healthy body to one in which that organ is failing. In spite of its cost, this operation is performed quite frequently in all countries of the industrialized world, taking tissue either from volunteers or from people recently killed in a way that left the relevant organ intact, such as an automobile accident. Transplants are done successfully with tissue from liver, kidney, pancreas, lung, bone marrow, parts of the eye, and even a complete heart. These transplantations do, however, encounter the problem of tissue rejection as a result of incompatibility of the MHC proteins in the donor and recipient. To minimize this problem, transplant recipients are given drugs that suppress their immune system and reduce the strength of their tissue rejection. People who take these drugs for a long time show a significant increase in their rate of developing tumors, indicating that their immune system had been preventing tumor development until it was suppressed.

Additional evidence comes from people who have become immunodeficient as a result of infection by the AIDS virus. A kind of cancer called Kaposi's sarcoma was rare until the dramatic increase in the number of people living with AIDS (now more than 40 million throughout the world). The AIDS virus inactivation of the immune system has effects analogous to the therapeutic reduction of immunity by drugs, as described for organ transplantation. Although AIDS does not cause Kaposi's sarcoma, this cancer is more likely to thrive when the immune system is working badly.

Another piece of evidence on the importance of adaptive immunity in cancer progression is that the stroma of tumors is now known to contain a significant number of lymphocytes, largely Tc cells. A tumor-specific function for these infection-fighting cells is not known, but observations on women with advanced and metastatic ovarian tumors suggest that their presence helps to check tumor growth. In recent clinical research, the major identifiable ovarian tumor was cut from a large number of women suffering from this cancer. The number of T cells per unit volume in each tumor sample was then counted. Five years later, 74% of the women whose tumors contained many Tc cells were still alive, whereas only 12% of those with low T-cell numbers survived. These observations support that idea that tumor-localized Tc cells inhibit tumor growth.

Oncogenic mutations can alter proteins enough to change the peptides displayed on a cell's surface.

Because the presentation of antigens by MHC proteins continues as cells go through cancerous transformation, the protein products of mutated genes are broken down into oligopeptides, transported to the plasma membrane, and displayed. If an oncogenic mutation causes a sufficient change in its protein product, that antigen may now be recognized as nonself and induce the action of Tc cells (FIGURE 9.23).

SIDE QUESTION 9.13

You have now read about several studies providing evidence that the immune system normally contributes to keeping naturally arising tumors in check. Which of these studies do you find the most convincing and which the least so? Given the weight of existing evidence, would you try to convince your grandmother that the immune system plays an important part in the control of cancer progression? What evidence do you think she would find most convincing? Why?

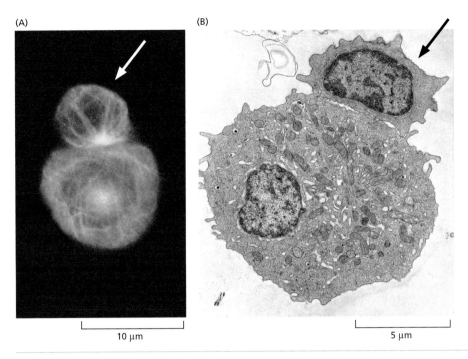

(A)

(B)

10 µm

5 µm

FIGURE 9.23 A cytotoxic T (Tc) cell interacting with a cancer cell to kill it. (A) A fluorescence light micrograph of a cancer cell interacting with a Tc cell (arrow) stained to show the microtubules in both cells. The centrosome and microtubules of the Tc cell have migrated to lie immediately next to the site of cell–cell interaction. This allows the cytoskeleton of the Tc to activate membrane traffic and bring cytoplasmic granules containing perforin and proteolytic enzymes to the place from which they can be secreted to interact with the target cell effectively. (B) A low-magnification electron micrograph of a thin slice cut from a cancer cell interacting with a T cell (arrow). Here, you can see the formation of tight associations between the plasma membranes of the two cells. Both images A and B show these interactions as they are just beginning to occur. (A, from Geiger B, Rosen D & Berke G [1982] *J Cell Biol* 95:137–143. With permission from Rockefeller University Press. B, from Zagury D, Bernard J, Thierness N et al. [1975] *Eur J Immunol* 5:818–822. With permission from John Wiley & Sons, Inc.)

Study of the proteins expressed by tumors has shown that the genetic instability of cancer cells frequently leads them to express proteins that are not part of a normal adult. For example, proteins that are normally made by embryos but not by adults have been observed in cancer cells. Cancer can also lead to the inappropriate expression of proteins that are normally expressed only in environments hidden from the immune system by a process called immune privilege. The eyes and the testicles are not protected by immunity. This condition probably arose to minimize the chances that inflammation might occur in these sensitive sites. Whatever the reason, proteins specific to these organs are not recognized as self by the immune system, but they do not normally evoke an immune response when expressed where they belong. If these antigens are expressed by genetically unstable cancer cells and presented on the cell surfaces, they can elicit an immune response, leading to the death of many tumor cells.

Developing cancers evolve mechanisms that allow them to evade immune surveillance and become a life-threatening growth.

One can imagine several routes by which cancer cells might evade the immune system. Carcinogenic mutations in key components of transformation, like *ras* (Chapter 6) and p53 (Chapter 7), can generate numerous alleles, many of which can induce cell proliferation and suppress apoptosis. There is probably a subset of those mutations that produces gene

From what you know about the immune system, list the ways in which it could be a factor in killing cancer cells and keeping tumors small.

SIDE QUESTION 9.14

products sufficiently similar to normal proteins that the process of antigen display does not evoke destructive attention from Tc cells. Such alleles will certainly be selected for in the development of cancer, because any allele that *does* stimulate an immune response will lead to the rapid elimination of that tumor cell.

Cancer cells can also go through additional mutations that block the presentation of antigens to their surfaces. Down-regulation of the mechanisms required for antigen presentation by MHC protein would do this, although as mentioned previously, this change alone can activate NK cells. There are, however, immune checkpoint proteins that block the actions of NK and Tc cells. If mutations in precancerous cells lead to an up-regulation in the expression of these proteins, the cells expressing these alleles will evade killing by both Tc and NK cells. Some carcinomas use yet another immune evasion strategy. They take advantage of a protein used by epithelial cells to attract NK cells when the epithelium is under stress, for example, from damage to the epithelial cell's DNA. These carcinoma cells synthesize this protein and traffic it to their surfaces, then secrete proteinases that cut the extracellular part of the protein off, allowing it to diffuse away as a soluble protein. Now the receptors on NK cells become saturated with the soluble form of the protein, and they can't tell which cell to kill. Diabolical!

Some cancer cells appear to be able to induce Treg cells to come to their rescue and inactivate both the B and T cells that might otherwise have led to cancer cell death. This scenario is currently a matter of conjecture, because no one yet knows for sure how cancer cells become so effective at suppressing lymphocyte activity. That cancer cells do inhibit immune responses is demonstrated by the fact that tumors do form and metastasize in people whose immune system is not compromised in any obvious way. Probably, multiple aspects of cancer biology are acting to block the activities of NK and Tc cells that would wipe out the cancer if they were not inhibited. Indeed, it is reasonable to say that we must add the ability to inhibit the immune system to the hallmarks of cancer, the properties that cells must acquire during a complete cancerous transformation.

Immune checkpoints are now receiving particular attention from clinical oncologists. Because these inhibitions work by means of specific signaling proteins that bind receptors on lymphocytes, they are obvious candidates for research on the ways that cancer is defeating the immune system. Procedures are in development to treat patients with antibodies to specific immune checkpoint proteins. This and related methods of experimental cancer therapy are described in more detail in Chapter 10. Indeed, medical science is now seeking multiple ways to stimulate the immune system to do a better job of killing cancerous cells. In the next section and in Chapter 10, you will see several approaches that cancer scholars are trying in efforts to bring the power of adaptive immunity to bear in cancer therapy.

THE IMMUNE SYSTEM AND CANCER TREATMENT

Given the remarkable properties of the immune system, it is no surprise that oncologists are seeking ways to take advantage of its power and specificity to develop treatments for cancer that would be specific for tumor parenchymal cells and therefore less damaging to normal cells. Collectively, these treatments are called **immunotherapies**. You are now in a position to understand how several of them work.

Some immunotherapies in current use are based on the idea that antibodies to tumor-specific proteins should lead to tumor regression and perhaps a cure.

Trastuzumab was mentioned in the previous chapter as a treatment for some breast cancers. This drug is an antibody that reacts with a growth factor receptor very similar to EGF-R, inhibiting its function. Two additional antibodies react with closely related growth factor receptors: cetuximab (Erbitux®) and pertuzumab (Perjeta®). Cetuximab is used to treat colorectal cancer, non-small-cell lung cancer, and head and neck cancers. Pertuzumab is another treatment for breast cancer.

These drugs are carefully engineered antibodies whose properties are a result of a deep knowledge of many aspects of adaptive immunity. They are **monoclonal antibodies**, meaning that only a single clone of B cells was used to make the protein. Thus, each antibody binds to only one antigenic determinant. They are so specific that each of them reacts with a different part of the receptor they target (FIGURE 9.24). In these cases, antibody binding inactivates the receptor and reduces the mitogenic signaling delivered to these cancer cells. About 15 monoclonal antibodies are now on the market, each designed to interact with a protein antigen that is important in some cancer: breast, stomach, colorectal, melanoma, lymphoma, and a kind of leukemia. Some of these antibodies work by blocking the action of a mitogen or growth factor, but some are used as cancer cell tags that elicit the activity of Tc cells, helping to reduce the tumor burden. For patients whose cancers express the relevant antigens in an accessible form, these drugs can be very effective. In combination with other treatments, like radiation, they have clinically significant effectiveness, for example, by increasing tumor cell sensitivity to radiation (FIGURE 9.25). Like most single anti-cancer drugs, however, monoclonal antibodies face the problem that a genetically unstable cancer will commonly change in ways that makes the drug lose its effectiveness.

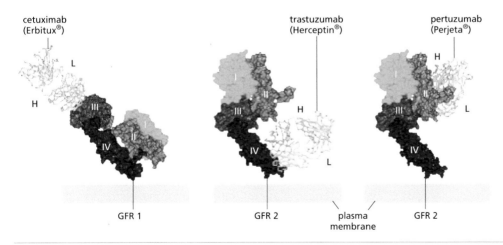

cetuximab (Erbitux®)   trastuzumab (Herceptin®)   pertuzumab (Perjeta®)

GFR 1   GFR 2   plasma membrane   GFR 2

FIGURE 9.24 Antibodies that bind to growth factor receptors (GFRs) can block their activity. Three commercially produced antibodies that react with two different GFRs are in use as drugs to block or slow tumor growth. Both the generic name for each drug (antibody) and its commercial name are given. Each antibody is shown as a dimer of both light (L) and heavy (H) chains. The constant regions of these antibodies (Fc) are not included. Each GFR is depicted with four structural domains in different colors and labeled with Roman numerals. The antibodies all react with different parts of the growth factor receptors because each antibody binds to a different antigenic determinant. These antibodies are all useful drugs for the treatment of specific cancers. (From Hubbard SR [2005] *Cancer Cell* 7:287–288. With permission from Elsevier.)

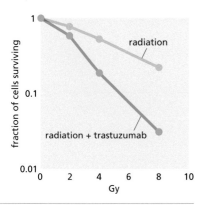

FIGURE 9.25 Antibodies to cell surface receptors sensitize tumor cells to radiation. The two lines on this graph compare the fraction of cells that can live after being exposed to a given amount of X-irradiation, shown on the horizontal axis in units called grays (Gy), which are defined as a given amount of radiation energy absorbed by a given mass of cells or tissue. Clearly, the addition of trastuzumab increases the sensitivity of these cells to radiation. (Adapted from Liang K, Lu Y, Jun W et al. [2003] *Mol Cancer Ther* 2:1113–1120. With permission from American Association for Cancer Research.)

Some cancer treatments use proteins that signal between cells of the immune system in an effort to stimulate the patient's immune response.

Cytokines are proteins used by cells to signal the need for an active immune response, as described previously. Several types of cytokines have been used, both in clinical trials and in standard patient therapies, to try to induce an effective immune response to a tumor. After a large amount of trial and error, interleukin 2 (IL2) has been identified as the most promising of the cytokines, and it has been used clinically for the treatment of melanoma and advanced kidney cancer. Although cytokine treatments are still in use, the data about their effectiveness are not encouraging. The difficulties with what looks like an excellent idea probably arise from the complexities of the signals conveyed by the multiple cytokines that immune cells use for communication. Simply increasing the concentration of one cytokine doesn't yield the desired result. We need to know more about how these signaling networks function in the human body before they can be manipulated to make effective therapies.

From this description, you can see that our bodies are protected by a wealth of cellular and molecular devices designed to identify cells and molecules that don't belong and then either kill them or otherwise eliminate them from our bodies. Our continued health relies on the action of these systems. A number of innovative uses of the immune system in cancer treatment are currently under development and testing. These are described in Chapter 10.

SUMMARY

Our innate immune system includes several structures and processes that help protect us from infection: (1) our skins (both outer and inner) help to keep pathogens away from our inner cells and tissues, (2) enzymes that are made and secreted by several cell types degrade macromolecules characteristic of pathogens, and (3) we harbor cells that can either eat pathogens (phagocytes) or kill them (NK cells).

Our adaptive immune system fights infections with both humoral and cell-mediated responses. The former includes antibodies, which are proteins specialized to bind molecules that do not come from our own cells. Antibodies are made and secreted by one kind of lymphocyte, a mature B cell, also called a plasma cell or effector B cell. These cells become antibody factories by going through a complex process of differentiation. After being born in bone marrow through the division of a blood-forming stem cell, a B cell first rearranges particular regions of its DNA, so it will make one and only one antibody molecule. While it is still in the bone marrow, the differentiating B cell puts a version of this antibody onto its surface, where it can bind to any molecules that pass by. If the antibody presented on the surface of the developing B cell does bind one of the body's own molecules, that cell initiates apoptosis and dies. If the presented antibody does not see an antigen it recognizes in the bone marrow, the B cell can then move out into lymph ducts and nodes, where it will encounter foreign molecules that might have come from pathogens or one of our own cells that is mutated, like a cancer cell. Now, binding of a molecule to the plasma membrane-localized antibody promotes further differentiation of that B cell. In combination with helper T cells, the B cell specializes to become a plasma cell, which will make and secrete large amounts of soluble antibody. The molecule to which the antibody binds is called its antigen, and the specific place on the antigen to which the antibody binds is an antigenic determinant.

T cells are other kinds of lymphocytes, ones that specializes in interacting with other cells rather than making and secreting antibodies. Like B cells, T cells go through a complex differentiation process that includes DNA rearrangements, so they can make one and only one antigen-binding protein, called a T-cell receptor. T cells are born in the bone marrow but migrate to the thymus, where each cell differentiates to make its own specific T-cell receptor, which is always situated in the plasma membrane of the cell that made it. In the thymus, by analogy with B cells, T cells are examined for the binding properties of their T-cell receptors. If the receptor binds with a molecule made in our own bodies, that T cell dies by apoptosis; if not, the T cell can differentiate further into one of three kinds of T lymphocytes: a helper T cell, which aids B cells in their differentiation process; a cytotoxic T cell, which can kill cells whose surfaces include an antigen they recognize as foreign; or a regulatory T cell, which works to limit the activity of B cells and other T cells, minimizing the chances that they will react with a self-produced antigen.

The immune system is important in limiting cancer growth. We know this from experiments, from the phenotypes of mutations in both animals and people, and from clinical treatments in which the immune system is weakened, whereupon cancers become more common. T cells are commonly found in tumors, but just what they are doing there and how they limit tumor growth are still matters for research. Antibodies to tumor-specific antigens, to immune checkpoints, and to cytokines (proteins used by the immune system for communication among its cells) are currently being explored as cancer treatments. However, all these potentially powerful therapies are limited in their effectiveness by unanticipated side effects and the rapidity with which cancer cells can mutate, as well as by the poorly understood complexity of the chemical signals by which immune cells communicate.

ESSENTIAL CONCEPTS

- The immune system protects us from disease, in part by keeping pathogens out and in part by killing pathogens that are infecting us. It also can limit the growth of cancer cells.

- We are born with some defenses against infection: our skin, molecules like RNAses and proteinases that disrupt pathogens, and cells that kill and eat invading pathogenic cells. Collectively, these processes are called innate immunity.

- We develop additional defenses through the action of specialized white blood cells called B and T lymphocytes (or simply B and T cells). Each of these cell types can rearrange the DNA of a particular genetic locus to make a vast range of proteins that can bind to other molecules. These are antibodies, made by B cells, and membrane-bound receptors, made by T cells.

- The variety of structures to which antibodies and T-cell receptors can bind is enormous, thanks to their great diversity and their varied structure.

- To avoid making antibodies or T-cell receptors that react with the body's own proteins and cells, our immune system includes a trial period for each kind of lymphocyte, during which these cells are exposed to our own molecules. If a lymphocyte reacts at this time, it dies. If not, it becomes a circulating lymphocyte. Thus, only lymphocytes that make immune proteins that do not react with the body's own molecules survive. They become an important part of our defensive repertoire for fighting infections and cancer.

- The immune system is greatly aided by a normal cellular process that works on all the proteins in our cells. This system uses a special proteinase to break proteins into fragments about nine amino acids long and a protein called the MHC protein to transfer those oligopeptide to the cell's surface. There the oligopeptide is exposed to the immune system.

- If a cell is infected by a pathogen, oligopeptide presentation assures that pieces of the pathogen will be shown or presented to the immune system, where they can be identified as nonself. This allows the immune system to kill pathogen-infected cells.

- The antigen presentation system operates in almost all our cells, continuously presenting self-oligopeptides to the immune system. This is part of the way the immune system learns what is a self-antigen.

- Cancer cells begin their lives as normal cells, so they are not perceived as foreign by the immune system. As they go through cancerous transformation and progression, however, they become less and less like self, and some of these altered cells can then be identified as foreign by the immune system. Thereupon, they can be killed.

- The emergence of a malignant cancer involves special tricks on the part of the cancer to evade detection by the immune system.

- One of the ways cancer cells evade detection is by making proteins that inhibit the immune response. These proteins mimic the action of molecules used by the immune system itself to regulate an immune response and limit the chances that the immune system will recognize self molecules and cells.

- We are still learning how all the component of the immune system work and how doctors might improve immune recognition and killing of cancer cells.

- Meanwhile, the specificity of antibodies is being used by doctors to devise numerous ways of attacking cancer cells. Some of these are already in clinical use. Their advantage is their specificity for cancer cells, meaning that they have few side effects. Their limitation, however, is that they are so specific that the rapid mutation rate of cancer cells can lead to the antibody becoming ineffective rather rapidly.

KEY TERMS

adaptive immunity
antibody
antigen
antigenic determinant
B cell
cellular adaptive immunity
complement system
constant region (Fc)
cytokine
cytotoxic T (Tc) lymphocyte
dendritic cell
effector B cell
Fas-ligand
Fas-receptor
helper T (Th) cell

humoral adaptive immunity
hypervariable region
immune checkpoint
immunoglobulin G (IgG)
immunotherapy
innate immunity
interferon
interleukin
leukocyte
lymphocyte
lymphokine
macrophage
major histocompatibility complex (MHC)
memory cell

monoclonal antibody
natural killer (NK) cell
neutrophil
oligopeptide
PD1
perforin
phagocyte
plasma cell
regulatory T (Treg) cell
T cell
T-cell receptor (TCR)
TCA4
thymus
variable region

FURTHER READINGS

For a more complete description of the immune system, see any of several good college-level texts in cell biology, for example, *Molecular Biology of the Cell* by Alberts et al. or *Molecular Cell Biology* by Lodish et al. These chapters are usually near the end of the book, so they assume a pretty good knowledge of cell biology, but they usually come just before a chapter on cancer, so the two go together.

Alberts B, Johnson AD, Lewis J et al (2015) Molecular Biology of the Cell, 6th ed. Garland Science.

Lodish H, Berk A, Kaiser CA et al. (2012) Molecular Cell Biology, 7th ed. W. H. Freeman Co.

For an authoritative but readable description of the immune system, see Parham P (2014) The Immune System, 4th ed. W. W. Norton & Company.

Pardoll DM (2012) The blockade of immune checkpoints in cancer immunotherapy. *Nat Rev Cancer* 12:252–264.

QUESTIONS FOR FURTHER THOUGHT

1. If you were able to make an antibody to the oncogenic allele of Ras protein that blocked its activity of stimulating cell growth and division, would it be useful as a new-style chemotherapeutic agent? Why or why not?

2. If you were able to make a function-blocking antibody to the cell membrane protein that transports the important sugar glucose across the plasma membrane, would it likely be a good new-style chemotherapeutic? Why or why not?

3. What is your personal assessment of the value of current research for ways to stimulate the immune system for an effective fight against cancer? If you were a member of a panel that judged grant proposals for funding, would you support such a proposal? Why or why not?

ANSWERS TO SIDE QUESTIONS

1. The breakdown products of proteins are either oligopeptides or amino acids, depending on how far the breakdown process goes. Those of nucleic acids are nucleotides. These are important nutrients and can be used by any cell to help them make their own proteins and nucleic acids. You will learn later in the chapter, though, that cells make additional use of oligopeptides that are made by partial digestion of foreign proteins. They are presented on the surface of the phagocyte for interaction with cells of the immune system.

2. A phagocyte, like any cell, is bounded by a plasma membrane. Some of the proteins in this membrane are receptors, molecules that protrude into the extracellular space where they can meet and bind molecules dissolved in extracellular fluid. Any signaling molecule released by a pathogen, a pathogen-infected cell, or any other cell that is under stress can be perceived by the receptors on phagocytes. The interaction between signaling molecules and their receptors can set off chains of events like the responses to growth factors, as seen in Chapter 6.

3. The larger number of receptor molecules with signals bound on the pathogen-facing side of the phagocyte does make sense, because the signals are being produced by the pathogen and diffusing through the extracellular space. The highest concentration of signaling molecules will be right next to the pathogen, and the concentration falls off with distance from that source. Thus, there will be a higher concentration of signaling molecules on the pathogen-facing side of the phagocyte, leading to a higher number of receptors with signal bound on that side. This difference could lead to an asymmetry of signal formation in the phagocyte, so it is more strongly stimulated on its pathogen-facing side, leading it to know which way to go in order to fight this threat.

4. An advantage of killing a pathogen-infected cell is that this will eliminate the environment in which the pathogen is thriving. A disadvantage of the process is that the organism loses a cell. If, however, the pathogen-infected cell would have produced many pathogens, stopping that growth is almost certainly to the host's advantage. You have seen in Figure 6.1 the very large number of virions that a single cell can produce, so stopping that production and perhaps degrading the viruses already made should be a big advantage. If the infected cell was killed simply by lysis, one can imagine the release of many infective virus particles, and that sounds bad.

If the infected cell is killed by apoptosis, which involves a great deal of proteolytic activity, many of the viruses or other intracellular pathogens will be destroyed before they get a chance to infect another cell, and this sounds like a good strategy.

5. When you read about apoptosis, you learned about caspases as proteinases. Caspase 9 was activated by the release of cytochrome *c* from mitochondria, and its action was to work as a proteinase to cleave and thereby activate a cascade of other caspases. If NK cells can put a proteinase into their target cell that initiates the same cascade of proteolytic caspases, this will greatly amplify the proteolytic activity the NK cell can provide, thereby making it a more efficient killer.

6. Any answer is of course an estimate, but it looks as if there are about six completely different surfaces that an antibody molecule could bind on the surface of this one, small protein from hen egg white. Since the surfaces seen by an antibody can overlap, the real number of possible antigenic determinants is larger than six, probably about 24.

7. It is clear from the previous descriptions that this would be very bad news. If the interaction were strong enough to set off the activities of either NK cells or cytotoxic T lymphocytes, the cells presenting that antigen would be killed. This could lead to the inactivation of an entire organ and major medical complications.

8. A great advantage of this ability for the pathogen is that the host is unable to develop memory B cells that are effective in a subsequent infection. Thus, this pathogen has multiple chances to invade the same host. A disadvantage of the ability (from the point of view of the pathogen) is that it must have a mechanism that allows this variation. Flu virus does it, in part, by having its genome broken into seven distinct pieces. If two different flu viruses are infecting one cell at the same time, new viruses can form by mixing the genomes from the two strains of flu. Malaria accomplishes the same goal by having a complex biochemical mechanism for changing the proteins that appear on the cell's surface.

9. The problem with the apparently clever idea of making antibodies and T-cell receptors that react only with nonself is figuring out how to do it. How can the cell know what pathogens it will encounter or what mutations

a genetically unstable cancer cell will experience? Cells can't read the future any better than we can. The immune system is a poignant example of the value of spending on defense. If you have evolved a system that can protect you from infection, it is worth a lot of synthetic effort (synthetic effort in this case is analogous to money) to keep that system fully functional.

10.   In preventive maintenance, a company invests money to send around repair people who change parts that are likely to wear out soon, for example, light bulbs or the bearings on a complicated machine. The process is expensive because of the person-hours needed to do it and because each part that is replaced might have lasted longer. If there were fewer replacement events over a long period of time, this would be an economy. On the other hand, the cost of a breakdown can be very high. If bearings wear out, a machine will probably shut down, so it is unproductive for a long period. Moreover, the failure of one part often leads to the destruction of others, so now the repair is time-consuming and expensive. Continued proteolysis is analogous because all proteins in a cell have only a limited lifetime. Most are turned over by proteolysis every few hours or days, depending on the protein. This costs the cell the expense of continued protein synthesis, but it saves the cell the potential disaster of a poorly functioning protein doing something that damages the cell.

11.   Your diagram should include similarities (motility, phagocytic ability, proteolytic digestion of foreign molecules, and presentation of oligopeptides on the cell surface). The principal difference between these cells is where

they go after accomplishing these tasks. Neutrophils commonly die as they ingest pathogens. Macrophages are more robust and can go on to consume additional pathogens or address future infections. Dendritic cells are programmed to go to the organs of the peripheral immune system—for example, the lymph nodes—where they will encounter many T and B cells, which can then interact with the antigens they are presenting.

12.   This is a kind of tumor promotion. Inflammation is not mutagenic, but it causes an irritation that induces cell division, increasing the likelihood of expanding clones of already mutated cells.

13.   Which study you find most convincing is a matter of opinion. You should certainly try to convince any relative of the importance of this phenomenon, because everyone should know about the roles the immune system plays in cancer and be aware that manipulation of this system is likely to be an important aspect of cancer control in the future. To be convincing, you might summarize each of the experiments described previously, ordered by the extent to which you found them convincing.

14.   This list should include the following: (1) NK cells recognizing a cancer cell because of its failure to have enough MHC molecules on its surface; (2) the binding of antibodies to oligopeptides derived from mutated proteins and presented on a cancer cell's surface by MHC protein, followed by the activation of phagocytes; and (3) the activation of Tc cells whose TCRs recognize mutant oligopeptides presented by MHC proteins.

A perspective on cancer futures

CHAPTER 10

The previous chapters of this book have described cancer as we currently understand it and treat it. This chapter builds on that material to show how modern research in human biology and medicine is opening new approaches to cancer screening, diagnosis, and therapy. This material is presented near the end of the book because understanding the relevant ideas and methods depends on many of the things you have learned in previous chapters. For example, cancer scientists and doctors are now sequencing patient DNA to identify the genes that have been mutated to become capable of serving as oncogenic drivers for cancer in particular patients. You will now be able to recognize those gene names and appreciate why their altered function is important for cancerous progression. Cancer scientists and oncologists are also using methods from immunology to identify and fight cancer cells specifically, rather than attacking cancer by poisoning all dividing cells. Likewise, they are using genetic, biochemical, and cell biological knowledge to design new drugs that should perturb or kill cancer cells more selectively. They are inserting carefully chosen DNA sequences into a patient's cancer cells, hoping thereby to stimulate immune responses in ways that you will now be able to understand. In short, the face of cancer medicine is evolving, and you can now understand the steps in that evolution.

Many of these innovations are new enough that their implementation as treatments is still experimental. Many of them are ferociously expensive, so the problem of cancer treatment at the level of public health is far from solved. Moreover, several of the inventive methods that sounded extremely promising are turning out to be less effective than was initially hoped by their inventors. Nonetheless, some of the novel methods now under study are sufficiently effective that we can expect better cancer treatments to be found. Foreseeing the future is, of course, impossible, but we can look at where we are, compare it with where we were, and project where we are likely to be in the years to come.

PROGRESS IN SCREENING FOR AND CHARACTERIZING CANCER

PROGRESS IN CANCER DIAGNOSIS

PROGRESS IN CANCER TREATMENT

NEW IMMUNOLOGICAL APPROACHES TO CANCER THERAPY

INTEGRATED APPROACHES

CAN THE COSTS FOR THIS KIND OF PROGRESS BE BORNE, AND WILL THEY BE WORTH IT?

LEARNING GOALS

1. Describe the ways in which new methods of screening for cancer may have a positive effect on early cancer detection and on the toll that cancer takes on both individuals and society at large.

2. Compare the new methods of cancer diagnosis with more traditional methods.

3. Give examples of antibodies to cancer-specific antigens that are in use as routes to interfere with the growth and/or health of cancer cells specifically.

4. Identify the cancers for which vaccines are currently available and describe reasons why vaccines to other kinds of cancer are hard to develop.

5. Describe what is involved in identifying a target for a new drug and why getting new drugs into the clinic is so costly.

6. Describe two kinds of currently available immunotherapies that are effective in treating cancer patients.

7. Explain why the development of additional, even more powerful immunotherapies is challenging.

8. Identify two examples in which natural products were the sources of new drugs or approaches for making effective cancer drugs.

PROGRESS IN SCREENING FOR AND CHARACTERIZING CANCER

Recent advances in medical imaging have revolutionized cancer screening. In the last 25 years, medical imaging has improved by giant steps. X-ray imaging, for example, in the preparation of mammograms, now commonly employs digital electronic detectors, which produce images that are available immediately, are of better quality than those recorded on film, and are obtained with lower X-ray doses than were necessary for film recording. Moreover, much X-ray imaging has moved from single images to computed tomography (CT) scans. X-ray scattering substances are now frequently administered to patients to enhance the visibility of blood vessels and other tubular structures in the body. The use of 3-D sonograms, magnetic resonance imaging (MRI), and positron emission tomography (PET) scans is now common, and the interpretation of the resulting images has become both sophisticated and precise (FIGURE 10.1).

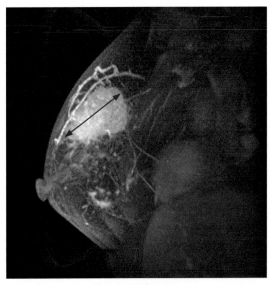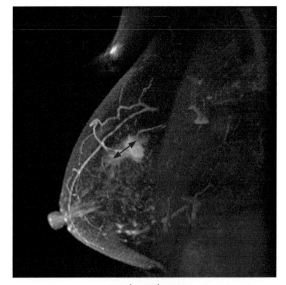

pre-chemotherapy
longest dimension = 47 mm

post-chemotherapy
longest dimension = 16 mm

FIGURE 10.1 **High-resolution imaging of tumors by magnetic resonance imaging (MRI).** When MRI is used in combination with stains that enter blood vessels, the identification and characterization of even rather small tumors can be done considerable precision. Here, this method is used to follow the behavior of a tumor as it responds to treatment. The images show how even a small tumor, whose length is indicated by the red arrows and quantified for each image, can be imaged with a sufficient signal-to-noise ratio to identify the lesion and either propose or evaluate a treatment. (From S.G. Partridge et al. [2002] AJR 179: 1193–1199.)

Part of this progress is the result of new imaging technologies, and part is due to quite sophisticated methods for computer image processing. These processing methods allow the combination of multiple images from different angles to generate 3-D models, as in the making of tomograms. They can also enhance edges, filter out noise (image contrast that conveys no information), boost the contrast that remains, and deliver quantitative information. These processes, taken together, have revolutionized the clarity and information content of images that describe what is going on inside a patient's body. When you add to these advances the techniques of endoscopy, which employ fiber optical instruments passed into almost any of the body's many tubes, the views of a living person's interior that are now available to help with diagnosis are truly remarkable. Their value is shown in the improvements of early cancer detection, changes that have had a big impact on the success of treatments for both breast and colon cancers, contributing to a significant reduction in the numbers of deaths per capita from these diseases.

SIDE QUESTION 10.1

Define image resolution, and describe in simple but accurate terms why improved image resolution is useful in the early detection and diagnosis of cancer.

Nanotubes may aid in the detection of cancer and in delivery of chemotherapeutic agents.

X-ray imaging, sonography, MRI, and PET have all come so far in the last 25 years that they may be at or near a plateau from which it will be hard for the existing instruments to go very much higher. The value of these instruments may, however, improve with new kinds of detection that provide useful information about tumors. The coming years will probably see improved stains for cancer cells, making CT scans and MRI more like PET scanning in their ability to distinguish cancer cells from normal ones. Just how this will be achieved is impossible to say right now, but an example of one approach under current development is the use of nanotubes to detect and characterize cancer cells.

A **nanotube** is a cylindrical array of carbon atoms, commonly one atom thick and with a diameter less than the width of a single protein molecule. These structures are fluorescent: they will absorb infrared light of one wavelength and emit light of a longer wavelength. Infrared light can enter and leave a patient's body reasonably well, so if nanotubes can be modified to bind antibodies to some cancer-related molecule, such as an oncogene product localized to the cell surface, it should be possible to use them for cancer detection in ways that would help clinicians to spot cancers early (FIGURE 10.2).

There is also hope for the use of nanotubes to deliver drugs to the cells they identify. In Figure 10.2, a nanotube is coupled to epidermal growth factor (EGF), so it will bind to cells that present that receptor (EGF-R) on their surfaces. If cancer cells are overexpressing EGF-R, such nanotubes will bind to those cells particularly well. The nanotubes in this figure are also coupled to Q-dots, tiny particles that fluoresce brightly with visible light, so their location can be followed by fluorescence microscopy. They have also been tagged with cisplatin, a DNA poison you read about in Chapter 8. Such nanotubes could not only mark cancer cells for identification by fluorescence microscopy but also poison them with some specificity. If the nanotubes are endocytosed by the cells to which they bind, they will deliver cisplatin to the cytoplasm of those cells, where it can be released and poison faithful DNA replication, increasing the frequency of mutation and inducing apoptosis.

This kind of high-tech approach to drug delivery is still very much under development. Its strengths are that it takes advantage of several recent innovations in both biology and materials science. Its limitations are that (1) cells other than cancer cells express EGF-R, so there may still be non-

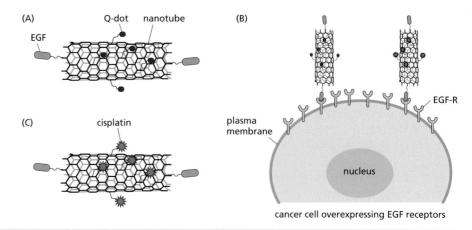

FIGURE 10.2 Nanotubes as tools for the localization of tumors and the delivery of chemotherapies. Nanotubes are commonly represented by a cylindrical cluster of hexagons in which there is a carbon atom at each vertex of every hexagon, and the lines connecting the vertices represent bonds between these carbon atoms. (A) These nanotubes have been coupled to both fluorescent particles (Q-dots) and to epidermal growth factor (EGF), so they will bind to EFG receptors (EGF-R) on cell surfaces (objects not shown to scale). (B) Some cancer cells overexpress EGF-R, so these nanotubes would bind particularly well to those cancer cells, marking them with Q-dots for identification by fluorescence microscopy (again, not to scale). (C) Nanotubes can also be coupled to a chemotherapeutic chemical, such as cisplatin. The binding of EGF to EGF-R often stimulates endocytosis, which would bring both the nanotube and the cisplatin into the cell, where this drug could act on DNA, modifying its structure and inducing mutations that could lead to apoptosis. This is an example of an effort to target chemotherapeutic agents specifically to cancer cells. (From Patel S, Bhirde AA, Rusling JF et al. [2011] *Pharmaceutics* 3:34–52. With permission from MDPI.)

specific, toxic effects; (2) nanotubes are usually quite long, so they may not be endocytosed efficiently; and (3) the release of cisplatin from the nanotube, once in the cell, may not be efficient, so the killing action may not be strong. Nonetheless, the kind of thinking represented by this approach to drug delivery reflects the imagination and creativity of scientists working to improve cancer treatments, and valuable things may come from it. Likewise, scientists may discover novel imaging methods that put tumor detection on a new level.

Scientific advances may lead to novel cancer screens.

Two tests that use a patient's blood to seek and characterize cancer are already in use. One of these is based on the fact that most cancer cells are significantly larger than the normal cells of blood: red blood cells, leukocytes, and platelets (cell fragments that contribute to the ability of blood to clot). A well-designed filter allows blood cells through but captures cancer cells, if they are present, isolating them for examination by more detailed cellular and molecular methods. Cancer cells in circulating blood are obviously a bad sign, because they imply that the parenchyma (the functional tissue of an organ) of an existing tumor contains cells that are escaping from their site of birth. Those cancer cells have the potential to establish sites of metastatic growth, making this cancer dangerous.

A more complex test, still under development, uses DNA sequencing to seek evidence for oncogenic mutations in DNA that circulates with the blood. The idea here is that some cells within a tumor die as a result of inadequate nutrition or oxygen supply. Dead cells break up into fragments. Phagocytes, like macrophages (Chapter 9), normally clean up this kind of debris, but some molecules escape. As a result, molecules from the dead cells can leak from the extracellular space into lymph ducts and from there into blood, where they circulate until they are degraded by enzymes in the blood. If the pieces of circulating DNA include sequences with oncogenic mutations, this information suggests that cancer is present, although the location of the cancer is completely unknown. Several companies are building devices that make the harvesting and characterization of such freely circulating DNA comparatively easy, sensitive, and informative.

Other, perhaps more valuable tests are still in the early stages of development. As mentioned in Chapter 2, blood tests with unprecedented sensitivity are a technology that may make significant contributions to cancer identification and characterization. The growth, division, and death of cancer cells can alter the chemistry of blood and/or urine. Such changes may provide evidence for the presence and nature of a cancer, once the right methods are found. This idea is plausible because mutations are characteristic of all cancer cells. Just as mutations alter cell behavior, they also alter the materials that cells secrete or leave behind as they die. Some of these cell products may become cancer signatures—that is, a particular substance or group of substances that only cancer cells make. A cancer signature might be a change in the balance of chemicals normally found in blood. It might even be a cancer-specific protein or nucleic acid that has not yet been identified.

Each of these methods would provide evidence for the presence of cancer. More precise determination of where the tumor is located and how far it has progressed would, of course, require medical imaging and probably biopsy, but the latter methods could be applied once the blood screen had shown that cancer was present. These approaches would avoid using radiation-based imaging for cancer detection, allowing its use only for cancer characterization. Sensitive and specific blood tests for the presence of cancer signatures would permit cost-effective use of the best forms of medical imaging to track down the location and nature of the problem, once you knew a problem was really there.

Where will these novel blood tests come from? Chemists are using ever more effective instruments to detect and analyze minute quantities of biological materials, so a highly refined and informative analysis of the chemicals in blood is not just a dream. However, two problems must be solved: (1) identifying specific molecules or conditions that are valid signatures for cancer and (2) learning how to find these molecules, when they will almost certainly be present in very small amounts, surrounded by the much more abundant molecules of normal blood. This will make it hard to detect the cancer signatures, the problem of a needle in a haystack. The task can probably be accomplished by identifying ways to bind the cancer signatures very tightly. This technology could then be used to pull the signature molecules away from all the other materials in blood, increasing their concentration and making it possible to study them.

A test based on cancer signature identification could be inexpensive and nondestructive, so its use could become widespread. If such tests were reliable, they would not cause people undue anxiety and concern from false positive and false negative identifications (**SIDEBAR 10.1**). Such tests would focus the use of other diagnostic methods, like imaging and biopsy, on patients who really needed them, rather than using them as screens.

SIDE QUESTION 10.2

Describe a false positive and a false negative in the case of a blood test for cancer.

Both cost and the increasing fraction of our population who are older present practical problems for cancer screening and treatment.

Modern medical imaging is not cheap, so people without insurance or without accumulated resources have trouble taking advantage of its benefits. Remember the experience of Woods (see Sidebar 2.1), who had great difficulty getting even an endoscopy to examine the possibility of cancer in his bladder. With MRI and PET scans, the costs are significantly

A CLOSER LOOK: Understanding the implications of results from a screen or test.

One of the most challenging issues facing anyone who tries to determine the value of a new screen or test is knowing the validity of the test's results. This issue came up in Chapter 2 in connection with the value of screening for cancer, but it is an important part of all new developments in screening and diagnosis (see the section in Chapter 2 entitled The quality of a cancer screen is measured by the frequency with which it gives a correct answer). When a test result says you do or do not have a particular condition, like a prostate tumor that needs treatment, what are the chances that the test is correct? Scientists address this issue with the idea of the **positive predictive value** (**PPV**) or **negative predictive value** (**NPV**) of the test. These two quantities are defined as follows:

PPV = number of true positives ÷ number of positive calls

NPV = number of true negatives ÷ number of negative calls

where the number of true positives and negatives is established by comparison with some completely reliable test, and the number of calls is in each case the number of positives or negatives that were given by the test under consideration. Stated simply, both the PPV and NPV are the fraction of times that your new test got the right answer.

It is intuitively obvious that these numbers (PPV and NPV) are key evaluators of a new test; if the number of times the test got things wrong is high, the test has no value. In each case, the ideal value is 1, which implies the test got it right every time. The hardest thing about getting accurate values of PPV and NPV is having a truly reliable test against which to compare the new test results. Such a test is usually called a gold standard.

An example of a test you already know about is the level of prostate-specific antigen (PSA) in a man's blood. If PSA is high, a physician might recommend a biopsy of prostate tissue to evaluate the character of the cells in this patient's prostate gland. The examination of a biopsy is fundamentally different from measuring the amount of PSA in the blood because a knowledgeable pathologist is now looking at cells, rather than simply the amount of one protein that those cells secreted. The cells in the biopsy could be used both for microscopy and for some molecular tests for the presence of abnormal cells and/or driving oncogenes. These would be the gold standard that would tell both doctor and patient whether a developing prostate tumor was really present. As stated in Chapter 2, the levels of PSA are no longer regarded as good predictors for the presence of prostate cancer, because quite a few men with high PSA have undergone biopsies, and the results have not found cancer. Likewise, some men who have clear evidence for prostate cancer have normal levels of PSA. The published value of the PPV for PSA testing is 30% (0.3), which means that for every three men with high PSA, only one will be found by biopsy to have indications of cancer. The concepts of PPV and NPV are important in thinking about any new screen, but you must always remember that finding a really reliable gold standard is difficult.

greater. Programs that increase the breadth of medical insurance coverage for all people can help with this problem, but there is an underlying difficulty that makes this issue even more difficult. All over the world, people are getting older. This sounds like a silly truism; of course they are. But changes in the fractions of people who are old or young are an important aspect of the modern world. Thanks to improved medicine and nutrition, to laws that have reduced workplace hazards, to moderately successful anti-smoking campaigns, and to our avoiding World War III (so far), the fraction of our population that can be described as old is increasing. In the hundred years from 1900 to 2000, the fraction of the United States population that was 65 or older increased about 3.5-fold faster than the population as a whole, reaching ~35 million people (FIGURE 10.3A). In the 50 years from 2000 to 2050, the fraction of the U.S. population ≥65 years old is again expected to rise faster than the population as a whole, reaching almost 80 million. Because cancer is far more common in older people than in the young (FIGURE 10.3B), it is not surprising that there has been an increase in the number of deaths from cancer. This implies that the incidence of cancer will go up, regardless of improvements in early detection and diagnosis.

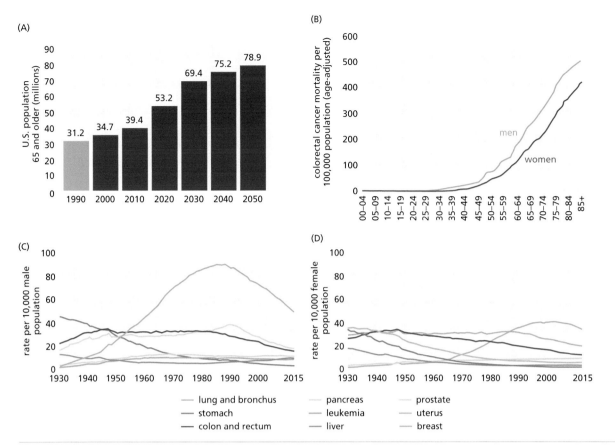

FIGURE 10.3 **Age distributions and frequencies of cancer for people living in the United States.** (A) The bar graph presents recent and expected data on the number of people in the United States who are aged 65 years or older. (B) This graph depicts the frequency with which people die of colon cancer as a function of their age, demonstrating the strong age dependence of cancer. When you know both the dependence of cancer on age and the age distributions in a population over the passage of time, you can correct data on cancer frequency collected over years for the effect of age on cancer frequency, carrying out age adjustment. (C, D) These graphs show age-adjusted death rates from 1930 to 2015 for several cancers in populations of (C) men and (D) women. Both medical and social changes have led to significant alterations in age-adjusted cancer frequencies. For example, better screenings for uterine and colorectal cancer have recently led to significant decreases in these diseases. The big upswing in lung cancer in men is attributed to the large number of men who became smokers during World Wars I and II. The increase in lung cancer for women occurred later than for men, probably for social reasons. The more recent downswing in lung cancer for both men and women may be due to increased awareness about the dangers of cigarette smoke. (A, courtesy of D. Singer and R. Hodes, U.S. Bureau of the Census, Projections of 1996. B, from Jemal A, Murray T, Ward E et al. [2005] *CA-Cancer J Clin* 55:10–30. With permission from Wiley. C and D, from Cancer Facts and Figures 2018, published by the American Cancer Society.)

If one takes the change in the age distribution of the population into account, using it to factor in the rapid increase in cancer's likelihood with age, one can obtain graphs that show the **age-adjusted incidence of cancer**. By this measure, medical science is making progress with the treatment of several cancers (FIGURE 10.3C,D). Most of this progress comes, however, from clearly identifiable factors, such as ways to prevent infection by a cancer-promoting virus or a big improvement in some screening technology, such as Pap smears, mammograms, and colonoscopies. Figure 10.3C also shows the dramatic changes in deaths from lung cancer in men. The big increase in this statistic and its more recent decrease are certainly due to American tobacco-smoking habits, so this cancer must be thought of as a special case. Note that the corresponding data for women (Figure 10.3D) show a later and smaller increase in lung cancer deaths with an encouraging more recent decrease. These changes too are probably due to smoking habits. The death rates from some other cancers are also encouraging: breast, prostate, colorectal, uterine, and stomach cancers have all shown some downturn, mostly due to improved methods for screening. Deaths from pancreatic and blood cancers, on the

Why is cancer more common in old people than in the young?

other hand, are quite stable. Clearly, we need to do better in the prevention, detection and treatment of cancer.

PROGRESS IN CANCER DIAGNOSIS

Some progress in cancer diagnosis has come simply from the thoughtful application of well-established methods, based on science you have read about in previous chapters of this book. Other progress is a result of more recent innovation. In the text that follows, you will find examples of both kinds of progress.

Antibodies to cell-type-specific antigens have refined aspects of cancer diagnosis.

Some of a cell's structural proteins are characteristic of the tissue where the cell resides. For example, intermediate filaments of the cytoskeleton are made from different proteins in fibroblasts, muscle cells, and various kinds of epithelia. In the case of metastatic cancer, antibodies specific for a particular class of intermediate filament can help to identify the tissue of origin for the wandering cells. This information can sometimes help doctors find an effective treatment.

As mentioned in Chapters 8 and 9, antibodies to specific membrane-bound receptors, such as the receptor for epidermal growth factor (EGF-R), can be used to help classify breast cancers; some cancer cells express this receptor in a recognizable form, while others do not. This is an example of ways in which immunological study of tissue biopsies can refine cancer diagnosis. If the results show that the cancer cells do display a known receptor, a cell-specific treatment can sometimes be employed. We will see that this approach is now an effective treatment for several cancers.

Advances in diagnosis are coming from improvements in methods that assess the levels of gene expression.

Mutations are, by definition, changes in the sequence of DNA. Some mutations are in the sequence that encodes the gene product, so these can alter both the structure and the function of the corresponding protein. Examples include both the change of a proto-oncogene into an oncogene and the inactivation of a tumor suppressor. Other carcinogenic mutations alter the segments of DNA that regulate the amount or timing of gene expression, changing the levels of expression of cancer-related mRNAs and the proteins they encode. Thus, methods that assess the levels of specific mRNAs in cancerous versus normal tissues are informative about the genetic and physiologic state of cancer cells. When this approach was first conceived, the methods for measuring relative amounts of RNA were quite primitive, but with recent developments in molecular biology, this kind of test has become quick, powerful, and wide reaching.

Scientists have learned how to attach small amounts of DNA with any chosen sequence to well-defined positions on a small transparent chip made of glass or a related material, like silicon. They have also developed methods to synthesize pieces of DNA, 20–50 nucleotides long, with any sequence they choose. Thus, they can make oligonucleotides with sequences that match each gene whose level of expression could be important for cancerous cell behavior: for example, genes that encode growth factor receptors or protein tyrosine kinases. When each of these DNA sequences is attached to a different position, the chip is called a **DNA microarray** (FIGURE 10.4). DNA microarrays can have many uses, depending on exactly which DNA sequences are attached to each of the thousands of locations

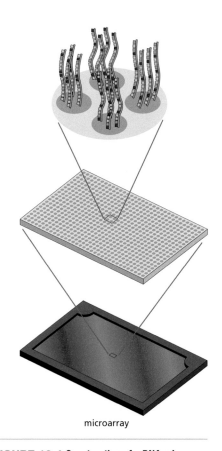

microarray

FIGURE 10.4 Construction of a DNA microarray. A DNA microarray is built on a thin wafer of glass or silicon (bottom), which is often called a chip. A robotic micromanipulator is used to generate a large array of spots to which DNA can be attached (middle). At each position of this array, appropriate chemical reactions can attach one end of several pieces of DNA, all with one selected sequence (top). At each other location, DNA with a different sequence is attached, filling all the locations on the chip with tens of thousands of different DNA sequences, all at known and well-defined positions. (Adapted from DNA Microarray, Learn.Genetics, Genetic Science Learning Center.)

on the chip. One such use is measuring the amount of mRNA expressed from each relevant gene in cancer cells and comparing those levels of expression with the levels of expression from the same genes in normal cells.

To accomplish this, a scientist will isolate all the mRNAs from the parenchymal cells of a tumor and from a comparable number of cells in a corresponding normal tissue. All the mRNAs in each sample are copied into DNA, by use of reverse transcriptase. These copy DNAs, commonly called cDNAs, are labeled with fluorescent dyes, one color for cDNA from tumor tissue and another color for cDNA from normal cells. The labeled cDNAs are then hybridized to two identical chips, one for the normal c-DNAs and one for the cancerous ones. By scanning such chips with a bright light that excites each color of fluorescence, a doctor can measure the relative amounts of cDNAs that have hybridized to the DNA sequences attached at each position. Such comparisons, which can be done rapidly by computer, produce **heat maps**, which are images that show whether the amount of a given mRNA is higher (red) or lower (green) in the tumor cells than in normal cells. If the levels on the two chips are the same, the heat map gives that position no color. Each position on the chip is now red, green, or black, and since the sequence of the DNA attached to each chip location is known, the heat maps tell the doctor the relative expression level of each gene. All these data can then be displayed in meaningful patterns by rearranging the colored spots into clusters that represent all the genes of a chosen kind (FIGURE 10.5).

Data and analyses like this may help to solve a very important problem that has emerged with improved medical imaging. In breast cancer, for example, high-quality imaging can reveal rather small tumors (Figure 10.1, right side). When such a tumor is seen during initial screening, as opposed to after treatment, the doctor cannot know for sure whether it is something that requires medical attention. If the condition were just left alone, would it grow and metastasize, would it simply stay as it is, or would the patient's immune system eliminate it with no medical intervention? From the X-ray images alone, there is not enough evidence to know the answer to this question, so most doctors act on the premise that it is better to treat and give unnecessary attention than not to treat and have a patient develop a fatal cancer. There is, however, a significant group of doctors

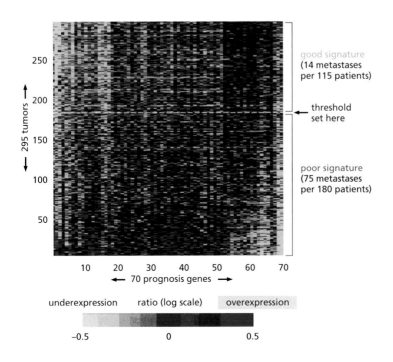

FIGURE 10.5 **A heat map of gene expression in 295 different breast tumors.** Seventy different genes were identified as ones whose level of expression might relate to the behavior of a tumor. The DNA sequences of these genes were used to make a DNA microarray containing 295 rows of 70 distinct spots. cDNAs of all the mRNAs from these 295 tumors were synthesized by use of reverse transcriptase. These pieces of DNA were labeled and allowed to hybridize to all the spots in a row of the DNA microarray, one row for each tumor. Hybridization was also done on a second copy of this microarray, using cDNA made from mRNA isolated from normal tissue from each of the same people, so measurements of fluorescence could be used to see whether each gene was expressed at a higher or lower level in each tumor. Patients were then sorted on the basis of their survival: good or bad. The expression signatures of the patients were then analyzed. Good outcomes are at the top of this map; poor outcomes are at the bottom. The clustering of green spots (decreased expression in the tumor) and red spots (increased expression in the tumor) shows that there are similarities in the expression signatures of patients with similar outcomes. (From van de Vijver MJ, He YD, van't Veer LJ et al [2002] *N Engl J Med* 347:1999–2009. With permission from Massachusetts Medical Society.)

and health advocates who say that screening by high-quality imaging is actually causing more trouble than it is saving by leading to aggressive treatment when the tumor might not have progressed or simply gone away without treatment. Data like those shown in Figure 10.5 display the kind of molecular information that can now be obtained through molecular analysis of biopsy material. This kind of information may become useful in guiding physicians and patients in making good choices about whether to treat or not. As we learn more about the molecular biology of cancer progression, the right choice about treatment may become obvious.

Information about the abnormality of gene expression in a given tumor may ultimately provide guidance about exactly how to treat, as well as whether to treat or not. For example, some transcription factors that regulate the expression of proteins controlling cell cycle progression are up-regulated in certain tumors. If such up-regulation is identified by a chip assay, like that shown in Figure 10.5, the knowledge of overexpression can guide treatment. Such refinements are already being carried out in some cases, as you read in Chapter 9; cancers that overexpress certain growth factor receptors can be treated effectively with monoclonal antibodies, like trastuzumab. Following further improvements in molecular methods, and with the accumulation of clinical experience in how to use the resulting knowledge, this kind of characterization should become more powerful and applicable to a wider range of cancers.

Assays based on DNA microarrays have also shown that one of the macromolecules whose expression can be altered in cancer is a kind of RNA that is not translated into protein; these are called noncoding RNAs. Some noncoding RNAs are quite short, so they have been dubbed **microRNAs**. A range of molecular biological experiments has shown that microRNAs bind to mRNAs and alter their translatability and/or stability, either up-regulating or downregulating expression, depending on the particular microRNA. Thus, expression levels of microRNAs can lead to changes in the expression of mRNAs that encode proteins key for carcinogenesis. This is a poignant example of the complexity of cells and the many reasons why a full understanding of carcinogenesis is taking many years to develop.

Improvements in methods for DNA sequencing are providing useful information about the character and properties of tumors.

As recently as the beginning of this millennium, the process of sequencing DNA was slow and expensive, and it required quite large amounts of DNA to be accurate. Recently, there has been dramatic progress in the technologies for determining DNA sequences (**SIDEBAR 10.2**). Progress here has been so great that it has altered the landscape of how sequence information can be used in cancer diagnosis and treatment. Now small amounts of DNA can be sequenced sufficiently rapidly and cheaply that it is realistic to use these methods to help characterize the changes that have occurred during the development of a particular cancer. Scientists now can get good DNA sequence information from one or only a very few cells. From earlier chapters in this book, you know that both proto-oncogenes and tumor suppressors are mutated into carcinogenic alleles as cells go through cancerous transformation. Thus, sequencing those genes and others in the DNA from a patient's cancer cells gives explicit information about just which genes have become mutated during this particular case of carcinogenesis. The significance of this progress is that the parenchymal cells of a tumor can now be characterized in completely new ways,

SIDE QUESTION 10.4

How can mutation of a transcription factor alter the level of expression of several genes? Do you expect changes in the levels of mRNAs to lead to alterations in the levels of the proteins they encode? Why or why not?

A CLOSER LOOK: DNA sequencing by the Sanger method.

Throughout this book, we have spoken of DNA sequence as something important for understanding cells and cancer, because the order of nucleotides along a strand of DNA defines the sequence of any RNA transcribed from that DNA and thus any protein that is translated from the RNA. We have also talked about how changes in DNA sequence are mutations that inheritably alter the structures of the RNAs that are transcribed and the proteins that are translated from that DNA. DNA sequences and their changes are therefore at the heart of carcinogenesis. Here, we describe the method invented by Frederick Sanger to determine the sequence of DNA, information that has become critically important in identifying the mutations that have actually occurred during the development of a particular cancer. This method, also called dideoxy sequencing, was not described previously because it involves a bit of chemistry.

Each strand of DNA is a chain of nucleotides. Each nucleotide contains a sugar (deoxyribose), a phosphate group (phosphorus atom surrounded by oxygens), and a base, either A (adenine), G (guanine), T (thymine), or C (cytosine); we have used the initials of the bases to distinguish the nucleotides (**Figure 1A**). When nucleoside triphosphates polymerize to make strands of DNA, two of the phosphates are broken off, a situation similar to the hydrolysis of ATP that cells use to power a wide range of process. The release of these phosphates provides the energy to drive DNA synthesis, and it leaves behind one phosphate group that forms the link between successive sugars in DNA's sugar–phosphate backbone (if this seems unfamiliar, see Figure 3.4 for a refresher on DNA structure). The bonds that build the sugar–phosphate backbone

are made by the phosphate-associated oxygens, which link the carbon atom (C) labeled 5′ (pronounced five prime) with the one labeled 3′ (pronounced three prime) on the next nucleotide of the chain. (The prime symbol indicates that we are talking about the sugar part of the nucleotide; the number specifies which carbon atom in the sugar ring is under consideration.) These links can form only when there is an oxygen atom connected to the 3′ C. When a normal nucleotide is used, as shown in Figure 1A, the chain can grow without limit. However, if a modified nucleotide with no oxygen on the 3′ carbon is used, as shown in **Figure 1B**, the chain can no longer grow. Such nucleotides are called chain terminators.

Fred Sanger, working in Cambridge, England, realized that he could use a mixture of normal and chain-terminating nucleotides to identify which nucleotide is present at each position along a chain of DNA. His method is described in **Figure 2**. This sequencing method is reliable and powerful; it has produced large amounts of useful data about the nucleotide sequences of different pieces of DNA. Its princi-

pal limitations, though, are that it is slow and expensive. As scientists have wanted more and more DNA sequence information, there has been strong motivation to improve on this method, and many advances have been made. The first to come was simply to automate Sanger sequencing by using fluorescent tags on the chain-terminating nucleotides. This trick produces the data shown in **Figure 3**. By combining the principles of the original Sanger sequencing with fancy devices for reading fluorescent signals emitted by tagged versions of the chain-terminating dideoxy nucleotides, the speed of this kind of sequencing increased, and its cost dropped by about a factor of 10.

Thanks to further technological improvements, it is now possible to sequence many pieces of DNA at the same time, in parallel, which has further sped up the sequencing process and significantly dropped its cost. Sequencing the first human genome cost many millions of dollars. Now an equivalent job is only about $4000, and even this comparatively low cost is dropping. Moreover, the sensitivity of sequencing is now so great that

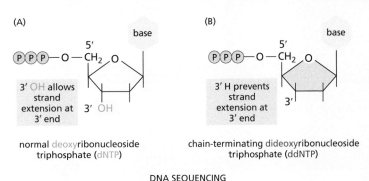

DNA SEQUENCING

Figure 1 The structures of normal and chain-terminating nucleotides. Dideoxy sequencing, or Sanger sequencing (named after the scientist who invented it), uses DNA polymerase, along with normal deoxyribonucleoside triphosphates (dNTPs, left) and special chain-terminating nucleotides called dideoxyribonucleoside triphosphates (ddNTPs, right), to make partial copies of the DNA fragment to be sequenced. These ddNTPs are derivatives of the normal deoxyribonucleoside triphosphates, but they lack the 3′ hydroxyl group. When incorporated into a growing DNA strand, they block further elongation of that strand. (Adapted from Alberts B, Johnson AD, Lewis J et al. [2015] Molecular Biology of the Cell, 6th ed. Garland Science.)

SIDEBAR 10.2

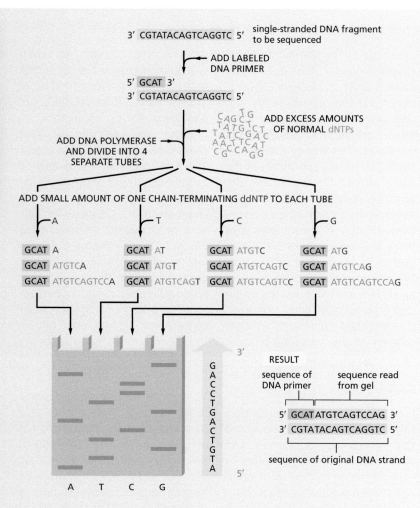

Figure 2 The use of mixtures of normal and altered nucleotides to determine DNA sequence. To determine the complete sequence of a single-stranded fragment of DNA (gray), the DNA is first hybridized with a short DNA primer (orange) that is labeled with a fluorescent dye or radioisotope. DNA polymerase and an excess of all four normal deoxyribonucleoside triphosphates (blue A, C, G, or T) are added to the primed DNA, which is then divided into four reaction tubes. Each of these tubes receives a small amount of a single chain-terminating dideoxyribonucleoside triphosphate (red A, C, G, or T). Because these ddNTPs are present at low concentration, they will be incorporated only occasionally into the growing oligonucleotide. Therefore, each reaction tube produces a set of DNA copies that terminate at different points in the sequence, thanks to the chance incorporation of a chain terminator. The products of these four reactions are separated by electrophoresis in four parallel lanes of a polyacrylamide gel (labeled here A, T, C, and G). In each lane, the bands represent fragments that have terminated at a given nucleotide but at different positions in the DNA. This works because there are many copies of the DNA being replicated in each of the wells, and only a small fraction of them is terminated by incorporation of the ddNTP at any one site along the DNA. By reading off the bands in order, starting at the bottom of the gel and reading across all lanes, the DNA sequence of the newly synthesized strand can be determined. The sequence, which is given in the green arrow to the right of the gel, is complementary to the sequence of the original, gray, single-stranded DNA. (Adapted from Alberts B, Johnson AD, Lewis J et al. [2015] Molecular Biology of the Cell, 6th ed. Garland Science.)

the DNA from a single cell provides sufficient material to obtain large amounts of sequence information. These methods for DNA sequencing have revolutionized many aspects of acquiring scientific knowledge. Their contributions to cancer biology and medicine are still evolving, as ever more uses for this detailed information are identified.

AUTOMATED DIDEOXY SEQUENCING

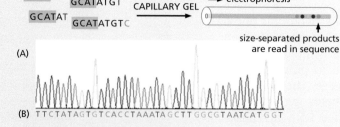

Figure 3 Diagram and image of automated Sanger (dideoxy) sequencing of DNA. Fully automated machines can run dideoxy sequencing reactions. (A) The automated method uses an excess amount of normal dNTPs plus a mixture of four different chain-terminating ddNTPs, each of which is labeled with a fluorescent tag of a different color. The reaction products are loaded onto a long, thin capillary gel and separated by electrophoresis. A camera (not shown) reads the color of each band as it moves through the gel and feeds the data to a computer that assembles the sequence. (B) A tiny part of the data from such an automated sequencing run. Each colored peak represents a nucleotide in the DNA sequence. (Adapted from Alberts B, Johnson AD, Lewis J et al. [2015] Molecular Biology of the Cell, 6th ed. Garland Science.)

based on the mutations that are leading to their cancerous behaviors and even assessing the genetic heterogeneity of a tumor.

In previous chapters, we have referred to cancers by the tissues from which they originated: lung, colon, prostate, and so forth. These classifications have been important to oncologists, both for localizing the primary tumor and for identifying effective treatments, based on previous clinical experience. Now, it is straightforward to identify the genes that are mutated and even to learn the specific mutations that have occurred. The characterization of a cancer can now include a description of the state of genes known to be important for the regulation of cell behavior: *ras*, *src*, *myc*, and the DNA that encodes p53, Rb, p21, and many others proteins that help to regulate the cell cycle, apoptotic behavior, and metastasis. This information facilitates a more refined approach to cancer diagnosis and treatment. Sometimes, this information can be obtained even faster and more cheaply by using a DNA microarray rather than by DNA sequencing. Microarrays are now available in which each position on the chip contains DNA sequences that match the sequence of a mutation known to be associated with cancer for each of the genes in question. Already, these classifications are leading to more focused treatments than were possible when all that was known was the tissue of tumor origin. This is an important example of the technologies that are enabling **personalized medicine**, the process of tailoring treatment to the specific problem in an individual patient, as described later in this chapter. In addition, there are at least five medical consequences of the increase in information available about the mutations in a tumor's parenchymal cells.

(1) Sequence information obtained from numerous examples of apparently similar cancers is allowing refined subclassifications that motivate specific treatments. For example, some cases of lung cancer are associated with up-regulation of the protein tyrosine kinase, EGF-R. This discovery has led doctors to treat these patients with inhibitors of this enzyme, for example, the drug gefitinib. This innovation has achieved significant patient improvement in some cases.

(2) DNA sequence can reveal which oncogenes may be driving cancerous behavior in a given tumor. For example, if *ras* is present in an oncogenic form, this identifies one of the important factors that can induce excess cell division. If the DNA encoding p53 reveals an allele known to reduce protein function, another key fact is available. Information like this is already guiding treatment, allowing doctors to take advantage of medicines known from previous clinical work to be effective when a given mutation is present in one kind of cancer and use them for treatment of other cancers in different tissues.

(3) DNA sequence information has revealed that carcinogenic mutations are not necessarily in the DNA that encodes the gene's protein product. The carcinogenic changes can also be in sequences that regulate the amount of gene expression. You will recall that viruses are sometimes oncogenic because they increase the level of expression of a proto-oncogene. Thanks to recently acquired sequence information from cancer cell DNA, scientists now know that analogous changes can occur through mutations that up- or down-regulation of expression cancer-related genes without the action of a virus. This finding is altering the way some doctors think about carcinogenesis.

(4) In a related discovery, information about DNA structure has shown that expression levels of tumor suppressors are often down-regulated by the addition of methyl groups (the atoms CH_3) to the nucleotide C in sequences that regulate expression of a gene in question. Changes in methylation levels have been known for years to down-regulate the expression of

specific genes during the course of human development, but currently there is a focus on methylation as a factor in altering gene expression during carcinogenesis. Changes in DNA methylation are not the genetic changes that we have seen as mutations, because they are not changes in DNA sequence. They are called **epigenetic** changes because they lie outside the genome itself. Nonetheless, when epigenetic changes occur in the promotor regions of genes important for cancer, be they proto-oncogenes or tumor suppressors, they can alter levels of expression of key regulatory proteins and bring on carcinogenic changes.

A remarkable thing about these epigenetic changes is that they are inheritable through cycles of DNA replication and mitosis. The enzymes that add methyl groups to the nucleotides of DNA are guided by the presence of existing methyl groups, so they repeat the pattern on newly synthesized DNA. Thus, although patterns of DNA methylation are not truly genetic changes, they are inheritable, and this is why they are now thought to be important in down-regulating the expression of tumor suppressors. Indeed, some cancer-promoting agents, like cigarette smoke, contain chemicals that increase DNA methylation; one of the ways smoking promotes cancer is through methylation of the DNA that controls the synthesis of tumor suppressors. The down-regulated expression of proteins that help to prevent cancerous transformation can definitely be carcinogenic, even if it is not a mutation.

(5) Scientists who have sequenced DNA from many tumors have found that a bacterium usually found in association with skin or mouth diseases is quite commonly found in the parenchymal cells of colon cancers. *Fusobacterium* would not normally infect colon cells, but when colon-derived cells are going through cancerous transformation, it appears that they become susceptible to such infections. The DNA of these bacteria is commonly found in colon cancers, suggesting that the presence of a bacterium may contribute to the pathology of the cancer. This is a new way of thinking about how cancers cause pathological changes in their hosts.

PROGRESS IN CANCER TREATMENT

Treatments of cancer are improving in at least three ways. The first improvement results from progress in cancer detection. Early detection means catching the disease at a stage when its parenchymal cells are not yet genetically so diverse, increasing the likelihood of effective treatment. The opinion that early diagnosis enables better treatment is based in part on the example of well-established treatments for acute lymphocytic leukemia (ALL) in children, as described in Chapter 8. The commonly used regimen of several conventional therapies leads to about 80% cures and many levels of remission among the remaining 20%. One reason for this success is probably that ALL in children involves a more consistent set of mutations than is found in other cancers; it therefore displays more consistent response to treatment. Thus, treatments that are effective in one patient are likely to be effective in other patients too. When doctors have an early diagnosis and detailed information about a given cancer, there is reason to hope that they will achieve similar levels of treatment effectiveness with other cancers as well.

The second source of progress in cancer treatment comes from more detailed diagnoses that characterize a tumor with greater precision, as described in the previous section. Better description of a tumor should enable more effective use of existing therapies. These therapies include targeted radiation and inhibitors of DNA synthesis and mitosis, plus inhibition of cell cycle-promoting enzymes and application of immunotherapies.

SIDE QUESTION 10.5

Imagine that studies sequencing the DNA of many cancer patients have identified the most common mutations in a gene that is important for tumor behavior, for example, *BRCA1*. How could you use DNA microarray technology to identify exactly which of those mutation(s) have occurred in a given patient's *BRCA1* gene?

The third path to improved cancer treatment is being built on the results from improved knowledge about individual cancers and about the human body more generally. We begin this discussion with a look at the fruits of personalized medicine.

Improved cancer diagnosis enables personalized medicine, which has both advantages and disadvantages.

The information about a cancer that can come from molecular analyses of tumor cells is impressive. DNA sequences from tumor cells help to describe what went wrong during this set of carcinogenic events; descriptions of RNA and protein expression levels characterize macromolecular syntheses, allowing a detailed comparison of gene expression in tumorous and normal tissue. Comparisons of cancerous cells with normal cells by light microscopy is, of course, what pathologists have been doing for years, but the additional information from molecular biology extends the description of a given cancer beyond what one can see in a microscope. Moreover, the use of antibodies to cancer-related antigens can provide additional useful information.

The upside of personalize medicine is the improved likelihood that an effective treatment can be found to help an individual patient. This benefit is likely to increase as more knowledge is obtained from experience treating patients whose cancer genotypes and RNA or protein expression levels have been determined. Such experience will reveal, for example, just how much of a particular chemotherapeutic agent is likely to be effective and minimally harmful in terms of side effects. In time, there should also be new immunological reagents (antibodies and cytokines) whose specificity for particular cancer-related cells and proteins can be focused by the knowledge of which genes have been mutated and in what ways they have changed. These aspects of personalized medicine are likely to develop into significant improvements in patient outcomes during the decades ahead.

However, the complexities of tumor cell biology pose significant hurdles in realizing the benefits of personalized medicine. Much evidence supports the view that each tumor starts as a clone of cells, formed by the multiple divisions of a single cell that had accumulated enough mutations to allow misbehavior. Often, however, these mutations also endow the parenchymal cells of the tumor with genetic instability, so daughter, granddaughter, and great-granddaughter cells are not identical to the founding tumor cell. Genetic instability leads to **tumor heterogeneity**, which shows up both in the diversity of tumor cell behavior and in variations in the protein products that are expressed in different cells of the tumor. Some daughters of the founding cell may become stem-cell-like in their behavior, leading them to be responsible for the observations that suggest the importance of **cancer stem cells** in tumor biology. Other cells may have become fast growing and account for much of the tumor mass, which in turn leads to much of the medical pathology that a tumor causes. Still other cells may acquire the mutations that lead them to become metastatic.

Tumor heterogeneity makes tumors hard to treat. It also means that the kinds of diagnostic methods that rely on sequencing DNA from a tumor biopsy may be of limited long-range value. When specific alleles of key genes are found by DNA sequencing, one cannot know the fraction of the tumor cells that contain those alleles. Moreover, the genotype of a tumor is a moving target. Thus, extensive sampling will be required to get reliable genetic information. Multiple samples will be needed to characterize

Why are young tumors likely to be less heterogeneous than old ones?

tumor heterogeneity, and samples taken over time will be necessary to follow tumor evolution, making this kind of diagnosis both costly and complex. Note, however, that the problem of tumor heterogeneity is often less pronounced in tumors that are relatively young. Thus, improved early diagnosis, together with molecular characterizations of the cancerous transformation up to that point, may enable powerful approaches to cancer treatment in the future.

An important downside of personalized medicine is its cost. Although the cost of DNA sequencing has come down dramatically, it is still high enough to add to the costs of cancer medicine; it currently costs about $4000 to obtain data on a complete human genome. Tools, like DNA microarrays, are being developed that will allow cheaper identification of specific mutations in specific oncogenes and tumor suppressors, but nonetheless, this kind of high-tech study is unlikely to become cheap. Both molecular diagnoses and the specific treatments they suggest, as will be described next, will put this kind of cancer care out of reach for most people in our country, unless we invest even more than we now do in public health.

Some emerging cancer treatments are based on our increasing knowledge of human biology.

Some progress in cancer treatments is coming from our increasing knowledge about the human body. For example, one class of compound that is already in use for certain cancers is based on chemicals that act to stimulate normal differentiation. A derivative of vitamin A called **all-trans retinoic acid** (**ATRA**) is used by our bodies as a signal to promote differentiation in some kinds of normal blood cells. One kind of leukemia leads to the release into circulating blood of large numbers of improperly differentiated white blood cells. This condition, called promyelocytic leukemia, can often be treated successfully by giving the patient ATRA. This chemical induces the immature leukocytes to complete their differentiation. As a result, the condition of the patient is significantly improved (FIGURE 10.6).

A different form of vitamin A is sometimes useful in comparable ways for the treatment of pre-malignant lesions in the mouth and throat. In both cases, the drug is overcoming the effect of a mutation that has inhibited normal cell differentiation. These vitamin A-based drugs cause cells to revert from a cancerous phenotype into a more differentiated structure

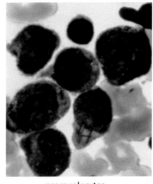
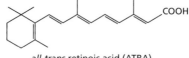
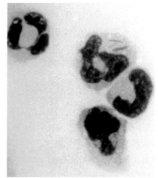

promyelocytes

neutrophils

FIGURE 10.6 Some poorly differentiated, cancerous white blood cells respond to *all-trans* retinoic acid (ATRA) by differentiating. ATRA is a relatively small biomolecule used by the human body as a signal during the differentiation of normal white blood cells. ATRA can be administered to patients whose white blood cells are not differentiating properly. In this kind of disease, the ill-differentiated white blood cells spill out from the bone marrow into the circulating blood. ATRA acts on these immature white cells, promoting the completion of their differentiation, as shown here with cells called promyelocytes differentiating into neutrophils. The cells that have differentiated do not proliferate, so the number of cancer cells will now increase at a much lower rate. The result of such a treatment is often a considerable improvement in the patient's condition. (Courtesy of P. P. Pandolfi.)

and behavior. This kind of treatment is currently valuable for only a limited number of cancers, but the idea of inducing a cancerous cell to revert to its normal path of differentiation is appealing. Other approaches like this may be found with further research.

Protein kinases are promising targets for novel cancer therapies.

In Chapter 8 you learned about imatinib (Gleevec®), a drug that inhibits the activity of an abnormal protein kinase that forms as the product of a particular oncogenic mutation, which is a common cause of chronic myeloid leukemia (CML). You will remember that this abnormal enzyme, the fusion of protein parts from Bcr and Abl, short-circuits normal cell cycle signaling pathways and drives cells into DNA replication and division without the action of growth factors. Imatinib blocks the action of this oncogenic enzyme and prevents its tumor-causing action. Treatment with imatinib is not a cure for this cancer, but it inhibits abnormal behavior in the cells that carry the mutation. The genetic defect is still there, but it no longer causes trouble, thanks to the enzyme inhibitor. It is likely that medical science will discover more drugs like imatinib that help to ameliorate multiple specific kinds of cancer.

Our cells contain over 500 different protein kinases, many of which are involved in regulation of the cell cycle and differentiation. It is easy to imagine that additional drugs will be found to inhibit protein kinases in ways that improve control of the cell cycle or of cell motility processes intrinsic to metastasis. A problem this work will face, however, is that many protein kinases have quite similar structures (FIGURE 10.7). Finding compounds that inhibit only the kinases important for a particular cancer phenotype will not be easy. Nonetheless, some useful inhibitors have already been found. Sunitinib inhibits several protein kinases that transfer phosphate from ATP to different growth factor receptors. Like imatinib, this drug binds to the ATP binding sites of kinases, the clefts marked with arrows in Figure 10.7. Although this inhibitor is not specific, it is effective in slowing tumor growth and helping to kill tumor cells (FIGURE 10.8).

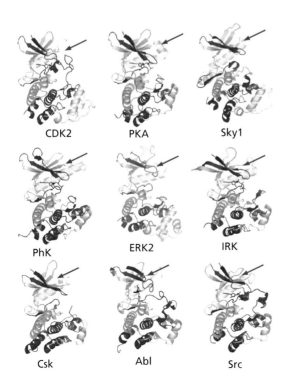

CDK2   PKA   Sky1

PhK   ERK2   IRK

Csk   Abl   Src

FIGURE 10.7 **Many protein kinases are built with similar structures.** The structures of nine protein kinases are shown, using ribbon diagrams to reveal the path of the amino acid chain in each. The red arrows indicate the cleft in each protein's fold where ATP binds. Clearly, these proteins all have similar designs. Thus, finding a small-molecule inhibitor that is specific for the activity of only one of these enzymes is a challenge for both chemists and drug companies. (Courtesy of N. M. Haste, S. S. Taylor, and the Protein Kinase Resource.)

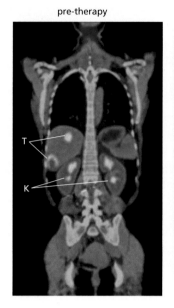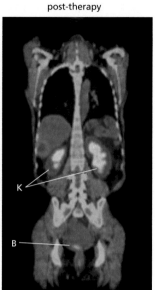

pre-therapy   post-therapy

FIGURE 10.8 The right protein kinase inhibitor can shrink tumors. A malignant sarcoma (cancer of a connective tissue) was located in a patient by PET scanning (marked T, for tumor, in the left image). To get these images, the patient was given a glucose analog that works as a stain for PET imaging (yellow), as described in Chapter 2. Kidneys (K) and the bladder (B) may be stained by this dye as well as the tumors, because the chemical is being eliminated from the body by urination. At right, a similar image was obtained after treatment of the patient with an inhibitor of multiple protein kinases (sunitinib). The kidneys are still stained, because the dye is again being eliminated from the body by urine flow. Now there is also staining in the bladder (B), showing that dye is on its way out of the body, but the staining of the tumor is gone; the drug has inhibited the rapid growth of these sarcoma cells and caused the tumors to shrink. (From Vander Heiden MG, Cantley LC & Thompson CB [2009] *Science* 324:1029–1033. With permission from AAAS.)

Key regulatory proteins may become valuable targets for new drugs.

You have read about c-Ras, the small, GTP-binding protein that is so important in cell cycle regulation. Oncogenic forms of Ras help to initiate DNA replication, even when signals from a growth factor are not present; they are found in many human tumors and are thought to be a major factor in driving cell proliferation in many human cancers. One can imagine a drug that inhibits this mutant protein without affecting the action of normal Ras. Such a drug might inhibit the growth of tumors that expressed oncogenic *ras* yet have few side effects. Like imatinib, this drug would not be a cure, but it might block tumor growth and alleviate both symptoms and further mutation or differentiation of the cells that carry the *ras* mutation. This inhibition might block the emergence of a metastatic cancer from the comparatively few tumor cells that were present at the time of diagnosis.

You will immediately recognize, though, that the genetic instability of tumors means that the benefit from any such a drug is likely to be short-lived. Additional mutations may arise that sidestep the inhibitory activity of each drug, rendering it ineffective. Indeed, experience from prolonged treatments with imatinib supports the validity of this concern. Thus, medical science will not be able to rely on a few magic bullets. It will need a whole arsenal of effective anti-cancer drugs. Each of these might provide only temporary patient benefit, but their successive use could provide prolonged cancer remission.

This kind of treatment is beginning to sound very expensive, but there is one feature of the cancer problem that may make the costs manageable. Every human cancer is formed by the mutation of normal, human DNA. Therefore, every patient who has a given carcinogenic mutation should be able to benefit from the drug that inactivates that particular oncogene. Thus, each drug that is developed should have widespread use, which will ultimately reduce its cost. Moreover, this kind of drug, used in patient after patient, is not likely to lead to a loss in the drug's effectiveness, as so often happens with antibiotics. After a given antibiotic has been used repeatedly to treat a particular bacterium, drug-resistant strains of that bacterium are almost certain to emerge. These pathogens can now go on to infect other people, and the resulting disease can no longer be cured by the same antibiotic. Because cancer is not passed from person to per-

SIDE QUESTION 10.7

Identify two more genes whose products are so important for cancerous transformation that they might be good targets for the kind of drug development already achieved with imatinib.

son, drug-resistant cancers would have to evolve in every patient that is treated. Each patient should therefore be able to benefit from the novel cancer treatment. In this way, treatment of cancer is fundamentally different from treatment of an infectious disease.

Some emerging cancer treatments focus on cell cycle arrest and cell death.

You have previously read about the tremendous importance of p53 in preserving genome integrity. It can act by arresting cell cycle progression to allow time for DNA repair, and when DNA damage cannot be repaired, this protein can eliminate cells by inducing apoptosis. p53 is down-regulated in many human cancers, either by mutation of the gene itself (loss of function through modification of the structural gene) or by down-regulation of its expression or stability (by modification of the system that controls the cellular levels of p53 protein). Recall that the protein MDM2 is responsible for keeping p53 levels down by promoting its degradation through proteolysis (the breakdown of proteins into polypeptides or simply amino acids). One drug that is important for effective treatment of some cancers inhibits MDM2 and therefore allows the levels of p53 to rise. The **nutlins** are small-molecule inhibitors of MDM2. Through their action, p53 concentrations are increased in some cancer cells, leading to slower cell cycle progression or the initiation of apoptosis. Other cancer cells are resistant to the nutlins for reasons that are not yet understood. Once again we find a drug that is useful in some cases but not all, but its strengths motivate a search for analogous drugs that could complement and extend the use of the nutlins we already know.

Another feature of some cancer cells that is key for disease progression is their ability to inhibit the cell's normal pathways for inducing apoptosis. Bcl-X is an anti-apoptotic protein; its function in normal cells is to balance the induction of apoptosis with factors that inhibit it, tending to keep the cell alive. Bcl-X is commonly overexpressed in cancer cells, which inhibits them from inducing apoptosis. The drug epigallocatechin gallate (EGCG) can inhibit Bcl-X function, allowing cancer cells to induce apoptosis after all. Here again is a cancer treatment that is useful in certain cells but not all. From these examples, you can see the importance of detailed molecular diagnosis for learning which drugs are likely to be effective in reducing the growth of a particular tumor. You can also see that effective treatment of cancer is going to require a large number of drugs to cover both the diversity of cancers and the abilities of cancer cells to mutate and acquire resistance to the pharmaceuticals used to block their growth.

One new approach to cancer treatment is based on early studies of cancer cell metabolism.

In the early twentieth century, a German scientist named Otto Warburg demonstrated that cancer cells process food and make ATP in ways that differ quite markedly from most normal cells in our bodies. Rather than using mitochondria and oxygen from the air we breathe to convert sugar efficiently into ATP, cancer cells tend to leave mitochondria out of the loop. They break down a sugar like glucose into pieces and capture some of the energy from its chemical bonds, but the process is only about one-sixth as efficient as the reactions that normally occur in mitochondria. A result is that cancer cells consume large amounts of sugar, a behavior that was mentioned when we discussed the staining of cancer cells for PET scanning (Chapter 2). The pathways for processing food are regulated by hormones, including insulin and insulin-like growth factor. The action of these hormones is mediated within cells by a protein kinase called AKT.

Debate with your neighbor, friend, or roommate the wisdom of investing further in novel cancer treatments versus seeking ways to prevent cancer, or at least to lower its prevalence.

Exactly which mutations occur to drive the changes in cancer cell metabolism is not yet known, but the dependence of most cancer cells on rapid glucose consumption is well established. Scientists are now trying to use this fact to find treatments that will damage cancer cells specifically. An advantage of this approach is that it should have some effect on many cancers. In this way, it is quite the opposite of personalized medicine. While it is unlikely to provide a cancer cure, this approach may lead to drugs that have a large effect on the rate at which many cancers grow.

Effective cancer drugs can come both from naturally occurring compounds and from those that are manmade.

The drug EGCG, which inhibits Bcl-X and allows cancer cells to induce apoptosis, is derived from a chemical found in green tea, which is brewed from the leaves of plants. Like EGCG, several valuable cancer drugs were first found as components of living things. Vinblastine and the taxanes, described in Chapter 8 as inhibitors of microtubule function and mitosis, are isolates from an ornamental plant and from the bark of a tree, respectively. Rapamycin is an inhibitor of a cell-regulating protein kinase called mTOR. This compound was originally isolated from the bacterium *Streptomyces*, which makes rapamycin to combat other microorganisms with which it competes for food. Rapamycin interferes with the biochemical pathways that control normal cell growth. As a result, it can give temporary inhibition of tumor growth.

Simply because a cancer-inhibiting compound is naturally occurring, however, does not mean that it is necessarily safe. Cyclopamine is a molecule made by a weed called the corn lily. This molecule resembles a normal steroid. When administered to animals, it interferes with an important signaling pathway based on a protein called hedgehog. This pathway is up-regulated in many human carcinomas, and cyclopamine can inhibit the resulting increase in hedgehog signaling, impeding the growth of this kind of cancer. Sadly, however, cyclopamine is so toxic that it is not a good drug. It is currently serving as a model for the chemical synthesis of related compounds that might be as effective in cancer treatment but not so lethal.

Some drugs now in use are the result of screening huge numbers of chemically synthesized compounds in a search for ones that have the desired activity. Chemists have devised ways of making vast numbers of similar but not identical molecules, literally hundreds of thousands of them. A biologist can then test these compounds, looking for ones that have inhibitory effects on a cellular process of interest. If an effective molecule is found, it is regarded as a **lead compound** that can focus a search for a related chemical that does the same job even better and with a minimum of side effects on so-called off-target processes. Both the construction of these chemical libraries and the testing of their many components are time-consuming and expensive operations. The number of useless chemicals that one must go through to find one that works is staggering. This is one factor that leads to the tremendous difference between the cost of making a molecule and of selling it as a drug.

Several groups of researchers interested in developing new cancer drugs have initiated programs called **rational drug design**. In this process, the scientists first identify a target protein, meaning one whose function is essential for cancerous cell behavior. They then learn the protein's structure in atomic detail by purifying it and using either electron microscopy or X-ray crystallography to image it at high magnification and resolution. Then the scientists try to invent a chemical that will fit into a cleft on the surface of the target protein and block its action. As you read previously, imatinib was much improved in this way, encouraging analogous efforts by other groups. Unfortunately, this approach has been less successful than one might hope, and many of the drugs currently in use were found by the straightforward, if time-consuming, approach of trial and error. Apparently, we don't yet understand protein structure and function well enough to make a rational approach work very well.

Discovery and testing of anti-cancer drugs requires both laboratory and clinical research to assess effectiveness and safety.

If a chemist finds that a particular compound inhibits a specific biochemical reaction, like the activity of a protein kinase, she must then ask, Will the compound work as a drug in a cell? To do so, it must be able to cross the plasma membrane and reach a concentration that blocks enzyme function in the part of the cell where the enzyme normally works. To assess the ability of a drug to do these things, cell culture is invaluable.

If the new compound does work on cells, researchers must then learn whether it will also work in an animal. Some chemicals are eliminated from the body so quickly they have little chance to have an effect. This can happen as a result of their being modified by body chemistry, for example, by enzymes in the liver, or because the chemical is passed rapidly out of the body in urine. Study of this issue is called **pharmacokinetics**, which is a scientific specialty in its own right. If a drug does work in an experimental animal, for instance a mouse, then one must determine a dose response curve; that is, how much drug is required and for how long that dose must be maintained to get the desired effect. Usually, many animals are used for this kind of experimentation, because of variability in the responses of living things to external stimuli. As described in Figure 8.11, one wants to know both the lowest drug concentration at which half the animals being tested show a positive response and also the concentration high enough that half the animals suffer bad effects or even die. The ratio of these two concentrations is the therapeutic index, an important metric in drug testing.

Once the therapeutic index is known for animals, doctors can consider how best to try the compound on people. These **clinical trials** are organized in three phases:

(1) A phase 1 clinical trial assesses the toxicity of the compound in people, a necessary step, because people don't always respond to a drug in the same way as mice or other experimental animals. With anti-cancer drugs, which may have significant toxicities, these tests are commonly done with a small number of volunteers, usually people who are already very sick with the disease the drug is supposed to cure. Commonly, the trials begin with a small dose of the drug and then increase the dose slowly, using the same or different subjects. The volunteers are willing to take the chance that it might help them, because nothing has helped them so far. If the dose they receive is too high, and they become even sicker, this is very sad, but it is the risk that these courageous people are willing to take. Although these clinical trials do include risks, they identify conditions for drug administration that are likely to be safe for other patients. In effect, one is obtaining the same sort of data shown in Figure 10.9 but getting it from people, rather than an experimental animal.

(2) In a phase 2 clinical trial, doctors use the knowledge acquired in phase 1 to give doses likely to be therapeutic but not dangerous. Now it is desirable to administer drugs with a double-blind design (see Sidebar 8.10), although for a particularly expensive drug or dangerous disease, the size

of the patient pool may be too small for this more time-consuming and expensive approach. At least a single-blind design is necessary, however, because of the importance of objective and unbiased data in drug evaluation. By the time of a phase 2 trial, the company that identified the compound under study will probably have invested well over $10,000,000 in its identification, development, and laboratory testing, so they are very keen for the drug to be successful. Getting successful compounds into the hands of physicians is important, but keeping useless or dangerous drugs off the market is equally important. Therefore, it is essential to have a trial structure in which the doctors treating the patients do not know who received the test drug and who received either the conventionally used drug or a placebo. Only then can one be confident that the results are an accurate reflection of drug effectiveness. Success in a phase 2 trial is a major breakthrough in the identification of a compound that is likely to be both helpful in human health and profitable for the company that made it.

(3) Phase 3 trials are conducted if the phase 2 trial indicated that the drug under study is safe and appears beneficial. In phase 3, researchers increase the numbers of patients under study, hoping to collect enough data to determine at least two things: the effectiveness of the drug compared with the currently available standard of care and the fraction of patients who respond positively to the drug's effect, or at least a distribution of its effectiveness. Both of these questions can be answered by administering the drug at proper doses to a fairly large number of people. Data on the effects of the drug are usually displayed on a graph designed by two statisticians, E. L. Kaplan and Paul Meier. These men realized the value of plotting the percentage of people still alive, or the percentage of subjects who are still cancer free, at different times after the initiation of treatment (FIGURE 10.9). This kind of graph is widely used to evaluate the efficacy (or not) of a new drug or treatment. Patients are organized into two groups, one of which is given a standard treatment, or perhaps a placebo or no treatment at all, while the other is given the new drug or other therapy. Careful records are kept, using a blind experimental design, so any differences in outcome can be seen as reliable indications of treatment efficacy.

It is easy to see that producing and testing a new drug is expensive. The difference between a chemical that is identified as promising by a laboratory scientist and a drug that can safely be marketed to the public is commonly an investment of more than a billion dollars. This staggering fact is one of the reasons that new pharmaceuticals are so expensive and that progress in cancer medicine is no faster than it now appears. The design and execution of clinical trials is a science in itself, and much work

SIDE QUESTION 10.9

Given the preceding description of the process for testing drugs, do you think it is morally acceptable to do preliminary testing on laboratory animals? Why or why not?

FIGURE 10.9 Results from a double-blind, phase 3 trial. One hundred twenty-six patients with pancreatic carcinoma were identified as subjects for this trial and placed in one of two groups. One group received a new experimental drug (GEM), and the other group was treated with the standard therapy of the day (5-FU). Success was measured by the length of time the patients survived. The results are plotted to show the percentage of patients still alive in each group after each month of treatment (a Kaplan–Meier plot). The new drug does prolong life, but the effect is small, much less than one would hope. Such a result is not uncommon, but if data like these are based on well-controlled, double-blind observations, they can lead to approval of the new drug for more widespread use. (From Burris HA III, Moore MJ, Andersen J et al [1997] *J Clin Oncol* 15:2403–2413. With permission from American Society of Clinical Oncology.)

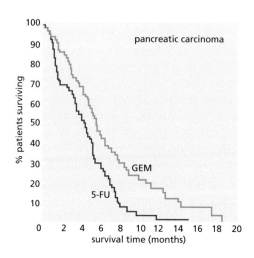

is ongoing to make them as efficient and effective as possible. You may find it interesting to see such trials in action, and this is possible, thanks to a Web site by the National Cancer Institute that keeps track of most trials ongoing throughout the country: https://clinicaltrials.gov/.

High-tech innovations hold promise of improving cancer surgery.

Scientific innovation and high-technology instrumentation are active in almost all fields of cancer medicine, including surgery. You will remember that a serious problem for a cancer surgeon is the question of how much tissue, beyond the obvious tumor, to remove in order to minimize the chances of leaving behind nearby micrometastases after surgery. Cutting out too little tissue can lead to a rapid reoccurrence of the cancer, whereas cutting out too much leads to longer patient recovery times. Surgeons in several hospitals are experimenting with a device, sometimes called an **iKnife**, that uses a powerful instrument for fast chemical analysis of the tissue currently being removed. This provides the surgeon with rapid feedback about the character of that tissue.

The iKnife uses heat and suction to vaporize tiny amounts of tissue as a surgeon moves the knife to different regions near a tumor. The suction pulls molecules from the region being cut and feeds them into a mass spectrometer. This instrument uses electric fields to charge up the molecules that come into it and to drive them at high speeds through a vacuum, where a magnetic field bends their path by an amount that depends on the ratio of their charge to their mass. A mass spectrometer measures this number with spectacular accuracy, so even though the result of the measurement is only one number, it is often sufficient to identify specific molecules, particularly if a computer associated with the instrument has previously been taught the values of this ratio that are diagnostic for the kinds of molecules the surgeon will encounter. The mass spectrometer works so fast, it can send signals back to the surgeon within seconds, telling him or her about the molecules in each region probed with the iKnife. Some molecules are characteristic of cancer cells, so when these signatures are no longer coming from the spectrometer, it is time to stop cutting there and try another place. In this way, the success of tumor removal is being improved.

NEW IMMUNOLOGICAL APPROACHES TO CANCER THERAPY

In Chapter 9, you learned about the remarkable specificity of adaptive immunity, and you read evidence that this part of the immune system is often active in identifying and eliminating cancers. You also learned some of the ways that cancer cells can secrete, or place upon their surface, proteins that inactivate cytotoxic T (Tc) cells, allowing the cancer to flourish. Biomedical scientists are now using a variety of methods to try to overcome this kind of inhibition and mobilize the patient's immune system to aid in defeating a life-threatening cancer.

Bacteria can be used to activate a patient's immune system, enabling it to fight cancer.

One class of well-established cancer immunotherapy employs a bacterium to stimulate the body's immune response, inducing it to interact with cancer cells at the same time. This approach is currently an established procedure for bladder cancer, as mentioned in Chapter 2. The bacterium

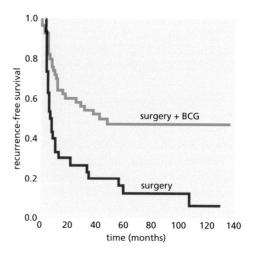

Bacillus Calmette-Guerin (BCG) can be injected directly into a patient's bladder, where it stimulates the immune system to bring in regulatory T (Treg) and Tc cells, plus natural killer (NK) cells and macrophages, creating a true inflammatory response. These cells not only eliminate the bacterial infection, they also kill tumor cells quite effectively. In combination with surgery, this method gives a significant improvement in the frequency with which patients survive without recurrence of their cancer (FIGURE 10.10).

Doctors are now experimenting with the use of other bacteria to accomplish the same trick that BCG does for bladder cancer, but under circumstances in which the microorganism is not as well segregated from the rest of the body as it is when injected into the bladder. Both *Salmonella* (a common cause of food poisoning) and *Listeria* (another food pathogen that causes listeriosis, a serious problem for people with weakened immune systems) have been used as tools to carry DNA that encodes a cancer-specific antigen into a patient's body. The idea here is similar to that described above for BCG: a real pathogen will activate the immune system strongly, turning on both innate and adaptive immunity. Doctors hope that if there is a cancer antigen around while the immune system is tuning up to fight the pathogen, effective antibodies will be made against the cancer antigen too. If there is a strong immune response, the patient may generate antibodies and T-cell receptors to fight both the pathogen and the cancer, with the result that both are killed.

Cytokines are being tried as stimulants for a cancer patient's immune system.

Several groups of doctors are pursuing the idea that a cancer patient already has T cells and antibodies that can combat their cancer, but these parts of the immune system need to be stimulated. You will recall that cytokines are proteins that signal from cell to cell, and many such signals contribute to the progress of a patient's immune response. One approach within this context is to treat patients with cytokines known to stimulate leukocytes in general, hoping to activate NK or Tc cells that are already present in the patient's tumor. One cytokine that has received particular attention is **granulocyte–macrophage colony-stimulating factor**, known more commonly as **GM-CSF**. This protein has been administered to patients with a severe cancer, such as metastatic melanoma, either alone or in combination with other treatments, hoping to activate any anti-cancer immune cells that are already present. While several studies published as recently as 2014 showed a prolongation of patient life in response to this treatment, none of these studies reported cures. This approach to the use of GM-CSF seems to need significant improvement.

Doctors have, however, tried more sophisticated ways to use GM-CSF in the treatment of cancer patients. In the 1990s, a group at Johns Hopkins Medical School took melanoma cells from several patients, transfected these cells in culture with the DNA needed to express and secrete GM-CSF at quite high levels, grew these cells up to make lots of them, and then irradiated them with X-rays, so they could no longer replicate their DNA and divide. These irradiated cells were then injected back into the patient from whom the cells had originally come. The idea was that the irradiated cells were too highly damaged to increase the number of tumor cells dividing in the patient. The cells would, nonetheless, secrete a cytokine, and this might stimulate the patient's T cells that recognized antigens on the surface of the injected tumor cells. Because the irradiated cells came from the patient into whom they were injected, there would be no problem with tissue rejection by the immune system. The stimulation of the patient's T cells, it was hoped, might initiate an immune response that acted not only on the GM-CSF-secreting cells but also on the patient's living tumor cells. Clever as this approach seems, it was not very effective.

Since 2010, several medical groups have been trying other ways to achieve essentially the same goal as that of the early experiments with GM-CSF. For example, an anti-cancer drug called talimogene laherparepvec (TVEC) is a modified herpesvirus that will infect melanoma cells. This virus has been engineered to express GM-CSF, which is then secreted by the infected melanoma cells, so it should, in principle, activate Tc and NK cells to act on the melanoma. Indeed, some patients treated in this way have shown a shrinking of their tumors. Strangely, however, many did not. Even with these high-tech approaches to the delivery of a cytokine, the desired response has not been achieved, so most researchers have turned to different approaches.

This kind of work with novel approaches to immunotherapy is usually done with people whose cancers are far advanced and who therefore have no reasonable hope of success from more conventional therapies. One example is a treatment for malignant melanoma that relies on a subset of the patient's own T cells that are already resident in the tumor. Recall from Chapter 9 that immunologists have good reason to believe that these cancer-fighting cells are responsible for helping to keep tumor growth in check, and it is easy to see why doctors interested in experimental cancer therapies might want to modify tumor-localized T cells in ways that would improve their anti-cancer performance. Experiments carried out at the National Institutes of Health have followed up on this idea by surgically removing tissue from one tumor of a metastatic melanoma, separating its T cells from the rest of the tumor cells, and then culturing them, a few at a time, in small vessels that allow assessments of the T cells' properties. Through study of the cells in each individual vessel, the scientists identified a subset of the tumor-localized T cells that was producing significant amounts of the cytokine **interferon** γ (**IFN-γ**). Because IFN-γ can stimulate T-cell function, these cells seemed likely to be more active than others in fighting the tumor effectively.

The cells expressing higher levels of IFN-γ were then increased in number by feeding them growth factors and combining them in cell culture with other cells that nurture T-cell growth. After many of the desired T cells had been grown up in the laboratory, the patient's bone marrow was exposed to a dose of X-rays sufficient to kill some but not all the cells resident there. The T cells that had been modified in the lab were then injected into the patient's bone marrow, allowing them to replace the cells that were dying

In one example cited above, tumor cells taken from a patient were engineered by DNA transfection to express a particular cytokine. These modified cells were then increased in number, rendered unable to divide further by X-irradiation, and injected back into the patient. In the last example described, tumor-resident T cells from the patient were selected for their synthesis of high levels of a particular cytokine. Those cells were then increased in number and injected back into the patient. What are the differences between these approaches, and which do you think is more likely to succeed? Why?

SIDE QUESTION 10.10

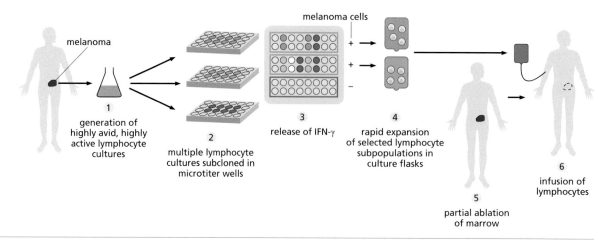

FIGURE 10.11 An experimental treatment for melanoma, based on T cells from the patients' tumors. A portion of a tumor was cut from each patient, and the T cells residing in that sample were isolated and cultured. These cells were put into many small vessels, often called microtiter wells. T cells in each well were assessed for their ability to make and secrete interferon γ (IFN-γ) when exposed to melanoma cells. The production and secretion of IFN-γ implied that the T cells recognized the melanoma cells and responded to them. The best cells by this criterion were induced to proliferate by adding growth factors and helper cells. The patients were prepared to receive extra T cells by reducing the number of active cells in their bone marrow, using X-ray irradiation. The tumor-recognizing T cells were then put back into the patients with the hope that those T cells would interact with the tumors and reduce their size (see Figure 10.13). (Adapted from Dudley ME & Rosenberg SA [2003] *Nat Rev Cancer* 3:666–675. With permission from Springer Nature.)

from radiation (FIGURE 10.11). The hope was that these T cells, which came from the tumor in the first place and were selected for their ability to secrete high levels of IFN-γ, would be particularly effective in controlling tumor growth and even in knocking the tumor back. The progress of patients given this treatment was monitored by X-ray tomography, and in some cases the results were most encouraging (FIGURE 10.12). Sadly, such a positive response has not been typical. About half of the patients treated in this way showed a prolonged life, but cases of real remission were few.

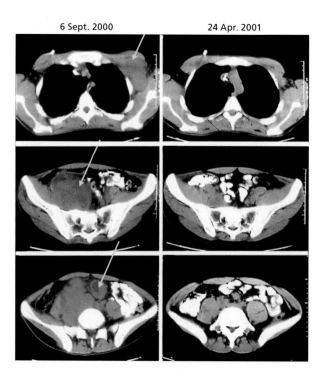

FIGURE 10.12 Assay for the effectiveness of an experimental treatment for melanoma. Images from computed tomography (CT) scans from a patient receiving the treatment described in Figure 10.11. The three images on the left show various regions of a patient's body where metastases have been detected (yellow arrows). Some 7 months following treatment, the tumors at all these locations had shrunk, showing a significant improvement (images on the right). Sadly, however, these positive results were not uniformly seen, so other methods must be sought. (From Rosenberg SA & Dudley ME [2004] *Proc Natl Acad Sci USA* 101:14639–14645. With permission from National Academy of Sciences.)

Several research groups are working to take advantage of specific antigen–antibody interactions to help the immune system fight cancer.

The problem that all scientists and doctors working to develop cancer immunotherapies must solve is how to activate the cell-killing powers of the immune system, so it recognizes tumor parenchymal cells as nonself. Much of the necessary experimentation has been carried out on patients with advanced melanoma, because until recently, there has been no effective treatment for this condition. The parenchymal cells in the tumors in some of these patients display antigens that are called melanoma-associated antigens (MAGE). The function of these proteins in normal humans is unknown, but some of them, such as MAGE-A3, are now quite well characterized. One approach to treating patients whose tumor cells display this antigen has been to make large quantities of the protein in the laboratory, then inject it into the patients, either as a soluble molecule or as an engineered part of the coat around a modified virus that won't cause an infection in humans. The immune system does respond to this kind of stimulation, because the number of Tc cells that bind the MAGE-A3 antigen goes up as much as 30-fold. This kind of response has led to remissions of the melanoma, some of which have lasted for 2 years. However, the fraction of patients who respond positively has thus far been only about 20%, for reasons that are not yet understood.

Cancer treatments that act on immunological checkpoints can reduce the down-regulation of an immune response against self. As described in Chapter 9 and Sidebar 9.2, doctors are injecting patients with antibodies to proteins in the T-cell membrane that are important for down-regulating a T cell's response to activating stimuli from an antigen-presenting cell (so-called immunological checkpoint proteins). These antibodies, like the drug pembrolizumab that binds to PD1, block the inhibition of T-cell responses and thereby stimulate the ability of a Tc to interact with a tumor antigen and kill the tumor cells. The idea of these treatments is to reduce the patient's natural down-regulation of immune responses against self. Large-scale tests of this approach have shown increased 2-year survival of patients with advanced melanoma, which is definitely encouraging. Indeed, this drug was given to former President Jimmy Carter; he attributed his significant improvement while dealing with metastatic cancer to the action of this antibody.

Perhaps even more encouraging, there are multiple membrane proteins now known to contribute to T-cell regulation, and several of these are already being tested in the clinic (TABLE 10.1). Probably more such molecules will be discovered soon. One can imagine a whole armory of antibodies to immune checkpoint proteins that could be used to combat different cancers, allowing doctors to beat very hard on any cancer in which several of these immune checkpoints are active. In support of this idea, recent work has identified OX40, which is a co-stimulatory molecule on T-cell surfaces; it interacts with the molecule OX80L on an antigen-presenting cell to enhance the T cell's response. Antibodies to OX40, like antibodies to PD1 and CTLA4, may contribute to a doctor's ability to activate the patient's immune system for fighting cancer. A note of caution must, however, be added. As these manipulations of the immune system are tried on more people, several are experiencing autoimmune disease, meaning that they are now mounting an immune response against their own tissues. Such conditions are hard to treat and are very unpleasant, so curing cancer by inducing autoimmune disease is no solution at all. Clearly, the biomedical community needs to understand the immune system better than it now does to use immunological treatments well.

Table 10.1 Clinical development of agents that target immune-checkpoint pathways

Target	Biological function	Antibody or Ig fusion protein	State of clinical development*
CTLA4	inhibitory receptor	ipilimumab	FDA approved for melanoma, phase II and phase III trials ongoing for multiple cancers
		tremelimumab	previously tested in a phase III trial of patients with melanoma; not currently active
PD1	inhibitory receptor	MDX-1106 (also known as BMS-936558)	phase I/II trials in patients with melanoma and renal lung cancers
		MK3475	phase I trial in multiple cancers
		CT-011[‡]	phase I trial in multiple cancers
		AMP-224[§]	phase I trial in multiple cancers
PDL1	ligand for PD1	MDX-1105	phase I trial in multiple cancers
		multiple mAbs	phase I trials planned for 2012
LAG3	inhibitory receptor	IMP321[†]	phase III trial in breast cancer
		multiple mAbs	preclinical development
B7-H3	inhibitory ligand	MGA271	phase I trial in multiple cancers
B7-H4	inhibitory ligand		preclinical development
TIM3	inhibitory receptor		preclinical development

CTLA4, cytotoxic T lymphocyte-associated antigen 4; FDA, U.S. Food and Drug Administration; Ig, immunoglobulin; LAG3, lymphocyte activation gene 3; mAbs, monoclonal antibodies; PD1, programmed cell death protein 1; PDL, PD1 ligand; TIM3, T-cell membrane protein 3. *As of January 2012. [‡]PD1 specificity not validated in any published material. [§]PDL2Ig fusion protein. [†]LAG3–Ig fusion protein. Data from Pardoll DM [2012] *Nat Rev Cancer* 12:252–264. With permission from Springer Nature.

There is progress in the development of cancer vaccines.

In the spirit of dealing with cancer by preventing its occurrence, doctors have long dreamed of a vaccine that would prevent cancers from forming in the first place. In Chapter 5, you read about the very significant success in cancer prevention that has been achieved with vaccines against human papilloma virus (HPV), an important risk factor in cervical and penile cancers and some other cancers as well. This is an excellent example of a vaccine helping people to avoid a serious cancer problem, but most human cancers are not virus-related. In these more common kinds of cancers, there is no external pathogen against which the immune system can mount a defensive response, so the dream of a general vaccine to prevent cancer has received only limited attention from cancer scholars. Recently, however, there has been enough progress to draw attention back to this appealing idea. Most researchers pursuing the idea of a cancer vaccine are seeking ways to increase the patient's immune response to an existing tumor. However, a tumor's stroma can inhibit T cells from getting at the tumor parenchyma to do the damage one wants. Another problem, of course, is that most tumor-specific antigens are self-antigens, as described in Chapter 9, so the immune system is programmed *not* to mount a response against them.

In spite of these problems, one currently promising approach is to identify a protein that is important for driving tumor cell proliferation, such as **insulin-like growth factor binding protein 2** (**IGF-BP-2**), a receptor that is important for the ability of some breast cancers to grow and thrive. IFG-BP-2 can be made in parts by transfecting segments of the gene that encodes it into bacteria, accompanied by the DNA sequences that will serve as promoters for bacterial RNA polymerase. Now, mRNA is made, and the bacterial ribosomes translate these messages into segments of this protein. It is easy to grow up large numbers of these bacterial cells and isolate the protein fragments of IGF-BP-2 from them, which makes available a large supply of these protein pieces. These can be injected, first into laboratory animals and then into people, to find out which pieces are good antigens in the sense that they elicit the kind of immune response that might help to fight a cancer that depends on this protein. Work in Seattle has shown that the C-terminal part of this protein elicits the activation and proliferation of T cells of the right kind, so these scientists are now exploring ways to administer this protein fragment to breast cancer patients in such a way that it will induce them to develop a successful immune response against their cancer.

Cancer doctors are also exploring the use of dendritic cells, which were mentioned in Chapter 9 as an important kind of antigen-presenting cell. Recall that all cells can use the major histocompatibility complex (MHC) proteins to convey oligopeptides from the cell's interior and display them on the cell surface. Dendritic cells are particularly good at antigen presentation, though their natural life cycle takes them through different stages where they are better at some kinds of antigen presentation than others, for reasons that are not yet well understood. Nonetheless, several groups of scientists and doctors are now exploring ways to use a patient's own dendritic cells, transformed with DNA that encodes a tumor antigen selected on the basis of known tumor features, to goad the immune system into developing the cytotoxic T cells that will eliminate the tumor parenchyma.

Some novel approaches to cancer therapeutics use membrane proteins.

In the preceding chapter, you saw approaches to cancer therapy that relied on antigens localized at the cell surface. Some monoclonal antibodies to cell surface receptors can inactivate those membrane molecules and prevent the cells from receiving growth-promoting signals (see Figure 9.24). Antibodies can also be used to localize a toxin to an antigen on a cancer cell's plasma membrane. The key idea here is to identify an antigen that is exposed specifically on a cancer cell's surface, for example, a protein that is sufficiently mutated to be distinguishable from the wild-type antigen on a normal cell's surface. Another possibility is an embryonic antigen that the tumor is now making but that is not found elsewhere in the adult body. Most studies following this approach are still experimental, but they too may develop into powerful therapies, especially as more cancer-specific cell surface molecules are identified and as better cellular poisons are brought into use. Thus, we can probably look forward to an ever-growing repertoire of anti-cancer treatments that are not simply poisons for all cells but are designed specifically to affect cancerous cells more than normal ones. Such treatments hold out the hope of doing serious damage to tumors while leaving the rest of the body in better shape, a most desirable change in treatment outcome.

In an innovative alternative to the approaches just described, a group of clever researchers has designed a novel membrane protein they call a **chimeric antigen receptor** (**CAR**). This protein has been engineered through the use of recombinant DNA technology to include an antibody to a tumor antigen on its extracellular part, a membrane-spanning domain, and an intracellular domain that can send stimulatory signals to the cell that expresses this protein as soon as the membrane-localized antibody interacts with its antigen (FIGURE 10.13A). The DNA for a CAR is then

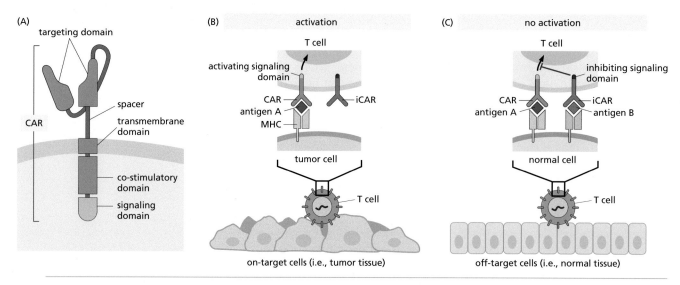

FIGURE 10.13 Membrane proteins engineered to enhance T-cell function. (A) An engineered chimeric antigen receptor (CAR) protein, illustrating that its extracellular domain is the part of an antibody that will bind to an antigen on the surface of a tumor cell (the targeting domain). Under this domain is a spacer, then a transmembrane domain. The rest of the protein is an intracellular signaling complex that can help to activate cytotoxic T cells. (B) At the bottom, an engineered T cell interacts with a tumor. At the top, a magnified view of the engineered CAR interacting with an antigen presented by a major histocompatibility complex (MHC) protein on a tumor cell's surface. The T cell becomes activated. (C) The same engineered T cell interacts with a cell from normal tissue. Now the inhibitory chimeric antigen receptor (iCAR) is activated, so the T cell does not respond and initiate cell killing.

transfected into cytotoxic T cells isolated from a cancer patient. As these Tc cells grow in culture, they express the CAR and insert it into their plasma membranes. This property should allow them to bind the tumor cells and initiate cell killing. The problem with such a construct, however, is making sure that it doesn't cause the T cells to interact with normal cells. To aid in this, scientists have added a second membrane protein called an inhibitory chimeric antigen receptor (iCAR), which is designed to recognize antigens on the surface of normal cells but not tumors. With both CAR and iCAR attached to the surface of the engineered Tc, cancer cells that lack the inhibitory antigen will stimulate the killing activity of the cytotoxic T cell, (FIGURE 10.13B) but normal cells will not (FIGURE 10.13C). This approach is again being evaluated with clinical trials, and its performance is somewhat better than that of the previously described methods. Further work may refine this approach into a really powerful cancer treatment. For an example of applying this strategy to the treatment of a B-cell lymphoma, look at the paper by Kochenderfer et al. cited in Further Readings.

In spite of progress with novel cancer treatments, much cancer therapy is still done with conventional methods.

In the context of the previous discussion, it is important to add that while immunotherapies for cancer sound attractive, many oncologists are still treating patients with focused radiation and familiar chemotherapeutic agents, such as inhibitors of DNA replication and mitosis. Why should doctors be so conservative? The answer at present is that, on average, the modern versions of the classic cancer treatments work better than most of the newly identified experimental treatments. Reasons for this situation are not entirely clear, but they have prompted some people from the business community to call for an end to the expensive experimentation in novel cancer therapies and switch to a focus on cancer prevention. This is not a foolish position, though from a long-term point of view it is probably self-defeating. Some research is identifying really novel ways to inhibit and kill cancer cells, and with further exploration, truly effective methods for treating cancer may be found.

INTEGRATED APPROACHES

One feature of modern science is the way that computers have influenced attitudes toward data. In times past, it has been common for structural biologists to examine the image they collected, for chemists to ponder their numbers about molecular composition and reaction rates, for molecular biologists to think about gene sequences, and for geneticists to consider the results of a given cross between sexually reproducing individuals. Data from each scientific source tended to remain in the context where it was collected, and groups working with different technologies rarely communicated. In recent times, these separations have broken down in many ways, some of which have a direct impact on the study of cancer.

Structural biologists now rely heavily on chemists and molecular biologists to help them develop novel stains that allow study of a particular molecule or bring out important image features. Chemists and molecular biologists work together to invent and assess the effects of new compounds on the many different reactions needed for life. Geneticists use molecular and chemical methods all the time to deepen their understanding of genes and how they work. Unifying all these approaches are the cell biologists, who are generally willing to use any method they can find to help answer questions about cell behavior. Enabling all these collaborations are the computer scientists, whose instruments and data storage devices have become essential for most aspects of modern science. The result is that doctors with a direct interest in the treatment of disease have quite amazing tools with which to observe their patients, learn details of the disease at hand, and then compare these results with those from comparable cases previously studied, all stored and analyzed by computer. This integration and analysis of information is one of the most promising aspects of a modern scientific approach to cancer, because it is allowing and will continue to allow a deeper, more analytical approach to the many diseases called cancer than has ever been possible before. Important advances may come simply from this more integrated approach to medical and scientific knowledge.

CAN THE COSTS FOR THIS KIND OF PROGRESS BE BORNE, AND WILL THEY BE WORTH IT?

Many people are deeply concerned about the cost of modern medicine and its tendency to increase without limit. The descriptions of possible paths to progress in cancer treatment given so far all sound expensive. When one thinks about the money that has already been spent on cancer research and the lack of effective modern treatments for many forms of cancer, one cannot help asking what lies ahead in terms of costs and frustrations.

There is no question that lots of money has been spent in this important human effort. Not only the National Cancer Institute, the largest institute at the National Institutes of Health, but also private charities, like the American Cancer Society and the Leukemia and Lymphoma Society, have been supporting high-quality cancer research in the lab and the clinic for many years. Given that there is not yet a cure for most cancers, we can legitimately ask whether this money has been wasted. Is the proposal to continue with cancer research simply throwing good money after bad?

This seems most unlikely. Remember the complexity of even a single cell, based on the pathways we have discussed. Remember that this book describes only a small fraction of a cell's true complexity. Remember that each cell is so tiny that about a million cells must be clustered to make an object as big as a pinhead. If you consider all these issues at once, you can see the real dimensions of the problems scientists and doctors are trying to solve. Yes, engineers have built moon rockets and supercomputers that are also very complex, and those machines work well, but in those cases, the complexity was man-made. The designers knew at each step exactly what they were adding to their device and could predict with considerable confidence how it was likely to work in combination with all the other components already assembled. With a cell, we are still trying to identify the parts from which it is made and how they interact among themselves.

The human genome project gave medical scientists information about all the sequences in human DNA. From this information, molecular biologists can predict some aspects of the structure and sometime even the function of many gene products. Now, many aspects of the control of gene expression are well understood, but we are still working to understand the full complexity of gene regulation. Moreover, many DNA sequences that are predicted to be protein-producing genes have never been studied; their products have not been isolated, and their roles are still mysterious. In addition, there are many genes that encode RNA that is never translated but instead interacts with DNA and other RNA molecules to contribute to gene regulation in ways that are just beginning to be sorted out. Thus, there is still a huge amount left to be learned about cells in their normal state. Until we have that base of information, correcting the pathological conditions that mutations can cause is likely to be impossible.

The author is optimistic about future progress in cancer research, but severe social and economic problems remain unsolved.

As I look back over the information presented in this book, realizing that it includes only a small fraction of the knowledge that has been accumulated about cells and cancer over the last 50 years, I am impressed with the amount that *has* been learned. Yes, it would be wonderful if biomedical science had done better, but considering the complexity of the subject and the difficulties of analyzing cells, it is remarkable how much we do know. I expect that the next 50 years of biomedical science will be at least as effective, and probably more so, because the technologies for cell study are now so much more powerful than they were when I began my scientific career in the 1960s. Thus, I think we can anticipate that our understanding of cells will continue to move forward, with new insights emerging all the time about both cellular processes and their control. This understanding will probably provide the foundations for novel cancer treatments that I can't even imagine as I consider the problem now.

Given all these considerations, I think there are many reasons for optimism about solutions to the cancer problem. Granted, they are not going to come quickly, and from what we now know, it is unlikely that there will ever be a single cancer cure. At every turn we have seen the individuality of cancers, so cancer treatments will probably require a focus on the problems of particular patients. This situation may make cancer treatment expensive and difficult, but I expect that those treatments will come. Certainly, the additional attention to cancer provided by efforts like former Vice President Biden's Cancer Moonshot is the kind of thing that helps to keep funds flowing for research. From a public health point of view, however, my biggest hopes are for cancer prevention and early diagnosis as ways to improve the human condition and keep health care costs down. Time will tell.

The optimistic tone of the previous paragraph fails to take into account a critically important dimension of the cancer problem: the inequities of cancer diagnosis and treatment that pervade the healthcare system of the United States. The improvements in both diagnosis and treatment reported in this book are not available to all residents of this country because of their cost and reliance on technologies that are not available in every medical facility. This is, of course, part of the broader problem of healthcare in general. It is ironic that the United States spends more money per capita on healthcare than any other country in the world, and yet markers of the effectiveness of our healthcare system, such as infant mortality, show that we are not getting good value on our investment. This is a vast problem. It is a subject to learn and think about, but it is a subject for another book.

SUMMARY

Scientific progress has led to improvements in three aspects of cancer medicine: screening, diagnosis, and treatment. Screening has improved largely through the use of advanced methods for medical imaging: X-ray imaging with electronic detectors, sonograms, CT scans, MRI, and PET scans, plus the diverse methods for endoscopy. Early detection of cancer is now more common, but this advance has provoked a controversy over the question of whether some of the cancers that are detected early, and therefore treated early, would have gone away on their own if they had not been found.

Methods for diagnosis, once a cancer has been found, have moved beyond light microscopy to the use of molecular methods that assess the levels of mRNA and protein expression from a wide range of genes. Current work is seeking correlations between patterns of gene expression and cancer outcomes; that is, patient responses to various treatments and their implications for long-range prognoses. Doctors hope that this molecular information will be useful in guiding more effective treatments in the future. Methods for DNA sequencing are beginning to be used to compare tumors and normal tissues, again seeking meaningful properties of a cancer that can guide treatment and lead to more consistently positive outcomes. All of these methods are faced, however, with two challenges: the heterogeneity that is common in large tumors, where genetic instability has led cells to diversify, making treatment difficult; and the costs of high-tech medicine. While the characterization of an individual patient's tumor can now be accomplished with great detail, opening up the possibility of personalized medical treatments, this advance may not have the hoped-for impact for treating cancer in our increasingly elderly population, given the heterogeneity of tumors, the rapid evolution of cancer cells, and the very substantial costs of this kind of personalized treatment. Nonetheless, a scientific approach is leading to new methods for treating the many diseases called cancer, some of which are already successful and some of which are promising.

Some cancers of the blood can be treated with a chemical that induces the dividing cells to differentiate and behave more like normal blood cells. Drugs that inhibit important signaling molecules can be used to impede cell growth and division. Other drugs can sometimes reactivate the cell's own p53, which can then signal cells with excess DNA damage to initiate apoptosis. Some of these drugs come from naturally occurring compounds while others are the products of great chemical ingenuity and effort. In both cases, however, they are fiercely expensive to bring to market, in large part because of the cost of a full regimen of testing that is required to assure their effectiveness and safety.

Some promising new treatments for cancer are based on immunology. Antibodies to cancer-specific cellular receptors can be used to silence these growth-stimulating molecules, slowing or blocking excess cell division. Antibodies can also be used to target poisons to cancer cells, but the number of antigens that are useful for this approach is still limited. Some doctors are experimenting with cells from the patient's own immune system, trying to activate their attack on cancer cells. One approach uses the cytokines normally employed for communication among lymphocytes. Other approaches try to engineer the patient's own lymphocytes to be more aggressive in their secretion of cytokines and thus their attack on tumor cells. These methods are not yet producing breakthrough successes, but there is still hope that a deeper understanding of both cancer and the immune system will lead to immunotherapies that are truly effective.

In brief, clever scientists and doctors are using a wide variety of scientific methods to try to improve all aspects of cancer medicine: screening, diagnosis, and treatment. There are reasons for hope of significant improvement in cancer medicine, but clearly the problem is difficult, and real successes are still few. Future researchers may be aided in their efforts by the intelligent use of effective treatments from traditional therapies, though this approach is still in its infancy. The power of human imagination to unravel the mysteries of nature is impressive, as our current understanding of cancer demonstrates. If allowed to continue working on this problem, scientists will certainly make progress, but it is not clear that society is willing to pay the very significant costs implied by this approach to cancer therapy.

ESSENTIAL CONCEPTS

- Advances in the science and engineering of imaging have led to vast improvements in the ways we can look inside the human body. This has resulted in major advances in the early identification of cancers.

- There is controversy about the value of early cancer diagnosis, because some doctors believe that the small tumors these technologies identify and treat would have gone away without treatment. If these doctors are right, at least some early diagnoses are a waste of money and cause patients unnecessary mental stress.

- Early diagnosis has the advantage that a tumor is identified before significant genetic instability has developed, and this makes treatments much more effective. Thus, pursuing methods for early diagnosis is almost certainly a valuable effort.

- Molecular methods for characterizing both genes (DNA sequences) and gene expression (amounts of mRNA and/or protein in a cell) have improved dramatically. The use of these approaches for characterizing tumors may provide sufficient detail about tumor character and prognosis that early detection will become more obviously useful.

- Tumors become heterogeneous as they grow, thanks to genetic instability. The resulting variation in tumor cell behavior is one reason for the difficulty of finding effective cancer treatments. It also means that molecular characterization of a tumor is not straightforward. Many samples of the tumor's cells may be required, and samples may have to be taken over time to follow cancer cell evolution.

- Deeper understanding of tumor biology is leading to several kinds of new cancer treatments. For certain cancers these are very effective. Drugs that induce differentiation can make tumor cells behave properly. Inhibitors of protein kinases can block or slow cancer growth. Drugs that modify protein behavior can sometimes push cancer cells into apoptosis.

- Finding new drugs is difficult and expensive, due to the complexity of cell signaling pathways and the lengthy process of clinical trials required to assure that a drug is both effective and safe for human use.

- Harnessing the immune system for cancer therapy is a potentially valuable approach, given the specificity of antibodies and T-cell receptors.

- There are antibodies that react with and inactivate cellular receptors for growth signals, and these can be useful for treating certain cancers. Other antibodies can target poisons specifically to cancer cells.

- Mobilizing the patient's own immune system to fight his/her cancer is a promising approach, but in practice the immune system is difficult to manipulate because of its complexity. Improving our understanding of cellular immunity is likely to help make immunotherapy more valuable.

KEY TERMS

age-adjusted incidence of cancer

all-trans retinoic acid (ATRA)

cancer stem cells

chimeric antigen receptor (CAR)

clinical trials

DNA microarray

epigenetic

granulocyte–macrophage colony-stimulating factor (GM-CSF)

heat map

iKnife

insulin-like growth factor binding protein 2 (IGF-BP-2)

interferon γ (IFN-γ)

lead compound

microRNA

nanotube

negative predictive value (NPV)

nutlins

personalized medicine

pharmacokinetics

positive predictive value (PPV)

rational drug design

therapeutic window

tumor heterogeneity

FURTHER READINGS

Weinberg R (2013) The Biology of Cancer, 2nd ed. Garland Science. The later chapters in this book describe many of the treatments presented here in considerably more detail.

Paterson Y, Guirnalda PD & and Wood LM (2010) *Listeria* and *Salmonella* Bacterial Vectors of Tumor-associated antigens for Cancer Immunotherapy. *Semin Immunol* 22:183–189. This is an interesting account of work that uses human bacterial pathogens as a pathway to introduce tumor antigens to the patient's immune system along with a bacterium that will stimulate the immune system.

Kochenderfer JN, Dudley ME, Feldman SA et al. (2012) *Blood* 119:2709–2720 (doi: 10.1182/blood-2011-10-384388).

QUESTIONS FOR FURTHER THOUGHT

1. If you were a member of a panel that allocated money for cancer research, where would you want to invest? Would you back innovative but expensive therapies? Would you favor efforts to improve existing immunological approaches over the identification of new drugs? Would you want to support research on naturally occurring chemicals? Would you want to nurture research into the cancer treatments suggested by traditional medicine? Form an opinion and then be prepared to defend it in a debate with your classmates.

ANSWERS TO SIDE QUESTIONS

1.   Resolution describes the fineness of the details that can be seen in an image of a given object. Image detail in a CT scan, an X-ray picture, or an MRI provides information about the features of a given region of the body, such as the size and shape of a suspicious growth. Such detail is invaluable in the decision that a doctor must make about whether to do a biopsy. An informed decision at this stage helps to make sure that the next stage in screening, such as the analysis of a biopsy by microscopy and/or molecular methods, is done whenever it is appropriate for the health of the patient.

2.   In any test for cancer, a false positive is a situation in which the test returns the answer that cancer is present, when in fact it is not. A false negative is the answer that no cancer is present, when in fact cancer is there. These principles apply for a blood test or another kind of cancer screen.

3.   Cancers arise as a result of carcinogenic mutations. Mutation is an essentially random process, so whatever the rate at which mutations are occurring, as a result of a combination of internal (genotype) and external (environmental) factors, the likelihood that multiple carcinogenic mutations have occurred increases with time (age).

4.   A transcription factor is a protein that binds to a promotor or other regulatory region of one or several genes and alters the rate at which RNA polymerase transcribes that gene (or those genes). A mutated transcription factor may be altered in structure and therefore bind differently to its normal transcription-regulating sites. The binding might be stronger than normal, perhaps up-regulating the expression of that gene (or those genes), or it might be weaker, leading to less gene expression. Changes in levels of mRNA are likely to lead to changes in levels of the proteins they encode. This is because the more copies of a given message there are in the cytoplasm, the greater the likelihood that one of them will interact with ribosomes and end up being translated, which will increase the amount of protein product. Note that this is not necessarily the case, because a given mRNA may be so unstable that although more of it is made, it breaks down before there has been a chance to translate it.

5.   A DNA microarray was described as short pieces of DNA with given sequences attached to specific places on a chip of glass or a similar material. If we know the mutations that are common in a given gene, for example, the common oncogenic mutation of the *BRCA1* gene, the corresponding sequences of DNA can be synthesized and attached to different places on the chip. The patient's DNA from that gene can be amplified in the lab, so there is enough of it to work with, and it can be labeled in a convenient way, for example, by attaching a fluorescent tag. The labeled DNA can then be applied to the chip under conditions where only identical DNA sequences will hybridize well. An investigator can now look to see which place on the chip bound the patient's DNA. Knowing the DNA sequence at each site on the chip will provide knowledge about the sequence in the patient's gene.

6.   Tumor heterogeneity arises over time and is the result of genetic instability in cancer cells. Early in the life of a cancer, there has not yet been sufficient time for the growing and dividing cells either to become unstable genetically or to manifest that instability in the acquisition of large numbers of mutations.

7.   There are quite a few possibilities among the genes discussed in this book: *EGF-R, src, myc,* genes that encode factors that inhibit apoptosis, and any other oncogene are all possibilities. A drug that inhibited a superactivated form of the product of such a gene might act to inhibit its action as a driver of cancerous behavior. Alternatively, one can imagine a drug that up-regulates the activity of a tumor suppressor that has been down-regulated by mutation. Activation by a drug is harder to achieve than inactivation, but it is possible.

8.   This is really an activity, not a question. Most people will argue that preventing cancer is preferable to treating it. The problem is figuring out how to do it. This conversation could take the form of a cost/benefit discussion for finding ways to alter people's behavior versus finding medicines that retard a cancer's growth. Consideration could follow along the lines of the cost of identifying cancer-inducing mutations in children (and finding ways to reduce the risk of those genes showing up in phenotype), versus the cost of developing more conventional cancer treatments. Think about it!

9.   This is actually a profound question, because it raises the issue of humanity's right to use animals for its own benefit. People who are not vegetarians do this all the time, usually without thinking about it, but the use of animals in drug evaluations raises some people's consciences to the point of resistance. American society used to allow the testing of drugs on human prisoners, but that is now illegal: an infamous test of malaria drugs on African-American prisoners put a stop to it. How you come down on this issue is a personal moral choice, but you should think things through well enough to be consistent in your thoughts and actions.

10.   In the first case, the level of cytokine expression by T cells was increased by the methods of molecular biology, leading to engineered T cells to put back into the patient. In the second case, the patient's own T cells were assessed for their rate of synthesis of that same cytokine, and the T cells that make and secrete high levels of the cytokine were selected, expanded, and injected back into the patient. As stated, these two methods are very similar. One difference between them, though, is that the patient's T cells that were good makers and secretors of the chosen cytokine might have been doing additional things that would help to fight the cancer, and by selecting these cells, the doctors might have done something with medical benefit that they didn't even know about at the time. In the case of the engineered cells, this is very much less likely to be true. For this reason, the second method is more likely to be successful than the first.

Minimizing cancer risk

11

CHAPTER 11

The best way to deal with any disease is not to get it. This is why immunizations against some serious viral infections, like smallpox and polio, have been so tremendously effective in reducing the worldwide impact of these ailments. The previous chapters have described cancer from the point of view of its causes, diagnoses, and treatment, presenting a pretty grim picture. We would all be happier if it were possible to avoid these issues completely. The question, then, is how best to prevent cancer from entering your life in the first place.

The nature of the many diseases called cancer means that there is no surefire way to avoid this malady, but some ideas are obvious. If cancer results from the accumulation of mutations that lead to cellular misbehavior, then one wants to avoid carcinogenic mutations. Minimizing one's exposure to mutagens is a realistic possibility, but DNA replication itself presents some chance of error, and these errors cause mutations. Since DNA replication and cell division are essential parts of the normal growth and maintenance of a healthy body, mutations will occur. It follows that some level of **cancer risk** is intrinsic to being alive.

If mutations occur at random along the DNA, which is a good approximation to reality, then there is an element of chance in who will get cancer and what kind of cancer will appear. As a poignant example, some heavy smokers live rich, full lives and die of heart disease, Alzheimer's, or an automobile accident, not lung cancer. Some people who are careful about their lifestyle and diet get cancer at a tragically early age. These manifestations of chance defy the kind of cause-and-effect explanation that most of us find satisfying. How, then, can we think about the problem of preventing cancer? There are three factors that define your risk of getting cancer: chance, heredity, and environment. An examination of these cancer risk factors goes a long way toward explaining why one person gets cancer while another person does not. We begin this chapter with a discussion of chance, which may appear quite arbitrary, but the role of chance in cancer is understandable in terms of the science you now know.

CHANCE AS A COMPONENT OF CANCER RISK

INHERITED CANCER RISK AND POSSIBLE ACTIONS TO REDUCE IT

REDUCING THE ENVIRONMENTAL COMPONENT OF CANCER RISK

DIET AND CANCER RISK

BEHAVIOR AND CANCER RISK

MEDICAL EFFORTS TO REDUCE CANCER RISK AND PROBLEMS WITH THEIR IMPLEMENTATION

LEARNING GOALS

1. Describe reasons for the elements of chance associated with the initiation and progression of cancer.

2. Give examples of the role that heredity can play in cancer risk.

3. List the aspects of your own lifestyle that affect cancer risk. For each factor, write a statement of what you could do to reduce your personal risk of cancer.

4. Describe the essential features of a responsible clinical trial for a novel drug that is supposed to reduce cancer risk.

5. Choose a case in which existing evidence shows a correlation between a particular behavior and cancer risk. Think of ways in which the correlation might be misleading. Identify an experiment whose results would help to determine whether this correlation reflects causality.

6. Use the Internet to find a Website that claims to have a cure for cancer, then use your knowledge of cancer to evaluate the legitimacy of the claim.

CHANCE AS A COMPONENT OF CANCER RISK

The idea that chance plays a role in human affairs is as old as history, and probably a good deal older than that. Accounts that come down to the present in the writings by ancient Egyptians, Greeks, Chinese, and Indians often talk about chance in the outcomes of war, business transactions, and love. There are elements of this kind of chance in cancer risk, such as the possibility that you were visiting the Three Mile Island or Chernobyl nuclear power plant at the time their reactor core melted down, with a resulting release of significant amounts of radioactive material. These are not, however, the kinds of chance events that play a role in day-to-day cancer risk, so they are not considered here. Instead, we look at chance in the behavior of biological molecules and the structures they form, examining the roles of random events in the development of cancer risk.

Some cancer risk is the unavoidable result of chance mutation.

A paper published during 2015 in *Science*, an important and reliable scientific journal, took a data-based approach to the evaluation of chance in the frequency of different kinds of cancer. The abstract of this paper states,

"Some tissue types give rise to human cancers millions of times more often than other tissue types. Although this has been recognized for more than a century, it has never been explained. Here, we show that the lifetime risk of cancers of many different [tissue] types is strongly correlated (0.81) with the total number of divisions of the normal self-renewing cells maintaining that tissue's homeostasis [i.e., maintaining cell number in the tissue, taking account of normal cell death]. These results suggest that only a third of the variation in cancer risk among tissues is attributable to environmental factors or inherited predispositions. The majority is due to 'bad luck,' that is, random mutations arising during DNA replication in normal, noncancerous stem cells. This is important not only for understanding the disease but also for designing strategies to limit the mortality it causes." (From Tomasetti C & Vogelstein B [2015] *Science* 347:78–81.)

The conclusions drawn in this paper were based on comparisons of cancer frequencies from 31 different tissues or genetic conditions (for example, people who inherited cancer risk, such as a carcinogenic allele of *APC*, the gene responsible for familial adenomatous polyposis). The correlations sought were between the frequencies with which cancer occurred in a given tissue (over the lifetimes of many people) and the number of stem cell divisions necessary to form and maintain that tissue in the face of cell death. The correlations observed were impressively good.

To restate the paper's conclusions more simply, these workers interpreted their findings to mean (1) the number of mutations that a somatic cell's DNA has experienced is proportional to the number of replications this

DNA has undergone since the fertilization of the egg from which it grew and (2) this number is a significant factor in the emergence of cancer because the number of carcinogenic mutations will be a a fraction of the number of all mutations. If this analysis is correct, a significant fraction of cancer risk is an inescapable result of being alive.

It is noteworthy, however, that this paper elicited a strong negative reaction from many highly regarded members of the biomedical community. Several of these scientists wrote to the journal, expressing their disagreement with the article's conclusion and giving reasons why they thought the paper wrong. These letters pointed out that the lifetime incidence of cancer in various tissues shows variation from one country to another, suggesting that lifestyle matters in the frequencies that were used to make the correlations in the paper. They pointed out that lifestyle-dependent factors, like obesity, increase cancer risk in general. They identified the importance of preventive measures in reducing cancer risk. We have here an excellent example of an important scientific issue, which is critically relevant to cancer causality and prevention, upon which reputable scholars disagree. This situation is certainly not unique; it reflects both the complexity of cancer and the difficulty of coming to a deep and legitimate understanding of cancer causality.

An article published only a few months later in the *New England Journal of Medicine* (see Further Readings) took a broader view of cancer causality and formulated a way to think about cancer risk from multiple points of view. All scholars of the subject accept that there is such a thing as **inherited cancer risk**, that is, inherited alleles of certain genes that increase the chance that an individual will get cancer. For example, no well-informed person would deny the significance of inheriting a loss-of-function allele of the genes that encode the retinoblastoma protein (*Rb*), certain breast cancer proteins (BRCA1 and BRCA2), or adenomatous polyposis coli (APC). Moreover, no one who has examined the data on increased cancer risk from exposure to ionizing radiation or the consumption of tobacco could claim that environment and behavior are not contributors to cancer risk. Add to these factors the inescapable evidence for increased cancer risk from certain viruses, such as human papilloma virus (HPV), and we have irrefutable evidence for contributions to cancer risk by both heredity and environment. On current evidence, then, we must conclude that chance, inheritance, and environment all play roles in the likelihood that an individual will confront cancer at some time during his/her life. As sensible people who want to minimize our own chance of getting cancer, we must ask which of the risk factors are under our own control, and what we can do to diminish them. Now we can take a closer look at the nature of cancer risk, which is really the **probability** that cancer will occur under a given set of conditions.

Chance and probability are important to cancer medicine in two distinct ways.

The idea of probability is used by people talking about cancer in two quite different ways, and it is worth some thought to understand the different meanings of the word. When a geneticist is talking about the probability of a mutation in a particular part of a human's DNA and says that the probability of mutation is essentially the same for all genes, s/he is really making a statement based on the simple but important idea of **symmetry**. From a chemical point of view, all stretches of DNA are essentially the same, so there is no reason to expect a higher frequency of mutation in one region than another. This is the same idea as the concept that a coin is equally likely to land heads up or heads down. It is the same as the fact that your chances of winning the lottery are given by the number 1 divided by the number of people who have bought lottery tickets for this drawing. If the lottery is fair, every person who bought a ticket has the same chance of being chosen: $1/N$, where N is the number of tickets. These expectations of chance are based on physical structure. A good coin is twofold symmetric in the sense that the two faces are mechanically equivalent, even though they are differently marked, so you can tell heads from tails. The mechanical equivalence means that the coin is equally likely to land heads-up or tails-up; this is its symmetry. Symmetry says that if there is a given chance that DNA polymerase will make a mistake when replicating each nucleotide, then the more times DNA is replicated, the more likely it is that the DNA will contain a mutation. Symmetry is at the heart of the argument presented above that the probability of cancer in different tissues depends on the number of cell divisions that go into making and maintaining that tissue.

A very different kind of probability is used when a doctor is giving a prognosis for a cancer patient who has a specific cancer, identified at a specific stage, and given a well-known treatment about which there are lots of data. As discussed briefly in Chapter 2, a doctor might say that the patient's chance of being alive after 5 years is 75%. Here, the situation has no symmetry to define the probability; the statement, which is a **prognosis**, is based on historical fact. Over the approximately 75 years that doctors have been keeping and publishing medical records about cancer outcomes, a large number of people have previously been diagnosed with the same cancer at roughly the same stage and then given the prescribed treatment. Publicly available information shows the average fate of people in that condition; 75% were still alive after 5 years. This historical fact becomes a prognosis for the patient, based on the unstated assumptions that neither the cancer nor its treatment has changed significantly during the years of data collection.

Historical probabilities and probabilities based on symmetry have a striking feature in common; they both depend on the amount and quality of data available. If a particular cancer has been studied for 5 years, allowing the accumulation of outcomes on 125 patients, one has some idea of likely patient outcome, but if a cancer has been studied for 50 years and there is outcome information for 10,000 patients, the reliability of predictions based on history will be significantly more reliable. In an analogous fashion, if you were to toss an unbiased coin four times and get heads four times, you would probably say that was pretty unlikely. If you tossed the coin 400 times, the likelihood of getting all heads is almost zero. With the greater number of tosses, you would be much more likely to get close to half heads and half tails, because that is a natural property of chance, based on the symmetry of the situation, when you are dealing with a large numbers of trials.

Ideas about probability allow one to work out the likelihood that you would get four heads in four tosses. It is small but nowhere near zero (the prediction is $\frac{1}{2} \times \frac{1}{2} \times \frac{1}{2} \times \frac{1}{2} = \frac{1}{16}$, which is 0.0625). The likelihood that you would get 40 heads in 40 tosses is *much* lower (roughly $1/110,000,000,000 = 0.00000000000091$). Yet if you tossed a coin 40 times, and repeated this tedious process $1,000,000,000,000$ times, it's reasonably likely that within all those tosses you *would* get 40 heads in a row, at least once. This is the nature of random events that are shaped by symmetry.If you are told that your chance of being alive in 5 years is 75%, you certainly cannot be assured that you won't die. This is the nature of historically defined probabilities. It is helpful to get symmetry-defined and historical probabilities straight in your mind when thinking about cancer and other medical problems.

INHERITED CANCER RISK AND POSSIBLE ACTIONS TO REDUCE IT

The nature of probabilities means that chance is something you cannot control. Similarly, you cannot control your risk for inheritance of cancer, but your inherited cancer risk can be a motivation for action. Additional vigilance or a more proactive approach to treatment can reduce this cancer risk.

If hereditary factors suggest you are at risk for a particular cancer, there are sometimes things you can do.

One of the best-documented examples of inheritable cancer risk is for breast cancer. Given the power and economy of modern methods for sequencing DNA, a person from a family with a history of this disease would be well advised to have a responsible lab examine the sequences of the relevant genes in their genomes, asking whether they are carrying alleles known to increase cancer risk. If so, then at some time in their life, removal of the tissues at risk (a mastectomy to remove breast tissue and/or an ovarian hysterectomy to remove ovaries, uterus, and related parts of the reproductive organs) could be a sensible course of action. Commonly, such cancers do not develop until midlife, so if a person at hereditary risk for breast cancer wants to have children, she might do so and then arrange to undergo one or more of the operations that will significantly reduce her cancer risk. In this case, timing is a significant consideration. This situation does, however, raise the question of whether or not to have children when you may pass your predisposition to cancer on to them. Is this a reasonable and honorable thing to do? The answers to such questions depend on many personal issues. Questions like this are worthy of thought and discussion, but the answers are not anything one can state quickly in a book. If you should encounter this dilemma in your own life, there are services in many communities that nurture informed discussion about genetic issues. For example, one option would be to resort to in vitro fertilization of your ova, so several embryos could all be cultured in a lab and one or a few cells could be taken from each embryo for genome sequencing. With the resulting data in hand, one could pick an embryo that had only wild-type alleles of the relevant genes. However, this undertaking is seriously expensive, time-consuming, and emotionally arduous. Given the complexity and personal significance of these problems, it is no surprise that the impact of genetically defined risk in the onset of breast cancer is the focus of ongoing **clinical trials**, that is, controlled experiments or observational studies on human subjects as opposed to animals.

SIDE QUESTION 11.1

You are working as a genetic counselor and have been approached by a 22-year-old woman who just learned that she carries an oncogenic allele of *BRCA1*. What would you suggest she think about in deciding whether or not to have children?

Even with no family history of cancer, breast and prostate cancers still pose a lifetime risk.

The comparatively high risk of cancer in breast tissue probably relates in part to the number of cell divisions necessary to develop normal breast tissue and to maintain its function during the changes that naturally occur with each menstrual cycle and with pregnancy. The effects of sex hormones during pregnancy and lactation (the time after birth when breasts produce milk) are also important, because extensive data demonstrate that the likelihood of breast cancer drops significantly for women who have had one or more children.

Both female and male tissues that are under the control of steroid sex hormones are common sites of cancer development, for reasons not yet fully understood. Nonetheless, checking for cancer in these tissues, as discussed in Chapter 2, is likely to reduce the risk of developing metastatic cancer. However, it is important to note that there is disagreement among oncologists and other concerned doctors about the age when mammography should begin and how frequently it should be done. There are still scientific meetings devoted to this issue, so even when people agree on the value of cancer prevention, there is disagreement about how best to do it. When does the value of a vigilant search for tumors, before they have gotten big, outweigh the added cancer risk from multiple X-ray exams? There is also the difficult issue of the medical futility associated with finding and treating a lump that would never have progressed to cancer, not to mention the emotional toll of such persistent reminders to patients of their cancer vulnerability. These are issues with no certain answers. Current evidence has led the U.S. Preventive Services Task Force to suggest beginning mammograms at around age 50 and having them every 2 years until about age 70, when their frequency can reasonably be increased. Until 2015, the American Cancer Society favored starting mammograms at age 40, but they have now raised that age to 45. For younger women with no family history of breast cancer, mammograms are not recommended because the chances of finding a growth that needs medical attention are low, and the added stress, risk, and expense are not warranted.

For prostate cancer, there is no simple and inexpensive test with value equivalent to a mammogram. However, regular physical exams by a good general practitioner, beginning at age 40, make sense for many reasons. As mentioned in Chapters 2 and 10, the value of measuring blood levels of prostate-specific antigen (PSA) is now in question, largely because the observation of high PSA has led to prostate biopsies that have produced mixed results. As with biopsies of breast lumps, the results from prostate biopsies can motivate treatments that were probably unnecessary. The use of magnetic resonance imaging (MRI) to look for cancerous growths on the prostate gland is increasing, but the expense of this imaging method makes the approach of questionable value for routine screening. Prostate cancer is a particularly elusive target because a study of the prostate's condition by autopsies on a large number of men who died of other causes has suggested that benign prostate tumors are remarkably common. Aggressive treatment of a suspicious prostate condition may often have no medical value, and the treatments can have side effects on quality of life. For example, an operation to remove the prostate gland can lead to impotence, though with modern surgical techniques, this problem is less common.

There are measures to manage inherited cancer risk, even when that organ cannot be removed.

If there is inherited risk for cancer in an essential organ, such as the intestine, complete tissue removal is not an option. However, one can still use DNA sequencing to determine whether an individual is carrying alleles known to increase risk of these kinds of cancer. If one is at risk, then close attention by effective medical imaging (a colonoscopy in the case of colon cancer, for example) is a valuable step to take. This approach may identify loci of precancerous or early cancerous development, which can be treated by small-scale surgery, radiation, and/or chemotherapy, reducing immediate concern through action that is neither drastic nor life-threatening.

Inherited risk will, however, almost certainly carry an emotional burden; it is hard to live with a genotype that feels like a death sentence. In this

SIDE QUESTION 11.2

The cost of using MRI to examine the prostate gland in a healthy man varies from state to state, but most hospitals will charge about $2000. With added fees from doctors who interpret the images, the cost is usually considerably higher (up to $12,000). The method has the advantages that image quality is excellent and the images do not involve ionizing radiation, but the cost of MRI makes it a questionable tool for routine screening. If insurance were to cover it, the cost would, of course, be passed along to everyone as an increase in the cost of insurance premiums. Given all these issues, do you think MRI should be used to screen men for prostate cancer, and if so, at what age should the process begin? Give reasons for your opinion.

situation, the key thing to remember is that *all* genotypes are death sentences; that is the nature of human life. A predisposition to cancer only changes the timing. This and related issues are discussed more thoroughly in the next chapter on living with cancer, a hard but common problem. Fortunately, there are numerous cancer support groups that can offer significant help in thinking about the issues of living with the disease and making peace with one's situation.

The other side of the inheritance coin is the possibility that one may be born with genes that are particularly effective in preventing cancer. Certainly, some people are genetically predisposed to be good athletes, exceptionally good-looking or fast-thinking, and so forth. It is almost certain that some people inherit alleles of genes encoding enzymes that are particularly good at repairing DNA, or preventing errors in mitosis, or some other aspect of avoiding the accumulation of mutations. These conditions may be factors that help to explain the good luck that some people have in avoiding cancer, even though they engage in risky behaviors. At the moment, the identity of those genes and the DNA sequences of the valuable alleles are unknown, but that situation may change as more information is collected about human DNA sequences.

REDUCING THE ENVIRONMENTAL COMPONENT OF CANCER RISK

Since there is little one can do to alter the fundamentals of inheritance and chance in cancer risk, any action plan for reducing one's life-time risk of cancer must focus on the environment in its broadest sense. Granted, some aspects of our environment are inescapable—for example, exposure to cosmic rays and ground-based radioactivity—but other environmental factors are under our control, or at least our influence. These are the topics that concern us here.

Avoiding aggressive carcinogens is an obvious approach to minimizing cancer risk.

Until quite recently, certain workplaces were the sites of potent and avoidable carcinogens. Uranium miners were exposed to ionizing radiation, asbestos miners to carcinogenic fibers, and dye workers to carcinogenic chemicals, to mention just a few. Thanks to improved knowledge, customs, and laws, the situation is now much better than it was prior to the mid-twentieth century. While there are still large numbers of dangerous chemicals being produced, thanks to the expansion of the organic chemistry industry in the period during and following World War II, the United States has seen a real reduction in environmental carcinogens, thanks to improved knowledge about the nature of carcinogenesis and scientific innovations, like the Ames test described in Chapter 5. These changes have provided effective and inexpensive ways to identify chemicals that are likely to be carcinogenic. Companies that make new chemicals can now flag the ones with mutagenic properties both quickly and economically. This is a major advance, and many companies now have policies in place that use this sort of test to minimize the cancer risk faced by both their employees and their customers.

However, there are situations, both in the United States and elsewhere throughout the world, where companies have ignored the carcinogenic impact of their waste products and dumped them in public places, minimizing their cleanup costs. America's Superfund sites (designated by the U.S. Environmental Protection Agency) include many examples of this

SIDE QUESTION 11.3

Given all the information that is widely available about the dangers of smoking tobacco, why do people still do it? Can you think of ways in which tobacco use could be reduced?

behavior, and countries in the developing world, such as those in sub-Saharan Africa, are sometimes used as marketplaces for goods that can no longer be sold in the developed world. This kind of corporate behavior deserves public condemnation, so people making the relevant decisions will experience negative consequences from such profiteering schemes.

The most obvious way to reduce your personal cancer risk is to minimize your exposure to physical and chemical carcinogens. This effort can take many forms. The most important is to avoid cigarettes and other tobacco products. The significance of this rather simple statement cannot be over-emphasized. Current data suggest that tobacco products are responsible for about 170,000 deaths per year in the United States alone, while carcinogens in U.S. workplaces contribute to only about 33,000 deaths annually. Thus, risk from tobacco is 5-fold (500%!) more important than risk from workplace exposure to carcinogens. Moreover, the number given for the workplace is probably affected by exposures to carcinogens that occurred many years ago, so the real ratio of risks between tobacco and workplace carcinogens may be even higher. Thus, giving up smoking and avoiding the smoke of people who still have this habit are wise steps in cancer risk reduction. While the data are still pending, the same principle will probably apply to e-cigarettes as well (**SIDEBAR 11.1**).

What about the carcinogenic effects of smoking or ingesting marijuana? Unfortunately, the evidence available on marijuana (leaves from the plant *Cannabis*) is too sketchy to make strong statements. There are studies suggesting that smoking pot causes cancer, but because recreational (as opposed to medical) marijuana is sometimes laced with tobacco, which is certainly carcinogenic, the evidence about cannabis itself is still wanting. Now that this substance has been made legal in some areas of the United States, there will probably be rigorous scientific studies that will answer this important question in the not-too-distant future. As a simple idea to help govern behavior, however, the temperatures reached by burning plant leaves are high enough that the process is likely to generate carcinogens. Inhaling these chemicals from any kind of smoke, whether from tobacco, marijuana, corn silk, or a forest fire, is not likely to do you good.

SIDEBAR 11.1

A CLOSER LOOK: E-cigarettes.

When the subject of cigarettes comes up these days, electronic cigarettes are commonly mentioned. What are they, and do they avoid the carcinogenic effects of regular tobacco products? An e-cigarette is a device made to convey the feeling of smoking a real cigarette without the inhalation of tobacco smoke. They look and feel a bit like a regular cigarette, because they are designed to make habitual smokers feel as if they are still engaging in their old habit, but no tobacco smoke is involved. E-cigarettes use an electric current to heat materials that allow you to breathe vapors of nicotine, the stimulant in tobacco that many people find enjoyable. At this time, there is controversy about the health effects of e-cigarettes, based on three not-yet-answered questions: (1) What does an e-cigarette smoker inhale besides nicotine?

Different brands have different levels of contaminants in the vapors that a smoker breathes in. Some of these compounds may be dangerous. (2) What are the long-term consequences of vaporized nicotine for the health of one's lungs or other body organs? There simply are not yet data to answer this question. (3) Does the switch from regular tobacco to e-cigarettes help people to give up smoking, or do e-cigarettes serve more frequently as a pathway by which young people go from not smoking to becoming tobacco users? Again, the relevant data are not yet available. In short, these tobacco substitutes are a questionable, unproven alternative to real cigarettes. They may be less harmful, but you would be better off not smoking anything!

Work to establish habits that are consistently healthy.

One's own behavior and lifestyle are matters where one can exercise some measure of control, so incorporating knowledge about carcinogenesis into your daily habits makes a lot of sense. For example, some older studies showed that smoking tobacco while consuming alcohol, particularly strong drink, increases the risk of throat cancer. Alcohol itself is now categorized as a Class1 carcinogen, meaning it has been demonstrated to be cancer-inducing in humans. This is probably a result of the fact that the liver treats ethanol (the chemical name for the active ingredient in what we call alcohol) as a poison; it uses enzymes to alter this chemical's structure, detoxifying your blood. However, one product of this reaction is acetaldehyde, which is carcinogenic (FIGURE 11.1). It then takes a while for the liver to carry out a second reaction and get rid of the acetaldehyde it has made. Thus, after a bout of drinking, while your liver is getting your blood alcohol levels back down, you are experiencing the action of a proven carcinogen. In combination with tobacco, this has clinically demonstrated carcinogenic effects.

Excessive weight gain is another manifestation of human behavior with clear implications for cancer risk. A person is said to be obese when their **body mass index** (BMI) is greater than 30. BMI is defined as weight in kilograms (that is, your weight in pounds, divided by 2.2) divided by the square of your height in meters (that is, your height in feet, divided by 3.3). No one has identified a direct line of causality between obesity and cancer, but the following correlations make a fairly strong case. According to data collected by both the National Cancer Institute and the American Cancer Society, women who are obese have a 2–4-fold greater frequency of endometrial cancer than women with a normal BMI. Obese people of both sexes are about 2-fold more likely to get either esophageal, stomach, liver, or kidney cancer.

There is also a link between long-term obesity and an increased likelihood of breast cancer after menopause. One possible reason for this correlation is that after the ovaries stop making estrogen, at the time in middle age when fertility decreases, a major source of estrogen is the cells in fat, called **adipocytes**. More fat cells probably mean more postmenopausal estrogen. As described in Chapter 8, the parenchymal cells of many tumors that arise in the breast are dependent on estrogen for their survival, so extra production of this hormone may increase the tendency of breast-derived cells to proliferate. Interestingly, obesity in older men is well correlated with an increased risk of colon cancer. Here, the mechanism is less obvious, but it may be a result of the fact that belly fat is commonly associated with inflammation, which can serve as a tumor promotor. Whatever the mechanisms, it follows that refraining from becoming obese can be added to the list of things under one's own control that can reduce cancer risk, regardless of gender. Moreover, keeping one's weight down is worthwhile for other aspects of health, such as avoiding diabetes, high blood pressure, cardiovascular disease, and stroke.

FIGURE 11.1 Reactions by which the liver detoxifies alcohol. The chemical name for the active material in alcohol is ethanol, a simple molecule whose atomic structure is given. In the human liver, the enzyme alcohol dehydrogenase (ADH) acts on ethanol and removes both an electron and a hydrogen atom by transferring them to a different small molecule. This reaction converts ethanol into acetaldehyde, a carcinogen. A second enzyme, acetaldehyde dehydrogenase (ALDH), acts on acetaldehyde, removing another hydrogen atom and an electron to make acetate, which is harmless. The transient presence of the acetaldehyde made by this set of reactions is probably the reason that ethanol has been shown to have carcinogenic effects in people.

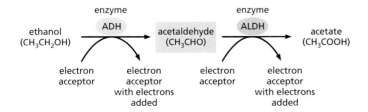

Sometimes, avoiding environmental carcinogens gets complicated.

A good example of the complexity in avoiding carcinogens is seen with the chemical formaldehyde (FIGURE 11.2). This simple substance is used for many beneficial processes—for instance, making ink, glue, some cosmetics, clothing, plastics, dyes, disinfectants, and household insulation. Formaldehyde is also used by some scientists in their experiments and by some undertakers for embalming dead bodies to preserve a lifelike appearance. In light of all these uses, it is surprising to learn that high doses of formaldehyde were shown in 1979 to be carcinogenic in mice. Strong evidence about formaldehyde's carcinogenicity for humans was obtained around 2000. Shortly thereafter, concerned scientists and government officials met to discuss what should be done about the widespread use of this dangerous chemical. One point of view said, Take this chemical and others like it out of the human environment immediately, so the potential of their negative impact will disappear. The other point of view maintained that since formaldehyde has so many uses, it would be very hard to replace. Furthermore, it is likely that chemicals used to replace formaldehyde would also be carcinogenic, although it might take years to learn this with any certainty. Removing formaldehyde from our industrial and manufacturing sectors would probably slow productivity and block the synthesis of many valuable products, which we would then have to do without.

A group of diverse people, convened by the Environmental Protection Agency to think about these issues, identified important factors to consider beyond simply use or no use. For example, there is a significant reduction in risk achieved through proper handling of any dangerous substance, such as the use of waterproof gloves and protective clothing. Formaldehyde poses an additional risk, however, since it forms vapors that are also carcinogenic. Again, with proper handling, the vapors can be removed from the workers' environment by a chemical fume hood that pulls the formaldehyde vapors away from the workplace and disperses them into the open air, diluting them to the point that they are no longer carcinogenic. Moreover, when formaldehyde is used to make things, it usually reacts with other chemicals and loses its carcinogenic properties. Cosmetics made with formaldehyde need not be carcinogenic. Thus, the knowledge that formaldehyde is carcinogenic can be used to establish procedures and behaviors that neutralize or diminish its damaging effects.

So, is it right to continue the use of formaldehyde? In 2011, the National Toxicology Program, an interagency program of the Department of Health and Human Services, named formaldehyde as a known human carcinogen, but sensible people have recommended its continued use, because of the considerations just discussed. The formaldehyde example shows

FIGURE 11.2 The structure of formaldehyde and the distribution of its many uses.
(A) Formaldehyde is composed of a central carbon atom that is doubly bound to an atom of oxygen (O) and singly bound to each of two hydrogen atoms (H). (B) The relative amounts of formaldehyde employed for different uses in the United States as of 2007. Resins are any of several hard polymers that are used in finishing surfaces and building construction. Chemical intermediates refers to a wide range of chemical structures with diverse uses that can all be synthesized by starting with formaldehyde. Agriculture refers to fertilizers and herbicides. Paraformaldehyde is a polymer of formaldehyde that has its own many uses, and chelators are additional kinds of chemicals. (B, data from Bizarri SN [2007] Chemical Economics Handbook Marketing Research Report: Formaldehyde. SRI International.)

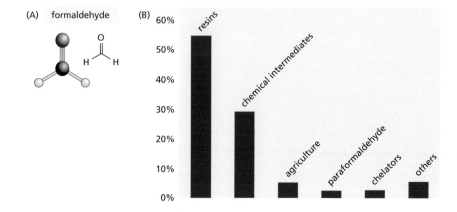

that decisions about reducing cancer risk are not always straightforward, but with thought, care, and investment, they can be effective.

Some environmental cancer risks are harder to get out of your life than you might think.

Tobacco smoke contains a cluster of chemicals that are much more carcinogenic than formaldehyde. (Look back at Table 5.2 in Chapter 5.) Giving up smoking is a surefire way of reducing cancer risk, both for yourself and for others near you when you smoke. Extensive tests have demonstrated that secondhand smoke (smoke from other people's cigarettes, cigars, or pipes) is carcinogenic. The Surgeon General of the United States issued a report in 1964 stating that 2.5 million adult Americans who were nonsmokers had died because they breathed secondhand smoke. Sadly, the secondhand smoke problem seems to be greatest for infants and children, if they are exposed to carcinogens at an early age when they are particularly vulnerable to respiratory and ear infections. A more recent report from the U.S. Department of Health and Human Services, published in 2017, found that 34,000 premature deaths from heart disease are related to secondhand smoke, so giving up cigarettes is important in ways other than reducing cancer risk.

In spite of this kind of evidence, some people find it very hard to stop smoking, partly because the nicotine in tobacco smoke is highly addictive. In the nineteenth century, a remark attributed to the great American author Mark Twain bragged that, "Giving up smoking is the easiest thing in the world. I know, because I've done it thousands of times." But the fact is that giving up smoking for good is worth the effort. The resulting reduction in cancer risk is substantial, and there are organizations available to help (see Websites mentioned in Further Readings).

SIDE QUESTION 11.4

If you were serving on a panel to decide about continued mining of asbestos (a fibrous mineral used in fireproofing material, but a known human carcinogen), what three factors would you consider to be the most important in making a wise and constructive decision?

DIET AND CANCER RISK

The food we eat can influence our cancer risk in two ways: reducing cancer risk, by acting against the effects of carcinogens we encounter from other sources, or increasing cancer risk, by serving as either a carcinogen or a tumor promoter. You have seen that carcinogenic chemicals are classified in two categories, direct and indirect. As described in Chapter 5, a **direct carcinogen** is carcinogenic in its own right; for example, it may interact directly with DNA to form an adduct. An **indirect carcinogen** is a chemical that is turned into a carcinogen by chemistry that goes on inside our bodies. Maximizing one's intake of cancer-inhibiting substances and minimizing one's intake of both direct and indirect carcinogens is obviously important for reducing one's risk of accumulating somatic mutations.

Some foods and food additives are suspected indirect carcinogens and should be avoided.

Certain food preservatives are indirect carcinogens. Many processed meats, like bacon and sausage, contain sodium nitrate or sodium nitrite, added as preservatives. These molecules are altered by cooking and/or by chemical reactions in the human stomach and liver to form nitrosamine, a proven carcinogen. However, the amount of nitrosamine produced by your body from a few slices of bacon is small, and there is no direct evidence that such foods are carcinogenic. Nonetheless, minimizing your consumption of substances that expose your body to nitrosamine is probably a good idea. Some foods contain indirect carcinogens but are also truly good

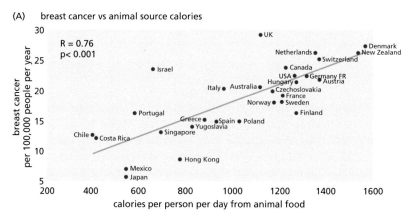

(A)   breast cancer vs animal source calories

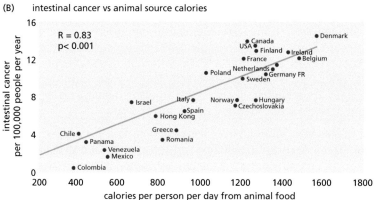

(B)   intestinal cancer vs animal source calories

for you. The common grocery-store mushroom *Agaricus bisporus*, which is widely used in salads and sauces, contains a fairly high concentration of hydrazine, a proven carcinogen. Laboratory experiments have shown a carcinogenic effect in rodents from high doses of these mushrooms. However, a cancer risk from enjoying a few of these mushrooms, in salads or other foods, has not been demonstrated and probably never will be. This difference can be attributed to the issue of dose; the experimental evidence for the carcinogenicity of hydrazine was based on much higher amounts of the chemical than one ingests from a few mushrooms. Moreover, these mushrooms also contain chemicals that work as antioxidants, which should reduce cellular amounts of reactive oxygen species (ROS) and be good for you. Other mushrooms, like shiitakes, are thought by some people to *reduce* cancer risk, although objective evidence is again wanting. One must admit that, for this sort of food issue, there is not yet enough evidence available to help you make clearly reasoned decisions.

There is, however, considerable evidence that a vegetarian diet reduces cancer risk. Several religions advocate or require vegetarian diets, for example, the Seventh-day Adventist Church. People who follow this faith have a lower incidence of most cancers than the American population as a whole. This difference may, of course, result from several aspects of their lifestyle, such as abstaining from alcohol, and not simply their reduced intake of meat. However, studies comparing cancer incidence in people from different countries that consume different amounts of meat are consistent with the idea that vegetarian diets reduce cancer risk (FIGURE 11.3).

When evaluating claims and data, it is important to recognize that correlations do not demonstrate causality.

Figure 11.3 clearly shows a strong positive correlation between the incidence of breast and intestinal cancers and the amounts of meat consumed in different countries. Such data are, however, less convincing than they first appear; they demonstrate correlation, not direct causality. The data are clear in showing that people who live in countries where meat is commonly eaten have a higher risk of several kinds of cancer than people who live where meat is rare. However, they do not show that meat consumption is the cause of the higher cancer rates. That issue is not settled by correlation, even when the correlation is very good (as indicated by the low p values given on the graphs, which mean the probability that there is no relationship between meat eating and cancer). Perhaps countries where meat is commonly eaten are richer and therefore have better diagnostic facilities, so cancer is identified more frequently than in poorer countries. Perhaps people who eat more meat are wealthier, and they use their wealth to do things other than buy meat, things that may actually be the cause of higher cancer rates. Perhaps these people don't do as much physical work and thus get less exercise. It could be that they can afford to buy expensive substances, like tobacco and recreational drugs. What if their wealthier environments include carcinogenic chemicals that have not yet been identified? These kinds of alternative explanations are the reason that correlation does not demonstrate causality.

While the above cautions are all legitimate concerns, the consumption of meat could have an impact on cancer risk. The combination of (1) national correlations, (2) low cancer incidence among groups of people who do not eat meat, and (3) carcinogenic chemicals that are generated by cooking meat fat at high temperatures (as described in Chapter 5) makes a moderately strong case. It is nowhere near so strong a case as for the carcinogenic effects of tobacco smoke, but it is still worthy of your consideration.

Reasons why a vegetarian diet might reduce cancer risk.

There are two plausible reasons why vegetarians are less prone to cancer than other people: (1) with less meat in the diet, there is less intake of meat fat, and (2) a vegetarian diet increases the consumption of fruits and vegetables. The latter is important because some plants contain chemicals that actually reduce the effects of some carcinogenic chemicals. When talking about carcinogenesis in Chapter 5, we discussed reactive oxygen species (ROS), which are chemically reactive molecules generated by ionizing radiation and some chemical reactions. ROS can initiate several forms of DNA damage, all of which we want to avoid (FIGURE 11.4). However, some ROS in your cells is inevitable, given that most of our cells use oxygen to allow mitochondria to make ATP from food efficiently. These chemical reactions generate ROS without any external agent, like ionizing radiation. It is not surprising, then, that our cells contain chemical mechanisms to combat ROS, commonly known as antioxidants; these help to inactivate ROS and prevent its destructive action. Relevant molecules include vitamins A, C, and E (FIGURE 11.5) as well as some chemical reaction systems that use enzymes to inactivate ROS. The latter include reactions that employ a small molecule called glutathione (Figure 11.5), in which there is a sulfur atom attached to a hydrogen that can inactivate ROS.

Plants also contain substances that inactivate ROS. Some of their ways of fighting ROS are based on chemicals different from the ones in animal cells. (If you are interested in this subject, see the article about it in Further Readings). Some of these chemicals can move from our intestine into our bloodstream and thence into our cells. Therefore, eating plants probably gives your cells additional ways of reducing ROS: the ways used by plants, in addition to the ways intrinsic to your own cells. Food obtained from animals, on the other hand, doesn't add to the ROS-reducing mechanisms that our own bodies can produce because we, as animals, already possess those mechanisms.

SIDE QUESTION 11.5

Look up the words correlation and causality in a good dictionary. Describe the differences between them in your own words. If two factors are correlated, how could you determine whether there is a causal relationship between them?

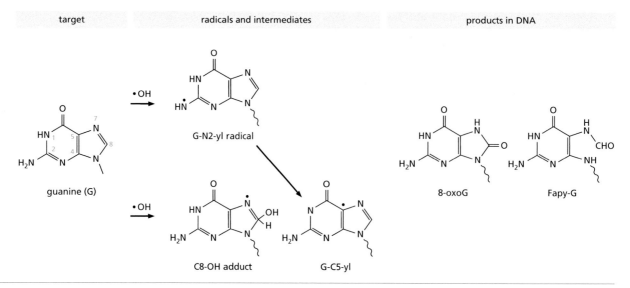

target | radicals and intermediates | products in DNA

FIGURE 11.4 Effects of reactive oxygen species (ROS) on the base guanine (G). There are several ways in which different forms of ROS can react with guanine, one of the bases found in the nucleotides that make up DNA. •OH is the hydroxyl radical that forms from water under the action of ionizing radiation. Analogous reactions can occur with the other three base founds in DNA. Most of these reactions have been observed in cells, showing that ROS can alter the components of DNA, which in turn can lead to a mutation. (Adapted from Cadet J & Wagner JR [2013] *Cold Spring Harbor Perspect Biol* 5:a012559.)

What are the chemicals in plants that help to protect us from ROS? This issue is not yet well understood. Several chemicals have been identified as likely candidates and tried by doctors as specific dietary supplements, but this work has not yet provided evidence that any of these compounds protects people from getting cancer. A useful word for thinking about this issue is **phytochemicals**, substances we ingest by eating plants (*phyton* is the Greek word for plant). This name captures the idea that plants make substances, probably for their own protection against ROS and other bio-

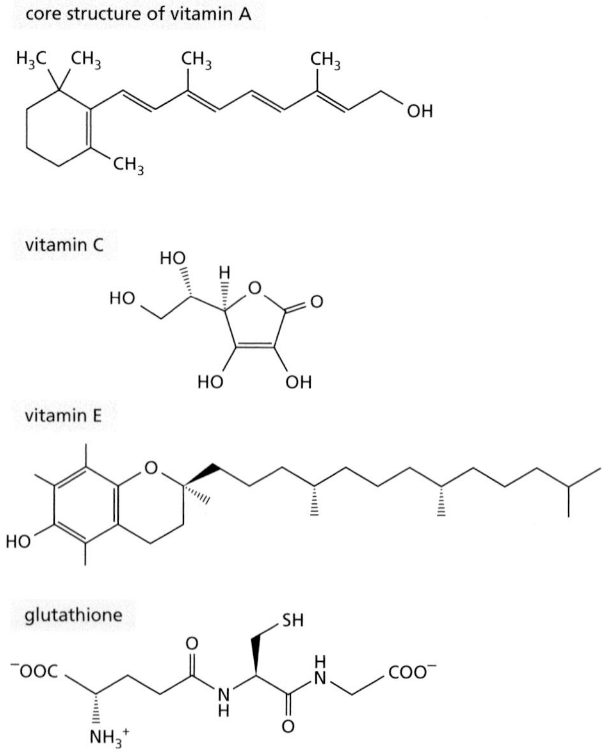

FIGURE 11.5 Molecules that help our cells to eliminate ROS. The four molecules shown (vitamins A, C, and E, followed by glutathione, which is made from three amino acids linked by peptide bonds) are all capable of reacting with ROS, such as the hydroxyl radical, •OH. These reactions inactivate the radical. In each case, the inactivation of ROS is accomplished by adding an electron to the molecule of ROS, which makes it unable to react with other molecules, such as DNA. Although each of the vitamins is destroyed by reacting with ROS, glutathione is regenerated by energy and electrons that come from food, so it can function repeatedly as a destroyer of ROS.

chemical hazards, which are also good for us. We may not yet know what they are, but eating them conveys health benefits. Certainly, many herbs have been used for uncounted generations as medicines for human healing. Some of these traditional remedies have been traced to specific chemicals, resulting in important medicines, like aspirin and the important new drug for treating malaria called artemisinin. Our knowledge of nutrition and biochemistry is simply not yet good enough to understand all these complex issues.

There is also talk about the value of fiber in our diet for reducing the risk of colon cancer, and there is some support for this view. The word fiber, as used here, means naturally occurring, water-soluble polymers in our diet that are indigestible but harmless. Most plants cells are surrounded by both soluble and insoluble fibers—the polymers that make up their cell walls (FIGURE 11.6). Some of these polymers are indigestible, so they contribute to the bulk in your feces, that is, the stuff left over from food that you eliminate from your body with a bowel movement. Fiber in the

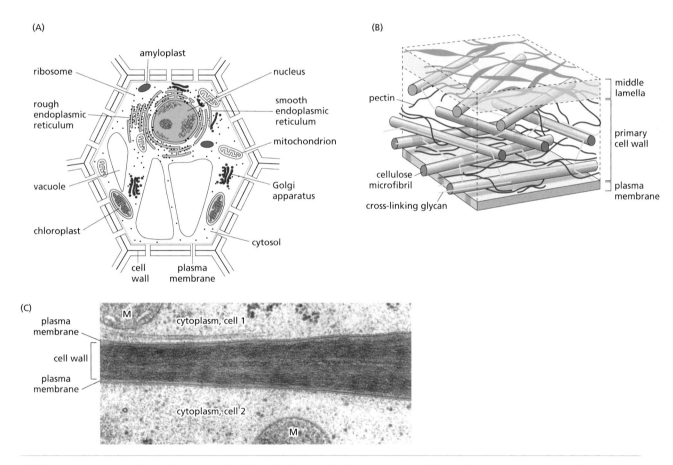

FIGURE 11.6 A plant cell and the components of its walls that provide fiber to our diets. (A) Plant cells have much in common with animal cells. They are surrounded by a plasma membrane, and they contain a nucleus that harbors their DNA. Between these structures is the cytoplasm, which contains a cytosol rich in enzymes and the many small molecules produced by metabolism. Also in the cytoplasm are membranes of the endoplasmic reticulum (ER), mitochondria, many ribosomes, and several Golgi apparatuses. Plant cells also include large, membrane-bound compartments that are not found in animal cells, called vacuoles. Vacuoles contain water and many of the chemicals a plant cell needs to discard. Outside the plasma membrane of a plant cell there is a cell wall, composed of several kinds of polymers. (B) Diagram of a plant cell wall. Its component polymers sit right on the outer surface of the plasma membrane. The principal polymer is cellulose (blue), a polysaccharide that forms bundles that are tough and hard to digest. Cross-linking the cellulose fibrils are smaller polysaccharides called glycans (yellow) and pectins (red) that are more readily digested. The middle lamella is an outer layer of the wall; it abuts the wall of the adjacent plant cell. Together these polysaccharides make up much of the fiber in our diets. (C) Electron micrograph of a thin section of a plant. The dark material in the middle is the fibers of the cell wall. Adjacent to the wall, top and bottom, one can see the cytoplasm of two cells that are separated by the wall. These cells made the wall by synthesizing and secreting the polysaccharides from which it is built. M indicates mitochondria in the cytoplasm of these two cells. (B, Courtesy of Michael W. Davidson and The Florida State University. C, from Albersheim P, Darvill A, Roberts K et al [2011] Plant Cell Walls: From Chemistry to Biology. Garland Science.)

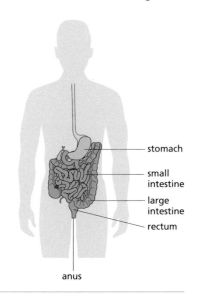

stomach

small intestine

large intestine

rectum

anus

FIGURE 11.7 The human intestine is long and shows variation in structure. The stomach drains into the upper part of the small intestine, a convoluted tube whose principal function is to pump small molecules generated by the digestion of food out from the intestinal lumen and into the surrounding blood vessels. The principal function of the large intestine is to bring water from its lumen into the surrounding blood vessels. This function serves both to concentrate the remaining waste products into the feces we pass with a bowel movement and to retain water for use in the body, so we are not always thirsty.

diet helps to keep feces bulky and soft, so they move through the intestines easily. If something causes the feces to get stuck in our intestines (which is called constipation), the chemicals we have ingested but not absorbed into our bloodstream are held in the intestine for a longer time. If, by chance, we have eaten something carcinogenic, it then has a longer time in contact with the cells of our lower intestine before it leaves our bodies. All this suggests that a high-fiber diet should reduce the risk of colon cancer, but recent clinical tests do not support this view with any strength. Eating plenty of fiber certainly won't hurt you, and it is probably good for your digestion, but it doesn't seem to reduce cancer risk.

The issues just raised do, however, provide a way to understand why cancer is more common in the epithelium of the lower intestine (colon) than the upper intestine. As ingested food moves through your digestive system, water is added, for example as saliva, to aid the enzymes that break down food polymers (proteins, nucleic acids, polysaccharides, and fats). In the lower intestine, much of this water is pumped back into the body, retaining it so you are not always thirsty (FIGURE 11.7). The condition called diarrhea, when your stools are runny or loose, is the result of failure in the epithelium of the lower intestine to pump water back into the body. Because water is normally removed from fecal material in the lower intestine, the concentration of any molecules not passed into your bloodstream increases there. If some of these chemicals are carcinogens, their concentration will be higher in the lower intestine than in the upper, thus increasing the likelihood of cancer in your colon.

Taking all these facts together, it seems that sensible choices about what you eat should have a positive impact on your risk of cancer. The case for eating a healthy diet is good, but it is nowhere near as strong as the case for giving up smoking or minimizing your exposure to ionizing radiation. An additional consideration that makes decisions about diet difficult is that some of the consumables associated with a modest cancer risk are good for you in other ways. Meat is an excellent nutrient. It provides all the amino acids you need to make your own proteins, whereas most plants are lacking some of the amino acids we need but can't make. Constructing a healthy vegetarian diet requires knowledge of or experience with how to combine vegetables, so you get all the nutrients you need. Even beverages that contain alcohol, cited above as a Class 1 carcinogen, have some beneficial effects. One glass of red wine a day, even two, has been shown to benefit heart health in older people. The question, then, is do you care enough about reducing cancer risk to change your habits by giving up nutritious food you might enjoy, like a barbecued steak? This is a highly personal matter, but the data encourage you to eat lots of fruits and vegetables and at least reduce your consumption of potentially harmful foods. In short, eat with balance and good sense.

There is no good evidence that supplemental antioxidants and/or vitamins reduce cancer risk.

The health food industry talks a lot about antioxidants and about boosting your immune system. Do they or other vendors have substances that really help to prevent cancer? If one asks for well-documented evidence, the answer is simply no, not at this time. There is even some evidence that specific antioxidants and other substances commonly presented as beneficial for health do *not* work to reduce cancer risk. Doctors have studied the effects of two natural antioxidants, vitamins A and E, on populations of people who were at serious risk for cancer. One such study on vitamin E was carried out on tens of thousands of middle-aged, male smokers. The rate of cancer in the vitamin E-treated group was no different from the group that was given a placebo. Similar studies have been done with β-carotene, a chemical in orange vegetables that is used by our bodies to make vitamin A. Again, when taken as an essentially pure substance, this important molecule had no positive effect on cancer rates in the 22,000 people studied.

It appears, from these studies and some others, that supplements either have no effect or make so little difference that their impact is not measurable. There are, however, serious limitations to the design of studies like the ones described. For example, the study on middle-aged smokers was carried out on people who were already at serious risk of cancer. What would the results have been if the study had been carried out on young, healthy people who took the supplement for 40 years instead of 5 years? Would such a study find a measurably reduced incidence of cancer? This work has not been done and probably never will be, because it would be so expensive and slow that no agency supporting scientific or medical research is ever likely to pay for it. Thus, there *may* be some reduction in cancer risk as a result of ingesting certain pure chemicals, but at the moment, we don't have evidence to demonstrate it.

Methods for gathering and evaluating data from clinical tests.

Many doctors, scientists, and pharmaceutical companies have ideas for ways to reduce cancer risk or drugs and procedures for treating cancer. For example, a researcher might have invented a new procedure for treating early-stage prostate cancer and want to evaluate the effectiveness of this treatment for reducing the risk of serious, metastatic disease. Cancer biology and cancer medicine are filled with situations where the interested parties must find objective and reliable ways to evaluate their ideas and treatments. Previously, you have read about the importance of double-blind tests, where neither the patient nor the experimenter knows which individuals in a test population have received the experimental or the conventional treatment or perhaps a control, meaning a placebo chemical or a benign alternative treatment. Even when a double-blind test is used, however, there is still the question of whether the results support the idea that the treatment is worthwhile. In evaluating the test results, one must consider whether the data gleaned from the test demonstrate an effective treatment or not.

If a result is clear (for example, 90% of 1000 people in the experimental group got better versus only 10% of 1000 people in the control group), then fancy methods for evaluating the results are not necessary, though they are often done. More commonly, however, the result is something like 25% of the experimental group got better and 20% and the control group did, too. Then what can you conclude? Do the data indicate that the treatment is better than the control, or was the difference observed the kind that would have been likely to arise by chance? (Remember, tossing a coin might give you heads four times in a row, even though this result is unlikely.) Do the data suggest that the treatment is worth its cost? These are the questions that researchers conducting clinical tests must wrestle with.

To understand how evaluations of treatments are done, we need some definitions and ideas. First, most researchers doing this kind of work consider the possibilities in terms of hypotheses; for example, the supposition that treatment with a given drug at a particular dose for a given period of time will produce a positive effect, such as shrinking a tumor. An alternative is that the drug does nothing; this is called the **null hypothesis**. It is also possible, of course, that the drug will actually do harm, but usually a combination of preliminary work in the lab and initial clinical work, called a phase one trial, have shown that this is not the case for the doses being used, so researchers conducting subsequent clinical trials do not need to consider harm as an issue. Now the hypotheses are tested by seeing which is better supported by the data.

Clearly, a trial of this kind require at least two groups of subjects, one to receive the experimental treatment and one to get either the currently used treatment or a control. These groups are often called **cohorts**. For a test to be valid, a double-blind design is important, and the cohorts should be equivalent. If the members of one cohort are more prone to cancer than members of the other, then the results of a test on a potential cancer-preventing substance would obviously be misleading. One strategy for obtaining equivalent cohorts is to use all data available about the participants (their age, gender, eating and smoking habits, and so forth) to distribute equal numbers of equivalent individuals into each group. The problem with this apparently rational strategy is that it assumes one knows the factors that are important to consider in distributing the subjects equally. Commonly, this is not a good assumption, because our understanding of all the issues relevant to the test is limited, no matter how much we know about the participants.

An alternative approach is to assign individuals to the cohorts randomly. For example, as each individual enrolls in the study, she or he is assigned a number taken sequentially from a list of numbers that are truly random. (Computers can generate such lists.) When it comes time to assign individuals to cohort A or B, one can put all people with odd numbers into one cohort and those with even numbers into the other. If the cohorts are large, the number of odd and even numbers assigned should be almost the same (by symmetry and the properties of large numbers, as previously discussed). The assignments to cohorts have now been made with no bias in age, gender, eating habits, or anything else because cohort assignment was done at random. This approach is described as use of **randomized samples**. The strength of this approach is its lack of bias. Its principal limitation is that if the numbers of individuals in the cohorts are not large, the groups may not be equivalent. Remember that it is not uncommon to get all heads when you toss a coin only four times. This is why many reliable studies involve thousands of individuals, and it is one reason that responsible testing is expensive.

With subjects assigned to cohorts randomly and with a double-blind test design, we are on our way to an unbiased assessment of a drug or protocol. We can collect data—for example, on the fraction of people in each group who do or do not get cancer, or the fraction of people still alive after some period of treatment—but how do we learn whether our results show a meaningful difference? The graphs commonly used to describe the resulting data are the Kaplan–Meier graphs presented in previous chapters. This form of presentation is useful and informative because it deals in a simple way with the issue of subjects who leave the test for some reason not related to the test itself (die of some other cause, move away, or withdraw from the test for any number of reasons). All one needs to do is to calculate the fraction of people who are still cancer free or still alive, using the number of people still present in the cohort as the denominator of the fraction.

We still need, however, to assess the likelihood that any differences we see might have arisen by chance, not as a result of our treatment. This issue is more difficult. To treat it well, some mathematics is necessary, so we will leave the subject for those of you who want to take a course on the statistical analysis of data. In overview, tests for the significance of a result look at the number of people in each cohort and the size of the difference between the cohorts at each time in the study. They then use the theories of probability to estimate the likelihood that such differences would have arisen by chance. The greater the differences observed and the greater the numbers of people in each cohort, the greater the likelihood that the null hypothesis is wrong and the interesting hypothesis is correct. For example, when the likelihood of the null hypothesis is low (say, <1%), the value for the treatment is high, and the process or drug is validated. If the likelihood of a null hypothesis is >1% but <10%, most scientists would still see the result as significant, validating the process or drug. If the probability of the null hypothesis is greater than 10%, the result is usually deemed more likely to be due to chance and thus less convincing. In this case, the process or drug has not been validated by the data.

Some countries are combining the most promising of their traditional remedies with the methods of modern clinical investigation.

Many people recognize the value of cancer prevention versus cancer treatment, so scientists all over the world are using traditional knowledge and/or modern science to seek foods and other substances that are effective in reducing cancer risk. This quest turns out to be difficult, in part because of the complexity of carcinogenesis and in part because of the complexity of valid testing, as just described. Nonetheless, there is some progress to report. A study in China has sought to evaluate the role of green tea in cancer prevention. More than 18,000 people were subjects, and the results showed that green tea drinkers were about half as likely as others to develop stomach or esophageal cancer as people in the control group. This sounds very encouraging, but a study carried out in the Netherlands got potentially different results. This study sought a link between the consumption of black tea and the subsequent incidence of stomach, colorectal, lung, and breast cancer. The study involved 58,279 men and 62,573 women, ages 55–69, and it took into account such factors as smoking and overall diet in establishing the cohorts. It found no link between tea consumption and protection against cancer.

Perhaps the difference in the results of these two studies is green versus black tea? Research at the U.S. National Cancer Institute suggests not. One of their studies looked at the therapeutic value of green tea among prostate cancer patients. Forty-two patients drank about 4 cups of green tea daily for 4 months. Only one patient experienced a short-lived improvement, and nearly 70% of the group experienced unpleasant side effects, such as nausea and diarrhea. The study concluded that drinking green tea had limited anti-tumor benefit for prostate cancer patients. Note, however, that this was asking green tea to *cure* cancer, as opposed to preventing it, and these are very different tasks. Ongoing studies are testing the components of green tea as preventive agents against skin cancer. One study is investigating the protective effects of green tea compounds, in pill form, against sun-induced skin damage; another is exploring the topical application of green tea compounds in shrinking precancerous skin changes. Thus, many scientists are interested in the possibility that traditional medicines or forms of food can have a significant anti-cancer effect, and reliable results are bound to come along soon.

Medical scientists in the United States are also looking for cancer-reducing diets and chemicals.

A nice example of both the promise and difficulty of using natural products in cancer prevention is found in studies that have revealed the capacity of broccoli sprouts to protect mice against the action of a chem-

SIDE QUESTION 11.6

What are the differences between testing a procedure or chemical for its value in reducing cancer risk versus its effectiveness as a cure of an already developed cancer? Think this through carefully, because it is a very important question.

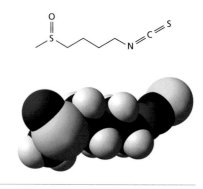

FIGURE 11.8 The structure of sulforaphane.
The structure of sulforaphane is shown with two
different representations of its atoms. (From Erkoç S
& Erkoç F [2005] *J Mol Struct: THEOCHEM* 714:
81–85. With permission from Elsevier. Structure
calculated in Spartan '04 Student Edition using
Hartree-Fock/6-31G*. Image generated in Accelrys
DS Visualizer.)

ical carcinogen. Chemists at Johns Hopkins School of Medicine, led by
Paul Talalay, were applying carcinogens to the skin of laboratory mice and
looking for foods that would reduce the subjects' subsequent frequency
of cancer. They succeeded with extracts from cruciferous vegetables,
most notably broccoli. Broccoli sprouts were even better than broccoli
itself. A chemical subsequently purified from the sprouts had all the beneficial activity of the sprouts themselves; its structure was determined,
identifying it as a molecule called sulforaphane (FIGURE 11.8). Even a
synthetic version of sulforaphane worked to protect mice against the skin
carcinogen, so a molecule with this structure is a cancer preventer, under
some circumstances.

The Johns Hopkins group and scientists in other labs have been trying
to figure out how this chemical works to prevent cancer, but the results
have turned out to be complicated. Sulforaphane inhibits the division of
human cells grown in culture, at least in part by interfering with formation of the mitotic spindle. Sulforaphane is also an antioxidant, so it might
be working by reducing ROS. Moreover, it is recognized by the liver as a
chemical likely to be damaging to cells, so liver cells turn on the synthesis
of enzymes that can interact with foreign substances and change their
chemistry. Sulforaphane may, therefore, be acting indirectly in the experimental mice, reducing ROS and inducing their cells to protect themselves
against the applied carcinogen. This interpretation suggests that sulforaphane would only reduce cancer risk if you were being treated at the same
time with certain chemical carcinogens. In sum, taking sulforaphane as a
drug or a dietary supplement is not likely, on current evidence, to reduce
your general cancer risk. It may be, however, that the ability of this phyto-
chemical to induce our body's chemical protection systems is part of why
a vegetarian diet is healthy; it stimulates your liver's enzymes to protect
other cells in the body from toxins, including carcinogens. Note that if
you look up sulforaphane on the Web, you will find that it has now been
studied in many labs. It is thought by some researchers to work by activating specific transcription factors. Other investigators have found that
it alters the histones that bind DNA, and still others have shown that it
alters the activity of MAP kinase. In short, understanding what specific
chemicals do in the context of a human body is a very challenging subject!

Butylated hydroxytoluene, usually called BHT, is an antioxidant widely
used as a food preservative. Interestingly, it too is a good inducer of the
liver's protective enzymes. Thus, not all chemicals that have unanticipated beneficial effects come from plants and other natural sources; we
may be getting some anti-cancer benefit from unexpected sources, like
chemical additives to our food!

Summarizing the above statements about food, dietary supplements, and
cancer, no single substance has yet been demonstrated to reduce the
everyday cancer risk of people, even though there are many that are
claimed to do so. Some classes of chemicals, like antioxidants, may reduce
the harmful effects of naturally occurring carcinogens, like ROS, but even
here, where the idea is highly plausible, evidence is generally lacking.
Taking such substances in modest quantities is not likely to harm you, but
it may not do you much good either. However, eating a diet that includes
lots of fruits and vegetables is very likely to be good for you in more ways
than simply reducing your cancer risk.

BEHAVIOR AND CANCER RISK

There are some behaviors that dramatically increase cancer risk, like
smoking cigarettes. Other behaviors increase risk in a demonstrable but

less extreme way, such as becoming obese or experiencing significant stress, as discussed below. This poses the question, Are there also behaviors that can reduce cancer risk? Fortunately, the answer is yes.

Exercise appears to reduce cancer risk.

Studies have compared the incidence of breast cancer in women who either did or did not do sports in college. The results showed clearly that women who were athletic during their youth had a lower incidence of breast cancer later in life. Is this correlation or causality? Perhaps women who do athletics in college are healthier than those who do not, and it is their good general health that gives them a reduced cancer risk. These are the kinds of factors that make it hard to pinpoint cancer causes through studies of populations, but such work does draw our attention to potential cancer risk factors that can be tested by experiment. Indeed, several scientists are now investigating the relationship between exercise and cancer risk. The data show clearly that repeated, mild exercise over a long period reduces the risk of breast and colon cancer. For other cancers, the evidence is less complete.

Is there any way to understand this result, given what we know about cancer initiation and progression? One plausible idea is that when you exercise, you breathe hard to get enough oxygen to keep your muscles working properly. Extra oxygen going into our cells and mitochondria means more ROS, which should make trouble and actually increase cancer risk, but it does not. This suggests that our cells have found ways to deal with the extra ROS. When we make ROS through natural behaviors, doing things that animals have been doing throughout their existence on planet Earth (running around to catch food or avoid being eaten, for example), the body's systems for maintaining internal balance are activated, and perhaps they more than compensate for the additional ROS. This idea suggests that even though exercise leads our bodies to make more ROS, the concentration of ROS in our cells is reduced, thanks to the cell's natural compensation systems. In any event, getting regular exercise seems to be good for reducing cancer risk, and it may also help to prevent other aspects of aging.

Stress could be a factor in cancer.

Some people worry that the stress of their lives is increasing their cancer risk, and some data support this concern. There is a twofold increase in the incidence of breast cancer among women who have recently suffered a disruption of their marriage through divorce or the death of their spouse, which is consistent with the idea that stress can be a factor in cancer biology. Clinical studies have shown effects of stress on the rates of wound healing and the risk of both heart problems and infection, so why not cancer? These issues are reviewed in an article cited in Further Readings.

Just how stress affects biological functions is not yet well understood. Several hypothetical pathways that link the impact of stress to levels of signaling molecules in humans are diagrammed in FIGURE 11.9. These pathways are, however, so complex that it is very hard to get convincing data about relationships, even correlations. It is, therefore, currently impossible to give a clear description of how stress might work to enhance cancer. It seems unlikely that stress is mutagenic, but stress does change the levels of numerous hormones. Altering the concentrations of these signaling molecules probably affects the ways in which cells respond to growth factors and other biological signals. Stress may even have an influence on the repair of damaged DNA, the development of local blood supplies, and the ways our cells cope with ROS.

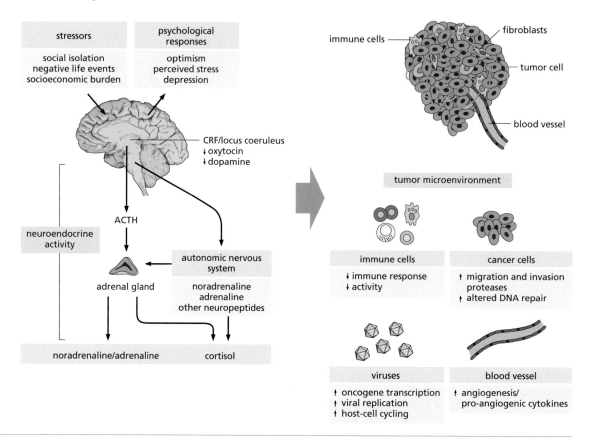

FIGURE 11.9 Possible effects of stress-associated factors on a tumor's microenvironment. The body's responses to stress involve the perception of a threat by the central nervous system and subsequent activation of both the body's hormonal systems and the autonomic nervous system (the part of your nervous system that functions automatically and over which you have very little control). Stress hormones (the numerous chemicals named in the figure, such as adrenaline, noradrenaline, oxytocin, dopamine, and cortisol) can be released from the adrenal gland, brain, and some of the body's nerve terminals. These chemicals can modulate the activity of multiple components of a tumor's microenvironment. Effects may include promotion of tumor-cell growth, increased capacity for migration and invasion, and stimulation of angiogenesis by inducing the production of pro-angiogenic cytokines. Stress hormones could also activate oncogenic viruses and alter several aspects of immune function, including antibody production, cytokine production, and cell trafficking. Collectively, these downstream effects could contribute to an environment that is permissive for tumor initiation, growth, and progression. ACTH, adrenocorticotropic hormone; CRF, corticotropin-releasing factor. (Adapted from Antoni MH, Lutgendorf SK, Cole SW et al [2006] *Nat Rev Cancer* 6:240–248. With permission from Springer Nature.)

Stress may also have an impact on cancer incidence through its action on our immune systems. Many of us have experienced the ability to keep working intensely for a prolonged period with no distracting illness, but immediately upon finishing the job, we become sick. Thus, one can imagine that stress and/or the end of stress affects the ability of our immune system to identify rogue cells early in their life and kill them before they develop into full-blown cancer. There are indications that stress plays a bigger role in cancer progression than in cancer incidence, but much remains to be learned about this complex issue. Scientists are working to understand such things; an article that describes one such project is cited in Further Readings. It is unlikely, though, that we will have clear answers about these issues any time soon.

MEDICAL EFFORTS TO REDUCE CANCER RISK AND PROBLEMS WITH THEIR IMPLEMENTATION

Although the majority of cancer-related medical research has been devoted to the discovery and optimization of cancer treatments, there are two kinds of work whose results have reduced cancer incidence: methods for early diagnosis and vaccines. These issues were discussed in previous chapters, but they are reconsidered here in the context of actions one can undertake to reduce personal cancer risk and the difficulties of achieving risk reduction.

Timely cancer screening has value but is not always available.

Chapter 2 includes a quite extensive account of the value of screening for some kinds of cancer, given that early detection is important for reducing the risk of metastasis. At several places in this book you have read about the potential dangers and costs of excessive screening, but screening at levels recommended by a good doctor is definitely a sensible course of action. However, many people throughout the world do not have the possibility of cancer screening, either because the necessary facilities are not available or because cost puts such examinations out of reach. This is part of a very important aspect of cancer medicine: the disparity in access to treatment that is faced by people with different levels of wealth and education. This problem is pervasive, even in wealthy countries, like the United States, because access to health care is so different, depending in part on insurance coverage and place of residence. Some health providers are working to minimize this disparity, and the American Cancer Society is seeking paths that might eliminate it. But even with recent efforts to improve access to health insurance in the United States, there are still very significant differences in the chances a cancer will be caught early and treated effectively, based on patient wealth and education. Sidebars 2.1, 8.3, and 8.4, describing patient Woods' experience with the diagnosis and treatment of his bladder cancer, are examples of this problem. Finding a solution to this pervasive problem should command attention and effort from us all.

There are vaccines that reduce the risk of certain cancers.

Cervical cancer, which is promoted by human papilloma virus (HPV), is the best example of a cancer where vaccine intervention is causing a significant reduction in the incidence of a specific cancer. As described in Chapters 5 and 6, HPV is a sexually transmitted disease, and the virus is oncogenic. Vaccination that enhances a person's immune resistance to this virus has been demonstrated to reduce subsequent risk of cervical cancer. The effect is strong enough that it is now common to inoculate both girls and boys before they become sexually active. The results are most encouraging, even though a vaccine that is effective against all types of HPV is not yet available. This situation is an excellent example of prevention being a sound strategy for avoiding the ravages of cancer.

Research has also shown that cancer can follow infections by the liver-infecting hepatitis B virus (HBV) or hepatitis C virus (HCV). There is an excellent vaccine to prevent disease from HBV (and the related, noncarcinogenic hepatitis A virus, HAV). Indeed, a program of universal HBV vaccination for infants and children has been active in the United States since 1982. HCV, on the other hand, has the ability to mutate the gene encoding its coat protein rather quickly, so no useful immunization to prevent this infection is yet available. Both HBV and HCV are commonly transmitted through routes like those that transmit the AIDS virus: unprotected sex and sharing of needles for drug injection, so prevention of liver cancer from HCV is partly a matter of avoiding risky behaviors and partly an issue of getting the treatment for this infection that is now available.

Several biomedical groups are working on more general cancer vaccines, but there is nothing very promising on the immediate horizon, and there are no pills that prevent the accumulation of mutations. Cancer risk is not a situation that a clever doctor can simply fix.

SUMMARY

Since cancer results from the accumulation of certain mutations, an effective way to avoid cancer would be to avoid mutations. Unfortunately, this is impossible. Some level of mutation is characteristic of the DNA replication and repair that are essential for cell growth and division, so some level of mutation is a consequence simply of being alive. Furthermore, everyone inherits genes that are important for the processes that govern the faithful transmission of genetic information from one cell generation to the next. The alleles you happened to receive from your parents may not be the best for accurate maintenance of your genome. However, we all must live with the genes we inherit, so some aspects of your personal cancer risk are defined by your heredity.

Nonetheless, you can reduce your cancer risk by minimizing your exposure to factors that increase the likelihood of mutation. Ionizing radiation should be avoided when possible, though some level of exposure is acceptable; for example, the occasional use of X-rays recommended by a responsible physician, thanks to the effectiveness of our cellular systems for DNA repair. Ultraviolet light, too, is mutagenic, but with appropriate protection, being outside is not a danger. Because sunlight generates vitamin D from molecules we eat or can make, complete sunscreen protection is a bad idea. A sensible balance of competing environmental factors is often the mindset needed to treat the issues of carcinogenesis with wisdom.

Chemical carcinogens in the environment are now a minor problem for most of us, in large part because science and technology have made possible both better identification of those chemicals and better procedures for limiting our exposure to them. The major exception is tobacco, which persists as a serious public health problem, even though an inexpensive and effective solution is immediately available through changes in human behavior (stop smoking!). Carcinogenic molecules in some components of food or food additives are of much lesser importance, but a few things or processes should be minimized, for example, frequent consumption of meat cooked at high temperature. Indeed, a vegetarian diet is likely to be valuable in reducing cancer risk, both because of reduced meat intake and because of some of the chemicals found in plants. There is no hard evidence that the consumption of particular chemicals thought to be cancer-inhibiting, like antioxidants, will actually reduce cancer risk, but the studies so far conducted may not have been the right ones for identifying effects that are detectable only over a long period of treatment. It is likely that a diet low in meat and high in fruits and vegetables, combined with regular mild exercise, will reduce your cancer risk and help to prevent obesity. Most doctors will recommend the same sorts of activities for many reasons, but prevention of cancer can go onto the list.

The problem with these recommendations is that they sound like advice from a grandparent: old-fashioned, boring, and no fun. That does not, however, make them bad advice. Thinking through what they mean to you now, whatever your age and situation, and thinking seriously about getting a diagnosis of cancer, are good ways of seeing what the stakes and the costs really are. It would probably not be wise to try to take on many lifestyle changes all at once, but if you can work your way toward the kind of lifestyle suggested, it will certainly reduce your lifetime risk of cancer, and it may even help you to feel better along the way.

It is important to add that the courses of action available may depend on economic factors that are not under your control. In some neighborhoods, there are no stores that sell fruits and vegetables, and in urban areas, fast food that is rich in fat and low in vegetables seems to be everywhere. Moreover, it may be unsafe to exercise outdoors in ways that dwarf the risk of getting cancer. It may also be that all your friends are behaving in ways that you know include self-destructive activities, but if you behave differently than them, the social costs will be high. In the face of these kinds of constraints, you may have to go slowly in changing your patterns of living. But do keep all that you have learned about carcinogenesis in mind; it might allow you to enjoy lifelong health.

ESSENTIAL CONCEPTS

- Each round of DNA replication includes some chance of a carcinogenic mutation, so some cancer risk is unavoidable; it is simply a result of being alive.

- Because some tissues go through more cell divisions than others, the risk of cancer varies from one tissue to another. Probably, however, there are additional factors that affect tissue-specific cancer incidence, such as the complex action of sex hormones (for example, testosterone and estrogen).

- Inheritance of loss-of-function alleles of tumor-suppressing genes contributes significantly to cancer risk, but inheritance of good alleles in these and other genes may be an important part of why some people with dangerous habits, like smoking tobacco, never get cancer.

- Environment plays a significant role in cancer, and many of the relevant environmental factors are under your control: smoking, contact with other proven carcinogens, exposure to ionizing radiation and ultraviolet light, diet, and exercise are all things you can manage in such a way as to reduce your personal cancer risk.

KEY TERMS

adipocytes	direct carcinogen	probability
body mass index	indirect carcinogen	prognosis
cancer risk	inherited cancer risk	randomized samples
cohort	null hypothesis	symmetry
clinical trial	phytochemicals	

FURTHER READINGS

Luzzatto L & Pandolfi PP (2015) Causality and chance in the development of cancer. *N Engl J Med* 373:84–88.

The National Cancer Institute's branch on cancer prevention has a Website at http://prevention.cancer.gov/.

Ames BN & Gold LS (1997) Environmental pollution, pesticides, and the prevention of cancer: Misconceptions. *FASEB J* 11:1041–1052; http://www.fasebj.org/content/11/13/1041.long.

Thompson HJ (2006) Pre-clinical investigations of physical activity and cancer: a brief review and analysis. *Carcinogenesis* 27:1946–1949.

Ji LL (2002) Exercise-induced modulation of antioxidant defense. *Ann NY Acad Sci* 959:82–92.

Schiller JT & Lowy DR (2012) Understanding and learning from the success of prophylactic human papillomavirus vaccines. *Nat Rev Microbiol* 10:681–692.

Gersch ED, Gissmann L & Garcea RL (2012) New approaches to prophylactic human papillomavirus vaccines for cervical cancer prevention. *Antiviral Ther* 17:425–434 .

Armaiz-Pena GN, Lutgendorf SK, Cole SW & Sood AK (2009) Neuroendocrine modulation of cancer progression. *Brain, Behav, Immun* 23:10–15.

A reliable Website that describes clinical trials now ongoing to evaluate the impact of inheritance in the emergence of breast cancer is Search the Studies: Find NIH Clinical Center Trials at http://clinicalstudies.info.nih.gov/search.html.

A report on the importance of nutrition and exercise by the World Cancer Research Fund and the American Institute for Cancer Research (2007) is available at http://www.aicr.org/assets/docs/pdf/reports/Second_Expert_Report.pdf.

Antoni MH, Lutgendorf SK, Cole SW et al (2006) The influence of bio-behavioural factors on tumour biology: pathways and mechanisms. *Nat Rev Cancer* 6:240–248. This is a review article on biological behaviors and cancer incidence.

An opinion piece on the health value of vegetarianism is available at http://www.vegsource.com/harris/cancer_vegdiet.htm.

Gill SS & Tuteja N (2010) Reactive oxygen species and antioxidant machinery in abiotic stress tolerance in crop plants. *Plant Physiol Biochem* 48:909–930. This is a description of antioxidants in plant cells.

QUESTIONS FOR FURTHER THOUGHT

1. There are clearly people who engage in risky behavior, such as smoking (pack-a-day regulars), who don't come down with lung cancer. You hypothesize that these people have unusually effective enzymes for DNA repair. What strategy would you pursue to test your hypothesis?

2. A college friend has become a real couch potato. You are worried that his behavior will lead to serious health problems in the future. What arguments would you marshal to try to convince him to eat less and better, as well as to get more exercise?

3. An estimated 5% of human cancers are caused by HPV infections, and most of these are cancers of the cervix. Worldwide, about 500,000 new cases of cervical cancer are diagnosed each year—80% in developing countries—resulting in more than 250,000 deaths. Two HPV vaccines that target the two most oncogenic virus types, HPV16 and HPV18, are now commercially available. In controlled clinical trials, these vaccines have proven to be effective at preventing infection and the associated cancer that is induced by these virus types. If you were investing American taxpayer's money in public health, what fraction of your cancer prevention budget would you be willing to invest in a worldwide immunization program? In a U.S.-specific immunization program? Describe your reasoning.

ANSWERS TO SIDE QUESTIONS

1. This question is difficult, thought-provoking, and has no right answer. It asks you to draw on your new scientific knowledge about cancer, about inheritance of cancer-related genes, and about the relationships between them in the tissues important for a woman's ability to bear children. Issues to think about include her genetics (is she homozygous or heterozygous for the cancer-causing allele?), her interest in having children, whether she is pregnant now, and if so, whether she has consulted a doctor about the sex of the child (if it's a boy, inheritance of the *BRCA1* gene is less threatening for his future cancer risk). How many children she is thinking of having, and how does she think and feel about operations that would remove the tissues that might put her at-risk for a serious cancer? Has she considered in vitro fertilization, and would she think it moral to test the cells of embryos for their genotypes? Could she afford such treatments?

2. This is another question with no right answer, but it is an interesting one to consider in the context of the relationship between preventive medicine and health care costs. You could use your knowledge about cancer to identify a group that is at a sufficiently high risk for prostate cancer (say, obese men 60 years and older) and propose a test that would ask the question, Does screening in this group identify cancers whose early treatments prolong life in a demonstrable way? There are other sensible proposals you could make, but they all require thought.

3. People smoke tobacco for a wide range of social and physiological reasons. These include desire for a particular self-image, peer pressure, responsiveness to advertising, enjoyment of the slight high that comes from nicotine, and sometimes an enjoyment of tobacco's taste. One can hope that education about the dangers of smoking will be helpful in reducing the habit, but experience over the last 20 years or so suggest that it will take more than facts to alter people's behavior. Since the 1920s, coffin nails has been a common slang name for cigarettes, and yet usage has increased very significantly since then. So how would you reduce tobacco use? Would you propose higher taxes on cigarettes, or would you simply let people smoke without cost penalties? Would you make it impossible for people who smoke to buy health insurance? Would you make them pay higher premiums? Interesting to think about.

4. Three possible factors to consider are (1) the magnitude of the continued need for this mineral in the face of many substitutes that have been identified, (2) the safety measures that could be introduced at an acceptable cost to assure a healthy working environment, and (3) the procedures now available for early diagnosis that might reduce the medical hazards of some asbestos exposure. Can you think of others?

5. Correlation means a mutual relationship between two or more things. Causality means that one thing is a result of the other, which is a very specific kind of relationship. Correlation means any kind of relationship, whereas causality means that when A happens, B will result. If two correlated issues can be studied by experiment, then you can vary A and see what happens to B, and vice versa. From these kinds of observation, it is often possible to determine whether A and B are causally related.

6. Reducing cancer risk can be achieved by reducing the effectiveness of a carcinogen or tumor promoter. This can be accomplished by removing the trouble-causing factor or by countering its activity. For example, if the carcinogen is producing ROS, providing antioxidants that reduce ROS will lower the strength of the carcinogenic activity and reduce the likelihood that it will lead to cancer. If cancer is already established, mutations have occurred, cells are now misbehaving, and metastasis may even have occurred. Reducing the rate of mutation or of tumor promotion at this point is like the proverbial problem of closing the barn door after the horse has already escaped. Instead, one must find ways to kill or stop cell division in the rogue cells, a process that is far more difficult than reducing carcinogenesis. You are trying to cure cancer rather than prevent it.

Living with cancer

12

Living with cancer includes many complexities. Foremost is coping with the disease itself, a burden that differs greatly depending on the kind of cancer, its stage of progression, and the patient's response to it. Second is the issue of patient care. Even if the cancer is advanced, both patient and family may prefer to arrange care at home, although medical problems make this impossible in some cases. Providing home care for someone who is increasingly ill is a significant amount of work, so the people giving this care are often heavily burdened by the chores that must be done. Thus, caring for the caregivers can also become a significant issue. A companion issue of great importance is cost. Some of the medicines and/or treatments for cancer are ferociously expensive, and if solid medical insurance is not available, this burden can be crippling. On top of these practical matters, there are the responses of everyone concerned to a very serious disease that can be life-threatening. Such responses are highly personal, so it is often hard to find useful things to say about them. In what follows, I, the author of this book, am trying to lay out the objective problems in an informative way. I also put into words a number of opinions and ideas I encountered through living with the cancer death of my son, the cancer experiences of friends, and the views I heard in interviews with additional cancer patients and their families. The fact that I am here, writing this book, means that in some significant way I cannot understand the full depth of the cancer problem, but I have seen it from close at hand. I hope that what I say will help to promote your own thinking and understanding of how people cope with this disease.

POINT OF VIEW

A FOCUS ON LIFE, NOT DEATH

SEEING CANCER TREATMENTS FOR WHAT THEY ARE

REMISSION AND THE RENEWAL OF HOPE

COPING WITH PAIN

DEALING WITH PHYSICAL WEAKNESS

CHOOSING AN ENVIRONMENT

CARING FOR CANCER CAREGIVERS

LEARNING GOALS

1. When a patient is terminally ill with cancer, there are often additional treatments that medical science can give to prolong life. Many of these treatments have costs, not only in fees to the doctors and hospitals but also in their negative effects on quality of life. Identify two treatments of this kind, and make up your own table of the pluses and minuses about receiving such medical attention.

2. A prognosis of death in a relatively short time often makes people think about what they really value in life. Imagine that you have been given 6 months to live. Now, identify the things you would want to do and not do during that time.

3. Compare the care offered to a terminally ill patient by a good-quality hospital and a good hospice program.

4. Look up the laws concerning medically assisted suicide in the state where you live. Do the same for the state of Oregon, which was the first state to allow such actions. Compare the two sets of laws, and make a decision on which set of laws you prefer. Describe the reasons for your opinion.

POINT OF VIEW

People's attitudes in the wake of a cancer diagnosis differ widely. It is common to feel despair and to see this moment as the beginning of life's end. Many questions are sure to come to mind, such as Why me?, Why now?, How will I cope?, and How will my family cope? There are themes that run through such feelings and questions, giving them a kind of unity. The foremost issue to define is the severity of one's bout with cancer, based on the prognoses given by a responsible doctor, both with and without recommended treatment. If you receive the suggested treatment and all has gone very well, it may be that a cure has been accomplished, and this is wonderful news. For these lucky people, cancer can become a thing of the past. For some individuals in this position, cancer is considered no more attentively than by most of us who have never had the disease. Other people are quite changed by their experience. A common response in this group is to devote a large amount of time and energy to raising money for important cancer causes, a generous response. I won't say more about these kinds of happy results because they speak for themselves.

For people who are less fortunate, the time ahead can seem dark. Death is one of the very few certainties of life. The relevant questions we can rarely answer, however, are how and when. Modern Americans seem to be better than any other people who have ever lived at delaying and ignoring death. This luxury probably comes both from our generally high standard of living and from our not having had a major war on our own soil for over 150 years. Perhaps as a result, we have developed a youth-oriented culture that considers old age and/or infirmity to be a bit obscene, something that should be hidden. Many people hide age with cosmetics, operations, or drugs; others keep the elderly out of sight by arranging institutional care. Thus, proximity to death is foreign to many of us, and we don't know how to deal with it.

Over recorded history, and even well before it, most societies have relied on the disciplines and frameworks of organized religions to help deal with birth, emergence from childhood, marriage, and death. For some people, the tenets of their religion give life meaning. However, many citizens of the United States no longer place the same reliance on religion as their ancestors did, and these people are left coping with life's major transitions more or less on their own. Something like a life-threatening disease can make that independence feel very lonely, urging one back towards traditional ways of understanding, particularly at life's last transition. One of our interviewees, however, found the approach of his death comparatively easy to accept. Woods, the homeless man, was remarkably thoughtful and well-read (**SIDEBAR 12.1**).

I will say very little about religion as a factor in living with cancer because it is too personal an issue for general treatment; each person will have his or her own views. For everyone, though, the experience of having to look squarely at the reality of one's own death, or the death of a loved one, provokes deep thought. It may be depressing, but it can also be a powerful encouragement to think about the things one values most deeply; if time is limited, it is not to be wasted. This realization gives some people who are living with cancer an amazing joy in living. From my perspective,

SIDE QUESTION 12.1

What are the things you would want to know if you were diagnosed with a serious cancer? Describe them and rank them by their importance.

CASE STUDY: Woods

Woods described his feelings about dying from cancer in a remarkably clear and balanced way:

"I didn't have as much trouble dealing with the reality of cancer as many people do. I've spent many years preparing for my own death. It occurred to me, really when I was a small child, that the time to reconcile yourself to death is as you go through life. I've been thinking about this through reading world theology for a long time. About 25 years ago, I came to terms with my own death, and ever since then it hasn't bothered me. There are some things in life that you simply have to accept; railing against them is just an exercise in futility. When I got to the moment of that cancer diagnosis, the work I'd been doing for the past 25 years really helped me. I could say to myself and mean it, OK, this is the problem, what are the potential solutions to it? You go through it and hope for the best. That's all you can do. However, I'm quite convinced by people I have known and things I have read that you have to keep a positive mental outlook. If you don't, cancer will kill you, and it'll kill you in real short order."

this is an admirable approach to take. All of us live with the prospect of a limited future, because life is a fatal condition. Most of us ignore our ultimate death and think only about getting on with life. When death is brought unavoidably to our attention, it can be life-affirming by helping us to appreciate what we have. This can be a positive aspect of dealing with a cancer diagnosis.

On the other hand, it is almost impossible not to feel the darkness of life's transience. This issue calls to mind many questions that have occupied and puzzled serious thinkers for time out of mind. Why was I born, and why am I here? Where am I going? What happens after death? In short, what is the meaning of life? These questions can take many forms, some of which sound profound and some mundane. However, they are all important, and they are worth thinking and talking about. Although answers may be elusive, just the act of thinking about them is an affirmation of humanity and of the importance of being alive. The great French philosopher René Descartes said, "I think, therefore I am." A consideration of these fundamental issues is one of the potential values of coping with a cancer diagnosis. If one were not confronted with one's own mortality, one might never pay attention to such deeply human questions.

My own views on such matters have been shaped by my training as a biologist. I see death as an essential part of life and the cycles upon which life is based. Obviously, one hopes for the chance to lead a long and fulfilling life, but death must come, and its inevitability must be accepted. Imagine what the world would be like if people never died!

A FOCUS ON LIFE, NOT DEATH

If one's time is limited, it makes sense to think about the life one has, not the death that will come to everyone. What are the things that you want most to do, or do again? Are there people you want to see, places you want to visit or revisit? Are there things you always wish you had done? These thoughts and their answers may make a strong, positive difference to you and the people you care about, so they merit immediate and focused thought. John, the roofer, was an extraordinarily insightful man, so his responses to these sorts of questions are well worth reading (SIDEBAR 12.2).

A problem for many people living with cancer is a sense of loneliness. It is common that some friends start to avoid you, fearing to let your illness into their lives. Even when there are friends or love ones around to talk

SIDEBAR 12.2

CASE STUDY: John

Part of John's strength in dealing with his cancer diagnosis came from his father, who was also in construction and set a great example for leading life with an open mind and an eye on what was going on. His dad used to help him figure out how to do jobs in the way that was the easiest, but would really work. This was the attitude that John brought to cancer. It helped him listen to what the doctors said, ask questions when he didn't understand, and then make up his own mind about the best way to go forward. He knew he had made lots of mistakes, but he had learned from most of them and had no regrets. As he said, "Lots of problems are bumps in the road, but you just keep going. Don't worry about it, just get on with it."

Many people told John that he should join a support group, but he always said no. "I have terminal cancer, and I know I'm going to die. So what I want to do is lead my life as best I can and enjoy what I can. As long as I keep a healthy attitude, I figure I'll be OK. I figure that the mental part is about 50%, the physical part is about 50%, and you can tip the scales with the way you approach things. The natural action of the tumor is going to take over eventually, but until then, the way you think about things makes a big difference."

When asked about how he made decisions, John said,

"In fact, the decisions have been easy, because there weren't many options. The hardest part of all this has been telling other people what was going on. I've never asked Why me? because it doesn't matter. You throw away all the garbage and then you get back to living. I wake up in the morning, and I think, I'm alive! Lee [his wife] and I have been able to joke about it along the way, and that has made it OK.

When the doctors told me in the first place that I had 3–10 months, I said, We'll see about that; I'm going to push the envelope. My surgeon has made things as easy as he could. Right now, I'm going to see him next week, taking reports of all the tests and scans I've just had, and he's going to help me decide what to do next. He knows that I have already outlived the odds, and he wants to be helpful. That helps me feel good. He's also really honest, so I feel I can trust him. For example, he said, 'John, we could do an operation, and it might give you a year, but it's going to take you 5 months to recover from the knife. What do you want to do?' That's the kind of treatment I can deal with. If I can't get up and running pretty soon, I'm not going to do another surgery. When I first heard that I had cancer, I set myself a goal of 1 year. Now I'm thinking 5 years. I know this may not be realistic, but it helps me to keep going. I've already made it for 2 years since my initial diagnosis, but I know that I had the cancer for a long time before it was diagnosed. I'm really a 7- or 10-year survivor already, and so maybe I can push it farther.

The hard part, though, is the effect of all this on other people. It's my friends who have the trouble. I continue to joke about things, especially with the doctors. I say things like, What's the worst option? What am I going to do, die? They hate hearing that, but it keeps things in perspective. This kind of humor is my release. If I can't have fun out of it, I'm not interested. I get excited just talking about it. Hey, I'm fighting death, I'm beating death.

There are several things I really enjoy, that I'd like to do more of, like travel. But one of the problems with cancer is the expense. I'm not a rich man, and I'm balancing a bunch of problem: mortgage, credit card debt, medical expenses, and the costs of the few things I'd like to do while I'm still kicking. I'm very lucky, because I have people I really care about: Lee, five or six good friends. So in a way I'm very rich, but I have to be a bit careful about what I take on at this point. I don't want to leave a lot of debts behind me, but with a bit of care and some planning, I think it's all going to work out fine. Money comes and goes, but friendships last forever. I'm still in touch with friends I made 40 years ago, so that makes me rich, even though the economy is not doing good things to the money I've saved in IRAs. I'm on disability now, and it doesn't pay enough even for my mortgage payments. Because I own my own company I can call the shots there, and I haven't paid myself for over 2 years. I'm getting a bit in debt, but the life insurance policies I have will cover everything, so it will all work out in the long run.

I haven't had any big problem with pain. I get nicks and twinges, but not what you'd call pain. I haven't had any need for painkillers. But some of it is attitude, I think. On the phase 3 trial I went into, I was ranked 23 out of 24 people [in likelihood of success]. When I came off the trial, because one of my tumors was growing again, the doctor rated me number 1, just because of my attitude. One of the worries about the upcoming surgery, though, is that if things go wrong, I'd lose the action of my one remaining kidney, and then I'd have to be on dialysis for the rest of my life. I told my doctors, 'If that happens, don't let me wake up. If I'm on dialysis, I'll torture all you doctors for the rest of my life,' and I think they believed me."

CASE STUDY: Rosemary

Rosemary never wanted to go to a support group. She said, "All along, the doctors and nurses mentioned support groups that I could join if I wanted company, but I never did. I'm just not a group sort of person. When the chemotherapy was over, though, when my hair was beginning to grow back in, that was when I actually felt the worst. One day as I was seeing the oncologist, I mentioned that I was feeling down-hearted. She said 'Oh, I should have mentioned. We have some-body just down the hall who might be able to help.' This was a psychiatrist who was seeing lots of people with cancer, so I made an appointment and described my symptoms to her: everything I used to enjoy was no longer pleas-ant, or tasty, or desirable. It was as if a light in my life had been switched off. I felt as if the dumb-est kid in my classes was smarter than I was, and it took four times more effort than previously, just to get a lecture ready. The woman heard me out and then said, 'These are classic symptoms of depres-sion. We can fix this in a couple of months. If you combine some pills, like Prozac, with some "talk ther-apy," you will almost certainly feel better soon.' So I followed her advice, and sure enough, in a few months I really felt much better. I did think, though, that someone might have suggested this solu-tion sooner. About 80% of women who are treated for breast cancer have a similar reaction, but per-haps none of us asks for help in the right way.

I think the factors that combine to make these feeling of depression so severe include the obvious fear that comes with cancer, the prob-lems of an operation, however mild, and the impact of chemo-therapy. But they also probably include the impact of changing hormone levels. I had been taking hormone replacement therapy for a few years before this all hap-pened, so going off those hormone doses 'cold turkey' probably added to my blue moods. I can't say, of course, whether my feeling better was a result of the wisdom of the woman I talked with, a year's treat-ment with Prozac, or simply the fact that a year had gone by since my diagnosis. Probably it was a combination of all three. Well, before the end of the year, I was feeling pretty much myself again.

My only real regret is that I didn't realize sooner how bad I was going to feel as I was getting bet-ter; I waited too long to get some psychological treatment. That is something people who are learn-ing about cancer should know. Some books about cancer talk about how survivors are suffused with a sense of hope and joy, but I sure didn't feel that way, and many people I know didn't either. Just when you think you're sup-posed to be feeling better, you feel worse than you've ever felt, and that's hard. There seems to be very little in the popular literature that prepares you for that, and the doc-tors aren't very interested in it either. Given that it is depression, and that depression is treatable, people really ought to know about it, so they don't waste time in useless misery."

with, it can seem as if they just don't understand, which in turn makes you feel isolated. This situation has several solutions. The American Cancer Society and some other cancer-focused organizations have phone lines and support groups in many parts of the country; these can pro-vide information and companionship from people who are either cancer survivors themselves or have lived with cancer in other ways. Many communities have a cancer support center where volunteers and/or pro-fessionals have organized books, Websites, and other people to help by providing both information and companionship to anyone who wants it. Talking with concerned, well-informed people is heartening for many people, especially if the combination of cancer and its treatment are lead-ing to depression (SIDEBAR 12.3).

Perhaps the most effective way to focus on life is to do things you enjoy with people you like (SIDEBAR 12.4). The beauty available from nature, from music or other forms of art, from friends, or from an interest in books or films that dig deeply into something you care about are all ways to strengthen or perhaps rekindle an appreciation of life. These statements may sound like platitudes, but there is truth in them.

In spite of good attitudes and sound information, however, the toll that a progressing cancer takes on the patient's body can be severe. With time, these physical problems can weigh heavily on the way a person feels.

Describe and rank in order the things you would want to do if you had only a few months to live.

SIDEBAR 12.4

CASE STUDY: Rob

While Rob was still strong enough to get around, he and Suzy turned their attention to visiting the people they really wanted to see. They went to California to see Rob's brother and some friends. Thanks to the generous gift of frequent flier miles from a business colleague, they also went back to Europe to see places and people they loved. This included a trip to Venice to visit a college roommate, accompanied by the purchase of a spectacular pink tie and a natty felt hat, which decorated Rob's now bald head.

Over the next couple of months Rob received visits from many of his best friends. Though he was now housebound, their company was a source of deep pleasure. Together, they reminisced, talked about books, films, and music they had all enjoyed, and basked in the glow of affection. Suzy was remarkable in the care and love she brought to this time in Rob's life. Every day she thought of something new that Rob could either do alone or that they could do together. Meanwhile, Suzy and Rob's mother worked like troupers to maintain a household that ran smoothly and could support an invalid. I was in and out, spending some time with them outside New York and some time at my lab in Colorado. It was a time full of stress for us all (my blood pressure went through the roof), but because of the loving care that Rob got at home, there were also some very rich moments.

Toward the end of his life, Woods, who was handling his bladder cancer without enough money to make life easy, began to feel the weight of his situation. For him, learning to take things one day at a time was very important (SIDEBAR 12.5).

SIDEBAR 12.5

CASE STUDY: Woods

Only a few weeks before his death, Woods described his situation like this:

"In these last few weeks, my energy has gone down again. I'm not experiencing a lot of pain, but I do take methadone [an opiate similar to heroin, but less addictive], one pill in the morning and one in the evening, and that probably helps. It appears to me as if this tumor may actually be shrinking now. My oncologist said that might happen, because she said this tumor is not behaving like a normal metastatic tumor. Usually they form in some other organ, but this one in my groin seems to be just sitting there. She was pretty positive, given the situation, but I have recently gone onto hospice care [read more about hospice care below], because that's really the way the land lies. It may be that with the radiation and all, I'm actually buying some more time. You know, in a way, that's what this is all about; buying more time to be alive.

I have to say, though, that without the Affordable Care Act, I'd already be dead, and it would probably have been a very painful death. These medications you need are not cheap, you know, and living on $175/month, there are not many of those you can buy. It just astounds me that people don't get this around the country.

I've accepted for years that I was probably going to die of cancer. I have it on both sides of my family: cancer and Alzheimer's. Maybe cancer is the better choice. So in some ways, I'm really getting off easy. I'm getting first-class medical care. I'm now getting a compassionate allowance with my social security, so I'm making $721/month, which as far as I'm concerned makes me a bit of a high roller. I have to pay rent out of that, but that's a fair charge, $191/month. I can handle that.

You know, the single most important thing in living with cancer is maintaining a positive mental attitude, not letting your mind catastrophize the situation. If I let my mind do that, I'd be under ground in about 5 minutes. While I can't prove this, I've been very conscious of working at this ever since my diagnosis. I read a lot, and that has helped me tremendously. It's not easy to keep at it, but that and meditation are big things in my recovery. I try to do each of these things for a certain time each day. One of the things my urologist said to me was that I had to look at this like a triathlon. You have to take each stage, one at a time. It's really all about taking it one day at a time, which I have already practiced for well over 20 years. For me, that was really good advice. Be in the present. That's really an important thing about life. Yesterday's a canceled check, tomorrow's a promissory note, but today, today is cash."

SEEING CANCER TREATMENTS FOR WHAT THEY ARE

If a patient has received a clear diagnosis of metastatic cancer, there are usually several treatment options available. Chapter 10 described a few of the experimental treatments now under development or in clinical trials. Some people with a terminal cancer seek out such trials, partly because of the chance that the new method might be helpful and sometimes to make a contribution through offering themselves as a test specimen. Moreover, new treatments continue to be developed. Antibodies that suppress immune checkpoints were described in Chapter 10, but even more recent work has led to the development of procedures that engineer a patient's immune cells to express proteins that accomplish an analogous job. A chimeric antigen receptor (CAR) approach to enhancing the ability of a patient's T cells to combat a tumor was approved for treatment of acute lymphoblastic leukemia in children and young adults as of August 2017. This is an exciting advance, but the cost of the treatment is about half a million dollars, even more than the cost of a bone marrow transplant. Barriers like this put some treatments out of reach for many. This often means that more conventional treatments, like X-rays, are the treatments of choice. John the roofer experienced several treatments that were in clinical trials, but it was the persistent use of radiation that did the most to slow the growth of his metastases. After John's death, his wife, Lee, described the situation in **SIDEBAR 12.6**.

When you are living with cancer, many treatments are available to you.

All relevant scientific treatment options will probably be mentioned to you by an oncologist. There are, however, alternative treatment options, some of which were sketched in previous chapters, and more are available on the Websites of professional institutions and organizations, such as the U.S. National Library of Medicine, or from journals, such as *Alternative Therapies in Health and Medicine*. In fact, there are so many possible ways to treat cancer that you could spend a large amount of time sorting through them and deciding if any is right for you. You could also spend a lot of money trying them out. It is likely, though, that none will do much better than the best treatment that evidence-based medicine can offer. Note, however, that there have been positive effects, now under careful study, of alternative medicines that make the effects of radiation treatments less bothersome.

As described in Chapter 8, a doctor's prognosis for a given condition, that is, a statement about the cancer patient's likely future, is based on the current symptoms and data about what happened to other people who previously had a similar cancer. Prognosis is a summary of what happened in the past to people with a similar condition. Knowledge of this history is the sort of thing that doctors can and should know. They cannot, however, foresee the future. Thus, taking advice from an oncologist is bound to feel unsatisfying, and it may increase a patient's feelings of uncertainty. This situation makes some people want to reach out to other health care providers who will give them more certain statements about the future, whether they are true or not.

Over the last 30 years, 5-year survival rates for most cancers have improved significantly. Part of this can be attributed to early diagnosis and part to improved treatments, including the use of highly targeted therapies. The statistics are particularly encouraging for people with ages ranging from 50 to 65; people older than that have not seen such marked improvement. However, the discrepancies in 5-year survival rates between

CASE STUDY: John

A few months after John's death, I talked with his wife Lee about his final months. She said, "When you last talked to John, about 3 years ago, he was still doing great, but it was about then that he started to need radiation treatment. It really worked, but the long-range side effects were hard. The first radiation they did was to treat a fast-growing tumor on his neck; in 6 weeks it had gone from the size of a pea to a golf ball. For that treatment, we hooked up with a doctor at the medical center who was a great radiation oncologist. He said, 'I want to do this hard and fast, and get rid of it,' and he did. John had no radiation burn, no obvious problems at all from that radiation, but the doctor warned that there might be damage, and there was. John developed trouble with his brachial plexus [a cluster of nerves that goes from the spine into the arm]; that problem probably came from the first bout of radiation. The doc really blasted it, and it totally worked. John didn't need another treatment for a year or two, but that was enough radiation to make trouble.

After that first treatment we were really hopeful. John was doing fine. He was driving himself to his doctor's appointments, but that first radiation was probably what really started his final slide. After that, John had a couple of tumors in his belly. The insurance company wouldn't approve him for radiation of those tumors, because they said it was too hard to treat tumors there without damaging other organs, but the doctor did it anyway. There were two tumors, so the doc irradiated them both and never charged us for it. Here was a radiation oncologist who was really working with you. He was amazing.

After the tumors grew in John's belly, there was another tumor in his adrenal gland. That one was really hard because it hurt so much. John said it felt as if someone was stabbing him. For each radiation of that tumor the doctors did tomotherapy [bringing in the radiation from many angles to do maximum damage to the tumor itself and a minimum of damage to surrounding tissues]. This radiation definitely gave him a longer life. We had to push for it, but it really made a difference. Of course there were problems. The damage to his brachial plexus was a problem, because it limited his use of that arm. He could still carry a ladder, but it's hard to be a roofer when you don't have the free use of both hands.

A while later, John got both a brain tumor and a spot on his hip-bone. Both of these showed up in CAT scans, which the doctors did every 3 months. For the brain tumor, they followed up the CAT scan with an MRI to get a better idea of where to administer radiation. All in all, he had six tumors treated by radiation. The insurance companies were mostly really great about this. There was a $100 copay, but that was it. In fact, the radiation John had at the medical center was free; we never paid for that. During this time, insurance covered thousands of dollars of work, and before that, when John was taking chemo, they covered most of that too. $40 copay, and that was it. Sometimes we would see the bills for the drugs, and it would be $6000 for a month of this, $7000 for that, and all we had to cover was the copays.

In fact even taking care of the copays was a problem, given the several hundred dollars monthly costs for the insurance itself, plus the copays. But without insurance, it would have been impossible. We were very lucky on that score. We paid $3000–4000 a year, which was hard, but it wasn't impossible. When insurance doesn't pay, you get a bill for something like $40,000, and that's just impossible. What it really meant was that someone had to be the advocate. That was my job, making sure it all worked. John never saw a single bill. He was the one who was sick, and I took care of everything else."

cancers at different sites are quite dramatic (FIGURE 12.1). Much of this is due to the difficulty of identifying cancers early when there is no useful screening procedure, for example, for cancers in liver, ovaries, esophagus, and pancreas. For breast and colorectal cancers, scientific medicine is doing quite well. Prognostic data, like those shown in Figure 12.1, are of obvious importance to a person living with a cancer diagnosis.

Outside the realm of evidence-based medicine, there are healers who are confident that they can cure cancer through faith. Their confidence can be inspiring and can lead a patient to faith in the healer's powers. In some cases, these treatments lead to a demonstrable improvement in well-being, but sadly, not in all (SIDEBAR 12.7). A faith healer can be in a very strong position, because if the patient does not get better, the healer can believe that the patient's faith was insufficient; the fault is with the patient,

estimated new cases of cancer

males				females		
prostate	164,690	19%		breast	266,120	30%
lung and bronchus	121,680	14%		lung and bronchus	112,350	13%
colon and rectum	75,610	9%		colon and rectum	64,640	7%
urinary bladder	62,380	7%		uterine corpus	63,230	7%
melanoma of the skin	55,150	6%		thyroid	40,900	5%
kidney and renal pelvis	42,680	5%		melanoma of the skin	36,120	4%
non-Hodgkin's lymphoma	41,730	5%		non-Hodgkin's lymphoma	32,950	4%
oral cavity and pharynx	37,160	4%		pancreas	26,240	3%
leukemia	35,030	4%		leukemia	25,270	3%
liver and intrahepatic bile duct	30,610	4%		kidney and renal pelvis	22,660	3%
ALL SITES	856,370	100%		ALL SITES	878,980	100%

estimated cancer deaths

males				females		
lung and bronchus	83,550	26%		lung and bronchus	70,500	25%
prostate	29,430	9%		breast	40,920	14%
colon and rectum	27,390	8%		colon and rectum	23,240	8%
pancreas	23,020	7%		pancreas	21,310	7%
liver and intrahepatic bile duct	20,540	6%		ovary	14,070	5%
leukemia	14,270	4%		uterine corpus	11,350	4%
esophagus	12,850	4%		leukemia	10,100	4%
urinary bladder	12,520	4%		liver and intrahepatic bile duct	9,660	3%
non-Hodgkin's lymphoma	11,510	4%		non-Hodgkin's lymphoma	8,400	3%
kidney and renal pelvis	10,010	3%		brain and other nervous system	7,340	3%
ALL SITES	323,630	100%		ALL SITES	286,010	100%

FIGURE 12.1 **Frequencies of diagnosis and death for 10 common types of cancer.** (Top) Numbers of cases of cancer at multiple sites for the population of the United States during 2015 and published in 2018, broken down by cancer site and patient gender. (Bottom) Number of cancer deaths in the same populations during the same period. The differences between the number of cancers identified and number of deaths reflect the effectiveness of treatment for each cancer. For example, cancer of the lung and bronchus (tube leading to the lung) accounts for only 14% of cases recognized in males but 26% of cancer deaths. (Data from Siegel RL, Miller KD & Jemal A [2018] *CA—Cancer J Clin* 68:7–30.)

not the healer. This situation can be quite upsetting and even destructive to people in poor health, because the last thing they need is a guilt trip about insufficient faith. This situation points out one of the difficulties in turning away from evidence-based medicine to alternative treatments. Getting the best data possible and letting a well-trained physician choose the most effective treatment may be your best course of action. Even here, though, faith can provide a framework within which to think about your prospects. Faith can also bring an acceptance of your situation, even if it includes the proximity of death, so it can be a very positive thing. Meanwhile, the pursuit of ever-better treatments may be a source of hope. It is always worth hoping that either a refined way of using a traditional remedy or a novel invention by scientific medicine will bring the cure you need.

The costs of cancer care are large, which is a factor in the inequities of cancer treatment.

The costs of treating and caring for cancer patients in the United States are significant, and they are going up (FIGURE 12.2A). These expenses derive from the costs of drugs, treatments, and hospital care. Commonly, the costs are separated into categories, such as the initial treatments that

SIDEBAR 12.7

CASE STUDY: Rob

Rob's wife, Suzy, had always loved the unconventional, both in her life as an artist and as a person of style. While she and Rob were living in Budapest, Hungary, Suzy came to know and admire a remarkable faith healer. He specialized in treating people with cancer through a combination of Reiki healing, a Japanese technique that induces relaxation and promotes healing, and his own particular form of psychic power. Suzy wanted to take Rob back to Budapest for treatment, but we agreed as a family that keeping him comfortably at home was important. The conclusion was to invite Károly to New York, to stay in Rob and Suzy's house, and see what he could do.

Upon arrival, he wanted to see some of Rob's medical records, particularly the X-rays of his head, so he would know where to concentrate his healing efforts; we were able to get this information from Rob's doctors at the Columbia University Irving Medical Center in New York City. Károly had brought a young Hungarian woman as an assistant, but when he saw the X-ray pictures, he felt he needed even more help. He undertook to train both Suzy and Rob's mother in Reiki healing. Suzy already had some experience, but Rob's mom was a novice. She turned out, however, to be very good at it, so the four of them formed a team that worked over Rob for hours every day, employing a mixture of laying on of hands, psychic support, and Reiki.

This was a huge effort for all. Rob was not predisposed to think it was going to help, but partly for Suzy's sake, since she was such a strong believer, and partly because it seemed as if medical science had effectively given up on him, he was game to give it a try. The treatment also included a rigid control of diet: beets in large quantities and beyond that mostly other vegetables and fruits. This was supposed to help Rob eliminate toxic substances from his body and aid the healing process. All the workers in this psychic team spent many hours conveying whatever healing powers they could to the affected parts of his body: his aching knees, his sore back and spine, and particularly his head, where the cancer was obviously widespread. The course of treatment was about a week, and by good fortune, Rob was due for another CT scan at the end of that time. Thus, there would be a good test of progress in beating back the cancer.

Just at the end of this week, Rob had a seizure, not an uncommon event for people with tumors in their brains. He was rushed for treatment to a local hospital and then taken to the Columbia University Irving Medical Center for closer examination and the scheduled CT scan. The sad news came back that the tumors had grown, in spite of what was certainly an intense treatment by an expert faith healer. Everyone was disappointed. It was Suzy, though, who came up with a truly remarkable next step, which showed what a strong and resourceful person she was. She went shopping and bought the creams, salves, and ointments to give Rob a first-class facial in his hospital bed. A family friend, who was a radiologist at Columbia, described his amazement as he came with dragging steps to speak with Rob about the not-encouraging results from the new CT scan and heard gales of laughter pouring from his room. Rob, Suzy, and Rob's brother were all in giggles as they applied the beauty treatments to his potentially sad face. It was a grand example of mind over matter, even if the faith healing had not worked.

SIDE QUESTION 12.3

To get some specific information about the costs of cancer care, look up the charge for a one-night stay in your nearest hospital. Now look up the cost of five cancer drugs and two methods for diagnostic imaging as performed in your neighborhood, so you can estimate what it would cost you in out-of-pocket expenses if you were diagnosed with a severe cancer and did not have medical insurance.

follow diagnosis, patient maintenance during continued treatment or remission, and the final costs incurred at the time of death. In Figure 12.2A, all these costs are pooled, leading to national health care costs in the tens of billions of dollars. FIGURE 12.2B shows a cost breakdown for women with cancer who were 65 or older at the time of initial diagnosis. Again, the costs are almost staggeringly high. They emphasize the importance of medical insurance for obtaining quality cancer care.

Both the expense of some cancer treatments and variation in the quality of medical services from place to place have led to considerable differences in the quality of medicine that people in the United States can afford, based on their different levels of income. These factors lead to inequities in the quality of cancer treatments, and these inequities in turn correlate with race. Such differences show up in studies of the distribution by race of the likelihood of 5-year survival for a wide range of cancers.

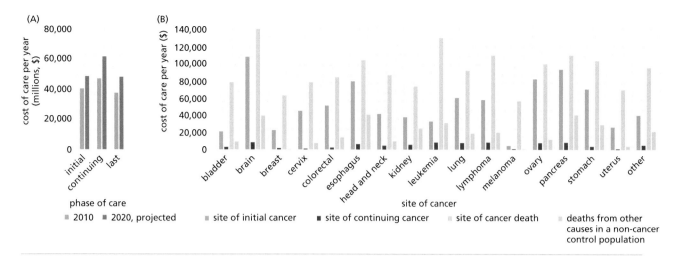

FIGURE 12.2 Costs of cancer patient care. (A) Costs of cancer care organized by phase of care: initial (at the time of diagnosis), continuing (continued treatment), and last (treatments and care at the end of life). The data combine information for all cancer sites, ages, and sexes. Projections to the year 2020 are based on the most recently available data about cancer incidence, survival, and cost of care, expressed in millions of 2010 dollars, to take account of inflation. (B) Costs of cancer care during the last year of life for women, aged 65 or older, broken down by cancer site. The heights of the green bars relative to the other bars demonstrate the costs of treating terminally ill cancer patients. (A, data from the National Cancer Institute, https://costprojections.cancer.gov/. B, data from Mariotto AB, Yabroff KR, Shao Y et al [2011] *JNCI, J Natl Cancer Inst* 103:117–128.)

The data available in a free, online journal called *CA—A Cancer Journal for Clinicians* show that there is essentially no difference by race in the stage at which cancer is diagnosed: (1) When is the cancer first identified? (2) Is it still localized? (3) Has it already become regional? (4) Has it developed distant metastases? There are, however, differences by race in the 5-year survival rates for a variety of cancers. White citizens of the United States are significantly more likely than black citizens to still be alive 5 years after diagnosis (FIGURE 12.3). This issue is of great concern to many health-care providers, but the solution will probably depend on significant changes in many aspects of race relations in our society.

REMISSION AND THE RENEWAL OF HOPE

Many cancer patients are lucky enough to find a treatment that works, even if it is painful or difficult to deal with; this leads to the state called **remission**. Medically speaking, remission can include a range of conditions that extend from a tumor's getting smaller all the way to its disappearance. The latter is a complete remission, and it does happen. With some cancers, like breast cancer caught reasonably early, complete remission is now common. A partial remission, in which cancer cells are fewer in number but still present, is likely to lead to a return as a full-blown medical problem, but only time will tell. Taking the good news of a remission in stride makes excellent sense; it is something to be happy about and enjoy to the fullest, but it is well to be realistic about the possibility of recurrence. One of the best aspects of remission is that it can focus one's attention on life and doing valuable and enjoyable things. All in all, remission is to be strived for and appreciated.

The opposite side of this coin, however, is the loss of hope and/or the development of anger that can accompany a bad time in any cycle of cancer treatment. Often, these dark times are associated with guilt, a natural but potentially destructive human emotion. It is surprising in a way that a cancer patient should feel guilty about their disease, but it is common. One origin of guilt can be the sense that one did not do all one might have

FIGURE 12.3 Data on the likelihood of 5-year survival, separated by cancer site and patient race. Each panel presents data for a different cancer site. Each bar in each graph gives the 5-year survival percentage for that group of people, depending on how much the cancer had spread at the time of initial diagnosis. The cancer sites with good prognoses are ones where early cancer detection is relatively common. Low percentages are seen with cancers of internal organs, where early detection is still difficult. The fact that the blue bars are shorter than the green bars in most of these panels is a direct reflection of inequities in cancer survival, according to race. (Data from Siegel RL, Miller KD & Jemal A [2016] *CA—Cancer J Clin* 66:7–30.)

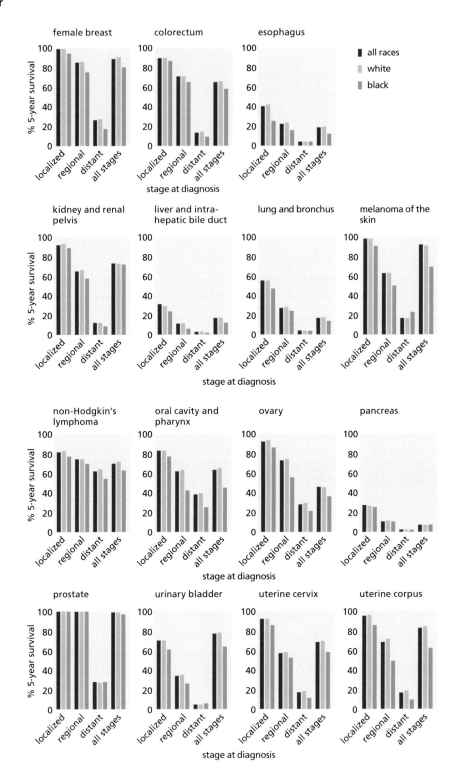

done to avoid carcinogenic circumstances: smoking, bad eating habits, excessive stress, and so forth. However, the discussions in Chapter 11 show that two major reasons for incurring cancer are inheritance and bad luck. One should not feel guilty about those aspects of cancer risk. But guilt is a sticky thing. Even if it is irrational, guilt can plague your life and erode your sense of hope. Under these circumstances, it may be well to seek counseling from someone experienced in helping patients with cancer. An oncologist, a psychologist, your religion, or the American Cancer Society can help here.

COPING WITH PAIN

Pain is one of the serious issues that can emerge in living with advanced cancer. The extent of pain depends on many things, but its origin is consistent: growth of the primary tumor or of metastatic sites can push on surrounding tissues and either hurt them or create the sensation of pain by pushing on nerves or a nerve network. In both bone and some soft tissues, there are nerves called **nociceptive receptors** that are slow to respond, but when greatly stimulated, they produce a strong sensation of pain. Cancer can also result in **referred pain**, also called reflective pain, which appears to be coming from a place other than the site of the painful stimulus (FIGURE 12.4). For example, a tumor in the liver can give rise to pain in the shoulder as a result of pushing on the relevant nerves. Reducing tumor size can reduce the cause of pain, but if tumor growth is not responding to therapy, the best course of action is to treat the pain itself.

There are many medications for reducing the sensation of pain, some of which are part of daily American life: aspirin, acetaminophen (sold as Tylenol), ibuprofen (sold as Advil and Motrin), and naproxen (sold as Aleve). Two sources of pain relief from traditional medicine are the spices ginger and turmeric. The latter contains the anti-inflammatory drug known to scientific medicine as curcumin. These are all useful medicines for the treatment of cancer pain, and they are frequently used. However, cancer pain can become too strong for these agents to dull, whereupon more potent medicines are required. The World Health Organization has defined a set of steps in pain management that begins with the drugs just mentioned and then moves up to **opioids** of various strengths. These are drugs that bind to receptor proteins in the membranes of neurons and block their response to pain, numbing any sensation of pain, despite strong pain stimuli (FIGURE 12.5). Many opioids are derived from the natural narcotic opium, which is obtained from a kind of poppy. The opioid **morphine** is

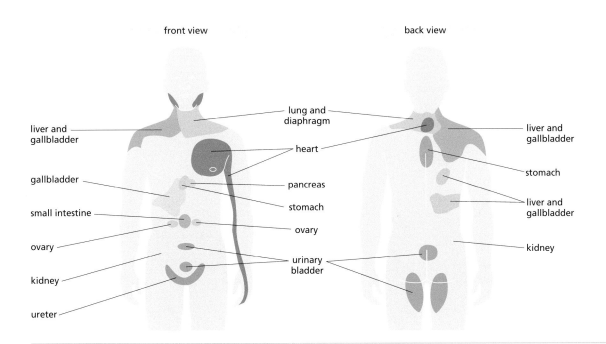

FIGURE 12.4 Sites of referred pain that can be important in cancer. The colored areas on the diagrams show where patients can feel pain as a result of cancer-related growths in the organs indicated by the labels. (Adapted from Tortora GJ & Derrickson BH [2013] Principles of Anatomy and Physiology, 14th ed. John Wiley & Sons.)

FIGURE 12.5 Strategies for management of cancer pain. Diagram of a three-step ladder developed by the World Health Organization for the relief of cancer pain in adults. They prescribe, "If pain occurs, there should be prompt oral administration of drugs in the following order: non-opioids (like aspirin), then mild opioids (like codeine), then stronger opioids (like morphine) until the patient is pain free. The term adjuvant is used here to mean additional drugs that can be used in case of patient anxiety." (Adapted from the World Health Organization.)

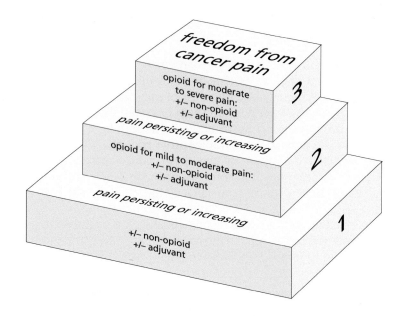

among the strongest and most generally effective pain medicines known. Modern chemistry has either isolated or designed and made a range of related chemicals that differ in their strength and duration of action (FIGURE 12.6). Morphine and related chemicals can do a lot to dull pain, but they also dull other mental processes, making a person foggy-headed and sluggish. They also can cause constipation, confusion, and depression. Unfortunately, opioids quickly reduce the body's sensitivity to them, so higher and higher doses are required to get a similar effect, a phenomenon called **tolerance**. Opioids are also notoriously addictive, meaning that one becomes accustomed to taking them and suffers when they are not available. Thus, there are losses associated with taking opioids, leading most patients and caregivers to look for some kind of balance in the medicines used.

One route to balance is through the careful choice of dose. Patients can now control the rate at which morphine is administered, allowing them to find the lowest dose that keeps them comfortable, thus leaving them as

FIGURE 12.6 Duration of effects for several opioid drugs used to treat cancer pain. Across the bottom of this graph are the names of several opioid drugs; asterisks indicate patented trade names. The vertical axis shows the duration of pain relief that each drug offers. When two values are present for a single drug, they reveal the range of time over which relief was experienced by different patients. (Data from Nersesyan H & Slavin KV [2007] *Ther Clin Risk Manage* 3:381–400. With permission from Dove Press Ltd.)

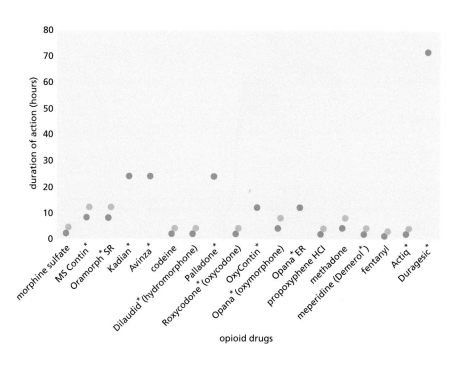

mentally active as possible. In addition, other drugs can be used in combination with morphine to help relieve some kinds of pain with a lower dose of morphine. For example, naproxen, a **nonsteroidal anti-inflammatory drug** (sometimes abbreviated NSAID), helps significantly with bone pain. Pain is a major issue for many cancer patients, which means that reliable information from a professional is important. An experienced cancer nurse may be more helpful with pain management than an oncologist, who may not have studied pain management. Whatever the source of information, getting good advice on pain management, one aspect of what doctors called **palliative care**, is very worthwhile. You can become better informed about the issues of pain management by reading materials on responsible Websites and from the many publicly available articles and books on the subject.

DEALING WITH PHYSICAL WEAKNESS

Some cancers have a significant effect on one particular bodily function, making it grow weak or fail while the rest of the body is still strong. Examples of this effect include loss of vision for an ocular tumor, loss of the ability to swallow from throat cancer, and loss of sexual drive and potency from a prostate tumor or its treatment. In each case, modern medicine is likely to be one's best source of information and treatment for dealing with these kinds of situations. Some treatments will alleviate the conditions, but sometimes a physical device, called a **prosthesis**, can provide the function that is either lost or is going fast. A good physician can make a big difference by providing access to these sorts of patches that may have nothing to do with a cure but a lot to do with the quality of day-to-day life.

As tumors grow, the rest of the body tends to weaken more generally. The cancer cells that have flourished are particularly good at getting nourishment. In a sense, they have been chosen by natural selection to command your body's resources, even if there is not enough to go around. In addition, advanced cancers often deregulate the body's controls that normally handle food, resulting in a loss of weight and muscle mass, followed by a general loss of vigor, even when ample nutrition is available. Doctors call this condition **cachexia** (ka-kex'-ia), a combination of two Greek words that mean bad condition. People react very differently to cachexia. Some become depressed while others keep their spirits up; once again, this is a personal thing without rights or wrongs. It is generally a good idea to anticipate this kind of weakening, both practically and psychologically, by identifying some things you will still be able to do, even when you are no longer vigorous.

CHOOSING AN ENVIRONMENT

Many people living in the industrialized world are sure that a serious illness means going to a hospital for treatment. In some cases, this is an excellent course of action, but for terminal cancer it may not be the best option. If a hospital can help you to get better, it is a good place to be. If it is a place to languish, it is an expensive and depressing environment; moreover, it can be all too good a source of serious infections. It is now evident that widespread use of antibiotics has led to the emergence of numerous antibiotic-resistant strains of common bacteria, like *Salmonella*, *Klebsiella*, and *Staphylococcus*. Getting rid of such an infection can be very difficult, so sticking to a home environment can, in some cases, be the healthier course of action.

SIDE QUESTION 12.4

Look up the rules about how hospitals in your neighborhood must care for terminally ill patients.

Another important issue is that hospitals are obliged by law to keep you alive if they possibly can. At some point, this may not be in your best interest, either personally or financially. There may not be much joy in the prospect of continued life, and the costs can be horrific. For an account of just how much cancer care costs, particularly at the end of life, see Figure 12.2B, but short of committing suicide, is there an alternative? Fortunately, the answer is yes.

Many years ago, a group of caregivers in England realized that the ability of scientific medicine to keep the body alive could lead to seriously inhumane situations. They started the **hospice** program, which is designed to provide good care for terminally ill people and to help them live the time that is left to them with as much dignity and comfort as possible. One policy of hospice programs is not to prolong patient's lives by artificial means. As a poignant example, if a patient can no longer eat or drink, a hospital will provide food and water through intravenous administration of nourishing fluids. After an automobile accident or the like, this can contribute significantly to care and recovery; it can be the difference between life and death. For patients who are terminally ill with an incurable disease, however, such care is probably just a prolongation of the status quo.

If one is in constant pain, prolonged life is a questionable proposition. One can legitimately ask, Is this what I want? Many people will easily say no, but it can be difficult to get a hospital *not* to administer life-sustaining care. One needs strongly written documents that conform to the legal standards of the state where you live. These include a living will, which states your wishes about how you want to live and die; a medical power of attorney, which conveys to another person, such as a close relative, the legal right to make medical decisions if you can't do it yourself; and a document called an order not to resuscitate, if you should become unconscious. This trio of legal papers is commonly needed to convince the administrators of a hospital that they will not be sued for medical malpractice if they stop administering every available treatment. From the hospital's point of view, continued treatment makes excellent sense. It uses their resources in a constructive and profitable way, but the costs to the patient can be astounding (see Figure 12.2). It is commonly said that people spend more money on healthcare during their last 6 weeks

SIDEBAR 12.8

CASE STUDY: Woods

Woods' experience with hospice was tremendously important for his quality of life during the months before his death. For about 5 weeks, they made it possible for him to stay in his own apartment, living as independently as possible. Hospice arranged for a nurse from the city's health care program to come daily and help him with taking his medicines, which by this time were numerous, strong, and a bit confusing. They got him in touch with a community of Buddhists who came regularly to talk with him about his views on death and to meditate with him. When his condition worsened, hospice helped him get into a resident-care nursing home, where he could receive the round-the-clock care he now needed. Even here, though, he was tended by the hospice nurses and doctors who could help him keep a level of medi-

cation that eased his pain but left him reasonably clear-headed.

When Woods moved into this nursing facility, he took with him many of his books on religion and his meditation cushion. He was well educated about Judaism, Christianity (the religion in which he had been raised), and Islam as well as Taoism and Buddhism. He wanted to talk with both his hospice caregivers and the other nurses about the different views of death that he took from all these texts. Up until the day before he died, he took pleasure and solace from these readings, thoughts, and conversations. His opportunity to do this was a direct result of the philosophy of the hospice program and the way they interacted with the nursing home to which he went.

CASE STUDY: Rob

As Rob's condition continued to deteriorate, we got in touch with hospice, so we could continue to care for him at home. The hospice program in that county was excellent, and they helped in many important ways. One of the most significant was the management of pain. Lung cancer commonly metastasizes to bones and joints, where the presence of cells that don't belong can cause extreme discomfort. The oncologist knew this and prescribed morphine, which was certainly a help, but it is such a strong narcotic that its effects on general mental function became a problem.

The hospice nurse who came to help care for Rob knew that naproxen, available as an over-the-counter medicine, was particularly good for pain in bones. By following her advice, Rob was able to cut down significantly on the amount of morphine he needed to dull the pain, so his mind became more acute, and he regained an interested in many things around him. The hospice nurse also knew about a special mattress, whose surface resembled an egg carton; this helped slow the development of bedsores. She also knew about Bag Balm, the old-fashioned salve used by farmers to ease pain in the udders of cows who are having difficulty being milked; this turns out to be a cheap and effective remedy for bedsores that have already developed. She provided a hospital bed that allowed Rob to adjust his own positions with motor-driven lifters, and she recommended a commode, which reduced the distance he had to walk to relieve himself. The sum of all these little things was a big improvement in his comfort and state of mind. They made it not only possible but advantageous for Rob to stay at home as he weakened.

As Rob's cancer progressed, it became harder for him to swallow, so his normal eating slowed. That was probably not a bad thing. By this time, his progression toward death was evident, but some of his medical friends were concerned about how he might actually die. If an old person were to have a cancer so advanced, death would probably come soon, perhaps from pneumonia, which is easy to catch in a hospital. But when you start with the physique of an athlete and the constitution of an energetic youth, there is concern about how you can actually die from a wasting disease.

Toward the end, Rob lived essentially without food for about 3 weeks, and he took in only a little water. Surely, he would have lived many weeks longer if he had been hospitalized and kept alive by fluids administered intravenously, a feeding tube, and additional tools of life support. His choice to die at home was truly a blessing for us all, but especially for him. The hospice nurse was able to keep him comfortable with good, simple remedies and the medical help of morphine. He had family with him throughout. Near the end this included his brother and sister too. He was loved and tended, and all of us felt that we were doing what could be done, little as it was. There was time to say goodbye and to accept the passing of a dear child in a way that would not have been possible, for me at least, in the more sterile environment of a hospital. I think hospice is a truly great institution.

than during the entire rest of their lives. Hospice programs offer a way to avoid this situation, because they take a different approach to patient care (SIDEBAR 12.8).

The founders of hospice recognized that we do not have to cure to heal. Terminally ill people can be cared for, talked with, and tended, but when their time has come, the most natural and appropriate thing for them is to die. Hospice, therefore, focuses on patient comfort and dignity, including provision of the best beds and mattresses, and on the availability of the best drugs to ease pain (SIDEBAR 12.9). Hospice does not provide treatments that will prolong life, such as intravenous feeding, and indeed most hospice programs will not take patients who are receiving such treatments elsewhere. In many parts of the country, hospice runs care centers analogous to hospitals where a patient can go to stay, but hospice can also enable patients to stay in their own homes. Although this care costs money, most insurance programs will pay for it, since it reduces the burdens of those extremely costly last few weeks in a hospital. More information about hospice care is available from the American Academy of Hospice and Palliative Medicine and the National Hospice and Palliative Care Organization, among others.

Find out about the hospice care services available in your neighborhood.

Compare the services that hospice offers with the kind of care a terminally ill patient would receive in your nearby hospital.

The society and culture of the United States, like those in many developed countries, put tremendous value on human life. Likewise, many religious traditions support life and strongly criticize any activity that would lead a person to take his or her own life. However, all these views arose long before the invention of intravenous feeding to nourish people who cannot eat, before dialysis machines for kidney failure, and before respirators for loss of the ability to breathe. These devices are all examples of advanced medical treatments that are hardly natural, even though they have become familiar. In the right situation, they can be of tremendous value for the patient, but in other situations they are quite the opposite. Some serious thought about these end-of-life issues *before* they become central to your own life is an excellent idea. The legal routes to making your intentions known are clearly defined, as described above, but following these routes correctly takes both time and care. It is also worth noting that some states in the United States have already enacted laws that permit physician-assisted dying (commonly called death with dignity or assisted suicide) under conditions like terminal cancer: Oregon was first, followed by Washington, Vermont, Montana, and California. Laws allowing this significant step commonly require input from more than one doctor, from family if available, and from lawyers. In spite of this complexity, however, assisted death can be a wise and humane path to take.

CARING FOR CANCER CAREGIVERS

When one member of a family is desperately ill, that person's needs naturally command attention. Commonly, the time and effort to care for the sick person will fall on only a few other people. Until one has been in the position of being a caregiver, it is hard to know and understand the strain that it poses, even for a healthy and well-balanced person. John's wife, Lee, took care of him right up to the end (SIDEBAR 12.10).

Time spent tending for the sick can mean loss of work, loss of sleep, and loss of personal freedom, as well as the possibility of a loss of joy in life. There are also more subtle problems, including a sense of guilt, even when the caregiver is working flat-out to provide all the help they possibly can. When someone else is in serious trouble, like a patient with terminal cancer, it is natural to feel fortunate that you are not gravely ill yourself. At some level, you might even feel secretly glad that this is their problem, not yours. This very human feeling can, however, inspire guilt that one is not suffering when another person is. If one is working hard to care for the patient, a feeling of guilt on top of all that effort can make the care-

CASE STUDY: Lee

Lee described the way she dealt with the practical problems surrounding John's death as follows:

"Through all of John's final health problems, I didn't take a minute off from work. I didn't even take funeral leave. My boss offered that I could take time off, but there was no one to take my place, so I just kept at it. In part I kept on coming to work because work was a really good outlet for me. Even from day 1, I just kept coming in, because it took my mind off of John's situation. I really had good support from friends at my job. Now and then, I had to make a lot of decisions, where the description of the options included a whole lot of words I didn't even know. But the doctors at the Health Center were a great group. There was a whole lot of camaraderie, and all of us bonded together."

CASE STUDY: Scott

As Scott began to heal, Mary was able to think of other important things. She said, "Once Scott was getting better, I began volunteering in the oncology branch of Children's Hospital. There were kids coming in from all over to get treated there: Wyoming, Kansas, and so forth. It was a good thing for me to talk with some of those parents about what was in store, because we had had such a good experience.

We were lucky that our medical insurance covered a lot of Scott's costs. The hospital bill itself was covered 100%, but there were a lot of other bills that kept coming in, like the anesthesiologist, the general surgeon, and so forth; they were only covered to something like 80%. So there were costs, but they were not crippling. I certainly wouldn't have done anything differently. We were able to trust the whole team in everything they did.

It was also very good to have people around, my friends and family, who cared and who wanted to help. It was not easy for me to take that, because I'm used to doing things myself, but my friends and family were feeling really frustrated by not being able to do anything, so accepting their help turned out to be a really good thing to do."

---

giver angry. Organizations that provide companionship for cancer patients can often offer help to their caregivers as well. Certainly, people on the helplines from the American Cancer Society will do their best to counsel any caregiver. Talking with people who have lived through a similar difficulty can sometimes be a great release.

Caregivers must also cope with sadness. As they figure out how to live and go forward in the face of an incipient death, they sometimes face problems as hard as, or even harder than, the person who will die. Living on can be tough, especially if there are serious financial problems, so caregivers have every right, and indeed even an obligation, to think of themselves and their own future. Making sure that one knows about legal and financial issues, so they are dealt with properly, can be an important part of coping with someone else's cancer. It is not wrong to ask about life insurance, bank balances, lawyers, and other things that may not have previously been on one's radar. It only makes sense to become as well informed as possible, so as to be able to deal with the practical problems that are almost unavoidable when one is dealing with a patient who has terminal cancer. But despite these negative aspects, being a caregiver can be a strongly positive experience, providing a sense of coping successfully with difficult problems and helping to make life work. Mary's care for her son, Scott, followed this pattern, as described in SIDEBAR 12.11.

SUMMARY

Cancer is a medical challenge that is not about to go away, and the issues of living with cancer are ongoing. A diagnosis of cancer can carry a wide range of difficulties, depending on the severity of the condition and the effectiveness of the treatments available. Thus, generalizations about the disease and how to live with it are hard to come by. The words in this chapter, and particularly the interviews with cancer patients and their caregivers, which are provided in sidebars throughout the book, should help you to understand the many dimensions of this disease. Moreover, the scientific understanding you have acquired from this book should equip you to deal with cancer if it should ever come into your life. The essence of this book's message about living with cancer is to become as well informed as you can and use this knowledge to help yourself and others make the best of the life ahead, whatever it may be. Knowledge can help you to see cancer with a broader view. A focus on the things you care about can help to put the dark side of cancer into perspective.

ESSENTIAL CONCEPTS

- There are many ways of reacting to a diagnosis of cancer, but some approaches to the disease are easier to live with than others.

- Modern Western culture is strongly oriented toward youth and health, so older people and sick people, particularly those with a very serious medical problem like terminal cancer, tend to be pushed out of the way or ignored.

- The religions of the world have traditionally provided strong help for people facing death, but to many people living with cancer in the industrialized world, this source of support is either unimportant or not available.

- Although there are no universally useful approaches for people living with a diagnosis of terminal cancer, it seems from the experience of many that keeping a positive attitude and focusing on both life and the things you enjoy can make the experience less difficult than it would otherwise be.

- Identifying activities you like to do and arranging your life so you can do them is an excellent way to bring positivity into the difficulties that cancer presents.

- For some people touched by cancer, discussing their feelings with others is a consolation. For others, however, this course of action is unappealing or even unpleasant. Whichever is the case, cancer organizations provide information on how to cope with the disease.

- A pervasive problem of living with cancer is the expense. Medicines, treatments, and care can all cost more than the patient or the family can afford. Medical insurance is therefore of great importance.

- There are many alternative forms of treatment for cancer, and there are many people willing, even eager, to provide them. It is unlikely that any of these treatments will be better than the recommendations of medical science, but some cancer patients want to try alternative treatments anyway. Caution is necessary, though, because expensive treatments by unscrupulous people can cost a great deal of time and money and accomplish nothing.

- Living with cancer commonly means living with pain. A nurse, well trained in pain management, may be a better source of help for this condition than an oncologist. Nurses also know many coping strategies that will simply make life easier and/or more pleasant for the cancer patient.

- End-of-life care can often be provided better at home than in a hospital, an institution that is organized around cures. If there is no cure for a patient's cancer, a hospital can be an unnecessary discomfort and a significant waste of money.

- Hospice is an organization committed to providing end-of-life care without attempting a cure or prolongation of life. Under some circumstances, hospice can provide a terminal cancer patient with the best possible option.

- People who care for very ill patients, such as those with terminal cancer, are themselves at considerable risk of exhaustion, depression, and even breakdown. Caregivers need to care for themselves as well as for their patient, which can sometimes be difficult to do.

KEY TERMS

cachexia

hospice

morphine

nociceptive receptors

nonsteroidal anti-inflammatory drug

opioids

palliative care

prosthesis

referred pain

tolerance

FURTHER READINGS

Mukherjee S (2010) *The Emperor of All Maladies: A Biography of Cancer.* Scribner.

Gawande A (2014) *Being Mortal; Medicine and What Matters at the End.* Metropolitan Books.

The following resources offer descriptions of ongoing efforts to help people who are living with cancer: The NIH National Library of Medicine's MedlinePlus (http://www.nlm.nih.gov/medlineplus/cancerlivingwithcancer.html), the American Cancer Society's MyLifeline (https://acs.mylifeline.org/getstarted), and the American Society of Clinical Oncology's Cancer.Net (https://www.cancer.net/coping -with-cancer).

Mariotto AB, Yabroff KR, Shao Y et al. (2011) Projections of the Cost of Cancer Care in the United States: 2010–2020. JNCI, *J Natl Cancer Inst* 103:117–128. This resource has detailed information on the costs of cancer care.

The NIH National Cancer Institute's Office of Cancer Complementary and Alternative Medicine (https://cam.cancer.gov/) has reliable information about many aspects of alternative medicines and cancer care.

ANSWERS TO SIDE QUESTIONS

Given the character of the questions in this chapter, no model answers are offered, and there are no questions for further thought.

Glossary

actinomycin D

A chemical produced by certain bacteria that binds to DNA and blocks the passage of **RNA polymerase** along the DNA, thereby blocking **transcription**. It was originally used as an effective antibiotic, but for many years it has also been used as an anti-cancer drug.

adaptive immunity

The part of our body's immune system that provides a huge variety of defensive molecules, both **antibodies** and **T-cell receptors**, and then eliminates the molecules that react with our own **antigens** and makes many more of the ones that react with pathogens when they appear in our bodies. Under some circumstances, these molecules and cells can also react with and help to eliminate cancer cells.

adenomatous polyposis coli (*APC*)

A **tumor suppressor** gene important for terminating the continued proliferation of **transit amplifying cells** in the crypts of the colon.

adipose cells

The principal cells of fat tissue. Adipose cells express enzymes that convert nourishment from the diet into a simple, stable fatty molecule that separates from water and can be stored as a large, intracellular droplet. They are the body's principal way of coping with excess calories in the diet.

adjuvant therapy

Adjuvant means serving to assist, coming from the Latin word meaning to help. Adjuvant therapies are additional treatments, like radiation, that may assist the initial treatment, such as surgery, to deal effectively with a cancer.

age-adjusted incidence of cancer

A simple mathematical process in which a statistician uses data on both the increased frequency of cancer in older people and the differences in age distributions of two populations to compare cancer incidence across different populations or times.

allele

A form of a particular gene. Different alleles of a gene can encode different characteristics for the same trait, such as eye or hair color.

allogeneic transplant

The insertion of cell or tissues from one individual into another individual. This is distinct from an **autologous transplant**, in which the cells or tissues that are transplanted come from the person into which they will be placed. These terms are important in distinguishing different types of bone marrow and other transplants used in cancer therapy.

***all-trans* retinoic acid**

A molecule closely related to vitamin A that is used as a signal to stimulate cell differentiation during normal human development. It is an effective chemotherapeutic agent for certain blood cancers, because it promotes the final differentiation of cells that have been prematurely released from the bone marrow into the circulating blood.

alpha rays

A form of **radioactivity**. An alpha ray, also called an alpha particle, is a fast-moving helium nucleus (two protons and two neutrons), released in some radioactive events, for example, when an unstable atomic nucleus rearranges.

alternative medicine

Any of the many forms of medical treatment and understanding that are not based on the scientific approach to medicine developed in Europe, the United States, and Japan over the last ~100 years. Other terms used to refer to these medical practices are **complementary medicine** and **traditional medicine**.

amino acid

A kind of small molecule that is very important for the life of a cell. Twenty different amino acids are used to make **proteins**, and some others are used both in **metabolism** and in conveying **information** from one cell to another.

anaplastic

A term used by pathologists examining a biopsy to say that the cells under study have lost most of their differentiated structure. This implies that they are far down the path of transformation from normal to cancerous.

aneuploidy

A condition in which cells do not have the correct number of chromosomes. This is common among cancer cells.

angiogenesis

The process by which blood vessels increase in number and length to bring increased blood flow to a particular region of the body.

angiogenesis inhibitors

Chemicals or processes that block an increase in the number or length of blood vessels.

antibody

Proteins made by our adaptive immune system that bind to molecules that are not a normal part of our bodies (**antigens**), such as proteins on the surface of a virus or a bacterium. When the antibody binds the foreign protein, it initiates a series of processes leading to inactivation of the invader, thereby protecting our body from infection.

antigen

Any molecule to which an animal can make an antibody.

antigenic determinant

The part of an **antigen** that binds to a particular antibody molecule.

apoptosis

Programmed cell death, a natural process in which an existing pathway for cell behavior is turned on, leading to the activation of enzymes that destroy the cell.

aqueous

An adjective meaning made mostly from water.

aromatase

One of the enzymes that can help to make estrogen.

aromatase inhibitor

A chemical that blocks **aromatase**, a key enzyme in one of the pathways by which the body normally produces estrogen.

atom

A single copy of any of the elements from which all things in the known universe are made.

atomic number

A way to describe and classify atoms. It is defined as the number of protons in the atom's nucleus and thus the number of electrons that normally circulate around the nucleus.

ATP

The commonly used name for adenosine triphosphate, a compound widely used by cells as the molecule to store and carry chemical energy from one process to another. It also contains one of the bases used to make both DNA and RNA.

ATRA

Abbreviation for *all-trans* **retinoic acid**.

autologous transplant

The insertion of cell or tissues from an individual back into that same individual. This is distinct from an **allogeneic transplant**, in which the cells or tissues that are transplanted come from another person. These terms are important in distinguishing different types of bone marrow and other transplants used in cancer therapy.

B cell

Also known as B lymphocyte; a cell that can make **antibody** molecules. B cells are born in the bone marrow, where they go through an initial differentiation process in which they choose exactly what antibody molecule they will make. They are then examined to make sure that the antibody they make does not bind to one of the body's own proteins, that is, to self. They then migrate to the peripheral immune organs, like **lymph nodes**, where they are present in case a pathogen appears and presents an antigen that binds strongly to the antibody they make. Then the B cell differentiates into a **plasma cell**, which functions as a factory to produce large amounts of that antibody.

bacteriophage

Viruses that infect bacteria.

basal cell carcinoma

A cancer that arises comparatively frequently in cells situated at the base of the epithelium that covers skin. Commonly, it is a result of the mutagenic action of ultraviolet radiation, like in sunlight.

basal lamina

A fibrous extracellular layer that separates all epithelia from the underlying layer of connective tissue.

base

A word with two biological meanings: (1) A substance that is alkaline, meaning the opposite of acid. (2) A specific group of molecules, used to make the nucleotides that polymerize to form nucleic acids. Two kinds of bases are used to make nucleotides: purines (adenine, A, and guanine, G) and pyrimidines (cytosine, C, and either thymine, T, in DNA or uracil, U, in RNA).

BAX

A family of proteins that helps to open the pores of the outer mitochondrial membrane, allowing **cytochrome *c*** to leak out and initiate **apoptosis**. These proteins are **tumor suppressors**.

Bcl2

A protein that helps to keep pores in the outer mitochondrial membrane closed, thereby preventing the release of cytochrome *c*, which is an important initiator of apoptotic cell death. Up-regulation of Bcl2 is oncogenic because it blocks **apoptosis** in cancer cells, allowing them to proliferate without inducing apoptosis.

benign

A word whose common meaning is gentle or well-meaning, but in cancer biology it means that a tumor is not likely to invade adjacent structures or to metastasize. Thus, it will probably not become life-threatening, or **malignant**.

beta rays

A form of **radioactivity**. Beta rays, or beta particles, are fast-moving electrons released when certain kinds of unstable atomic nuclei rearrange.

biochemistry

The study of biological molecules by the methods of chemistry.

biomolecules

A general name for any of the molecules characteristic of cells. This includes biological macromolecules, such as nucleic acids and proteins but also lipids, sugars, amino acids, and vitamins. It does not include minerals like sodium, calcium, potassium, and iron, all of which are essential for life but which also play important roles in nature outside of cells.

biopsy

A medical procedure in which a small amount of tissue is removed for diagnostic study, commonly with a microscope. Four kinds of biopsies are common: (1) surgical biopsy, or removal of a small tissue sample; (2) core biopsy, in which a cutting needle is inserted into the tissue of interest to remove a slender cylinder of tissue; (3) aspiration biopsy, in which a needle thinner than a cutting needle is used to suck cells from the tissue of interest; and (4) punch biopsy, in which a cylindrical blade 1–8 mm in diameter is used to cut a sample from the patient's skin.

body mass index

A number, commonly called BMI, that expresses the ratio of one's weight (measured in kilograms) to the square of one's height (measured in meters). It is a useful number because it gives a quick assessment of whether an individual is underweight, normal, overweight, or obese.

brachytherapy

A cancer treatment in which a small piece of radioactive material is implanted into a tumor so the radiation emitted will affect the well-being of the tumor cells in that vicinity.

BRCA1 and BRCA2

The names given to two genes that serve as **tumor suppressors**. The genes were identified by genetic analysis of families in which breast cancer (hence BRCA) was common. The products of these genes are involved in DNA repair.

cachexia

Severe wasting away of the body as a result of a chronic illness, such as advanced cancer.

cancer

A disease caused by the uncontrolled growth of cells from our own body.

cancer risk

The likelihood that cancer will occur. This term is commonly used in association with conditions in the environment (environmental risk factors) or inheritance.

cancer stage

A way of rating cancers that tells how far the cancer has progressed. The cancer stages are based on the size of the primary tumor, the presence of nearby tumors, including in neighboring lymph nodes, and whether there are already distant metastases.

cancer stem cells

Cells in a tumor that have the ability to initiate a new tumor when they are transplanted into a different organism. They are thought to be analogous to normal adult stem cells in that they can grow and divide to give rise to the **parenchymal** cells found in the cancerous tissue.

cancerous transformation

The changes by which a normal cell is altered to become cancerous. This kind of transformation is brought about by the accumulation of multiple behavior-altering mutations in the same cell. Also known as carcinogenic transformation.

CAR

Abbreviation for **chimeric antigen receptor**.

carbohydrate

A class of molecule made by the chemical combination of the element carbon (C) and water (H_2O). Sugars are examples of simple carbohydrates. Many carbohydrates are used by cells to store energy for other uses. Complex carbohydrates are **polymers** of sugars, sometimes connected to other kinds of molecules, like amino acids. Complex carbohydrates are one of the four main classes of **macromolecules** found in cells. They too are used to store energy, but cells also use them for structural purposes and sometime as ways to store information.

carboplatin

A platinum-containing compound that is highly reactive. It is used as a chemotherapeutic agent because it will react with DNA to create **DNA adducts** and induce mutations, causing cells to enter **apoptosis**.

carcinogen

A chemical or a physical process that increases the likelihood of cancer.

carcinogenic

The property of promoting cancer.

carcinoma

A cancer that arises in an epithelial tissue.

Cas9

A nuclease that cuts double-stranded DNA at a sequence defined by its guide RNA as described under **CRISPR-Cas**.

caspase

Any of a group of cytoplasmic proteases that control and execute **apoptosis**.

catalyst

A chemical or object that can speed up (catalyze) a chemical reaction.

Cdk

Abbreviation for **cyclin-dependent kinase**.

cell biology

The study of cells. This modern subject combines information from microscopy, chemistry, and the study of genes to analyze the structures and processes that allow cells to accomplish their multiple tasks.

cell culture

A laboratory method for growing cells under controlled conditions, making it easier to manipulate them and experiment upon them.

cell transformation

A change in the behavior of cultured cells brought about by **carcinogenic** agents, such as a **tumor virus**. This process is thought to be a good model for the transformation that occurs in a human body as a cell changes from normal to cancerous.

cellular adaptive immunity

The part of the immune system that fights pathogens with **T cells**.

centrosome

An organelle that serves in most animal cells as the principal organizer of the cell's **microtubules**.

checkpoint

A process that helps to regulate the cell's growth and division cycle by slowing or stopping an essential process, such as cell division, until a process that is prerequisite, such as DNA replication or DNA damage repair, has been properly completed.

chemokine

A specific kind of **cytokine** that induces the migration of white blood cells.

chemotherapy

A form of cancer treatment in which the patient is administered chemicals that are harmful to rapidly growing cells, such as those in a cancer.

chimeric antigen receptor

The name given to an innovative approach for engineering **T cells** from a cancer patient so they will be more effective in killing cancer cells. Antigen receptor means that a CAR uses an antibody as a cell surface receptor. Chimeric is an adjective coming from the ancient Greek word *chimera*, which means a mythical beast composed from parts of multiple creatures, such as a lion, an eagle, and a snake. Molecular biologists use this word to refer to genes and proteins that they engineer by putting parts from different molecules together to form a novel molecule.

chromatin

The nuclear material composed of DNA wrapped around histone core particles, which provide the DNA with both compaction and regulation of gene expression.

chromosome

A piece of cellular DNA. Chromosomes at mitosis can be stained with dyes and become visible as colored bodies, which is what their name means. During interphase, the chromosomes are decondensed and mingle with one another in the cell's nucleus, so they can no longer be seen as distinct objects.

chronic myeloid leukemia

A kind of leukemia that develops more slowly than acute leukemias but is still life-threatening. Fortunately, modern chemotherapies, such as the use of the drug **imatinib**, also known as **Gleevec**®, are particularly effective with this cancer, and they will generally hold it in check for many years.

cisterna

(plural cisternae) A flattened sac, like the ones seen in the **Golgi apparatus**.

cisplatin

A platinum-containing compound that is highly reactive. It is used as a chemotherapeutic agent because it reacts with DNA to create **DNA adducts** and induce mutations, causing the cells to enter **apoptosis**.

clinical trials

Experiments on patients that are designed to evaluate a new drug or treatment.

clone

A group of cells that have all been formed by the growth and division of one cell and its daughter cells. Many tumors are clones. Bacterial cells can also form clones, and since bacteria are used by laboratory biologists to grow up particular pieces of DNA that they want to study, the term clone has also come to be used as a verb for the act of making a large number of copies of a specific piece of DNA.

CML

Abbreviation for **chronic myeloid leukemia**.

codon

A sequence of three **nucleotides** in either DNA or RNA that spells one genetic word.

cohort

Any group of people that is being treated as a group. The term is commonly used in studies of the effects of chemicals or procedures on people or animals, where the goal is to assess the effect on people or animals in general, as opposed to an individual in particular.

colon

The part of the large intestine between the small intestine and the rectum and anus.

colonoscopy

A diagnostic procedure that uses light-carrying fibers to look inside the large intestine, starting from the anus and reaching up into the colon.

complement system

Part of **innate immunity** that complements other parts of the immune system and can help with the body's defenses against pathogens by building holes in the plasma membrane of invading cells.

complementary medicine

See **alternative medicine**.

compound microscope

A light microscope that uses two lenses in series to obtain increased magnification.

computed tomography

A process for generating a 3-D image of a part of the body. For example, X-ray pictures of a given region are taken over a wide range of angles, and a computer uses this information to calculate the 3-D structure that must be present to have produced all those views. It is an important tool for cancer diagnosis and evaluation of cancer treatments.

connective tissue

A kind of tissue that lies beneath an **epithelium** and surrounds most other tissues, providing mechanical links between them. To a student of tissue, connective tissue includes bones, ligaments that connect them, tendons that connect muscle to bone, loose connective tissue that underlies epidermis and is part of skin, and even blood.

constant region

The portion of an **antibody** molecule that is the same in all antibodies of a given kind. This is distinct from the **variable region** of the antibody, which includes the part of the antibody that binds to the **antigenic determinant** recognized by that antibody. The constant region binds to and activates phagocytic cells and NK cells, drawing the activity of these parts of **cellular inate immunity** into play in dealing with the presence of a foreign antigen.

contact inhibition of growth

A behavior of normal cells grown in culture by which they stop their growth and division cycles when they make contact with neighboring cells on all sides.

contractile ring

A belt made from microfilaments that the cell builds just under the plasma membrane at the plane of the metaphase plate. After mitosis is over, this ring contracts to bring in the cell membrane, dividing the cell in two.

contrast

The property of an image that describes the range of light and dark or of different colors that allows the image to convey information to the viewer.

contrast agent

A drug administered during a diagnostic study that gives better definition of a structure of interest. An oral contrast agent helps define the inside of the gastrointestinal tract, whereas an intravenous contrast agent helps to define the inside of blood vessels.

CRISPR-Cas

A set of processes by which bacterial cells cut DNA sequences from foreign DNA, for example, from an infecting virus, and put them into a special part of their own DNA. RNA transcribed from these sequences then serves as a guide to control the activity of a **nuclease** called Cas9, which will now cleave any double-stranded DNA that contains the same sequence as its guide RNA.

CT scan

Abbreviation for **computed tomography** scan.

CTLA4

A protein that resides in the membrane of a T cell and can interact with proteins on the surface of an antigen-presenting cell, when that cell is expressing and presenting CD80 or 86. When this interaction occurs, the T cell is inhibited from becoming active, that is, from killing its target cell. This interaction is one example of an **immune checkpoint**.

culture medium

An aqueous solution that contains all nutrients and conditions necessary for the growth of cells in the laboratory.

cyclin

A protein that serves as activator for **cyclin-dependent kinase**.

cyclin-dependent kinase

One of a set of enzymes that play key roles in controlling progression of the cell growth and division cycle.

cyst

A closed sac formed in the body by cells from nearby tissue. Cysts are abnormal, but usually they are not dangerous. They can contain liquid, air, or harder material so they appear as a lump, not unlike a tumor, but they do not contain cancerous cells.

cytochrome *c*

A small mitochondrial protein that participates in the conversion of food energy into synthesis of ATP. It is also the primary intracellular signal for the initiation of **apoptosis**. If cytochrome *c* leaks out of mitochondria, it initiates protease activity by **caspases**, which take the cell apart, leading to its death.

cytokine

A class of small proteins secreted by cells that are in trouble. Cytokines induce other cells, particularly white blood cells (leukocytes), to migrate toward the secreting cell and become active in ways that will allow them to improve a bad situation. For example, cytokines can induce a white blood cell to kill a virus-infected cell that is secreting a cytokine.

cytoplasm

The region of a cell inside the **plasma membrane** and outside the **nucleus**. The cytoplasm includes many organelles, such as **mitochondria** and the **endoplasmic reticulum**, as well as the aqueous solution that includes many of the cell's molecules, called the **cytosol**.

cytoskeleton

The fibrous structures in cells that contribute to their shape and ability to change shape or move. The principal fiber systems of the cytoskeleton are **microtubules**, **microfilaments**, and **intermediate filaments**.

cytosol

The liquid part of **cytoplasm**.

cytotoxic T cell

A kind of T cell that can kill cells whose surface antigens it identifies as foreign. Also called a cytotoxic T lymphocyte (CTL).

daughter cells

The two cells formed when a cell (the parent or mother cell) divides.

dendritic cell

A dedicated **phagocyte** that is attracted to the site of an infection, where it engulfs pathogens. It then migrates to **lymph nodes**, where it presents **oligopeptides** cut from the pathogen's proteins to B and T cells, stimulating them to differentiate and fight the infection.

development

The complex series of biological processes by which an organism at a simple stage of its life cycle, for example, a fertilized egg, changes to become a more complex form, such as an embryo, a fetus, or an adult. Development is a whole field of biology, because each transformation in structure or function involves many processes, involving changes in gene expression, altered cell structure, and changed cell–cell interactions.

differentiation

The biological process by which cells turn on the expression of certain genes and turn off the expression of others and thereby become different from their sister cells.

diffusion

The physical process of molecular motion that results from the motion of all atoms and molecules that we call heat. Since all molecules are always in motion, particularly when they are in a liquid or gaseous condition, diffusion means that each molecule will tend to move from where there are many copies of it to a place where there are fewer.

diploid

A genetic condition of many cells and organisms in which they contain two copies of every gene, usually as a result of sexual inheritance in which they got one set of genes from one parent and another set of genes from the other parent.

direct carcinogen

A chemical that acts directly on an organism to induce cancer; no modification of the chemical is necessary to achieve this effect.

DNA

The acronym for deoxyribonucleic acid, the double-helical polymer of **nucleotides** that stores genetic information in all living things.

DNA adduct

The result of a chemical reaction in which a molecule binds strongly to a base in DNA and thereby blocks accurate base-pairing between complementary bases during subsequent rounds of DNA synthesis. DNA adducts induce **mutations**.

DNA damage checkpoint

The set of processes used by cells to prevent the passage of DNA damage from parent to daughter cells. Thanks to this checkpoint, DNA damage is sensed, and that information is used to stall the cell growth and division cycle, allowing more time for accurate repair. If satisfactory repair is not achieved, **apoptosis** is normally induced.

DNA hybridization

A process in which single-stranded DNA molecules of identical or very similar sequence are allowed to bind to one another by Watson–Crick base-pairing.

DNA microarray

A flat piece of glass or similar material onto which a very large number of specific DNA oligomers have been attached, each oligomer in a different position. These arrays can be used to measure the amounts of various RNAs that are present in a cellular sample or even to assess the sequence of specific pieces of DNA, thanks to the accuracy of **DNA hybridization**.

DNA polymerase

An enzyme complex that participates in the synthesis of DNA by binding to a strand of DNA, running along it in a single direction, and taking soluble bases from solution to synthesize a new strand of DNA that is complementary to the old one.

dominant

A term used to describe a situation in which **diploid** individuals who are **heterozygous** at a given genetic locus display the properties of only one **allele** in their **phenotype**.

double-blind test

A form of test in which the subjects are coded in some way, so neither the researchers carrying out the test nor the subjects themselves will know whether they are receiving a new experimental treatment, a standard treatment, or no treatment at all (being given a **placebo**). Thus, when the results are collected, neither the investigators nor the subjects know which treatment was given. Only after all data are in are the results decoded and put into their proper categories, allowing interpretation of the results. This mode of study is the most objective for obtaining information about the efficacy of a given treatment.

double-strand break

A form of DNA damage in which both strands of the DNA double helix are broken at nearly the same site. These breaks can be repaired by either of two processes: **homology-directed repair** or **nonhomologous end joining**.

driver mutation

A mutation that occurs during cancerous transformation that contributes significantly to the cancerous behavior of that cell and its progeny.

dysplastic

A term used by pathologists examining a biopsy to say that the cells under study have lost some of their normal differentiation and therefore show signs of **cancerous transformation**. They are not as fully transformed as cells that are **anaplastic**, but transformation is underway.

EBV

Abbreviation for **Epstein–Barr virus**.

ECM

Abbreviation for **extracellular matrix**.

effector B cell

Another name for a **plasma cell**.

EGF

Abbreviation for **epidermal growth factor**.

EGF-R

Abbreviation for **epidermal growth factor receptor**.

electron microscope

An instrument that uses a beam of electrons, rather than light, to probe the structure of a specimen. Though expensive to buy and operate, these microscopes provide an improvement in resolution, relative to the light microscope, of about 100-fold. They can visualize individual molecules, but unless special procedures are used, they cannot see much detail within them.

endocrine

An adjective that refers to glands that secrete molecules, like hormones, that help to regulate the behavior of other cells.

endocytosis

A cellular process in which the plasma membrane bends inwards to form a pit that can pinch off to form a vesicle that contains material that used to be outside the cell.

endoplasmic reticulum

A system of cytoplasmic membranes that is the site of synthesis of lipids and proteins that will either become part of a membrane or be secreted from the cell.

endoscope

An instrument for medical imaging that uses **fiber optics** to illuminate an interior organ and to transmit an image of that organ for examination by a doctor. Most endoscopes are used to slide into one of the body's tubes and thereby to look inside without any destruction of the body, for example, a surgical slit.

energy

An important physical concept that resembles the nonscientific meaning of the word. Energy is a property of a state or thing that describes its ability to do work. There are many forms of energy: electrical, magnetic, thermal, mechanical, etc. These forms are interconvertible, but with certain costs. The most basic form of energy is heat, and as energy changes from one form to another, some energy is commonly converted into heat. Heat energy can be converted back into other forms of energy only when it flows from a higher temperature to a lower one, so heat energy can become unavailable; if there is no place around with a lower temperature to which the heat can flow, heat energy cannot do work.

enhancer

A segment of DNA whose sequence allows it to bind an activator protein and thereby increase the level of expression of a particular gene.

enzyme

A protein that can catalyze (increase the speed of) a particular chemical reaction.

epidermal growth factor

A protein that circulates in blood and lymph to convey signals to the cells making the right receptor, telling them that they should grow and divide.

epidermal growth factor receptor

A membrane protein expressed in some cells that extends one part into the extracellular space, where it can bind **epidermal growth factor**. EGF binding induces a dimerization of the receptor, which allows the cytoplasmic portions of the proteins to phosphorylate one another. Once this phosphorylation has occurred, the receptor can activate a cascade of signaling events that initiate DNA synthesis and cell division.

epigenetic

Factors that affect inheritance (one cell generation to the next) but are not written directly into the sequence of a cell's DNA. An important example of an epigenetic factor is the state of **methylation** of DNA (the addition of a CH_3 group to a nucleotide, generally C). These changes commonly down-regulate the expression of a gene. This modification is commonly inherited as a cell goes through its growth and division cycle, because the enzymes that put methyl groups on DNA are guided by preexisting patterns of methylation. Epigenetic marks are generally removed during sexual reproduction, so they are not inherited from parent to child, though there is some recent, not yet rigorous evidence that questions this view.

epithelial to mesenchymal transition (EMT)

A transition in cell shape and behavior that is characteristic of the progression to malignancy in carcinomas.

epithelium

Tissues that cover other parts of the body and certain other tissues that are more subtly exposed to space outside the body. Skin and intestine are both covered by epithelia. Liver, too, is an epithelium, because one side of each liver cell faces a duct that carries bile from the liver to the lumen of the intestine, which is connected with the outside of the body.

Epstein–Barr virus

A human virus, often called EBV, which causes infectious mononucleosis (also known as mono or kissing disease) and which collaborates with the pathogen that causes malaria to induce a form of cancer called Burkitt's lymphoma. It may also be involved in the formation of some other cancers in ways that are not yet well defined.

ER

Abbreviation for **endoplasmic reticulum**.

Erbitux®

A monoclonal antibody that reacts with a growth factor receptor. It is used to treat some cancers.

erythrocyte

A small, hemoglobin-containing cell in the blood, also known as a red blood cell. Erythrocytes carry oxygen (O_2) from the lungs to our tissues and carbon dioxide (CO_2) from our tissues to the lungs, thereby allowing most cells in the body to burn sugars efficiently and make ATP.

estrogen

A steroid hormone that serves as a growth and survival factor for many of the cells in mammary glands and female reproductive organs. It also conveys less well understood signals involved in the way the body handles calcium and other atoms and molecules important for health.

eukaryotic

An adjective used to describe cells, like human cells, that contain a **nucleus**. This distinguishes them from **prokaryotic** cells that lack nuclei, such as bacteria.

exocytosis

A process by which cell secrete materials they have made (commonly proteins) by causing a vesicle that contains those materials to fuse with the **plasma membrane**, so the vesicle's contents are released into the extracellular space.

extirpate

To cut out, as a tumor, so as to remove it or destroy it completely.

extracellular matrix (ECM)

The collection of fibers found outside cells that gives many tissues their texture. This noncellular component of the body contributes to the environment of all cells and is important for tissue structure and some aspects of cell differentiation.

extravasation

The process by which a cell, such as a white blood cell or a metastatic cell, passes through the wall of a blood vessel to leave the circulation.

false negative

An event that can occur during testing of a chemical or procedure in which the sample is scored as negative (for example, the disease under study is said to be absent) when in fact the result is positive (disease is in fact present). These are also called type 2 errors.

false positive

An event that can occur during testing of a chemical or procedure in which the sample is scored as positive (the disease under study is said to be present) when in fact the result is negative (disease is in fact absent). These are also called type 1 errors.

familial adenomatous polyposis (**FAP**)

An inherited predisposition to cancer in which the gene for **adenomatous polyposis coli** (**APC**) is mutated to loss of function, leading to a high incidence of intestinal **polyps** and a high likelihood of colon cancer.

Fas-ligand

See **Fas-receptor**.

Fas-receptor

A protein inserted into the **plasma membrane** of cells that are in trouble. Its extracellular domain serves as a binding partner for the Fas-ligand, which is on the surface of specific white blood cells. When these proteins bind to one another, they induce the cell that had put the Fas-receptor on its surface to initiate **apoptosis**.

Fc

Abbreviation for fragment, constant, meaning the part of an antibody molecule that does not vary in structure from one immunoglobulin molecule to another. See **constant region**.

fiber optics

A bundle of thin, transparent fibers that can carry the light necessary to convey an image from one place to another. These optics are used in **endoscopes**.

fluorescence

A property of some small molecules and proteins that allows them to absorb light of one color and emit light of a different color (always with a longer wavelength). Fluorescent molecules are now widely used in both cell biology and chemistry for the study of biological structures and processes.

fluorescence microscopy

A special kind of light microscopy that uses molecules that are fluorescent to provide contrast for making images of cells that reveal the position and/or behavior of specific molecules.

fluoroscope

A medical imaging device for visualizing **X-rays** that have been used to look at structures inside a patient. A fluoroscope uses a screen of fluorescent material that absorbs X-rays and emits visible light, allowing a doctor to view the results of X-ray imaging in real time.

fluorouracil

A molecule that resembles the base called T (thymine) used to make DNA. Its similarity lets it be used by DNA polymerase to synthesize DNA, but its differences from T mean that it cannot base-pair properly with the complementary base A, so mutations are introduced, and cells often go on to initiate apoptosis. It is therefore a good chemotherapeutic agent for blocking cancer cell growth.

free radical

A molecule containing an unpaired and therefore highly reactive electron. Free radicals can react with almost any other molecule, so they are damaging to cells.

FU

Abbreviation for **fluorouracil**.

G1

A phase of the cell growth and division cycle that occurs between mitosis and **S phase**.

G2

A phase of the cell growth and division cycle that occurs between **S phase** and mitosis.

G-protein

There are two important kinds of G-proteins, both of which participate in regulation of cell function. Both kinds get their name because they bind GTP to become active. When in their active state, G-proteins bind to another protein, such as a **protein kinase**, and activate it. Ironically, G-proteins are GTPases, meaning that they hydrolyze the bound GTP to GDP and inorganic phosphate, whereupon they become inactive. In this way, the activation of G-proteins is self-terminating. How quickly they become activated (bind GTP) or inactivated (hydrolyze GTP) is controlled by companion proteins that bind to them. These include GTP exchange factors (GEFs) and GTPase activating proteins (GAPs).

gadolinium

An element used in intravenously injected solutions for medical magnetic resonance imaging to enhance the contrast of blood vessels.

gamma rays

A form of electromagnetic radiation in which the emission has very high energy and short wavelength, analogous to X-rays but even more energetic.

GAP

Abbreviation for GTPase activating protein; see **G-protein**.

GEF

Abbreviation for GTP exchange factor; see **G-protein**.

gene

The unit of biological inheritance. Each gene corresponds to a segment of DNA. It commonly includes a structural part, which is the segment of DNA that is transcribed into RNA when the gene is expressed, and a regulatory part, which helps to govern how often and at what level that gene is expressed.

gene expression

The process by which a gene is read, which includes transcription of the DNA sequence into RNA and then usually the translation of the RNA into protein.

gene product

The **macromolecule** whose structure is defined by a particular **gene**. This can be the RNA molecule that is transcribed from a gene or the protein that is translated from the RNA. Commonly the term is used for whichever of these molecules expresses the **phenotype** of that particular gene.

general anesthesia

A medical procedure in which a patient is put to sleep by administering chemicals (called anesthetics) that numb the nervous system and allow doctors to operate on the patient without causing pain.

genetic instability

A condition in which cells are unable to grow and divide to produce cells that have genomes exactly like the parent cells. This condition can result from poor error correction in either DNA replication or mitosis.

genome

The set of all genetic information possessed by a cell or organism.

genotype

The genetic characteristics of an individual, based on the DNA in their cells, as opposed to the **phenotype**, which constitutes the visible characteristics of that individual or cell, thanks to the genes they have expressed.

germ-line cell

A cell that can differentiate to become one of the cells that participates in fertilization to form the next generation, for example, a sperm or egg cell.

germ line

All the cells in an individual that are, or are in the process of making, cells that can participate in fertilization. Germ-line cells are distinguished from **somatic** cells, which are part of an individual's body and never participate in reproduction of the organism.

Gleevec®

A drug, also known as **imatinib**, that was designed to bind specifically with the product of the Abl oncogene, a protein tyrosine kinase, making it inactive. Imatinib is an effective drug for the treatment of chronic myeloid leukemia.

glucose

A sugar built from six carbon atoms and a comparable number of oxygens and hydrogens. It is half of the sugar known as sucrose, or normal table sugar. The other half is a similar six-carbon sugar called fructose.

GM-CSF

Abbreviation for **granulocyte–macrophage colony-stimulating factor**.

Golgi apparatus

An organelle within the cytoplasmic membrane system. It receives both membrane-enclosed vesicles and their protein contents that bud from the **endoplasmic reticulum**. The Golgi contains enzymes that modify these proteins by adding or removing sugars from their surfaces, preparing them to take up residence and function in different parts of the cell. Membranes and proteins leave the Golgi by budding off as vesicles from the membranes on the side of the Golgi that is far from the ER.

granulocyte–macrophage colony-stimulating factor

An important **cytokine** that can stimulate lymphocyte activity.

growth factor

A protein secreted by one cell and perceived by another cell that stimulates the recipient cell to enter the cell cycle and divide. A growth factor is like a protein hormone, but it is specialized for regulating cell growth.

GTP

The commonly used name for guanosine triphosphate, a compound that is used in making both DNA and RNA. GTP is also part of key regulatory systems that control cell metabolism and behavior, based on **G-proteins**, which are proteins that can bind GTP to become activated and then hydrolyze GTP to become inactive again.

half-life

The time during which half of all nuclei of a given atomic isotope will change and emit a radioactive ray. The shorter the half-life of an isotope, the greater the fraction of nuclei that will change in a given period of time, and therefore the more radiation they emit per atom present and per unit time.

heat map

A form of graphic data display in which color reflects the strength of an effect. Usually red means an increase in effect while green means a decrease. Black commonly means no change.

helper T cell

A kind of T cell that interacts with B cells and foreign antigens to aid in maturation of a B cell into an antibody-secreting **plasma cell**. Also called a Th cell.

Herceptin®

A drug, also known as **trastuzumab**, for the treatment of certain kinds of breast cancer. It is a **monoclonal antibody** that binds to a form of the **epidermal growth factor receptor** and inactivates it. For patients whose cancers express this receptor, Herceptin is a very effective chemotherapeutic agent.

heterozygous

A condition found in **diploid** organisms in which the two copies of a given gene are different.

histology

The science of the study of tissues.

homeostasis

A biological process that helps to maintain the status quo.

homology-directed repair (HDR)

A mode of DNA repair in which an identical or homologous chromosome is used both to bring the two pieces generated by a double-strand break into close proximity and then to guide the repair process through the use of its DNA sequences to guide the choice of bases to be added to the damaged strands during the DNA synthesis necessary for repair.

hospice

A program for care of terminally ill people, following the premise that continuation of life is not necessarily in the patient's best interests but attention to quality of life is key. Hospice programs specialize in home health care for terminally ill people, and they can provide essential services for end-of-life care.

human papilloma virus (HPV)

A small DNA virus that causes warts and other lumps on human skin. This virus has been identified as an important factor in the development of cervical and penile cancers.

humoral adaptive immunity

The parts of the immune system that work through molecules acting in solution. **Antibodies** are an example of humoral adaptive immunity, as distinct from **cellular adaptive immunity**.

hybridize

The process of allowing a piece of single-stranded DNA or RNA to interact with another single-stranded nucleic acid to see whether the two have complementary sequences and will form many Watson–Crick bonds and therefore bind to one another.

hydroxyl group

A molecule or part of a molecule composed of one oxygen atom and one hydrogen atom (OH).

hyperplasia

Growth of a structure to larger than normal size.

hypervariable region

A subregion in the variable parts of both antibodies and T-cell receptors that shows even more variation from one immunoglobulin to the next than is seen in the variable regions. These structures are one reason for the high specificity of binding shown by antibodies and T-cell receptors for specific antigenic determinants.

hysterectomy

An operation that removes the uterus. A related operation, called an ovarian hysterectomy, removes the ovaries as well.

IFN-γ

Abbreviation for **interferon γ**.

IGF-BP-2

Abbreviation for **insulin-like growth factor binding protein 2**.

IgG

Abbreviation for **immunoglobulin G**.

iKnife

The name for a recent invention that may aid surgeons in distinguishing cancerous from noncancerous tissues. This instrument combines a heated surgical knife with a suction device that vaporizes molecules from tissue and feeds them into a mass spectrometer, which returns data to the surgeon almost immediately, giving information about the chemical properties of the tissue being cut.

IL-2

Abbreviation for **interleukin 2**.

imatinib

The generic name for the drug **Gleevec®**, designed to bind to a mutant form of Abl protein tyrosine kinase and render it inactive.

immune checkpoint

Any of several protein–protein interactions by which the membranes of a **T cell** and an antigen-presenting cell can interact, conveying information to the T cell about how it should behave. There are both inhibitory and excitatory interactions in the immune checkpoint system.

immune system

The set of all cells, molecules, and devices the body uses to protect itself from infection.

immunoglobulin G

A shorter name for γ-immunoglobulin (IgG), the kind of antibody circulating in the blood that is discussed in this book. There are several other kinds of antibodies.

immunoglobulin locus

The region of DNA in any individual that contains the genes that encode antibodies and T-cell receptors.

incision

A cut in the surface of the body, made to get access to something inside.

indirect carcinogen

A chemical that can promote cancerous progression after it has been modified by chemistry carried out by some part of a patient's body, usually the liver and/or the bladder.

inflammation

A response by our bodies to infection or injury in which many changes occur locally in an effort to limit or remove harmful activity. Inflammation includes the migration of several kinds of white blood cells to the place in question, an expansion of blood vessels in the nearby region, and the secretion of numerous **cytokines** and other signaling molecules. The response helps to contain and treat the problem area, but it is often accompanied by swelling and soreness. Excess inflammation can make a variety of troubles for the body, so the process needs rigorous controls.

inherited cancer risk

The chance that an individual will get cancer as a result of genes inherited from parents.

innate immunity

The features of our bodies that naturally help to protect us from infection and need no additional development or adaption. These include skin and our intestinal epithelium; enzymes at these locations, such as ribonucleases, that cut up macromolecules from pathogens; **macrophages** and

other **phagocytes**; and our **natural killer cells**, which can identify cells that don't belong in our bodies and execute them.

insulin-like growth factor binding protein 2

A membrane-bound receptor that is important for the ability of some breast cancers to grow and thrive.

interferon

One of a class of proteins made by cells that are infected by a virus or other pathogen. They serve as signals, both within and between cells, to alert the body's immune defenses.

interferon γ

A particular kind of **interferon** that has been tried extensively as a chemical that might stimulate the patient's immune system to respond to a tumor.

interleukin

A kind of **cytokine** that participates in communication between white blood cells.

interleukin 2 (IL-2)

A cytokine that has limited value in stimulating some white blood cells for improved resistance to tumor growth.

intermediate filament

One class of fibers that comprise the cytoskeleton. These fibers are thinner than a microtubule but thicker than an actin-based microfilament, hence their name. They are made from some form of the protein keratin, which also makes hair. They are tough and flexible but not very stretchable.

interphase

The period of the cell cycle that occurs between cell division. It includes **G1**, **S phase**, and **G2**.

intravasation

The process by which a white blood cell or a metastatic cancer cell enters a blood vessel.

invasive

A property acquired by normal cells during cancerous transformation that allows them to move away from their normal position and wander locally.

ion

An atom or molecule that has an unequal number of protons and electrons and is therefore electrically charged. Ions can be positively or negatively charged.

isotope

Any form of a naturally occurring element. Many elements are found to have multiple isotopes as a result of having different numbers of neutrons in their atomic nuclei, though they have the same number of protons.

lagging strand

The strand of double-stranded DNA that is being replicated by polymerases that must continually re-initiate synthesis, because the enzymes are running away from the replication fork. This is in contrast with **leading strand** DNA synthesis.

lead compound

A small molecule discovered in a search for a new drug that is the first to be identified with the desired properties. It points the way towards finding structurally related molecules that may be even better at accomplishing the desired function and/or show fewer or weaker side effects.

leading strand

The strand of double-stranded DNA that is being replicated by polymerases that run continuously, advancing the position of the replication fork. This is in contrast with **lagging strand** DNA synthesis.

lens

A device that bends the path of a probe, like light or electrons, allowing the formation of an image, as in a microscope.

leukemia

A cancer of **leukocytes**, or white blood cells.

leukocyte

Another name for a white blood cell, one class of cells in blood. The other class is red blood cells or **erythrocytes**, which carry oxygen and carbon dioxide. There are many kinds of white blood cells; most of them are involved in the body's defenses again infection.

ligation

The molecular process of joining two adjacent pieces of DNA together, end to end.

light microscope

An instrument that uses light to form enlarged and useful images of small objects, like cells. This is a powerful tool for biologists.

lipids

The word scientists use for fats. These are biological molecules that are poorly soluble in water. Lipids have a texture like oil or wax; their physical properties change with temperature, so a solid lipid can usually be melted at a higher temperature. There are many kinds of lipids important for cells. Some are used to store food energy, some serve a structural role as part of a membrane, and some convey information within or between cells.

local metastases

Small tumors situated near the **primary tumor**.

loss of heterozygosity

A kind of mutation in which one **allele** of a gene that is **heterozygous** is lost, so the resulting cell now contains only the allele that was not lost. This is an important mechanism in the process by which a recessive allele, such as a **tumor suppressor**, can become expressed.

lumen

The interior of any tube or compartment.

lumpectomy

Surgical removal, usually from a breast, of a lump that might be cancerous. It is distinguished from a **mastectomy**, in which most or all of the breast is removed.

lymph

A liquid derived from blood as a result of the seepage of blood liquids (without blood cells) through the walls of the blood vessels called capillaries. Lymph bathes many of the cells in our body but drains slowly back into the vasculature, into either another capillary or a **lymphatic vessel**.

lymph nodes

Nodules, or enlarged areas, that form along **lymphatic vessels** that are specialized to let white blood cells encounter many other cell types, examining them to make sure they are parts of the body and not some sort of invading cell, such as a bacterium. Lymph nodes are important in cancer biology because they are places where wandering tumor cells will often dwell for a while. For example, breast cancer commonly spreads through lymphatic vessels to lymph nodes in the armpit. The presence of cancer cells in lymph nodes indicates the tendency of those cells to wander, making metastasis likely.

lymphatic system

The set of all lymphatic vessels and nodes, as well as the cells they contain.

lymphatic vessel

A branching tube that serves to collect lymph from many parts of the body and drain it back into the bloodstream via the veins that return blood to the heart.

lymphedema

A condition in which **lymph** does not drain properly into the circulating blood, resulting in swelling. It is a common problem following the removal of many lymph nodes.

lymphocyte

One of a special group of white blood cells commonly found in lymph and lymph nodes. These are cells of the immune system, including **B cells**, **T cells**, and **NK cells**.

lymphokine

A specific kind of **cytokine** that is used for communication between lymphocytes.

lymphomas

Blood cell tumors that develop from lymphatic cells.

lysosome

A membrane-bound, cytoplasmic organelle formed by the fusion of vesicles that budded off from the **Golgi apparatus** and in from the plasma membrane. Lysosomes contain digestive enzymes that can degrade old cellular structures that should be replaced and other materials brought into the cell by **endocytosis**.

macromolecular complexes

Assemblies of two or more macromolecules that are brought together to perform a function more complex than either of them could do alone.

macromolecule

A large **polymer** of smaller molecules. There are three kinds of macromolecules important in biology: proteins, carbohydrates, and nucleic acids.

macrophage

A kind of white blood cell that is specialized for phagocytosis.

magnetic resonance imaging (MRI)

A method of modern medical imaging that uses a strong magnet and radio waves to form images of tissues inside the body. Combined with the right contrast agent (a material introduced into blood) and the mathematics of axial tomography, it produces 3-D images of excellent resolution and contrast for diagnosis and characterization of tumors.

major histocompatibility (MHC) proteins

Cellular proteins made on the rough **endoplasmic reticulum**, where they bind small polypeptides generated by proteolysis from any protein in the cell. MHC proteins are then transferred to the **plasma membrane**, where they sit with their peptide-binding domain projecting into extracellular space. In this way, MHC proteins display fragments of any protein on the surface of the cell where they were made. When peptides cut from a body's own proteins are displayed, these are identified as self. When peptides cut from foreign proteins are displayed, for example, from a virus that is infecting that cell or a carcinogenic protein produced by expression of an oncogene, they can attract the action of **cytotoxic T cells**. MHC proteins are also responsible for marking cells as self or nonself in the case of tissue transplants. The fact that they can be recognized as nonself proteins contributes to the rejection of tissue or cell transplants from unrelated individuals.

malignant

A word whose common meaning is evil but whose meaning in cancer biology is life-threatening. A malignant growth, or malignancy, is a dangerous **cancer**.

mammogram

An X-ray study of the breast used to see if there is any evidence of cancer.

MAP kinase

A protein kinase in the signaling pathway from **epidermal growth factor receptor** for the regulation of gene expression, helping to activate **transcription factors** that induce the synthesis of proteins necessary for the onset of DNA synthesis.

mastectomy

Surgical removal of a part of one or both breasts. Radical mastectomy is the removal of the entire breast. This procedure is now comparatively rare, in contrast to a **lumpectomy**.

matrix metalloproteinase

One of a group of secreted enzymes that can cleave peptide bonds in proteins of the **extracellular matrix**. These proteases use a metal (usually zinc) in their active sites. Their proteolytic activity modifies the texture of the extracellular space. They are important for the metastatic behavior of cancer cells.

MDM2

A protein that serves as a negative regulator of **p53**.

medical imaging

The process of creating images of structures inside the body.

melanin

A biological molecule that absorbs visible light and therefore appears black. It is made by specific cells in skin and serves to modulate skin color. It is produced in large amounts by people of African descent and in lesser amounts by others, though exposure to sunlight increases its production. It protects skin cells against the harmful effects of ultraviolet light in sunlight.

melanoma

Cancer of **melanin**-producing cells. These cells are intrinsically migratory, given their origins in embryonic life, so melanomas tend to become metastatic and dangerous.

membrane traffic

The set of processes by which membrane made in the **endoplasmic reticulum** or brought in from the **plasma membrane** is delivered to different cellular locations.

memory cell

A special kind of **B cell** that is set aside as each B cell differentiates. It has the ability to make the same antibody as the rest of the B cells in that clone, but it becomes dormant and neither replicates nor makes antibody during the initial immune response. However, it does not die as that immune response becomes no longer necessary, so it is present in **lymph nodes** and can become active, replicate, and differentiate into many **plasma cells** if the same antigen should reappear.

mesenchymal cells

Cells of embryonic connective tissue. During progression to malignancy, epithelial cells come to look more and more like mesenchymal cells, through a change called the **epithelial to mesenchymal transition**.

mesothelioma

A cancer that formed in a specific lung tissue called the mesothelium. It is most commonly found as a result of prolonged exposure to asbestos.

messenger RNA (mRNA)

The processed transcript of a gene that is ready to enter the cytoplasm and be translated into protein.

metabolism

The collection of all the chemical reactions that are used by cells to process food, converting its molecules and energy into forms that are useful for cell growth, division, and differentiation.

metaplastic

A term used by pathologists examining a biopsy to say that the cells under study have lost their normal differentiated appearance and changed to look like a different kind of cell. This conditions usually suggests that cells in the biopsy are more transformed than **dysplastic** cells but less so than **anaplastic** cells.

metastasis

The process by which some cancer cells set up sites of growth at places distant from the initial tumor. To do this, the cancer cells migrate from the site where they were born to some other site in the body, where they can initiate the growth of a new tumor. Cells that do this are said to metastasize.

metastatic growth

A growth of cancer cells, usually leading to a tumor, that have migrated to a site that is distant from their initial site of growth.

metastasize

The verb that implies the movement of cancer cells away from their site of origin to form a tumor elsewhere in the body.

methotrexate

A drug often used as a chemotherapeutic agent for cancer because it interferes with the synthesis of thymidine. If cells run out of thymidine, they cannot make DNA.

methylation

The name for a chemical process in which one carbon atom and three hydrogens (a methyl group) are added to another molecule. Methylation of DNA is a way in which cells can modify their genome to alter the control of gene expression.

MHC

Abbreviation for **major histocompatibility complex**.

microenvironment

A word used to describe the special environment within a tumor that is generated by the **stromal** cells of the tumor. Tumor microenvironments can be important for the way the **parenchymal** cells of the tumor behave.

microfilament

A thin cytoplasmic fiber made from the muscle protein actin. Microfilaments are an important part of the **cytoskeleton**, because they contribute to several aspects of cell motion and the mechanisms by which cells change their shape.

microRNA

A small RNA that is transcribed from DNA but is not used to make protein. Instead, it helps to regulate the level of protein expression from mRNAs encoded by other genes. Commonly microRNAs work by binding to mRNA near its tail and altering either its stability or the rate at which it binds to a **ribosome** and is **translated**.

microtubule

A component of the **cytoskeleton**, made by polymerization of the protein tubulin and controlled by many cellular factors, such as the **centrosome**. Microtubules are particularly important for chromosome segregations during **mitosis**.

mismatch repair

A form of DNA repair in which a damaged nucleotide is cut out and replaced. Genes that encode the enzymes that accomplish this kind of DNA repair are important in preventing the emergence of colon cancer.

mitochondrion (plural mitochondria)

The **organelle** that burns food molecules in the presence of oxygen to gather the energy with which to make a cell's **ATP**. Mitochondria also play an important role in the process by which cells commit suicide, known as **apoptosis**.

mitogen

A signaling molecule that helps to send a cell into its growth and division cycle.

mitosis

The process in which a cell organizes and segregates its chromosomes.

mitotic spindle

The cellular machine composed of **microtubules** and many microtubule-associated proteins that organizes and segregates the duplicated chromosomes in preparation for cell division.

MMP

Abbreviation for **matrix metalloproteinase**.

molecular biology

A modern scientific discipline that uses the chemistry of nucleic acids and a knowledge of the enzymes that modify them to design and analyze the macromolecules that store information in cells.

molecule

An ordered assembly of atoms that defines a specific chemical object.

monoclonal antibody

An antibody that is made by a single clone of **B cells**.

morphine

A chemical derived from opium, a natural product isolated from certain kinds of poppies that is a powerful narcotic for people. Morphine is widely used for relief from pain, but it is a sedative that makes the recipient feel groggy, and it is addictive, meaning that the more one uses it, the more one needs to get the same effect. Many derivatives of morphine, such as heroin, are also addictive.

motor enzymes

Enzymes that use ATP to provide the energy for them to walk along a cytoskeletal filament (a microtubule or a microfilament), pulling along a cargo, such as a vesicle or a chromosome.

MRI

Abbreviation for **magnetic resonance imaging**.

multi-drug resistance

A phenomenon observed in cancer cells that have been treated with several chemotherapeutic agents. These cancer cells develop resistance to many drugs by up-regulating the expression of a **plasma membrane**-associated protein that can transport molecules of multiple shapes across the membrane and out of the cell, thereby making drugs ineffective.

mutagen

A chemical or process that promotes the formation of **mutations**.

mutation

An inheritable change in the structure of a cell's DNA. Most commonly, mutations are changes in the sequence of bases along the DNA backbone. Such changes alter the form of a **gene**, the basic unit of heredity. Mutations often have no detectable effect on the cell that hosts them, whereupon they are called silent mutations.

Myc

A transcription factor important for regulating the expression of genes whose products are involved in DNA replication. The *myc* gene is often mutated to a hyperactive form in cancer, in which case it acts as an **oncogene**.

nanotube

A cylindrical array of carbon atoms linked by strong chemical bonds. Nanotubes with a single wall are no greater in diameter than a single protein, but they can be long and very strong. They are a comparatively recent invention of material science. Uses for them in cancer diagnosis and treatment are currently under investigation.

natural killer cell

A kind of white blood cell that is capable of recognizing cells that don't belong in our bodies, by sensing molecules on their surface that are not part of a normal healthy body, and then killing these cells, either by making holes in their **plasma membranes** or by inducing them to initiate **apoptosis**.

negative predictive value (NPV)

A parameter used in evaluating a new test. It is defined as the number of true negatives divided by the number of negative calls, which is the number of times the test gave a negative result. It is to be distinguished from a **positive predictive value**.

neoplasm

A new growth that occurs as a result of cells dividing when they should not.

neuroblastoma

A tumor that forms from cells that are differentiated far enough to be committed to making nerve cells but still lack any specific neuronal function. Neuroblastomas are more common in children than adults, and they are a serious form of cancer.

neutrophil

A kind of white blood cell that works as a dedicated **phagocyte**, migrating to places where an injury has occurred and engulfing and then digesting any foreign cells, such as bacteria, that have gotten into the body.

NHEJ

Abbreviation for **nonhomologous end joining**.

NK cell

Abbreviation for **natural killer cell**.

nociceptive receptors

Nerves found in bone and some soft tissues that are slow to respond, but when greatly stimulated, they produce a strong sensation of pain.

nonhomologous end joining

One mechanism for the repair of **double-strand breaks** in DNA. During this process, the ends generated by a double-strand break are simply brought together and ligated to form a continuous piece of double-stranded DNA.

nonsteroidal anti-inflammatory drug

Any compound, like naproxen, that prevents swelling (inflammation) but is not chemically related to steroids, like cholesterol, estrogen, and testosterone.

nuclear envelope

A pair of membranes that surrounds the nucleus and separates its interior from the cytoplasm.

nuclear pores

Holes in the nuclear envelope that allow the passage of molecules in and out of the nucleus. The pores are filled with a protein machine called the nuclear pore complex, which serves as a filter to let only certain molecules pass through the pore. **mRNA** and **ribosomes** can pass out though the pores, and proteins necessary for nuclear function can pass in.

nuclease

An enzyme that cuts nucleic acids, such as DNA or RNA.

nucleic acids

The class of biological **macromolecules** that includes DNA and RNA. These are **polymers** of small molecules called **nucleotides**. Nucleic acids are important to cells as ways to store and convey information. RNA is a more complicated and diverse group of molecules than DNA because different RNAs can undertake many different functions.

nucleotide

A class of small molecules used to make DNA and RNA and for other important cellular jobs. A nucleotide is made from a **base** combined with a sugar, either ribose (for RNA) or deoxyribose (for DNA), and a phosphate group. The four bases used in DNA are A (which also occurs in ATP), G (which also occurs in GTP), T, and C. In RNA, T is replaced by U. The nucleotides containing these bases are known by the same one-letter abbreviations.

nucleotide excision repair

A pathway to repair DNA damage that involves several adjacent nucleotides, so the simple removal of one nucleotide is not sufficient to achieve repair. Enzymes that accomplish this kind of repair are important for correcting damage done by sunlight, so people who lack these enzymes are prone to the condition called xeroderma pigmentosum.

nucleus

The **organelle** in which cells store DNA, regulate expression of genes, make RNA, and replicate DNA in preparation for cell division.

null hypothesis

The postulate that nothing has happened as result of a particular treatment or procedure. This term is commonly used when statistical methods are used to assess the likelihood that the result of a given treatment or chemical might have occurred simply by chance.

nutlins

A group of small-molecule inhibitors of the regulatory molecule **MDM2**. By blocking MDM2 action, nutlins increase a cell's concentration of **p53** and can help reverse aspects of a cancerous phenotype.

oligopeptide

A string of amino acids of moderate length (oligo means a few, 7–12), connected together by the same peptide bonds that connect amino acids in protein molecules.

oncogene

A mutant gene whose product drives cells into a cancerous state. The wild-type allele of each oncogene is a **proto-oncogene**.

oncogenesis

The process of cancerous transformation.

oncogenic virus

A virus that is capable of promoting the formation of cancer.

oncologist

A medical doctor who specializes in diagnosing and treating cancer.

opioid

Any chemical derived from opium, a natural narcotic obtained from the opium poppy.

organelle

A substructure within a cell, big enough to be seen in a light microscope, that performs one or more complex tasks.

outpatient

A kind of medical procedure that can be administered without admitting the patient to a hospital.

p53

A protein with a molecular mass of 53,000 units, which works as a **transcription factor** to regulate the synthesis of proteins that help to slow the cell cycle or to induce **apoptosis**. It is one of the major players in helping cells to repair broken DNA or, if they cannot, to die. p53 is mutated to reduced function or loss of function in many human cancers, so it is often called the guardian of the genome.

palliative care

Treatments designed to help a patient be more comfortable but not to cure him/her.

palpation

The use of touch to characterize a structure. Palpation is used to identify structures that don't belong in the body, such as tumors.

Pap smear

A simple test for the normality of cells in the uterine cervix. It is a safe and cheap test that begins with a spreading of the vagina, which allows a doctor to reach the cervix and remove a small sample of tissue. These cells are transferred to a glass slide for staining and light microscopy. A skilled pathologist can then recognize normal or abnormal cells and either say that all is well or recommend further tests or treatment.

paracrine

A kind of signaling between cells in which the sender and receiver are near enough for the signal to pass without the help of circulating blood, but not so near as to allow the cells to touch one another.

parenchymal cells

Cells that are characteristic of a given tissue. Parenchymal cells of a tumor are the cancer cells themselves, as distinct from stromal cells that have joined the parenchyma to provide both blood and strength. The term is also used to describe normal tissues, so the parenchymal cells of the liver are hepatocytes that are designed to carry out liver function. The blood vessels and fatty tissues in liver are not liver parenchyma.

passenger mutation

A mutation that occurs during cancerous transformation but has no effect on the cancerous behavior of cells. It is simply a result, not a cause, of the cell's genetic instability and cancerous behavior.

pathogen

Any organism that can infect our bodies and cause harm.

pathologist

A medical doctor who is trained to identify and characterize abnormalities in cells and tissues.

PD1

A protein that can be expressed on the surface of a **T cell** when that cell is activated. It binds to a protein called PDL1, which is expressed and presented on the surface of an antigen-presenting cell. This binding serves to turn down the activity of the T cell, so PD1 is an **immune checkpoint** component.

peptide bond

The bond formed between two amino acids when they are being joined together to form a chain that will make a polypeptide or protein.

perforin

A protein made by both **cytotoxic T cells** and **NK cells**. It is stored in cytoplasmic granules until either of these cells is stimulated to kill a target cell, whereupon it is secreted, along with enzymes that stimulate the apoptotic pathway in the target cell. Once perforin has been secreted, it incorporates into the **plasma membrane** of the target cell and forms pores large enough for proteolytic enzymes from the cytotoxic cell to enter the cytoplasm of the target cell and cleave **caspases** there, initiating the apoptotic pathway.

personalized medicine

The name given to a new trend in cancer medicine in which diagnostic tools are used to characterize a patient's tumor as completely as possible, and then treatment is prescribed on the basis of that diagnostic information.

PET scan

Abbreviation for **positron emission tomography** scan.

phage

A shortened form of the term **bacteriophage**.

phagocyte

A cell that is differentiated to be effective and efficient at phagocytosis, a cellular process by which one cell engulfs and then kills and digests other cells or particles that have gotten into the body.

pharmacokinetics

The science that studies the rate at which a drug takes effect after being administered to an animal or patient and then the rate at which it is inactivated, by being degraded or eliminated from the body, for example, by urination.

phase microscopy

A special kind of light microscope that uses optical tricks to cause light that has interacted with the specimen to interact with light that has not, leading to the development of contrast, even if the specimen is transparent.

phenotype

A visible or otherwise detectable trait that is the result of a particular **genotype**.

Philadelphia chromosome

An abnormal human chromosome formed by the translocation of part of chromosome 9 to one arm of chromosome 22. This mutation is commonly found in the white blood cells of people who have chronic myeloid leukemia. The name for the chromosome comes from the city in which this abnormality was first recognized.

phosphatase

An enzyme that removes a phosphate group from another molecule, such as a protein.

phosphorylate

The action of adding a phosphate group to another molecule, such as a protein.

phytochemicals

Substances made by plants, particularly those that may contribute constructively to the human diet.

placebo

A benign and innocuous substance given to some of the subjects in a test. It is not the chemical under study, although it might look like it, so the recipient will not know whether they have been given a tested drug or not. Placebos are chosen to have no effect on the patient, for example, a sugar pill.

plasma cell

A fully differentiated **B cell** that circulates in the blood and makes large amounts of the **antibody** molecule that was selected during the differentiation of its precursor B cell. A plasma cell is also called an **effector B cell**.

plasma membrane

The membrane that surrounds each cell, defining what is inside and what is outside the cell.

plasmid

A small, circular piece of DNA that normally lives in a bacterium. Plasmids are important tools for transferring genes between cells, both naturally in bacteria and in many laboratory processes that involve recombinant DNA.

plasmin

A protease that normally inhabits blood plasma and other extracellular fluids in an inactive form, called plasminogen. It is activated to become a protease by plasminogen activator, a factor sometimes released by cells.

polymer

A **macromolecule** formed by the attachment of many small molecules, joined end to end in a chain.

polyoma virus

A small DNA virus that causes tumors in a variety of animal tissues.

polyp

An overgrowth of cells that forms an identifiable protrusion from the surface of a tissue. Polyps are usually precancerous, meaning they are not malignant, but they result from an excess of cell division, which indicates a loss of normal growth control. Their presence in cancer diagnosis is used as an indication of some level of **cancerous transformation**.

polypeptide

A chain of amino acids connected by peptide bonds, the bonds found in proteins.

positive predictive value (PPV)

A parameter used to evaluate a new test. It is defined as the number of true positives divided by the number of positive calls, which is the number of times the test gave a positive answer. It is to be contrasted with the **negative predictive value** of the test.

positron emission tomography (PET)

A method for medical imaging that uses a positron-emitting sugar to label cells that are also using lots of regular sugars, for example, any rapidly growing and dividing cells in a tumor. The positrons generate **gamma rays**, which can be detected from outside the body and used with **computed tomography** to find the places in the body that have concentrated the labeled sugar. This method is good for identifying sites that have high levels of metabolic activity, including primary tumors and metastatic growths.

post-translational modifications (PTMs)

Changes in protein structure made by adding a chemical group to an already synthesized protein. The most common modification is protein phosphorylation, in which an enzyme called a **protein kinase** binds the protein to be modified and ATP, then transfers one phosphate group from ATP to the protein, releasing ADP and a **phosphorylated** protein. The negative charges on the phosphate alter the local structure of the protein and can either activate it (for example, make an enzyme work faster) or inactivate it (for example, block the ability of that protein to polymerize).

primary tumor

The tumor that grows from cells that went through initial cancerous transformation. This is to be distinguished from secondary tumors, which can lie near the primary tumor or at a distant site, which is then called a metastatic growth.

probability

A term that means likelihood. In mathematics, a probability is a number between 0 (completely unlikely) and 1 (certain to happen).

prognosis

A doctor's estimate of how a disease will progress in the future. A prognosis is based on a careful examination of the patient's condition, the application of a recommended treatment, and then a projection of what will happen to the patient, based on information about many patients with a similar disease who have been similarly treated in the past. A prognosis can change in light of different treatments.

projection

A term borrowed from geometry that describes a way of mapping an object into a plane. A projection is generated by passing straight lines through the object and onto the plane. All the contrast along the line as it passes through the object is added up to give the value projected into the corresponding point at the plane, which generates an image of the object. **X-ray** images are projections of structures inside the body.

prokaryotic

An adjective used to describe bacteria and other cells that lack a **nucleus**, as distinguished from **eukaryotic** cells.

promoter

Part of the regulatory section of a gene. The promoter helps to bind **RNA polymerase** and initiate the synthesis of an RNA transcript that is complementary to one DNA strand and identical to the other through the region of the adjacent structural gene.

prostate gland

The part of the male reproductive system that makes most of the fluid that carries sperm during an ejaculation.

prostate-specific antigen (PSA)

A protein made by the prostate gland and secreted into blood. Its function is unknown, but it is easy to detect by a clinical test. High levels of PSA in blood have been thought to correlate with an enlarged and potentially cancerous prostate gland, so this test is sometimes used in screening for prostate cancer, in the hope of identifying the condition early in its progression.

prosthesis

A device that takes the place of a lost body part or function.

protease

A synonym for **proteinase.**

protein

One of the four important kinds of macromolecules found in cells. Proteins are linear **polymers** of amino acids, attached end to end in a specific order defined by the sequence of the **gene** that encodes that protein. The choice of amino acids and their position in the sequence of this polymer defines the structure of the protein, and thus its function. Many proteins are **enzymes**, meaning they are **catalysts** that speed up a specific chemical reaction. The activity of enzymes controls much of cellular chemistry. Other proteins are not catalysts but serve structural roles, either as parts of **membranes** or **chromatin** or as fibers of the **cytoskeleton** that help to give cells their shape and strength.

protein kinase

A kind of enzyme that takes one phosphate group from **ATP** and transfers it to a protein, changing that protein's shape and function. Protein kinases are important regulators of

protein activity in cells. There are more than 500 distinct protein kinases in humans, and many of them are important regulators of cell behaviors, such as growth and division. These enzymes accomplish one kind of **post-translational modification** of proteins, contributing to their regulation.

proteinase

An enzyme that cleaves the chain of amino acids that makes up a protein. Some proteinases break proteins down into small pieces, usually short strings of amino acids called oligo-peptides. Other proteinases will carry the breakdown process all the way to single amino acids.

proto-oncogene

A normal gene whose product helps to advance the cell cycle. Mutations that produce gain-of-function alleles of proto-oncogenes create **oncogenes**. Both these mutations and the genes themselves are important factors in driving **cancerous transformation**.

radiation oncologist

A cancer doctor who specializes in the use of any kind of radiation for cancer diagnosis and/or therapy.

radioactivity

A physical phenomenon resulting from the fact that some naturally occurring atoms are made with unstable nuclei that spontaneously fall apart, emitting one or more kinds of radiation. These forms of radiation include **alpha rays** (two protons and two neutrons), **beta rays** (single, fast-moving electrons), and **gamma rays** (very high-energy electromagnetic waves, similar to those that make X-rays and visible light). Some unstable nuclei emit an unusual form of radioactivity called a positron. Positrons are used to generate signals in **positron emission tomography**.

radioiodine therapy

A treatment for thyroid cancer that takes advantage of the fact that the only iodine-containing molecule made by the body is thyroid hormone. Thus, one can administer a radioactive isotope of iodine to a thyroid cancer patient and deliver a significant dose of ionizing radiation to the cells of the thyroid gland in a specific and not very invasive way.

radiologist

A medical doctor with specialized training in the use of diagnostic medical imaging.

randomized samples

If a scientist is trying to assess the effect of a particular chemical or treatment on people or animals, s/he wants to have two equivalent groups of individuals, or **cohorts**. Then, when one group is treated and the other is not, any difference in outcome between the cohorts is likely to be due to the treatment, not to differences between the groups. One way to get equivalent cohorts for study is to assign individuals to the cohorts at random (with a complete lack of order), so no bias has been used to make one cohort different from the other. These cohorts are then known as randomized samples.

Ras

The name of a protein that participates in transduction of signals from **epidermal growth factor receptor** and several other signaling pathways. Ras is a small **G-protein**, which is often mutated to a gain of function in cancer.

rational drug design

The process of using knowledge of cancer cell biology to identify a good molecular target for pharmaceutical inhibition and then using that structure to design an effective inhibitor of the target's function.

Rb

Abbreviation for **retinoblastoma**.

Rb protein

Abbreviation for **retinoblastoma protein**.

reactive oxygen species

Molecules containing oxygen in a form that is promiscuously reactive. Most of these molecules are **free radicals**, such as the hydroxyl radical (OH•) and the hydroperoxy radical (HO$_2$•). Some are strong oxidants, such as hydrogen peroxide (H$_2$O$_2$).

receptors

Cellular proteins that bind signaling molecules and convey to the rest of the cell the message that the signal has arrived, so the cell can respond.

recessive

The property of some **alleles** of genes in a diploid organism that they are not expressed in the **phenotype**. This is to be distinguished from alleles that are **dominant**.

recombinant DNA

Any piece of DNA that has been constructed in the laboratory by combining pieces of DNA from different sources.

rectum

The lowest part of the large intestine, the segment immediately inside the anus.

red blood cell

See **erythrocyte**.

referred pain

Pain that is sensed at a position where there is no reason for that region to hurt. Also known as reflective pain.

regress

A tumor is said to regress when it becomes smaller or goes away. This will often happen in response to treatment. It may also happen spontaneously, although that is very rare.

regulatory T cell

A kind of T cell that interacts with other cells of the immune system to control their activity. Also called Treg cells, they are key players in preventing immune responses against the body's own antigens.

remission

A period of time following a cancer treatment in which the cancer appears to have gone away or at least gotten very much better.

replication origin

A site on a DNA double helix that separates to allow the binding of the enzyme complexes necessary for DNA replication. An origin initiates two replication forks that travel in opposite directions.

resolution

The property of an image that describes the fineness of the detail it can display.

restriction enzyme

A protein that cuts DNA as a specific sequence. Also called a restriction endonuclease.

retinoblastoma

A tumor of cells at the back of the eye (the retina), which is relatively common in children and is very difficult to treat.

retinoblastoma protein

A protein that serves as an inhibitor of initiating DNA replication and entering **S phase** until it is turned off by the action of mitogens. This protein is commonly mutated to loss of function in human cancers. It was discovered through a study of the childhood cancer called **retinoblastoma**, hence its name.

retrovirus

A virus that uses RNA for its genome but includes an enzyme, called **reverse transcriptase**, that can transcribe viral RNA into a complementary strand of DNA, making it possible for virus-derived DNA to incorporate into the host cell's genome. Some retroviruses are cancer viruses. All RNA tumor viruses are retroviruses, but not all RNA viruses are retroviruses.

reverse transcriptase

An enzyme characteristic of RNA tumor viruses or **retroviruses** that can use a molecule of RNA as a template to synthesize a strand of DNA.

ribosome

The macromolecular complex that can bind to **messenger RNA** and **translate** it into the corresponding sequence of amino acids that will make the protein encoded by that gene.

risk

The chance that a particular unwanted event will occur, for example, the onset of cancer.

RNA

Ribonucleic acid, the polymer of nucleotides made by transcribing a given stretch of DNA. The sequence of such a transcript is therefore complementary to the strand that was copied and identical to the sister strand that was not.

RNA polymerase

The enzyme complex that binds to the regulatory part of a gene and then synthesizes an RNA transcript, copying the sequence of one DNA strand by becoming complementary to the other. This enzyme is responsible for gene expression.

RNA primer

A short piece of RNA synthesized on DNA to help get DNA replication going.

ROS

Abbreviation for **reactive oxygen species**.

S phase

The part of **interphase** in the cell cycle during which DNA is replicated.

sarcoma

A cancer that has arisen in a connective tissue.

secondary tumor

A tumor that develops from metastatic cells. It is distinguished from a **primary tumor**, which is at the place the cancer cells first formed a tumor.

sentinel nodes

The **lymph nodes** positioned so they are most likely to receive and harbor metastatic cells from a given tumor. For tumors in some locations, selective biopsy and examination of sentinel nodes can make the examination of many nodes unnecessary.

serum

The liquid part of blood, which serves to carry circulating red and white blood cells. One makes serum by taking whole blood and letting it clot, then spinning out both the clot and all the blood cells with a centrifuge. The remaining liquid is serum.

signal

A general term meaning a message that can be detected and interpreted. The word is used with instrumentation, like an **MRI** or a **PET scanner**, but it is also used to refer to messages that a cell receives or sends out, providing useful communication.

signal transduction

The processes by which a signal is converted into a form in which the information it carries can be understood and acted upon.

signal-to-noise ratio

The strength of a signal that carries information, relative to all similar inputs that are meaningless and therefore called noise. Noise is a familiar term when you think about sound, but the idea of noise is applicable to any kind of information: visual (background glare or meaningless flashing lights), tactile (too many bumps to let your fingers sense the little thing they are looking for), or smell (too much surrounding stink to let you smell the roses).

somatic

An adjective that means from the body, as opposed to from egg or sperm cells (**germ-line cells**). A somatic cell is any cell other than the ones used to produce offspring.

sonogram

A 3D image of organs inside the body obtained with the use of **ultrasound**. The resolution of a sonogram is low compared with a **CT scan** or an **MRI**, but sonograms are benign, comparatively inexpensive, and useful.

specimen preparation

A general term for the processes used to get biological material ready for study with a particular instrument. The term is most commonly used for microscopy, but the idea is general.

spindle assembly checkpoint (SAC)

A set of processes that control the onset of mitotic chromosome segregation, delaying it until all chromosomes have become properly attached to the mitotic spindle.

spontaneous remission

A rare event in which an established cancer simply goes away. Its cause is unknown, but it is thought to be a result of action by the patient's immune system.

Src

A virus-encoded protein tyrosine kinase that drives cells into inappropriate cell division. The gene that encodes this protein is from the genome of the Rous sarcoma virus. The wild-type allele of this gene, c-src, is a **proto-oncogene** found in many vertebrate cells, where it is part of the network for signal transduction that regulates the onset of S phase.

stem cells

Cells with the ability to divide and produce daughter cells that can either be new stem cells or differentiate into cells with specific traits and functions.

sterile technique

A set of procedures designed to minimize the chances that unwanted organisms will get into the material being examined. Sterile technique is used both during surgery and when culturing cells in the laboratory.

steroid

A class of molecules that serve as hormones. Some, like estrogen and testosterone, are involved in the development of reproductive organs and behaviors in females and males. Other steroids serve as growth and maintenance factors for other cell types, such as muscle.

stroma

Tissues in a tumor that are not made from or by cancer cells. These include blood vessels that supply the tumor with nutrition and connective tissues that support the tumor's structure.

structural element of a gene

The portion of a gene that is read by **RNA polymerase** to make an RNA transcript, which is then processed to make **messenger RNA**.

sugars

A special kind of small carbohydrate that is rich in **hydroxyl** (OH) groups. Sugars are water-soluble and rich in energy. A particular sugar, either ribose or deoxyribose, is part of all the nucleotides that are used to make RNA and DNA, respectively. Another sugar called glucose is a key part of a cell's energy **metabolism**. Glucose can couple with another simple sugar called fructose to make sucrose, which is table sugar.

SV40

A small DNA virus (simian virus 40), originally found in monkey tissue culture cells, that can transform the cells of several other animals into cancerous cells in culture.

symmetry

A property of objects that describes the similarity of one part of the object relative to another. Objects can show symmetry about a plane when, for example, the structure on one side of the plane is the mirror image of the structure on the other. They may show symmetry about a line, for example if you rotate the structure by one-third of a turn, it looks just as it did before rotation. They may also show symmetry about a point, such as a cube, all of whose face, edges and vertices are identical on opposite sides of the cube's central point. In the context of cancer biology, symmetry implies that if there is a given chance that DNA polymerase will make a mistake when replicating each nucleotide, then the more times DNA is replicated, the more likely it is that the DNA will contain a mutation. Thus the probability of cancer in a particular tissue depends on the number of cell divisions that go into making and maintaining that tissue.

syndrome

A set of symptoms or medical conditions that commonly occur together, suggesting that they derive from a common cause.

synthesized

The biologists' and chemists' word for made or constructed. Proteins, nucleic acids, and other macromolecules are all said to be synthesized by cells.

T antigens

A set of several proteins produced in cells that have been transformed to a cancerlike state by infection with any of several DNA tumor viruses.

T cell

A kind of lymphocyte that is the central player in all aspects of **cellular adaptive immunity**. T cells are born in the bone marrow but mature in the **thymus**, from which they get their name. There are several kinds of T cells: **cytotoxic T cells**, **helper T cells**, and **regulatory T cells**.

tamoxifen

An artificial molecule that resembles estrogen. It binds to the estrogen receptor in cells but does not stimulate it to perform its normal roles. If both tamoxifen and estrogen are present in the same cell, they compete for binding to the estrogen receptor. Thus, an appropriate concentration of tamoxifen will decrease a cell's response to estrogen. Many breast cancer cells require estrogen, both to live and to grow and divide, so tamoxifen blocks their growth and leads them to turn on the **apoptotic** pathway.

TATA box

A sequence of DNA, rich in the bases T and A, in the regulatory part of a gene. It initiates the binding of **RNA**

polymerase, which can transcribe the adjacent structural gene into RNA.

taxanes
A group of chemicals related to paclitaxel, a chemical found in the bark of Western yew trees. Taxanes bind to microtubules and make them unnaturally stable, so structures that rely on the growth and shortening of microtubules, like the mitotic spindle, fail to work properly. Mitosis is blocked by taxanes, so these chemicals are used as cancer chemotherapeutic agents.

Tc cell
Abbreviation for **cytotoxic T cell** (or lymphocyte).

TCA4
Abbreviation for cytotoxic T lymphocyte-associated antigen 4. This is a T-cell membrane protein that is a component of the **immune checkpoint** system. It is an important part of the way an antigen-presenting cell inhibits an associated cytotoxic T cell from becoming activated.

T-cell receptor (TCR)
A kind of protein in the membrane of every T cell that shows the same sort of cell-to-cell variation that is seen with antibodies in B cells. These are the molecules that give T cells their specificity of interaction with **antigens**.

telomerase
An enzyme normally expressed in stem cells that uses a bound piece of RNA as template to add DNA sequences to the ends of chromosomes and thereby prevent their shortening during successive rounds of DNA replication.

telomere
The structure at each end of a chromosome that helps to protect the chromosome from degradation by enzymes and other processes that might alter its structure. Telomeres are based on the repetitive sequences of DNA added by **telomerase** and proteins that bind to them.

temperature sensitive mutants
Mutant strains of cells, viruses, or animals that will grow and behave normally at a lower temperature, e.g. 30° C for human cells, but misbehave in some consistent and inheritable way at an elevated temperature, e.g., 40° C. See also **cold sensitive mutants**.

testosterone
A steroid hormone important for development and maintenance of cells in the male reproductive organs. It also has effects on other cells in the body, such as muscle.

TF
Abbreviation for **transcription factor**.

Th cell
Abbreviation for **helper T cell**.

therapeutic index
A metric used to describe aspects of the effect of a drug on a patient. It is the ratio of the amount of drug that patients (on average) can tolerate without ill effects to the amount needed (on average) to have a positive effect on their condition. Also known as therapeutic ratio.

therapeutic window
The range of drug concentrations between a value that is toxic to 50% of the test group and one that has a detectable effect on 50% of the test group.

thymus
A gland located where the neck meets the chest that plays a central role in the differentiation and action of **T cells**.

thyroid gland
A gland located in the neck that makes hormones important for the regulation of the rate at which our body burns up the food we have eaten.

tight junction
A specialized region where two epithelial cells make contact. It forms a seal that will keep all materials in the extracellular space from passing over to the other side of the junction.

tolerance
A phenomenon that is common for both aspect of human perception and human responses to drugs; stronger stimulation or higher doses are required to get a similar effect.

traditional medicine
See **alternative medicine**.

transcript
The RNA molecule made from DNA by transcription.

transcription
The process of transcribing, that is making a piece of RNA from a particular stretch of DNA.

transcription factor
A protein that binds to a transcription-regulating element of DNA to regulate the binding of **RNA polymerase** and initiate **transcription** of a **gene**.

transcription terminator
A DNA sequence that instructs RNA polymerase to stop transcribing and fall off from DNA.

transfection
A method that allows the introduction of foreign DNA into cultured cells. This is not infection with DNA, which is accomplished by a DNA virus, because the process is not a natural one and doesn't require any viral proteins. Any of several tricks can be used to force the uptake of foreign DNA, including precipitating the DNA with crystals of calcium phosphate, which the cell takes in by phagocytosis. Transfection can also be accomplished by mixing DNA with the right lipids or detergents to open the plasma membrane a little, allowing DNA entry. Also effective is the application of fairly strong alternating electric fields.

transgene
A gene that has been inserted into a cell from another source, usually by the methods of **recombinant DNA**.

transit amplifying cell
One kind of cell produced by the division of a stem cell. When a stem cell divides, one of its daughters retains the stem cell identity. The other commonly becomes a transit amplifying cell that first divides several times, making more of its own kind. All of these cells then differentiate to become specialized and capable of doing specific jobs for the body.

translate
To read the sequence of a messenger RNA and convert it into the corresponding sequence of amino acids to make the protein product of that gene; this process is called translation.

translocation
An abnormal cellular process in which part of one chromosome is exchanged with part of another chromosome, so the content of both chromosomes is altered, but the content of the cell's genome is essentially unchanged.

trastuzumab
The generic name for the drug **Herceptin**®, a monoclonal antibody that binds a growth factor receptor and inactivates it, reducing the rate of growth of some tumors.

Treg cell

Abbreviation for **regulatory T cell**.

tumor

A collection of cancer cells, connective tissue, and blood vessels that is the result of cancerous growth.

tumor burden

The total amount of tumor in a patient. This can be measured by the number of cancer cells in a patient's body, but more commonly it is expressed as the weight, in grams, of all the tumors in a patient. This measure includes both parenchyma and stroma.

tumor heterogeneity

A property of many advanced tumors that they contain cells of various genotypes and behaviors. This is probably the result of genetic instability in the cells of the tumor at an earlier time. This property of tumors makes treating them effectively quite difficult.

tumor promoter

A chemical or process that promotes the formation of cancer, even though it is not mutagenic. Tumor promoters increase the amount of cell division and thereby expand any clones of already mutated cells, increasing the likelihood that a complete **cancerous transformation** will occur.

tumor suppressor

A gene whose normal function is to make a protein that retards progression through the cell cycle. Loss-of-function mutations of tumor suppressors are an important aspect of **carcinogenesis**.

tumor virus

Any virus that can transform the cell it infects into a tumor cell.

type 1 error

See **false positive**.

type 2 error

See **false negative**.

ultrasound

Sound whose pitch is so high the ear cannot hear it. Ultrasound is used by an instrument for medical imaging to generate **sonograms**.

ultraviolet light

Light whose wavelength is shorter than can be seen by our eyes.

variable region

The portion of an **antibody** or **T-cell receptor** that shows variation in amino acid sequence (and therefore structure) from one molecule to another. This is the part of the molecule that interacts with a specific **antigen**.

vinblastine

A small molecule, isolated from the Madagascar periwinkle, that is used as a chemotherapeutic agent for some kinds of leukemia. For mechanism, see **vincristine**.

vincristine

A small molecule made by the plant *Vinca rosea*, or periwinkle. It binds the protein tubulin and forces it into an unnatural polymer that is not a microtubule, thereby blocking the assembly of the mitotic spindle. It is therefore used as a cancer chemotherapeutic agent for blocking cell division.

virus

A biological structure composed of a nucleic acid (DNA or RNA) and protein, and sometimes also lipid and carbohydrate, that can infect a living cell and use the cell's many molecules to make more virus.

viscera

A general term for the internal organs of the human body.

vitamins

Small molecules that are essential for one or more of the chemical reactions that cells carry out but are built in such a way that cells are unable to make them. Thus, they must be acquired through food or some other external source.

white blood cell

Another name for **leukocyte**.

Wilms tumor

A specific kind of kidney tumor that is more common in children than adults.

X-rays

A form of short-wavelength electromagnetic radiation, useful for penetrating solid objects and producing images of the insides of those objects, as in medical diagnosis or airport security.

xeroderma pigmentosum (XP)

An inherited genetic condition in which at least one of the genes that encodes an enzyme necessary for **nucleotide excision repair** has been mutated to loss of function. When this kind of DNA repair cannot be done, the individual is very much more sensitive than normal to the damage that can be done by UV radiation in sunlight.

Index

Notes: Entries followed by 'f' refer to figures; those followed by 't' refer to tables; those followed by 's' refer to sidebars; those followed by 'sq' refer to side questions, and those followed by 'tq' refer to thought questions.